Lecture Notes in Computer Science 14329

FoLLI Publications on Logic, Language and Information

Subline of Lecture Notes in Computer Science

More information about this series at https://link.springer.com/bookseries/558

Natasha Alechina · Andreas Herzig ·
Fei Liang
Editors

Logic, Rationality, and Interaction

9th International Workshop, LORI 2023
Jinan, China, October 26–29, 2023
Proceedings

Springer

Editors
Natasha Alechina
Utrecht University
Utrecht, The Netherlands

Andreas Herzig
CNRS, IRIT
Toulouse, France

Fei Liang
Shandong University
Jinan, China

ISSN 0302-9743 ISSN 1611-3349 (electronic)
Lecture Notes in Computer Science
ISBN 978-3-031-45557-5 ISBN 978-3-031-45558-2 (eBook)
https://doi.org/10.1007/978-3-031-45558-2

This Springer imprint is published by the registered company Springer Nature Switzerland AG
The registered company address is: Gewerbestrasse 11, 6330 Cham, Switzerland

Preface

This volume contains the papers presented at the 9th International Conference on Logic, Rationality, and Interaction (LORI-IX 2023), held during October 26–29, 2023, in Jinan, China, and hosted by the School of Philosophy and Social Development, Institute of Concept and Reasoning, Shandong University.

The topics of the conference included: agency; argumentation and agreement; belief representation; probability and uncertainty; belief revision and belief merging; knowledge and action; dynamics of informational attitudes; intentions, plans, and goals; decision making and planning; preference and utility; cooperation; strategic reasoning and game theory; epistemology; social choice; social interaction; speech acts; knowledge representation; norms and normative systems; natural language; rationality; and philosophical logic.

We received 32 submissions. The papers were selected on the basis of at least two single-blind reviews; all but 5 papers had at least three reviews. Among them, 17 full papers and 7 short papers were accepted and included in the proceedings.

In addition, there were presentations by six keynote speakers:

Philippe Balbiani (CNRS, IRIT, University of Toulouse)
Patrick Blackburn (University of Roskilde)
Thomas Bolander (Technical University of Denmark)
Sujata Ghosh (Indian Statistical Institute)
Minghui Ma (Sun Yat-sen University)
Chinghui Su (Shandong University)

The first LORI event, LORI-I, took place in August 2007. Following the notable success of this initial meeting, the series continued with eight more outstanding events over the past 15 years. Beyond gathering researchers from East Asia, the LORI series succeeded in attracting scholars from outside the region working on topics related to logic, rationality, and interaction. A full history of the series can be found at http://www.golori.org/.

We would like to thank all the members of the Program Committee for their hard work and thorough reviews, Fenrong Liu and Johan van Benthem for their guidance, and Liwu Rong, Chinghui Su, Zhiqiang Sun, Wenfang Wang, and Lun Zhang for the local organisation. We also thank the School of Philosophy and Social Development, Institute of Concept and Reasoning, Shandong University for sponsoring the conference.

October 2023

Natasha Alechina
Andreas Herzig
Fei Liang

Organization

Organization Committee

Fei Liang (Chair)	Shandong University, China
Liwu Rong	Shandong University, China
Chinghui Su	Shandong University, China
Zhiqiang Sun	Shandong University, China
Wenfang Wang	Shandong University, China
Lun Zhang	Shandong University, China

Program Committee Chairs

Natasha Alechina	Utrecht University, The Netherlands
Andreas Herzig	CNRS, IRIT, Toulouse, France

Program Committee

Thomas Ågotnes	University of Bergen, Norway
Thomas Bolander	Technical University of Denmark, Denmark
Agata Ciabattoni	Vienna University of Technology, Austria
Hans van Ditmarsch	CNRS, IRIT, France
Rustam Galimullin	University of Bergen, Norway
Pietro Galliani	Free University of Bozen-Bolzano, Italy
Sujata Ghosh	Indian Statistical Institute, India
Valentin Goranko	Stockholm University, Sweden
Meiyun Guo	Southwest University, China
Wesley Holliday	University of California, Berkeley, USA
John Horty	University of Maryland, USA
Kohei Kishida	University of Illinois Urbana-Champaign, USA
Dominik Klein	Utrecht University, The Netherlands
Sophia Knight	University of Minnesota Duluth, USA
Alexander Kocurek	Cornell University, USA
Louwe B. Kuijer	University of Liverpool, UK
Jérôme Lang	CNRS, LAMSADE, University Paris-Dauphine, France
Fei Liang	Shandong University, China
Hu Liu	Sun Yat-sen University, China
Munyque Mittelmann	University of Naples Federico II, Italy
Laurent Perrussel	IRIT, University Toulouse Capitole, France
Gabriella Pigozzi	LAMSADE, University Paris-Dauphine, France
R. Ramanujam	Institute of Mathematical Sciences, India
Olivier Roy	University of Bayreuth, Germany

Katsuhiko Sano	Hokkaido University, Japan
Francois Schwarzentruber	École normale supérieure de Rennes, France
Chenwei Shi	Tsinghua University, China
Sonja Smets	University of Amsterdam, The Netherlands
Chinghui Su	Shandong University, China
Paolo Turrini	University of Warwick, UK
Fernando R. Velázquez-Quesada	University of Bergen, Norway
Wenfang Wang	Shandong University, China
Wei Wang	Xi'an Jiaotong University, China
Yì Nicholas Wáng	Sun Yat-sen University, China
Fan Yang	Utrecht University, The Netherlands
Francesca Zaffora Blando	Carnegie Mellon University, USA
Zhiguang Zhao	Taishan University, China

Additional Reviewers

Soham Banerjee
Fausto Barbero
John Barnden
Gaia Belardinelli
Kees van Berkel
Francesco Chiariello
Tiziano Dalmonte
Rustam Galimullin
Didier Galmiche
Dazhu Li
Xiaolong Liang
Anuj More
Dmitry Rozplokhas
Jonni Virtema
Jialiang Yan

Abstract of Invited Talks

Parametrized Modal Logic III: Applications to Social Epistemic Logics

Philippe Balbiani

Toulouse Institute of Computer Science Research, CNRS-INPT-UT3, Toulouse University, France
philippe.balbiani@irit.fr

Abstract. By means of a two-typed modal language where the modal connectives of one type are indexed by formulas of the other type, we axiomatically introduce different two-typed parametrized modal logics. Then, we prove their completeness by means of the use of a tableaux-based approach and an adaptation of the canonical model construction and we prove their decidability by means of the use of guarded fragments and an adaptation of the filtration method. Finally, we will review their possible applications to the so-called social epistemic logics.

Keywords: Parametrized modal logic · Completeness · Tableaux-based approach · Canonical model construction · Decidability · Guarded fragments · Filtration method · Social epistemic logics

For reasoning about knowledge, states and agents have been identified as the primitive entities of interest. In this respect, one usually considers relational structures of the form $(S, \equiv)$ where S is a nonempty set of states and $\equiv$ is a function associating an equivalence relation $\equiv_a$ on S to every element a of a fixed set A of agents [4, 5, 7]. For all $a \in A$, in that setting, two states s and t are equivalent modulo $\equiv_a$ exactly when a cannot distinguish between s and t. As for distributed knowledge, it is of interest to assume that $\equiv$ is also a function associating an equivalence relation $\equiv_B$ on S to every $B \in \wp(A)$ in such a way that for all $B \in \wp(A)$, $\equiv_B = \bigcap\{\equiv_a \colon a \in B\}$.

The modal language interpreted over such relational structures traditionally consists of one type of formulas, state-formulas, to be interpreted by sets of states. State-formulas are constructed over the Boolean connectives and the modal connectives $[B]$ — B ranging over $\wp(A)$. The state-formula $[B]\varphi$ is true in a state s of some model if the state-formula φ is true in every state of that model that is equivalent to s modulo $\equiv_B$.

In many situations, one would like to use relational structures of the form $(S, A, \equiv, \triangleright)$ where on top of the above-considered elements S and $\equiv$, one can find a nonempty set A of agents and a function $\triangleright$ associating a binary relation $\triangleright_s$ on A to every element s of S. Which situations? Situations where relationships between agents such as the following ones have to be taken into account: "agent a trusts agent b in state s", "agent a is a friend of agent b in state s", etc [3, 6, 8] . In these situations, for all $s \in S$, two agents a and b are related by $\triangleright_s$ exactly when a trusts b in state s, a is a friend of b in state s, etc. Moreover, on top of the assumption that $\equiv$ is also a function associating an equivalence relation $\equiv_B$ on S to every $B \in \wp(A)$ in such a way that for all $B \in \wp(A)$, $\equiv_B = \bigcap\{\equiv_a\colon\ a \in B\}$, one naturally assumes that $\rhd$ is also a function associating a binary relation $\rhd_T$ on A to every $T \in \wp(S)$ in such a way that for all $T \in \wp(S)$, $\rhd_T = \bigcap\{\rhd_s :\ s \in T\}$.

The modal language interpreted over relational structures of the form $(S, A, \equiv, \rhd)$ naturally consists of two types of formulas: state-formulas, to be interpreted by sets of states, and agent-formulas, to be interpreted by sets of agents [1, 2] . State-formulas are constructed over the Boolean connectives and the modal connectives $[\alpha]$ — α ranging over the set of all agent-formulas — whereas agent-formulas are constructed over the Boolean connectives and the modal connectives $[\varphi]$ — φ ranging over the set of all state-formulas. The state-formula $[\alpha]\varphi$ is true in a state s of some model if the state-formula φ is true in every state of that model that can be distinguished from state s by no α-agents, whereas the agent-formula $[\varphi]\alpha$ is true in an agent a of some model if the agent-formula α is true in every agent of that model that is trusted by agent a at all φ-states.

By means of a two-typed modal language where the modal connectives of one type are indexed by formulas of the other type, we axiomatically introduce different two-typed parametrized modal logics. Then, we prove their completeness by means of an adaptation of the canonical model construction and we prove their decidability by means of an adaptation of the filtration method. Finally, we will review their possible applications to the so-called social epistemic logics [9] .

References

1. Balbiani, P.: Parametrized modal logic II: the unidimensional case. In: Areces, C., Costa, D. (eds) Dynamic Logic: New Trends and Applications. LNCS. Springer, Cham, vol. 13780, pp. 17–36 (2023). https://doi.org/10.1007/978-3-031-26622-5_2
2. Balbiani, P., Fernández González, S.: Parametrized modal logic I: an introduction. In: Advances in Modal Logic, vol. 14, pp. 97–117 College Publications (2022)
3. Ben-Naim, J., Longin, D., Lorini, E.: Formalization of cognitive-agent systems, trust, and emotions. In: Marquis, P., Papini, O., Prade, H. (eds)., A Guided Tour of Artificial Intelligence Research. Springer, Cham (2020). https://doi.org/10.1007/978-3-030-06164-7_19
4. Van Ditmarsch, H., Kooi, B., van der Hoek, W.: Dynamic Epistemic Logic. Springer, Dordrecht (2008). https://doi.org/10.1007/978-1-4020-5839-4
5. Fagin, R., Halpern, J., Moses, Y., Vardi, M.: Reasoning about Knowledge. MIT Press (1995)
6. Liu, F., Lorini, E.: Reasoning about belief, evidence and trust in a multi-agent setting. In: An, B., Bazzan, A., Leite, J., Villata, S., van der Torre, L. (eds) PRIMA 2017: Principles and Practice of Multi-Agent Systems. PRIMA 2017. LNCS, vol 10621. Springer, Cham (2017). https://doi.org/10.1007/978-3-319-69131-2_5
7. Meyer, J.-J., van der Hoek, W.: Epistemic Logic for AI and Computer Science. Cambridge University Press (1995)

8. Perrotin, E., Galimullin, R., Canu, Q., Alechina, N.: Public group announcements and trust in doxastic logic. In: Blackburn, P., Lorini, E., Guo, M. (eds) Logic, Rationality, and Interaction. LORI 2019. LNCS, vol 11813. Springer, Heidelberg (2019). https://doi.org/10.1007/978-3-662-60292-8_15
9. Seligman, J., Liu, F., Girard, P.: Logic in the community. In: Banerjee, M., Seth, A. (eds) Logic and Its Applications. ICLA 2011. LNCS, vol 6521. Springer, Berlin, Heidelberg (2011). https://doi.org/10.1007/978-3-642-18026-2_15

Knowledge, Propositional Quantification and Hybrid Logic

Patrick Blackburn

Section for Philosophy and Science Studies, IKH, Roskilde University, Denmark
patrick.rowan.blackburn@gmail.com

Modal logics enriched with propositional quantification have been explored for at least a century now, but the literature about them remains relatively small. In some ways this is understandable: as Kit Fine showed fifty years ago, the seemingly simple act of allowing propositional symbols to be bound by existential and universal quantifiers typically leads to extremely complex logics. But if we are interested in exploring the logics of knowledge, it seems that this complexity must be embraced. Knowledge lives near both truth and paradox, and the expressivity offered by explicit propositional quantification, although computationally awkward, makes such logics a natural tool for mapping the many forms of reasoning it gives rise to.

In this talk I will discuss modal logics with propositional quantification, starting with early work by Arthur Prior on paradoxes involving knowledge, and moving towards knowledge representation for AI. Along the way I will present some recent joint work (with Torben Brauner and Julie Lundbak Kofod) on how to hybridize modal logics with propositional quantification with the help of general models.

From Dynamic Epistemic Logic to Socially Intelligent Robots

Thomas Bolander

Technical University of Denmark, Denmark
tobo@dtu.dk

Dynamic Epistemic Logic (DEL) can be used as a formalism for agents to represent the mental states of other agents: their beliefs and knowledge, and potentially even their plans and goals. Hence, the logic can be used as a formalism to give agents a Theory of Mind allowing them to take the perspective of other agents. In my research, I have combined DEL with techniques from automated planning in order to describe a theory of what I call Epistemic Planning: planning where agents explicitly reason about the mental states of others. One of the recurring themes is implicit coordination: how to successfully achieve joint goals in decentralised multi-agent systems without prior negotiation or coordination. The talk will first motivate the importance of Theory of Mind reasoning to achieve efficient agent interaction and coordination, will then give a brief introduction to epistemic planning based on DEL, address its (computational) complexity, address issues of implicit coordination and, finally, demonstrate applications of epistemic planning in human-robot collaboration.

On Epistemic Reasoning in Games on Graphs

Sujata Ghosh

Indian Statistical Institute, Chennai
sujata@isichennai.res.in

Let us start our discussion with two-player games on graphs, where we consider two players playing a turn-based game by moving a token through a directed graph, tracing out a finite or infinite path. Such simple games provide us with compelling tools for reasoning about diverse phenomena arising in areas like computer science, logic, linguistics, economics, mathematics, philosophy, and biology. One usually considers different variants of such graph games where such variations arise from different winning conditions (e.g., reachability, parity [6]), independent moves of players (e.g., cop and robber game [10]), one player obstructing moves of the others (e.g., sabotage game [2], poison game [5]) and similar others. In the subtle interactions of the distinct domains like game theory, logic, linguistics and computer science, these graph games provide natural representations for intelligent machine models that need to interact with the uncertain environment.

Both two-player and many-player versions of these games on graphs have been studied in some detail from the algorithmic and combinatorial perspectives, e.g., see [4, 7]. From the logical perspectives, such studies got a renewed focus with the corresponding study on sabotage games [1], and subsequently, a more general overview [3] on logics for graph games. However, in all these studies, it was assumed that players have perfect information at each and every step of the game and the discussion pivoted around the moves of the players. Following the lines of study in [3], this work starts off with the simplest of the graph games, namely, travel games, and then moves on to study the variants like cops and robber game and sabotage game, where we mainly deal with the information available to the players with respect to their positions and moves. Our primary focus of study is the cops and robber game, alongside which we briefly touch upon the other ones as well.

The cops and robber game is an ideal setting for modelling search missions and pursuit-evasion environments in the study of multi-agent systems. As mentioned earlier, extensive research has been done on these games, mostly from the algorithmic and combinatorial perspectives (see e.g., [4]), assuming a notion of perfect information for the players. What we intend to do in our logical study is to bring to the fore the interplay of the moves made by the players and the information available to them. A natural question that arises in this context is about the extent to which such information can be made a primary notion in the graph game model. To this end, we propose to add an epistemic dimension [8] to the existing logical frameworks of cops

Supported by Department of Science and Technology, Government of India for financial support vide Reference No DST/CSRI/2018/202 under Cognitive Science Research Initiative (CSRI) to carry out this work.

and robber games on graphs [9]. We consider $1 \leq i \in \mathbb{N}$ cops and a robber moving along a finite connected graph, and describe their knowledge in terms of limited resources.

While epistemic reasoning of the players playing the hide and seek game or the cops and robber game blends in quite naturally with the game scenarios in terms of winning conditions, for the general travel games or the sabotage games, such reasoning involves moves and positions of the players in a more structural manner, akin to the notion of imperfect information in game theory. Even in such cases, the framework developed for the cops and robber game may provide strategic insights into the action-information interplay amongst the concerned players. Consequently, the natural next steps for the study of epistemic reasoning in games on graphs may well be along these directions.

References

1. Aucher, G., van Benthem, J., Grossi, D.: Modal logics of sabotage revisited. J. Logic Comput. **28**(2), 269–303 (2018)
2. van Benthem, J.: An essay on sabotage and obstruction. In: Hutter, D., Stephan, W. (eds.) Mechanizing Mathematical Reasoning: Essays in Honor of Jörg H. Siekmann on the Occasion of His 60th Birthday, pp. 268–276. Springer, Heidelberg (2005). https://doi.org/10.1007/978-3-540-32254-2_16
3. van Benthem, J., Liu, F.: Graph games and logic design. In: Liu, F., Ono, H., Yu, J. (eds.) Knowledge, Proof and Dynamics, pp. 125–146. Springer, Singapore (2020). https://doi.org/10.1007/978-981-15-2221-5_7
4. Bonato, A., Nowakowski, R.J.: The game of cops and robbers on graphs. AMS (2011)
5. Duchet, P., Meyniel, H.: Kernels in directed graphs: a poison game. Discrete Math. **115**(1–3), 273–276 (1993)
6. Grädel, E.: Back and forth between logic and games. In: Apt, K.R., Grädel, E. (eds.) Lectures in Game Theory for Computer Scientists, pp. 99–145. Cambridge University Press (2011)
7. Kvasov, D.: On sabotage games. Oper. Res. Lett. **44**(2), 250–254 (2016)
8. Li, D., Ghosh, S., Liu, F.: Cops and robber game: A logic study. Manuscript (2023)
9. Li, D., Ghosh, S., Liu, F., Tu, Y.: A simple logic of the hide and seek game. Studia Logica (2023)
10. Nowakowski, R., Winkler, P.: Vertex-to-vertex pursuit in a graph. Discrete Math. **43**(2–3), 235–239 (1983)

Cut Elimination in Gentzen Sequent Calculi for Classical Tense Logics

Minghui Ma

Institute of Logic and Cognition, Department of Philosophy, Sun Yat-sen University, Guangzhou, China
mamh6@mail.sysu.edu.cn

The philosophical development of modern tense logic goes back to Prior's analysis of temporal modalities. Basic tense operators include $\Box$ (future necessity), $\Diamond$ (future possibility), $\blacksquare$ (past necessity) and $\blacklozenge$ (past possibility). From a mathematical perspective, the minimal (classical) tense logic K_t is the extension of classical propositional logic with a Galois connection. In general, a pair of unary maps $\langle f, g\rangle$ on a poset P is called a Galois connection if for all $x, y \in P$, $f(x) \leq y$ if and only if $x \leq g(y)$. Thus in a tense logic L the pairs $\langle \Diamond, \blacksquare\rangle$ and $\langle \blacklozenge, \Box\rangle$ form Galois connections, namely, (i) $\Diamond\varphi \vdash_L \psi$ if and only if $\varphi \vdash_L \blacksquare\psi$; and (ii) $\blacklozenge\varphi \vdash_L \psi$ if and only if $\varphi \vdash_L \Box\psi$. Thomason [9] formulated tense logics as extensions of classical propositional logic with modal axioms and rules.

Many efforts have been made by logicians to find cut-free sequent calculi for modal and tense logics. Difficulties are encountered already for many normal modal logics, and thus partial solutions including labelled sequent calculus, display calculus, hyper-sequent calculus and deep inference are tried so that they generalize Gentzen's sequent calculus (cf. e.g. [8]). In this talk, we start from display calculi for tense logics (cf. [1, 5]), and show a new type of cut-free sequent calculus for classical tense logic which has been developed in [6, 7] .

Belnap's display calculus for tense logic generalizes Gentzen's sequent calculus by introducing structural connectives for tense operators. A typical feature of such a calculus is the displaying property, namely, for every sequent $\Gamma \Rightarrow \Delta$ and every part Σ of the antecedent (resp. succedent), there exists a structurally equivalent sequent $\Sigma \Rightarrow \Theta$ (resp. $\Theta \Rightarrow \Sigma$) such that Σ alone is its antecedent (resp. succedent). Here Σ is displayed in $\Sigma \Rightarrow \Theta$ (resp. $\Theta \Rightarrow \Sigma$). Cerrato [2] uses structural operators $[.]$ and $\langle . \rangle$ for $\Box$ and $\Diamond$ respectively to give a cut-free sequent calculus for the minimal modal logic K, but it fails to show cut-elimination for very simple normal extensions of K. Kashima [3] introduced the structural operators ${}^P\{.\}$ and ${}^F\{.\}$ for $\blacklozenge$ and $\Diamond$ respectively, and showed cut-elimination in one-side sequent calculi for eight tense logics.

In the author's recent joint work [7], structural operators $\circ$ and $\bullet$ for $\Diamond$ and $\blacklozenge$ are used to construct single-formula sequents, and a Gentzen sequent calculus GK_t for the minimal tense logic K_t is developed. This sequent calculus has the left side displaying property, namely, for every sequent $\Gamma \Rightarrow \alpha$ and a part Δ of Γ, there exists an equivalent sequent $\Delta \Rightarrow \beta$ such that Δ is displayed on the left side. The cut-elimination of GK_t is shown through some intermediate sequent systems. Furthermore, we consider strictly positive axioms of the form $\varphi \to \psi$ where φ and ψ are strictly positive formulas which are built from propositional variables and $\bot$ using only $\wedge$, $\Diamond$ and $\blacklozenge$. Strictly positive modal logics are explored in [4]. All tense logics axiomatized over K_t by strictly positive axioms have cut-free sequent calculi which are obtained by enriching GK_t with structural rules calculated from strictly positive axioms. In this talk, we shall show further how some tense formulas can be calculated into an equivalent strictly positive axiom via an algebraic correspondence theory.

References

1. Belnap, N.: Display logic. J. Philos. Logic **11**, 375–417 (1982)
2. Cerrato, C.: Modal sequents for normal modal logics. Math. Logical Q. **39**, 231–240 (1993)
3. Kashima, R.: Cut-free sequent calculi for some tense logics. Studia Logica **53**, 119–135 (1994)
4. Kikot, S., Kurucz, A., Tanaka, Y., Wolter, F., Zakharyaschev, M.: Kripke completeness of strictly positive modal logics over meet-semilattices with operators. J. Symbolic Logic **84**(2), 533–588 (2019)
5. Kracht, M.: Power and weakness of the modal display calculus. In: Proof Theory of Modal Logic, H. Wansing (ed.), pp. 93–121. Kluwer Academic Publishers, Dordrecht (1996)
6. Lin, Z., Ma, M.: A proof-theoretic approach to negative translations in intuitionistic tense logics. Studia Logica **110**, 1255–1289 (2022)
7. Lin, Z., Ma, M.: Cut-free sequent calculi for tense logics. Axioms **12**, 620 (2023)
8. Poggiolesi, F.: Gentzen Calculi for Modal Propositional Logic. Springer, Heidelberg (2011). https://doi.org/10.1007/978-90-481-9670-8
9. Thomason, S. K.: Semantical analysis of tense logics. J. Symbolic Logic **37**(1), 150–158 (1972)

The Hindsight via the Consequent

Ching Hui Su

School of Philosophy and Social Development, Institute of Concept and Reasoning, Shandong University, Jinan, China
pcs0929@gmail.com

Abstract. The purpose of the present paper is to develop an account of counterfactuals that respect the following three ideas: the important role of the consequent when evaluating, the inferential connection between the antecedents and the consequents, and the gappy intuition about conditionals. The first step is to adopt Dov M. Gabbay's accessibility relation among possible worlds in his 1972 paper, where worlds relevant to both the antecedent and the consequent are accessible to the base world. The second step is to impose the probabilistic dependency between the antecedent and the consequent. The final step is to replace the truth condition of conditionals by the justification condition of conditionals, and the resulting account is a justification theory of counterfactuals.

In [4] , Dov M. Gabbay proposes to formalize counterfactuals as '$\Box_{\varphi\psi}(\varphi \rightarrow \psi)$', where '$\rightarrow$' stands for material implication, indicating two ideas: one is that the consequent, ψ, is important when evaluating counterfactuals, and the other is that there is some inferential connection between the antecedents and the consequents. Though many authors have addressed the latter since Nelson Goodman [5] , few addressed the former. Let's call the former idea '**Consequent**' and the latter idea '**Inferentialism**'. The motivation for Consequent pertains to **IC** pairs, i.e., two conditionals have the same antecedent and inconsistent consequents, but intuitively they both can be true. Here is his proposal: $w(\varphi > \psi) = 1$, if and only if, for all $w' \in W$, if $(w, \varphi, \psi, w') \in \mathcal{R}_>$, then $w'(\varphi \rightarrow \psi) = 1$, where '$>$' stands for counterfactuals. However, his proposal is not good enough, for we still have no idea which possible world should be accessible.

Based on Goodman [5] , when considering '$\varphi > \psi$', we propose that ψ should be derived from φ with the context set $\Delta_{\phi\psi w}$ (w. r. t. the antecedent φ, consequent ψ and the base world w). Now, we're confined to context set $\Delta \subseteq w$. Given that we can order the context sets in terms of their degrees of deviance from the actual world, here is our proposal: $\Delta_{\varphi\psi w}(\varphi > \psi) = 1$, if and only if, for $\Delta_{\varphi\psi w'}, \Delta_{\varphi\psi w''} \in \Upsilon_{\varphi\psi w}$, $\Delta_{\varphi\psi w'} \leq_w \Delta_{\varphi\psi w''}$, where $\Delta_{\varphi\psi w'}$ is defined as the context sets that instantiate $\Delta_{\varphi\psi w} \cup \{\varphi, \psi\}$ and $\Delta_{\varphi\psi w''}$ is defined as the context sets that instantiate $\Delta_{\varphi\psi w} \cup \{\varphi, \neg\psi\}$.

Another motivation for **Inferentialism** is spurred by many psychologists, for their findings, based on well-designed experiments, suggest that there is some kind of inferential connection between the antecedents and the consequents [3] . So we propose that $\Delta_{\varphi\psi w} := \{\chi | \varphi \Leftarrow \chi \text{ or } \psi \Leftarrow \chi\}$ where $\psi \Leftarrow \chi$ stands for χ and ψ are probabilistically dependent in w (i. e. $\mathrm{pr}(\psi|\chi) \neq \mathrm{pr}(\psi)$). Hence, the above proposal will become: $\Delta_{\phi\psi w}(\varphi > \psi) = 1$, if and only if, (a) $\mathrm{pr}(\psi|\varphi) \neq \mathrm{pr}(\psi)$, and (b) for

$\Delta_{\varphi\psi w'}, \Delta_{\varphi\psi w''} \in \Upsilon_{\varphi\psi w}$, $\Delta_{\varphi\psi w'} \leq_w \Delta_{\varphi\psi w''}$, where $\Delta_{\varphi\psi w'}$ is defined as the context sets that instantiate $\Delta_{\varphi\psi w} \cup \{\varphi, \psi\}$ and $\Delta_{\varphi\psi w''}$ is defined as the context sets that instantiate $\Delta_{\varphi\psi w} \cup \{\varphi, \neg\psi\}$.

In [1], Ernest Adams believes that we don't have clear intuition about how to apply the notion of truth to conditionals, and proposes that we should talk of the justification condition or assertability condition of conditionals, rather than truth condition of conditionals. To set apart these three ideas, let's consider the following conditional satisfying Inferentialism [2]:

1. (a) If Jones proposed, William turned her down.
   (b) If anyone constructs a proof of this theorem, I will buy him or her a big cake.

Most people will agree that conditionals will be true, if both the antecedent and the consequent are true, and be false, if the antecedent is true but the consequent is false. However, it seems unclear whether (1a) is true when Jones didn't propose, and whether (1b) is true when everyone fails to construct a proof. Let's call the underdetermination here 'Gappy'. Now consider (1b), which looks like making a promise, i.e., a speech act. As many will agree, speech acts are actions to which the notion of truth is not applicable, so that we can explain why people lack clear intuition about the truth value of conditionals. That is, when asserting a conditional, the speaker is making a speech act: supposing that the antecedent holds, then the consequent will follow. This line of thought can be applied to (1a) and many others.

The motivation for Gappy is due to the obscurity of truth-makers for conditionals. Many authors take possible worlds as truth-makers for conditionals, but one may wonder what they are. It seems that modal realism takes possible worlds as primitive, resisting any further explanation, but this primitiveness brings us nothing but obscurity. Therefore, it is proposed that we should replace the truth condition by the justification condition when considering conditionals. Starting with Gabbay's proposal, we end up with a justification theory of counterfactuals which respects Consequent, Inferentialism, and Gappy.

References

1. Adams, E.: The logic of conditionals. Inquiry Interdisciplinary J. Philos. **8**(1–4), 166–197 (1965)
2. Cantwell, J.: Indicative conditionals: Factual or Epistemic? Studia Logica **88**, 97–117 (2008)
3. Douven, I., Elqayam, S., Singmann, H., van Wijnbergen-Huitink, J.: Conditionals and inferential connections: toward a new semantics. Thinking Reasoning **26**(3), 311–351(2020).
4. Gabbay, Dov M.: A general theory of the conditional in terms of a ternary operator. Theoria. **38**(3), 97–104 (1972)
5. Goodman, N.: The problem of counterfactual conditionals. J. Philos. **44**(5), 113–128 (1947)
6. Krzyzanowska, K., Douven, I.: Missing-link conditionals: pragmatically infelicitous or semantically defective? Intercultural Pragmatics **15**(2), 191–211 (2018)
7. Nute, D.: Topics in Conditional Logic. D. Reidel Publishing Company (1980)

Contents

An Inferential Theory of Causal Reasoning

Alexander Bochman(✉)

School of Computer Science, Holon Institute of Technology (HIT), Holon, Israel
bochmana@hit.ac.il

Abstract. We present a general formalism of causal reasoning that encompasses both Pearl's approach to causality and a number of key systems of nonmonotonic reasoning in artificial intelligence.

Keywords: Causation · Rationality · Semantics · Inference · Defaults

The aim of the paper consists in providing a principle-based description for a particular theory of causal reasoning. This theory, called the causal calculus (see [14,16]) has been born as part of a general field of nonmonotonic reasoning in artificial intelligence, where it has been shown to cover important areas and applications of AI. A new stage in the development of this theory has emerged with the realization that it can also provide a formal representation for Pearl's approach to causality in the framework of structural equation models (see [9]). A detailed description of the causal calculus, as well as the range of its current applications in AI and beyond can be found in [7].

In this paper, we are going to show that this formalism is based on a profound rationality principle that provides foundations for a general theory of causal reasoning, a kind of reasoning that goes well beyond standard deduction and correspondence semantics.

The formalism of causal reasoning described below will have a language that consists of a set of (causal) inference rules on propositions. It will also have a semantics defined in terms of valuations on propositions that are in accord with the causal rules. This semantics, however, will be based on a radically different, causal principle of acceptance for propositions that will set the corresponding reasoning system apart from traditional representational approaches to language and meaning. In particular, causal rules cannot be defined in this formalism as rules that preserve acceptance. Moreover, though a causal theory will determine its associated rational semantics of acceptance, the latter does not and even cannot determine the source causal theory. This fact will have multiple consequences for the corresponding theory of causal reasoning. It will lead to an entirely new agenda and desiderata for such a reasoning.

1 Causal Theories and Their Semantics

As it is common for reasoning formalisms, our system of causal reasoning will have a language and an associated semantics. Its language will be a set of causal

N. Alechina et al. (Eds.): LORI 2023, LNCS 14329, pp. 1–16, 2023.
https://doi.org/10.1007/978-3-031-45558-2_1

rules that are defined on an underlying language of propositions, while its semantics will be a set of valuations on propositions that conform to the causal rules. At the first stage, our underlying language L will be defined simply as a set of (atomic) propositions.

A *causal rule* is an inference rule of the form $a \Rightarrow A$, where a is a finite set of propositions and A a proposition. The rule asserts that a set a of propositions *causes* proposition A. A set of causal rules will be called a *causal theory.* A causal theory will provide an ultimate basis of causal reasoning, mainly in the form of constraints it imposes on acceptance of propositions.

The basic principle of causal reasoning can be formulated as the following rationality postulate of acceptance for propositions:

Causal Acceptance Principle. A proposition A is accepted with respect to a causal theory Δ if and only if Δ contains a causal rule $a \Rightarrow A$ such that all propositions in a are accepted.

The principle states that (acceptance of) propositions can both serve as and stand in need of *reasons*. In this sense, it can be viewed as a constitutive principle of rationality in our causal context (see [11]). Sets of accepted propositions that conform to the above principle will form the *models* of the corresponding causal theory.

The two parts of the above principle could be expressed as two independent rationality postulates:

Preservation Principle. If all propositions in a are accepted, and a causes A, then A should be accepted.

Principle of Sufficient Reason. Any proposition should have a cause for its acceptance.

The Preservation Principle states that the very concept of an inference rule (however understood) presupposes that such a rule should preserve (or 'transmit') acceptance of propositions. On a normative reading, it states that existence of (sufficient) reason is sufficient for acceptance.

Leibniz' Principle of Sufficient Reason is again a normative principle of reasoning stating that propositions *require* reasons for their acceptance, and such reasons are provided by establishing their causes. The origins of this principle go back to the ancient law of causality, but it surfaces, for instance, as the principle of definitional reflection in proof-theoretic approaches (see, e.g., [22]).

Example 1. *The following causal theory provides a causal description of some well-known example originated in [17].*[1]

$$Rained \Rightarrow Grasswet$$
$$Sprinkler \Rightarrow Grasswet$$
$$Rained \Rightarrow Streetwet.$$

[1] We assume that the labels of associated propositions are self-explanatory.

Just as for ordinary deductive inference systems, if, for instance, Rained is accepted with respect to such a causal theory, then both Grasswet and Streetwet should also be accepted. However, in a causal reasoning with this causal theory, any acceptable set of propositions that contains Grasswet should contain either Rained or Sprinkler as its causes. Similarly, Streetwet implies in this sense acceptance of both its only possible cause Rained and a collateral effect Grasswet. Both derivations from causes to their effects and from effects to their possible causes constitute essential parts of causal reasoning.

Rational Semantics. The semantics of a causal theory will be defined in terms of *valuations* on propositions. A valuation is a function $v \in \{0,1\}^L$ that assigns either 1 ('truth') or 0 ('falsity') to every proposition of the language L. If $v(A) = 1$, we will say that proposition A is *accepted* ('taken-true') in the valuation v. A valuation can be safely identified with its associated set of accepted propositions, and we will often view v itself as a set of (accepted) propositions.

For any set u of propositions and a causal theory Δ, we will denote by $\Delta(u)$ the set of all propositions that are directly caused by u in Δ, that is,

$$\Delta(u) = \{A \mid a \Rightarrow A \in \Delta,\ a \subseteq u\}.$$

This notation will help in formulating the semantics for our causal language.

Definition 1. – *A* causal model *of a causal theory Δ is a valuation that satisfies the condition*

$$v = \Delta(v).$$

– *A* rational semantics *of a causal theory is the set of all its causal models.*

The notion of a causal model provides precise formal expression of the Causal Acceptance principle since it determines that a proposition is accepted in a model if and only if it has a cause in this model.

$\Delta(u)$ is a monotonic operator on the set of propositions, while causal models correspond to its fixed points. Consequently, any causal theory has at least one causal model, so it always has a rational semantics. As an important special case, a causal theory always has the least model. This model provides a representation of (deductive) *provability* in our causal framework. However, it expresses only a small part of the informational content embodied in the source causal theory. Moreover, this observation can actually be extended to the rational semantics itself. A causal model, viewed just as a set of (accepted) propositions, contains only purely categorical, *factual* information. In this respect, it provides only a possible factual instantiation (a "factual shadow," if you like) of the rich causal information embodied in the source causal theory, what causes what. For instance, the Preservation principle *cannot* be used as a sole principle of validity for the causal rules themselves. Namely, we cannot follow Tarski in *defining* causal rules as inference rules that preserve acceptance. This could be seen already from the fact that such a stipulation would sanction the Reflexivity postulate of deductive consequence (namely, all rules of the form $A \Rightarrow A$) and this would trivialize in turn the second part of our rationality postulate, the principle of sufficient reason: on a causal reading, rules $A \Rightarrow A$ will make all propositions self-justified (self-evident).[2] Incidentally, this observation indicates also that (absence of) Reflexivity constitutes one of the key differences between causal inference and deductive consequence.

2 Causal Inference

It turns out that there are *formal* derivations among causal rules that always preserve the rational semantics. Such derivations will be taken to constitute the underlying *logic* of causal inference.

Definition 2. *A* causal inference relation *is a set of causal rules that is closed with respect to the following derivation rules:*[3]

Monotonicity *If* $a \Rightarrow A$ *and* $a \subseteq b$, *then* $b \Rightarrow A$;
Cut *If* $a \Rightarrow A$ *and* $a, A \Rightarrow B$, *then* $a \Rightarrow B$.

The notion of causal inference incorporates two of the three basic postulates for ordinary Tarski consequence relations. It disavows, however, the first postulate of Tarski consequence, the Reflexivity postulate. It is precisely this 'omission' that creates the possibility of causal reasoning in this framework. Still, the remaining two postulates capture the notion of *derivability* among propositions that is determined by a given set of (causal) inference rules.

It should be noted that we do *not* require that our causal inference should be anti-reflexive. Reflexive rules $A \Rightarrow A$ can belong to a causal theory, but in the framework of causal reasoning they already acquire a nontrivial content. More precisely, such a rule says that A is a self-evident proposition that does not require further justification for its acceptance. Propositions that satisfy such rules will be called assumptions in what follows.

We will extend causal rules to rules having arbitrary sets of propositions as premises using a compactness recipe: for any set u of propositions, we define $u \Rightarrow A$ as follows:

$$u \Rightarrow A \equiv_{def} a \Rightarrow A, \text{ for some finite } a \subseteq u.$$

For a set u of propositions, $C(u)$ will denote the set of propositions caused by u with respect to a causal inference relation, that is, $C(u) = \{A \mid u \Rightarrow A\}$. This causal operator plays much the same role as the usual derivability operator for consequence relations. In particular, the above postulates of causal inference can be recast as the following properties of the causal operator:

Monotonicity If $u \subseteq v$, then $C(u) \subseteq C(v)$.

[2] See [20] for a similar point.
[3] In what follows, causal rules $a \Rightarrow A$ are used both as formal objects of our theory and as statements in the meta-language (saying that a causes A).

Cut $C(u \cup C(u)) \subseteq C(u)$.

Thus, C is monotonic. However, the inclusion $u \subseteq C(u)$ does not always hold. Also, C is not idempotent, that is, $C(C(u))$ can be distinct from $C(u)$.[4]

For a causal theory Δ, we will denote by $\Rightarrow_\Delta$ the least causal inference relation that includes Δ, while C_Δ will denote the associated causal operator.

2.1 Causal Vs. Semantic Equivalence

Now we will show that causal inference constitutes a logical framework for reasoning with causal models.

Definition 3. *Two causal theories will be called* semantically equivalent *if they determine the same rational semantics.*

If $\Rightarrow_\Delta$ is the least causal inference relation that contains a causal theory Δ, then we have:

Lemma 1. *Any causal theory Δ is semantically equivalent to $\Rightarrow_\Delta$.*

Definition 4. *Two causal theories Δ and Γ will be called* logically equivalent, *if each can be obtained from the other using derivation rules of causal inference. Or, equivalently, when $\Rightarrow_\Delta$ coincides with $\Rightarrow_\Gamma$.*

Now, as an immediate consequence of the previous lemma, we obtain:

Corollary 2. *Logically equivalent causal theories are semantically equivalent.*

The reverse implication in the above corollary does not hold, and a deep reason for this is that the rational semantics *does not* fully determine the content of the source causal theory. This means, in particular, that it may well happen that two essentially (i.e., informationally) different causal theories could determine the same rational semantics. This under-determination is closely related to the fact that semantic equivalence of causal theories is *nonmonotonic*; it is not preserved under extensions of causal theories with further causal rules. The following simple example illustrates this.

Example 2. *Let us consider two causal theories: $\{A \Rightarrow B\}$ and $\{C \Rightarrow D\}$. These causal theories are obviously different, though they are semantically equivalent since they determine the same rational semantics which contains a single model $\emptyset$ in which no proposition is accepted. Now let us add to these causal theories the same causal rule $A \Rightarrow A$. Then the first causal theory will already have an additional model $\{A, B\}$, while the semantics of second theory will include two models, $\emptyset$ and $\{A\}$.*

A stronger, logical counterpart of semantic equivalence that would be preserved under addition of new causal rules can be defined as follows.

[4] For instance, A can directly cause B, though there are no intermediate causes between A and B.

Definition 5. *Causal theories* Δ *and* Γ *are* strongly semantically equivalent *if, for any set* Φ *of causal rules,* $\Delta \cup \Phi$ *is semantically equivalent to* $\Gamma \cup \Phi$.

Strongly equivalent causal theories are "equivalent forever"-that is, they are interchangeable in any larger causal theory without changing the associated rational semantics. Accordingly, strong equivalence could be viewed as a kind of logical equivalence with respect to some background logic of causal rules. And the theorem below shows that this logic is precisely the logic of causal inference.

Theorem 3. *Two causal theories are strongly semantically equivalent iff they are logically equivalent.*

The main idea behind the proof of the above theorem is that if two causal theories are not logically equivalent, we can always find some further causal rules such that their addition to these two theories will produce new causal theories that will already have different rational semantics. This result implies, in particular, that causal inference relations are maximal inference relations that are adequate for causal reasoning with respect to the rational semantics.

2.2 Causal Inference Vs. Deductive Consequence

A further insight into the properties of causal inference can be obtained by comparing it with deductive consequence. Thus, for a causal inference relation $\Rightarrow$ we can define the associated consequence relation $\vdash_{\Rightarrow}$ as follows:

$$u \vdash_{\Rightarrow} A \ \equiv_{def} \ A \in u \text{ or } u \Rightarrow A.$$

Then the following fact can be easily verified.

Lemma 4. *If* $\Rightarrow$ *is a causal inference relation, then* $\vdash_{\Rightarrow}$ *is the least consequence relation containing* $\Rightarrow$.

Let $Cn_{\Rightarrow}$ denote the consequence operator corresponding to $\vdash_{\Rightarrow}$. Then we have the following equality, for any set u of propositions:

$$Cn_{\Rightarrow}(u) = u \cup C(u).$$

Now, the Cut postulate immediately implies the following equality:

$$C(u) = C(Cn_{\Rightarrow}(u)).$$

Actually, the same Cut postulate implies also $C(u) = Cn_{\Rightarrow}(C(u))$, so the causal operator absorbs $Cn_{\Rightarrow}$ on both sides:

$$Cn_{\Rightarrow} \circ C = C \circ Cn_{\Rightarrow} = C.$$

This shows that deductive consequences of a given causal theory can be used as intermediate premises and conclusions in causal inference. In a hindsight, this could explain why it has been so difficult to distinguish causal reasoning proper from general deductive reasoning. In fact, causal rules could even be seen as a special kind of deductive rules. This view naturally fits Aristotle's theory of reasoning in his *Analytics* where (causal) demonstrations were viewed as a species of syllogisms (deductions) (see [7]). It should be kept in mind, however, that deductive inference alone is insufficient for determining the *causal* consequences of a set of propositions. In other words, deductive consequence cannot be used for determining the corresponding causal inference associated with a given set of inference rules.

Axioms Versus Assumptions. The rational semantics requires that any accepted proposition should have a cause. Accordingly, justification of accepted propositions constitutes an essential part of this semantic framework. In fact, this is a common feature of many other formalisms of nonmonotonic reasoning in AI (see, e.g., [12] for an abstract theory of justifications in nonmonotonic reasoning).

The law of causality inevitably leads to a fundamental problem known already in antiquity as the *Agrippan trilemma*: if you do not want to accept infinite regress of causation (or justification), you should accept either uncaused or self-caused propositions. Now, in the framework of causal theories, there are two kinds of propositions that can play, respectively, these two roles:

Definition 6. – *A proposition A is an* axiom *of a causal theory Δ if the rule $\emptyset \Rightarrow A$ belongs to Δ;*
– *A proposition A will be called an* assumption *of a causal theory if the rule $A \Rightarrow A$ belongs to it.*

Example 3. *Let us return to Pearl's example (Example 1):*

$$Rained \Rightarrow Grasswet \quad Sprinkler \Rightarrow Grasswet \quad Rained \Rightarrow Streetwet$$

Note first that, taken by itself, this causal theory does not have causal models (more precisely, it has a single empty causal model), mainly because the causal status of Rained and Sprinkler are not determined. But now let's make Rained and Sprinkler causal assumptions of our theory:

$$Rained \Rightarrow Rained \qquad Sprinkler \Rightarrow Sprinkler.$$

As a result, the rational semantics of this causal theory will acquire three additional causal models:

$$\{Rained, Grasswet, Streetwet\} \quad \{Sprinkler, Grasswet\}$$
$$\{Rained, Sprinkler, Grasswet, Streetwet\}$$

These models display already some correlations *between the relevant propositions. For instance, that Rained is always accompanied by Grasswet and Streetwet in these models (deduction), but also that Streetwet is always accompanied by Rained (abduction).*

In contrast to deductive reasoning, both axioms and assumptions provide reasonable end-points of justification in causal reasoning: axioms do not require justification, while assumptions correspond in this sense to self-evident propositions. For causal inference relations, any axiom will also be an assumption, though not vice versa. The difference between the two can be described as follows. Every axiom *must* be accepted in any reasonable model, and hence it should belong to any causal model. In contrast, any assumption *can* be accepted when it is consistent with the rest of accepted propositions, but it does not have to be accepted. As a result, causal theories admit in general multiple causal models depending on the assumptions we actually accept. This functionality makes assumptions much similar to abducibles in a system of *abductive reasoning*. In fact, it has been shown in [5] that causal inference allows us to provide a uniform and syntax-independent description of abductive reasoning.

3 Supraclassical Causal Reasoning

Now we will raise our abstract theory to a full-fledged reasoning system that will subsume, in particular, both Pearl's causal formalism and a number of prominent systems of nonmonotonic reasoning in artificial intelligence.

A basic disederatum, or prerequisite, of such a full-fledged system of reasoning is the ability to use ordinary classical entailment as an integral part of causal reasoning. Technically, a solution to this task is quite straightforward. Recall that any causal theory has an associated consequence relation, and this consequence relation can be safely used inside causal derivations. Accordingly, we only need to require that this consequence relation should be *supraclassical*, that is, it should subsume classical entailment.

From now on, our language L of propositions will be a classical propositional language with the usual classical connectives and constants $\{\wedge, \vee, \neg, \rightarrow, \mathbf{t}, \mathbf{f}\}$. The symbol $\vDash$ will stand for the classical entailment while Th will denote the associated classical provability operator. Also, $p, g, r, \ldots$ will denote propositional atoms, $A, B, C, \ldots$ will denote arbitrary classical propositions, and $a, b, c, \ldots$ finite sets of propositions.

Definition 7. *A causal inference relation in a classical language will be called* supraclassical *if it satisfies the following additional rules:*

(Strengthening) *If* $b \Rightarrow C$ *and* $a \vDash B$*, for every* $B \in b$*, then* $a \Rightarrow C$*;*
(Weakening) *If* $a \Rightarrow B$ *and* $B \vDash C$*, then* $a \Rightarrow C$*;*
(And) *If* $a \Rightarrow B$ *and* $a \Rightarrow C$*, then* $a \Rightarrow B \wedge C$*;*
(Truth) $\mathbf{t} \Rightarrow \mathbf{t}$*;*
(Falsity) $\mathbf{f} \Rightarrow \mathbf{f}$.

As a matter of fact, the origins of the above postulates can be found in Input/Output logics of [15], the only difference being the last postulate, Falsity. The latter could be viewed as a causal version of *Ex Falso Quodlibet* ("from falsehood, anything"), and its role consists, in effect, in excluding classically inconsistent models.

Causal reasoning in this setting requires also an appropriate 'upgrade' of the rational semantics. Namely, it requires that causal models should also be closed with respect to classical entailment.

Definition 8. – *A* classical causal model *of a causal theory Δ is a consistent valuation that satisfies*

$$v = Th(\Delta(v)).$$

– *A* supraclassical rational semantics *of a causal theory is the set of all its classical causal models.*

A classical causal model is closed both with respect to the causal rules and classical entailment. The principle of sufficient reason in such models is generalized, however, to the principle that any accepted proposition should be a classical logical consequence of accepted propositions that are caused in the model.

It turns out that supraclassical causal inference provides an adequate logical framework for reasoning with respect to the supraclassical rational semantics.

Definition 9. *Causal theories Δ and Γ will be called* s-equivalent *if they determine the same supraclassical rational semantics, and* strongly s-equivalent *if, for any set Φ of causal rules, $\Delta \cup \Phi$ is semantically s-equivalent to $\Gamma \cup \Phi$.*

If $\Rightarrow^s_\Delta$ is the least supraclassical causal inference relation that contains a causal theory Δ, then Δ will be strongly s-equivalent to $\Rightarrow^s_\Delta$. Moreover, the following theorem shows that supraclassical causal inference constitutes a maximal logic for the supraclassical semantics.

Theorem 5. *Two causal theories are strongly s-equivalent if and only if they determine the same supraclassical causal inference relation.*

4 Defaults in Causal Reasoning

The causal calculus has been shown to cover significant parts of nonmonotonic reasoning such as abduction and diagnosis, logic programming, and reasoning about action and change. As a further illustration, we will describe now a 'causal counterpart' of one of the key, original formalisms of nonmonotonic reasoning, default logic of [21]. This causal representation will also allow us to clarify the meaning of the main notions associated with default logic and first of all of the concept of default itself.

Defaults can be viewed as a special kind of assumptions. Namely, they are assumptions that we *must* accept unless there are reasons to the contrary.

Let us say that a proposition A is *refuted* if there is a cause for the contrary proposition $\neg A$. Then we can formulate the following informal principle:

Default Acceptance. *A default is an assumption that is accepted whenever it is not refuted.*

Note, however, that $\neg A$ is accepted in a causal model only if it has a cause in this model (that is, when A is refuted). Accordingly, the principle of Default Acceptance in causal models boils down to the principle of Default Bivalence:

Default Bivalence. *For any causal model v and any default assumption A, either $A \in v$ or $\neg A \in v$.*

Default Bivalence can be viewed as a characteristic property of defaults. Again, this is in contrast with classical logical reasoning where *all* propositions are required to satisfy bivalence. Note also that any axiom of a causal theory will also be a default (namely a default that cannot be consistently refuted), so defaults can be viewed as an intermediate notion between axioms and assumptions in general.

Reasoning in default logic amounts to deriving justified conclusions from a default theory by using its inference rules and defaults. However, if the set of all defaults is jointly incompatible with the background theory, we must make a reasoned choice among the default assumptions. At this point, default reasoning requires that a reasonable set of defaults should refute all defaults that are left out. The appropriate choices of defaults (called stable sets) determine then the *extensions* of a default theory which are taken to constitute the nonmonotonic semantics of the latter. This notion of extension presupposes, in turn, that any proposition which is not a default is accepted only if it is grounded, ultimately, in the set of accepted defaults. In other words, once we choose an acceptable ("stable") set of defaults, the rest of acceptable propositions should be derived from this set. This stringent, 'puritan' understanding of acceptance creates, in effect, a *bipolar system* of reasoning that divides all propositions into two classes, factual propositions and defaults, with opposite principles of acceptance. It is this understanding that also makes default logic an instantiation of (assumption-based) argumentation [10] where defaults play the role of arguments.

The following definition provides a causal representation of default reasoning.

Definition 10. – *A* default causal theory *is a pair $(\Delta, \mathcal{D})$, where Δ is a causal theory, and $\mathcal{D}$ a distinguished subset of its assumptions, called* defaults.

– *A* default model *of a default causal theory is a classical causal model m of Δ that satisfies the following two conditions:*

(Default Grounding) *m is caused by its defaults:* $m = \mathcal{C}_\Delta(m \cap \mathcal{D})$.[5]

(Default Bivalence) *For any default $D \in \mathcal{D}$, either $D \in m$ or $\neg D \in m$.*

– *A* default semantics *of a default causal theory is the set of its default models.*

Default semantics can be viewed as a special case of the rational semantics of causal theories. Still, there are two reasons why the reverse inclusion does not hold. First, a causal model may be generated not only by defaults, but also by other assumptions (on our causal understanding of the latter). Second, even when a causal model is caused by some set of defaults, it may still not satisfy

[5] Here $\mathcal{C}_\Delta$ is a causal operator corresponding to the least supraclassical causal inference relation containing Δ.

the principle of default bivalence. This might happen, in particular, when the background causal theory lacks appropriate cancellation rules that would allow us to *refute* incompatible defaults. As an extreme case, a default causal theory may even lack default models at all (though it always has causal models).

The above formalism provides an adequate description of default logic in the sense that there exist back and forth translations between them (see [8]). Moreover, the formalism of logic programming in AI naturally corresponds to the special case of default causal theories in which classical negative literals are defaults (see [3]).

Among many other things, default logic provides a feasible and working account of *defeasible reasoning* while preserving monotonicity of inference rules themselves, in contrast to approaches that are based on a total rejection of monotonicity as a way of solving the problem of defeasible inference. However, the discussion of the relative merits and shortcomings of these two basic approaches to defeasibility in AI is beyond the scope of this study.

5 Pearl's Causal Models and Basic Inference

Pearl's approach to causal reasoning in the framework of structural equation models can be viewed as an important instantiation of our general theory. According to [18], a causal model is a triple $M = \langle U, V, F\rangle$, where U is a set of *exogenous* variables, V is a finite set of *endogenous* variables, and F is a set of functions that can be represented as *structural* equations $V_i = f_i(PA_i, U_i)$, where PA_i is the minimal set of variables in $V\backslash\{V_i\}$ (parents of V_i) sufficient for representing f_i, and similarly for the exogenous variables $U_i \subseteq U$. Each such equation stands for a set of "structural" equalities

$$v_i = f_i(pa_i, u_i) \quad i = 1, \ldots, n,$$

where v_i, pa_i and u_i are, respectively, particular realizations of V_i, PA_i and U_i. Such an equality assigns a specific value v_i to a variable V_i depending on the values of its parents and relevant exogenous variables.

In Pearl's account, every instantiation of the exogenous variables determines a particular "causal world" of the causal model. Such worlds stand in one-to-one correspondence with the solutions to the above equations in the ordinary mathematical sense. However, structural equations also encode causal information in their very syntax by treating the variable on the left-hand side of the $=$ as the effect and treating those on the right as causes. This causal reading plays a crucial role in determining the effect of external interventions and evaluation of counterfactual assertions with respect to such a model.

The representation of Pearl's causal models in the causal calculus, suggested in [9], amounted in effect to viewing each structural equality $v_i = f_i(pa_i, u_i)$ as a causal rule saying that the instantiation pa_i of the parent endogenous variables PA_i and the instantiation u_i of exogenous variables U_i *causes* the instantiation $f_i(pa_i, u_i)$ of V_i:

$$PA_i = pa_i, U_i = u_i \;\Rightarrow\; V_i = f_i(pa_i, u_i).$$

In the special case when all the relevant variables are Boolean, a Boolean structural equation $p = F$ (where F is classical logical formula) produces in this sense two causal rules

$$F \Rightarrow p \quad \text{and} \quad \neg F \Rightarrow \neg p.$$

Given this translation, it was shown in [9] that Pearl's causal worlds correspond precisely to classical causal models of the associated causal theory that are (classical) *worlds.*

Example 4. *The following set of (Boolean) structural equations provides a representation of Pearl's example (see Example 1) in structural models:*

$$Grasswet = Rained \vee Sprinkler \qquad Streetwet = Rained.$$

If Rained and Sprinkler are taken to be exogenous variables, while Grasswet and Streetwet are endogenous ones, then the corresponding Pearl's structural model will have the same causal worlds as the following causal theory:

$$Rained \Rightarrow Grasswet \quad Sprinkler \Rightarrow Grasswet \quad Rained \Rightarrow Streetwet$$
$$\neg Rained, \neg Sprinkler \Rightarrow \neg Grasswet \quad \neg Rained \Rightarrow \neg Streetwet$$

with an additional stipulation that Rained, ¬Rained, Sprinkler and ¬Sprinkler are assumptions:

$$Rained \Rightarrow Rained \quad \neg Rained \Rightarrow \neg Rained$$
$$Sprincler \Rightarrow Sprinkler \quad \neg Sprinkler \Rightarrow \neg Sprinkler$$

Compared with our previous causal description of this example (see Example 3), the above causal theory contains additional causal rules, namely causal rules for the corresponding negative literals. These negative causal rules can be reproduced, however, using a systematic procedure called negative causal completion - see [7] for further details.

5.1 Counterfactual Equivalence

In Pearl's framework, the relation between causal theories and their (rational) semantics reduces to the relation between causal and purely mathematical understanding of structural equations. Thus, as in the general case of causal theories discussed earlier, two conceptually different sets of structural equations may "accidentally" determine the same causal worlds. According to Pearl, however, the relevant differences between them can be revealed by performing the same interventions ("surgeries") on them.

In the Boolean case, the corresponding transformation of causal theories can be described as follows:

Definition 11. *For a causal theory Δ and a set L of literals, a* revision $\Delta{*}L$ *of Δ is a causal theory obtained from Δ by removing all 'contrary' rules $A \Rightarrow \neg l$ for $l \in L$ and adding the rules $\mathbf{t} \Rightarrow l$ for each $l \in L$.*

Revisions of causal theories correspond to *submodels* of causal models in the sense of Pearl. Now, according to Pearl, every causal model stands for a whole set of its submodels that embody interventional contingencies. These submodels determine, in a sense, the "causal content" of a given causal model. In accordance with that, we can introduce the following definition:

Definition 12. *Causal theories Γ and Δ are* intervention-equivalent *if, for every set L of literals, the revision $\Gamma{*}L$ has the same causal worlds as $\Delta{*}L$.*

It can be shown that intervention-equivalence of two causal theories amounts to coincidence of their associated causal counterfactuals (see [7]). Accordingly, the content of causal theories is fully determined by their 'counterfactual profile' in Pearl's approach. In this sense, it can even be viewed as a further development of the counterfactual approach to causal reasoning initiated by David Lewis in [13]. Taken in this perspective, the difference between our approach and that of Pearl amounts to taking intervention-equivalence instead of strong semantic equivalence as a basic information concept for causal theories. This counterfactual approach sanctions, however, a somewhat different logic for causal reasoning.

5.2 Basic Causal Inference

It turns out that the Cut rule of causal inference does *not* preserve intervention-equivalence. In order to cope with this situation, we will revise our definition of causal inference as follows.

Definition 13.
- *A set of causal rules in a classical language will be called a* causal production relation *if it satisfies all the postulates of supraclassical causal inference except Cut.*
- *A causal production relation will be called* basic *if it satisfies the following rule:*

  (Or) *If $A \Rightarrow C$ and $B \Rightarrow C$, then $A \vee B \Rightarrow C$.*

As follows from the definition, basic inference is obtained from supraclassical causal inference by replacing the Cut postulate with Or. A detailed description of this kind of inference has been given in [2,4]. It has been shown, in particular, that causal rules in this formalism can already be given a purely modal *logical* interpretation in possible worlds models; by this interpretation, a causal rule $A \Rightarrow B$ is representable as a modal conditional

$$A \to \Box B,$$

where $\Box$ is a standard necessity operator defined in relational Kripke frames (see also [23]).

It has been shown in [6] that basic inference constitutes, in effect, the internal logic of causal reasoning in Pearl's causal models. More precisely, it has been shown that basically equivalent causal theories are intervention equivalent. Moreover, the reverse implication has been shown to hold for the special case of *Pearl's* causal theories, that is, for causal theories obtained from structural equation models by the translation of [9]. Some consequences of this correspondence were discussed in [7] in the context of analyzing different approaches to the notion of actual causality.

6 Classical Causal Inference and Causal Worlds

The differences between Pearl's theory and our approach disappear once we restrict our rational semantics to causal models that are worlds (in the usual classical meaning of the term). Thus, it was shown already in [2] that the postulate Or becomes an admissible derivation rule with respect to the world-based rational semantics. Note, however, that the restriction of causal models to worlds amounts to imposing Bivalence on the set of accepted propositions.

Definition 14. *A causal inference relation will be called* classical *if it is supraclassical and satisfies Or.*

Classical causal inference combines the properties of both basic and supraclassical causal inference. In particular, the causal rules of such an inference inherit a logical semantics in the modal framework of possible worlds, in which they are interpreted as modal conditionals $A \to \Box B$. And its corresponding rational semantics can also be restricted to causal models that are worlds.

Definition 15.
- *A* causal world *of a causal theory Δ is a classical causal model of Δ which is also a world (maximal classically consistent set).*
- *A* classical rational semantics *of a causal theory is the set of all its causal worlds.*

Classical causal inference has been shown to provide an adequate logical framework of reasoning with respect to the classical rational semantics.

Classical rational semantics could be viewed as the closest causal counterpart of the traditional correspondence semantics. Nevertheless, even this rational semantics remains nonmonotonic with respect to the underlying causal theory.

7 Conclusions

The causal calculus has been shown to provide a formal basis for reasoning and problem-solving in many areas and applications of AI. Moreover, due to deep and natural connections of causes with reasons and explanations, causal reasoning brings with it the promise of Explainable AI, an approach to artificial intelligence that is not only practically successful but is also susceptible to rational explanation and justification. We have seen also that this theory provides a formal representation for Pearl's approach to causation, and in this sense it can be viewed as a natural basis for a unified approach to causal reasoning.

The theory of causal reasoning described in this paper poses, however, a lot of questions and challenges for a general theory of reasoning. To begin with, being a nonmonotonic formalism, it is based on a unidirectional connection between the language (of causal rules) and its (rational) semantics, and this forces us to reconsider the basic notions associated with denotational approaches, such as truth and meaning/reference of language expressions. It also puts into question the very possibility, or even desirability, of constructing causal reasoning or its semantics bottom up from propositional atoms. Thus, we have employed a global, holist approach to incorporating classical entailment into causal reasoning. This construction provides all that is needed for the use of such a reasoning in applications, including derivations of conclusions and computation of the corresponding models. Actual work with this formalism could defuse the suspicion that it is somehow deficient or flawed in this respect. This construction distinguishes our theory also from standard proof-theoretic approaches that attempt to provide a modular inferential description of logical connectives in terms of associated introduction and elimination rules.

In a more general perspective, the miracle of resurrection of causal reasoning in artificial intelligence and other important fields of science confirms once again that causation should be viewed as an essential part of our reasoning, a kind of reasoning that has deep, though almost forgotten, roots in human history. Our inferential approach to causation fully endorses Elizabeth Anscombe's claim that causality consists in the derivativeness of an effect from its causes (see [1]), and it goes back as far as to Aristotle's theory of causal demonstrations as a special kind of syllogisms (deductions), to Leibniz's obliteration of the distinction between reasons and causes, and to Hume's views of inference as an 'impression source' of causation. This view of causal reasoning provides also natural connections of our theory with a general approach of inferentialism (see, e.g., [19]), or at least with a version of it that (in contrast to Sellars and Brandom) does not put conceptual barriers between causal and inferential (normative). But all this could be a subject of an entirely different study.

References

1. Anscombe, G.E.M.: Causality and determination. In: The Collected Philosophical Papers of G.E.M. Anscombe, pp. 133–147. Basil Blackwell (1981)
2. Bochman, A.: A causal approach to nonmonotonic reasoning. Artif. Intell. **160**, 105–143 (2004)
3. Bochman, A.: A causal logic of logic programming. In: Principles of Knowledge Representation and Reasoning: Proceedings of the Ninth International Conference (KR2004), Whistler, Canada, 2–5 June 2004, pp. 427–437. AAAI Press (2004)
4. Bochman, A.: Explanatory Nonmonotonic Reasoning. World Scientific, Singapore (2005)
5. Bochman, A.: A causal theory of abduction. J. Log. Comput. **17**, 851–869 (2007)
6. Bochman, A.: On laws and counterfactuals in causal reasoning. In: Principles of Knowledge Representation and Reasoning: Proceedings of the Sixteenth International Conference, KR 2018, pp. 494–503. AAAI Press (2018)
7. Bochman, A.: A Logical Theory of Causality. MIT Press, Cambridge (2021)
8. Bochman, A.: Default logic as a species of causal reasoning. In: Proceedings of KR-2023 (2023). (to appear)

9. Bochman, A., Lifschitz, V.: Pearl's causality in a logical setting. In: Proceedings of the 29th AAAI Conference on Artificial Intelligence, pp. 1446–1452. AAAI Press (2015)
10. Bondarenko, A., Dung, P.M., Kowalski, R.A., Toni, F.: An abstract, argumentation-theoretic framework for default reasoning. Artif. Intell. **93**, 63–101 (1997)
11. Brandom, R.: Articulating Reasons: An Introduction to Inferentialism. Harvard University Press, Cambridge (2000)
12. Denecker, M., Brewka, G., Strass, H.: A formal theory of justifications. In: 13th International Conference on Logic Programming and Non-monotonic Reasoning (LPNMR), pp. 250–264 (2015)
13. Lewis, D.: Causation. J. Philos. **70**, 556–567 (1973)
14. Lifschitz, V.: On the logic of causal explanation. Artif. Intell. **96**, 451–465 (1997)
15. Makinson, D., van der Torre, L.: Input/output logics. J. Philos. Log. **29**, 383–408 (2000)
16. McCain, N., Turner, H.: Causal theories of action and change. In: Proceedings of the Fourteenth National Conference on Artificial Intelligence (AAAI-97), pp. 460–465 (1997)
17. Pearl, J.: Embracing causality in formal reasoning. In: Proceedings of the Sixth National Conference on Artificial Intelligence (AAAI-87), pp. 369–373 (1987)
18. Pearl, J.: Causality: Models, Reasoning and Inference, 2nd edn. Cambridge University Press, Cambridge (2009). (1st ed. 2000)
19. Peregrin, J.: Inferentialism: Why Rules Matter. Palgrave-Macmillan, London (2014)
20. Prawitz, D.: The seeming interdependence between the concepts of valid inference and proof. Topoi **38**(3), 493–503 (2019)
21. Reiter, R.: A logic for default reasoning. Artif. Intell. **13**, 81–132 (1980)
22. Schroeder-Heister, P.: The categorical and the hypothetical: a critique of some fundamental assumptions of standard semantics. Synthese **187**(3), 925–942 (2012)
23. Turner, H.: A logic of universal causation. Artif. Intell. **113**, 87–123 (1999)

Indicative Conditionals in Awareness Framework

Tianyi Chu(✉)

ILLC, University of Amsterdam, Amsterdam, The Netherlands
tianyi_chu@qq.com

Abstract. We propose a novel approach to capturing the acceptability conditions of indicative conditionals using awareness logic. Specifically, we posit that the relevance between the antecedent and consequent of a conditional is determined by the concomitant occurrence of their awareness. We provide a sound and complete axiomatization of the logic. We compare the properties of the indicative conditionals in our framework with other existing theories.

Keywords: Awareness · Relevance · Conditionals · Indicative Conditionals

1 Motivation

Conditionals are sentences that have the form "If ϕ, [then] ψ". While the classical method to deal with conditionals–the material conditional theory, is simple, it has been criticized for producing counterintuitive inferences [8,17,24]. For example, material conditionals with false antecedents or true consequents are always considered to be true. Empirical experiments have shown that people do not use the material conditional for daily inferences [8].

While there isn't a widely accepted theory for classifying conditionals, they generally fall into two categories: indicative and subjunctive conditionals. Subjunctive conditionals are used to talk about hypothetical, imagined, or unlikely situations. The subjunctive mood is often signaled by the word "would", as exemplified by sentences such as, "If I were you, I would win the game". In contrast, indicative conditionals are used to discuss real, factual, or likely situations. Typically, these sentences for indicative conditional employ the word "will", as in "If it rains tomorrow, I will not go to work". This thesis narrows its focus on the latter notion, indicative conditionals.

Various theories for indicative conditionals have been developed, many of which build upon the idea of the Ramsey test [27]. Ramsey argued that we determine whether to accept a given conditional by hypothesizing its antecedent and then evaluating the acceptability of the consequent. In other words, "If ϕ then ψ" implies that we should accept ψ given ϕ and our background knowledge [2,22].

N. Alechina et al. (Eds.): LORI 2023, LNCS 14329, pp. 17–30, 2023.
https://doi.org/10.1007/978-3-031-45558-2_2

This idea has led to the development of a related but distinct concept called inferentialism [5], which posits that for a conditional to be true, it should be possible to infer its consequent from its antecedent. Inferentialism has not yet provided a formal account of conditionals, as the inference between the antecedent and consequent can be inductive or abductive. However, inferentialists have emphasized the importance of relevance in inference and conditionals. We can only infer B from A when they are relevant, and a good theory of inference must take topic-relevance into account. Relevance has also played a key role in psychological theories of conditionals [10,25], and its importance has been argued by Krzyzanowska et al. [19] and experimentally tested by Douven et al. [9].

There are several existing logic theories that address the problem of conditionals through the use of relevance. One well-known example is Relevance Logic, which was initially created by Routley and Meyer and later simplified by Restall in 1995 [28]. However, the semantics of Relevance Logic itself fails provide a concrete definition of relevance, as the concept of relevance depends on a ternary relation in the model which is not determined. As a result, the burden of explaining relevance remains when interpreting the ternary relation. Another theory related to Relevance Logic is the Analytic Implication theory of William Parry. This theory requires that conditionals satisfy a strict variable-sharing criterion [13]. However, the variable-sharing property, while useful as a syntactic test for relevance, may not always be sufficient. In some cases, sentences that share the same variables may still be irrelevant. Recently, Özgün and Berto [4] developed a new approach for handling indicative conditionals that combines Adam's Thesis [1] with topic relevance [26]. However, similar to other theories for conditionals based on probability, their system does not address the issue of nested conditionals.

Based on the previous theories, this paper develops a novel framework for indicative conditionals based on awareness and relevance. Note that instead of the truth conditions, this paper focus on capturing the acceptability conditions for indicative conditionals. We propose that relevance is a subjective notion and it has a close connection to the notion of awareness. Given that awareness can be clearly represented in awareness logic [11], the notion of relevance can be modeled in a similar system and be used as a core requirement for accepting indicative conditionals.

2 From Awareness to Relevance

Relevance is a complex notion that can be interpreted both subjectively and objectively. Objective relevance generally refers to the relevance of sentences to a specific question or problem, regardless of individual interests or contexts. This kind of relevance is determined by causal relationships and can be evaluated independently of any particular individual's perspective. On the other hand, when we talk about subjective relevance, we are generally referring to the relevance of sentences to an individual's specific interests, needs, or context. This kind of relevance varies greatly from person to person and from situation to situation [6,16].

This paper leans towards the subjective interpretation of relevance, as it seeks to delineate the acceptability conditions rather than the truth conditions of indicative conditionals, in line with inferentialism. By adopting this stance, relevance becomes a subjective notion because the inferential connections that individuals draw can vary based on their personal knowledge, experiences, and interpretive frameworks. What seems like a valid or relevant to one person might not seem so to another, and vice versa.

Traditionally, relevance has been defined only as a relationship between sentences: two sentences are considered relevant if they share the same atoms or topics. This leads to the question of how to determine whether two sentences share the same topic, which can be a challenging task. For example, it is easy to see the relevance between sentences like "A and B" and "B", since they share the same sub-sentence. However, in some cases it may be more difficult to identify the relevance, such as with the sentences "Tom won the election" and "Jerry died". Without further context, it is difficult to see any connection between these sentences. But if we know the context that "Tom and Jerry were the only two candidates for a position", then the sentence "Jerry died" could be seen as a reason for "Tom winning the election". This suggests that contexts should also be taken into consideration when to determine whether two sentences are relevant. Before we determine the relevance between two sentences, we should check the relation between the sentences and the context in advance. We clarify the terms here to avoid misunderstanding: we use "connected" to express the relation between a sentence and a context and use "relevant" to describe the relation between two sentences.

An pressing problem is how we can tell which sentences are connected to a context. Perhaps it's simpler to identify which ones are definitely not connected to the context. One intuitive answer is that, if one does not even think about the sentence in this context, then the sentence cannot be connected. Consider, for instance, a scenario in which a group of colleagues are discussing a complex project at work. They are brainstorming ways to overcome a specific technical challenge. In this context, a sentence like "We could try implementing a new algorithm to solve this problem" is clearly connected, as it directly relates to the topic of discussion. However, a sentence such as "I'm planning to bake cookies this weekend" might not be considered connected to the context. Though it may be a truthful and meaningful statement for the speaker, it does not seem to bear any direct connection to the technical challenge being discussed. Here's where the idea of awareness becomes crucial. The perceived connection of a sentence can be influenced by the shared awareness and understanding of the individuals participating in the conversation. For instance, if the colleagues are aware that the speaker often uses baking as a metaphor for their problem-solving process, they might perceive a connection between the cookie-baking plan and the work discussion. Someone without this awareness, however, would likely miss the connection. This illustrate that, the relevance of a sentence is not just about its content or the topic at hand, but also about the awareness and interpretive frameworks of the individuals involved in the conversation. Therefore, relevance in this sense is subjective and dependent on awareness.

Given the above, we can then determine whether two sentences are relevant. Intuitively, if two sentences are relevant, then thinking about one sentence should lead us to think about the other. It seems that a sentence can guide our thoughts towards sentences of relevant topics. This becomes clearer when viewed from another angle: if there is a situation where one is contemplating a sentence without even being aware of another sentence, these two sentences should be deemed irrelevant. After all, we can't consider a sentence without being aware of the topics surrounding it.

Rather than delving into the essence of awareness, our actual aim here is a practical method for determining the relevance between sentences. To simplify the issue, we propose a minimal requirement for relevance: two sentences are relevant according to the agent (if and) only if the awareness of one sentence necessarily lead to the awareness of the other. In other words, we regard relevance as concomitant occurrence of awareness. This is also desired from the point of view of inferentialism, since the inference from antecedent to consequent requires that we can get to the consequent by thinking about the antecedent.

This paper employs the framework of awareness logic to define the relevance relation between sentences as the concomitant occurrence of their awareness. By means of this notion of relevance, we can get an apt theory for indicative conditionals.

3 Logic of Awareness Conditional L_{AC}

This paper proposes to model relevance by awareness logic in order to capture the acceptability conditions of indicative conditionals. Awareness logic was developed by Fagin and Halpern [11], which is a system that aims to describe the properties of awareness and unawareness. It is based on the model of normal modal propositional logic, with the addition of a function that assigns to each world a set of propositions or propositional atoms. These propositions represent the content that we can be aware of in a particular world. Using this model, we can define a modal operator of awareness, stating that the agent is aware of a proposition in a world if and only if it is contained in the set of propositions assigned to that world [14,15,29]. This allows us to establish a link between sentences and worlds by awareness.

The language of our logic of awareness conditional L_{AC} is similar to the language of basic awareness logic (without operators for knowledge), but it has been enriched with an extra connective $\rightarrow$ for indicative conditionals and an S5 global modal operator $\Box$. We fix a countable set $\mathbb{P}$ of proposition letters.

Definition 1 (Language L_{AC}). *The language of topic-relevance* L_{AC} is defined as follows:

$$\phi ::= p \mid \neg\phi \mid \phi \wedge \psi \mid \phi \vee \psi \mid \phi \rightarrow \psi \mid \phi \supset \psi \mid \Box\phi \mid A(\phi)$$

where $p \in \mathbb{P}$, $\Box$ is the standard S5 modal operator for necessity, $A(\phi)$ means that the agent is aware of ϕ , $\supset$ denotes material conditional and $\rightarrow$ denotes indicative conditional. We abbreviate the formula $(\phi \supset \psi) \wedge (\psi \supset \phi)$ as $\phi \equiv \psi$.

In addition, we denote the set of all atoms in a formula ϕ by $At(\phi)$.

Definition 2. *The set of all atoms in a formula* ϕ is defined inductively as follows:

$$At(p) = \{p\}$$
$$At(\neg\phi) = At(\Box\phi) = At(A(\phi)) = At(\phi)$$
$$At(\phi \wedge \psi) = At(\phi \vee \psi) = At(\phi \supset \psi) = At(\phi \rightarrow \psi) = At(\phi) \cup At(\psi)$$

Definition 3 (Frame). *A frame of* L_{AC} is a binary structure $\mathbb{F} = \langle W, C \rangle$, where W is a nonempty set of possible worlds and C is a function that assigns to each world w a set of propositional atoms $C(w)$ that we can be aware of in that world.

Here we adopt the kind of awareness logic of Halpern in [14,29] which stipulates that awareness is closed under subformulas and is generated by primitive propositions. The awareness function C describes the topics of the different worlds which represented by the propositional atoms, and the agent in any world w can only access the atoms in $C(w)$ which are atoms that the agent can be aware of.

Definition 4 (Model). *A model of* L_{AC} is a triple $\mathbb{M} = \langle W, C, V \rangle$. A frame becomes a model when it is endowed with a world-dependent valuation function $V : W \times \mathbb{P} \rightarrow \{1, 0\}$.

The semantic clauses here are exactly the same of those in awareness logic. In our logic, support $\models$ is defined to capture the acceptability instead of truth conditions of sentences.

Definition 5. Given a state s and a model $\mathbb{M}$ of L_{AC}, the *support* relation $\models$ is defined inductively as follows:

$$\begin{array}{lll}
\mathbb{M}, w \models p & \text{iff} & V(w,p) = 1 \\
\mathbb{M}, w \models A(\phi) & \text{iff} & At(\phi) \subseteq C(w) \\
\mathbb{M}, w \models \neg\phi & \text{iff} & \mathbb{M}, w \not\models \phi \\
\mathbb{M}, w \models \phi \vee \psi & \text{iff} & \mathbb{M}, w \models \phi \text{ or } \mathbb{M}, w \models \psi \\
\mathbb{M}, w \models \phi \wedge \psi & \text{iff} & \mathbb{M}, w \models \phi \text{ and } \mathbb{M}, w \models \psi \\
\mathbb{M}, s \models \phi \supset \psi & \text{iff} & \mathbb{M}, w \models \neg\phi \text{ or } \mathbb{M}, w \models \psi \\
\mathbb{M}, w \models \Box\phi & \text{iff} & \forall x \in W : \mathbb{M}, x \models \phi
\end{array}$$

Definition 6. $\mathbb{M}, w \models \phi \rightarrow \psi$ iff all of the three conditions are satisfied:

- Truth Functionality: $\mathbb{M}, w \models \phi \supset \psi$
- Global Relevance: $\mathbb{M}, w \models \Box(A\phi \supset A\psi)$
- Local Awareness: $\mathbb{M}, w \models A(\phi \supset \psi)$

All the clauses in definition 5 are the same as in standard awareness logic with an S5 modal operator. We use $A(\phi)$ to express that the agent is aware of ϕ in the world, and we use $\Box\phi$ to express that ϕ is necessary.

Definition 6 establishes the three conditions that define indicative conditionals. The first condition is Truth Functionality, which means that an indicative conditional entails the corresponding material conditional. This condition ensures that the truth is preserved from the antecedent to the consequent.

The second condition, Global Relevance, states that for an indicative conditional to be accepted, the awareness of the consequent must necessarily follow from the awareness of the antecedent in all possible worlds. In simpler terms, awareness of the antecedent necessarily leads to the awareness of the consequent. Global Relevance requires to check awareness in all possible worlds because an agent may be contingently aware of two irrelevant sentences in the actual world. Therefore, relevance is global, while awareness is local. It is important to note that this condition does not stipulate that if we are aware of the consequent, we should be aware of the antecedent, indicating that the indicative conditionals $\rightarrow$ do not satisfy Contraposition. However, we can, of course, add this constraint, which will be discussed in Sect. 7.

The final condition, Local Awareness, asserts that the agent should be aware of the entire conditional statement before it is accepted. This condition makes the indicative conditionals hyperintensional and also results in the desired invalidity of Monotonicity. Strengthening the antecedent may render us unable to be aware of the conditional anymore. These conditions collectively represent what we refer to as relevance requirements for indicative conditionals.

Validity and logical consequence of L_{AC} are defined in a classic way:

Definition 7. For any proposition ϕ,

- $\models \phi$ iff for any world w and model $\mathbb{M}$, $\mathbb{M}, w \models \phi$
- $\Sigma \models \phi$ iff for any world w and model $\mathbb{M}$, if $\mathbb{M}, s \models \psi$ for all $\psi \in \Sigma$, then $\mathbb{M}, s \models \phi$.

4 Principles Valid in L_{AC}

Firstly, we examine the valid principles for indicative conditionals in L_{AC}. The first principle is Modus Ponens: from a conditional and its antecedent, we can infer its consequent. This is a basic and fundamental rule of inference involving conditionals. However, it has been challenged by Mcgee [23], who provided a counterexample:

> "Opinion polls taken just before the 1980 election showed the Republican Ronald Reagan decisively ahead of the Democrat Jimmy Carter, with the other Republican in the race, John Anderson, a distant third. Those apprised of the poll results believed, with good reason: X: If a Republican wins the election, then Y: if it's not Reagan who wins, Z: it will be Anderson. A Republican will win the election. Yet they did not have reason to believe: Y$\rightarrow$ Z: If it's not Reagan who wins, it will be Anderson."

In this example, we can agree with "If X, then if Y, then Z", and we also agree with "X", but we do not agree with "If Y then Z" since we believe Carter will win the election if Reagan lose. Thus the Modus Ponens inference from 'If X, then if Y, then Z" and "X" to "If Y then Z" fails.

However, this is not the case in L_{AC} if we adopt another interpretation. It is important to note that in L_{AC}, $X \rightarrow (Y \rightarrow Z)$ is not equivalent to but strictly stronger than $(X \wedge Y) \rightarrow Z$. Consider the statement "X: If a Republican wins the election, then Y: if it's not Reagan who wins Z: it will be Anderson." This does not mean that $X \rightarrow (Y \rightarrow Z)$, since it is not the case that the speaker is aware of both Y and Z in all worlds in which the agent is aware of X. In this case, "a Republican wins the election" only makes the agent aware of Reagan, and the agent become aware of Anderson only upon receiving the information that "if it's not Reagan who wins". Therefore, the statement should be interpreted as $(X \wedge Y) \rightarrow Z$ rather than $X \rightarrow (Y \rightarrow Z)$, which means that Modus Ponens is not rejected.

Fact 1 (Modus Pones). $\phi \rightarrow \psi, \phi \vDash \psi$ is valid in L_{AC}.

Proof. For any model $\mathbb{M}$ and any world w, assume $\mathbb{M}, w \vDash \phi \rightarrow \psi$, then we get that if $\mathbb{M}, w \vDash \phi$ then $\mathbb{M}, w \vDash \psi$. Given that $\mathbb{M}, s \vDash \phi$, we get $\mathbb{M}, s \vDash \psi$.

The next principle is Restricted Identity. While identity may seem like a basic principle, stating that every conditional whose consequent is identical to the antecedent is true, the original version of Identity is not valid in L_{AC}. This is because we cannot guarantee that the consequent or the antecedent of the conditional is relevant to our current context. For example, we cannot say "If Tom is a Kakabuka, then Tom is a Kakabuka" if we do not even know the word "Kakabuka", even though the sentence is true. Therefore, in L_{AC} we only have the Restricted Identity, which states that every aware conditional whose consequent is identical to the antecedent is accepted.

Fact 2 (Restricted Identity). $A(\phi) \vDash \phi \rightarrow \phi$ is valid in L_{AC}.

Proof. For any model $\mathbb{M}$ and any world w, assume $\mathbb{M}, w \vDash A(\phi)$. Then 1) $\mathbb{M}, w \vDash A(\phi \supset \phi)$, 2) if $\mathbb{M}, w \vDash \phi$ then $\mathbb{M}, w \vDash \phi$, 3) for any $w' \in W$, if $\mathbb{M}, w' \vDash A\phi$ then $\mathbb{M}, w' \vDash A\phi$. Thus $\mathbb{M}, w \vDash \phi \rightarrow \phi$.

The next valid principle in L_{AC} is Transitivity, which states that a conditional with antecedent X and consequent Y, in combination with a conditional with antecedent Y and consequent Z, entails the conditional with antecedent X and consequent Z. This principle has been somewhat controversial, with some arguing that it should be rejected. A famous counterexample was provided by Adams [1]:

> "If Brown wins the election, Smith will retire to private life. If Smith dies before the election, Brown will win it. Therefore, if Smith dies before the election, then he will retire to private life."

In this example, it is completely normal for someone to accept that "X: If Brown wins the election, Y: Smith will retire to private life", while also accept "Z: If Smith dies before the election, X: Brown will win it". But the conditional from Z to Y seems absurd. Therefore Adam objected the validity of Transitivity for indicative conditional.

However, this is not the case in L_{AC} if we adopt another interpretation. When we say "X: If Brown wins the election, Y: Smith will retire to private life", we are actually omitting the possibility that "Z: If Smith dies before the election". Whenever X and Y are accepted, the agent must not be aware of Z. Therefore, in L_{AC}, the first two conditionals should be reformulated as "$X \wedge \neg A(Z) \rightarrow Y$" and "$Z \rightarrow X$". These two sentences cannot be accepted simultaneously, so the counterexample provided by Adams does not hold in L_{AC}.

Fact 3 (Transitivity). $\phi \rightarrow \psi, \psi \rightarrow \chi \vDash \phi \rightarrow \chi$ is valid in L_{AC}.

Proof. For any model $\mathbb{M}$ and any world w, assume $\mathbb{M}, w \vDash \phi \rightarrow \psi$ and $\mathbb{M}, w \vDash \psi \rightarrow \chi$. Then we have 1) $\mathbb{M}, w \vDash A(\phi \supset \chi)$ and $\mathbb{M}, w \vDash A(\psi \supset \chi)$. 2) For any $w' \in W$, if $\mathbb{M}, w' \vDash A(\phi)$ then $\mathbb{M}, w' \vDash A(\psi)$, and then $\mathbb{M}, s \vDash A(\chi)$. 3) if $\mathbb{M}, w \vDash \phi$ then $\mathbb{M}, w \vDash \psi$, and then $\mathbb{M}, w \vDash \chi$. Thus $\mathbb{M}, w \vDash \phi \rightarrow \psi$.

5 Principles Invalid in L_{AC}

Moreover, we examine the invalid principles for indicative conditionals in L_{AC}. The first one is False Antecedent, which states that the falsity of the antecedent is sufficient for the truth of the conditional. This principle should be rejected, as it is obviously not always true. Additionally, even its restricted version, which requires that the sentence is aware of, is also invalid.

Fact 4 (False Antecedent and Restricted False Antecedent). $\neg\phi \vDash \phi \rightarrow \psi$ and $\neg\phi, A(\phi \rightarrow \psi) \vDash \phi \rightarrow \psi$ are invalid in L_{AC}.

Proof. Counterexample: Set the model $\mathbb{M}$ and the only two worlds in the model w, w' such that $V(p) = \emptyset$, $C(w) = \{p, q\}$ and $C(w') = \{p\}$. Then we have $\mathbb{M}, w \vDash \neg p \wedge A(p \rightarrow q)$, but $\mathbb{M}, w \nvDash p \rightarrow q$ since $\mathbb{M}, w' \vDash A(p) \wedge \neg A(q)$.

Similarly, the principle of True Consequent, which states that the truth of the consequent alone is sufficient for the truth of the whole conditional, should also be rejected. Like False Antecedent, even its restricted version which requires that the sentence is aware of, is also invalid.

Fact 5 (True Consequent and Restricted True Consequent). $\psi \vDash \phi \rightarrow \psi$ and $\psi, A(\phi \rightarrow \psi) \vDash \phi \rightarrow \psi$ are invalid in L_{AC}.

Proof. Counterexample: Set the model $\mathbb{M}$ and the only two worlds in the model w, w' such that, $V(q) = \{w, w'\}$, $C(w) = \{p, q\}$ and $C(w') = \{p\}$. Then we have $\mathbb{M}, w \vDash q \wedge A(p \rightarrow q)$, but $\mathbb{M}, w \nvDash p \rightarrow q$ since $\mathbb{M}, w' \vDash A(p) \wedge \neg A(q)$.

As an obviously corollary, conditionals with necessary consequent and conditionals with impossible antecedent are also invalid, since necessity does not guarantee relevance.

Fact 6 (Necessary Consequent and Impossible Antecedent). $\Box\psi \vDash \phi \rightarrow \psi$ and $\Box\neg\phi \vDash \phi \rightarrow \psi$ are invalid in L_{AC}.

Proof. The counterexamples from the fact 4 and fact 5.

The next principle is Linearity: for any two formulas X and Y either "If X then Y" or "If Y then X" holds. It should be rejected since maybe X and Y are totally irrelevant. Even its restricted version requiring that the agent is aware of the sentence, is also invalid.

Fact 7 (Linearity and Restricted Linearity). $\vDash (\phi \rightarrow \psi) \vee (\psi \rightarrow \phi)$ and $A((\phi \rightarrow \psi) \vee (\psi \rightarrow \phi)) \vDash (\phi \rightarrow \psi) \vee (\psi \rightarrow \phi)$ are invalid in L_{AC}.

Proof. Counterexample: Set the model $\mathbb{M}$ and the only three worlds in the model w, w', w'' such that $C(w) = \{p, q\}$, $C(w') = \{p\}$ and $C(w'') = \{q\}$. Then $\mathbb{M}, w \vDash A((p \rightarrow q) \vee (q \rightarrow p))$, but $\mathbb{M}, w \nvDash p \rightarrow q$ and $\mathbb{M}, w \nvDash q \rightarrow p$.

The next invalid principle is Monotonicity. This principle states that strengthening the antecedent of a true conditional by adding any conjunct will not change its truth value. There are many counterexamples to Monotonicity, with the Sobel sequence being one of the most well-known:

- X: If Sophie had gone to the parade, Y: she would have seen Pedro dance; but of course,
- X: if Sophie had gone to the parade and Z: been stuck behind someone tall, ¬Y: she would not have seen Pedro dance;

In this scenario we have "If X then Y" but not "If X and Z then Y". Thus Monotonicity should be rejected.

Fact 8 (Monotonicity). $\phi \rightarrow \psi \vDash (\phi \wedge \chi) \rightarrow \psi$ is invalid in L_{AC}.

Proof. Counterexample: Set the model $\mathbb{M}$ and the only world in model w such that the agent is aware of both ϕ and ψ in w while the agent is not aware of χ.

The next invalid principle is And-to-If, which states that any conjunction entails the corresponding conditional. If the two conjunctions are completely irrelevant, it is unreasonable to suggest that they provide a reason for each other. Like the previous principles, even its restricted version, which requires that the sentence is relevant, is also invalid.

Fact 9 (And-to-If and Restricted And-to-If). $\phi \wedge \psi \vDash \phi \rightarrow \psi$ and $\phi \wedge \psi, A(\phi \wedge \psi) \vDash \phi \rightarrow \psi$ are invalid in L_{AC}.

Proof. Counterexample: Set the model $\mathbb{M}$ and the only two worlds w, w' such that $V(p) = \{w, w'\}$, $V(q) = \{w, w'\}$, $C(w) = \{p, q\}$ and $C(w') = \{p\}$. Then we have $\mathbb{M}, w \vDash p \wedge q \wedge A(p \wedge q)$, but $\mathbb{M}, w \nvDash p \rightarrow q$ since $\mathbb{M}, w' \vDash A(p) \wedge \neg A(q)$.

The same counterexample can also invalidate the necessary version of And-to-If, since necessary truth does not guarantee the relevance between antecedent and consequent.

Fact 10 (Necessary And-to-If). $\Box(\phi \wedge \psi) \vDash \phi \rightarrow \psi$ is invalid in L_{AC}.

Proof. The counterexample from fact 9.

Unlike the previous principles, the validity of Contraposition is highly controversial. Some theorists, such as Stalnaker, reject it as counterintuitive, while others, such as Lycan [21] and Bennett [3], tend to accept it. In L_{AC}, $\rightarrow$ fails Contraposition here because the fact that the awareness of the consequent cannot be deduced from awareness of the antecedent. But as we will see in Sect. 7, if we strengthen the Global Relevance condition to both direction, the new operator for indicative conditionals can satisfy Contraposition.

Fact 11 (Contraposition). $\phi \rightarrow \psi \vDash \neg\psi \rightarrow \neg\phi$ is invalid in L_{AC}.

Proof. Counterexample: Set the model $\mathbb{M}$ and the only two worlds in model w, w' such that $V(q) = \{w, w'\}$, $C(w) = \{p, q\}$ and $C(w') = \{q\}$. Then we have $\mathbb{M}, w \nvDash \neg\psi \rightarrow \neg\phi$, although $\mathbb{M}, s \vDash \phi \rightarrow \psi$.

6 Axiomatization

Despite L_{AC} aims to capture the acceptability of indicative conditionals, it is very similar to the S5 awareness logic in [14], and the only additional operator $\rightarrow$ for indicative conditionals can be deduced from basic operators. Thus, we have the axiomatization AIC for awareness indicative conditionals as follows:

	Axiomatization AIC
CPL	classic propositional tautologies and Modus Ponens
$S5_\Box$	$S5$ axioms and rules for $\Box$
	Axioms for A:
A_1	$A(\phi \wedge \psi) \equiv (A(\phi) \wedge A(\psi))$
A_2	$A(\neg\phi) \equiv A(\phi)$
A_3	$A(A(\phi)) \equiv A(\phi)$
A_4	$A(\Box\phi) \equiv A(\phi)$
A_5	$A(\phi \supset \psi) \equiv (A(\phi) \wedge A(\psi))$
A_6	$A(\phi \rightarrow \psi) \equiv (A(\phi) \wedge A(\psi))$
	Axiom for $\rightarrow$:
C	$((\phi \supset \psi) \wedge A(\phi \supset \psi) \wedge \Box(A\phi \supset A\psi)) \equiv (\phi \rightarrow \psi)$

AIC is similar to the S5 system but enriched with axioms for awareness and axioms for indicative conditionals. Its soundness and completeness can be proven.

Theorem 1. *AIC is a sound and complete axiomatization of L_{AC}: for any proposition ϕ, $\vdash_{AIC} \phi$ if and only if $\models \phi$.*

Proof. A straightforward modification of the proof of soundness and completeness of $S5$ awareness logic. We omit the details here. For similar proofs, see [12,14].

7 Variants, Comparison and Conclusion

In the previous section, we present a method for capturing the acceptability of indicative conditionals through relevance requirements. However, the relevance requirements in definition 6 can of course be further strengthened, and L_{AC} has at least two natural variants. The strengthening method for the first variant is to replace the single-direction Global Relevance condition $\mathbb{M}, w \models \Box(A\phi \supset A\psi)$ by double-direction Global Relevance $\mathbb{M}, w \models \Box(A\phi \subset\supset A\psi)$. This means that the awareness of the antecedent should also follows from the awareness of the consequent. In other words, the agent must be aware of the antecedent and consequent at the same time in all possible worlds. And we can also strengthen the Truth Functionality condition $\mathbb{M}, w \models \phi \supset \psi$ to Global Truth Functionality $\mathbb{M}, w \models \Box(\phi \supset \psi)$ to get the second variant. This actually says that an indicative conditional $\phi \to \psi$ should entail the corresponding strict interpretation of material implication [18] that is $\Box(\phi \supset \psi)$.

We use $\to_1$ to represent the first variant of indicative conditionals, and $\to_2$ for the second variant. Formally we can write their semantic clauses as:

Definition 8. The truth conditions for variant indicative conditionals are follows:

- $\mathbb{M}, w \models \phi \to_1 \psi$ iff $\mathbb{M}, w \models (\phi \supset \psi) \wedge \Box(A\phi \subset\supset A\psi) \wedge A(\phi \to \psi)$
- $\mathbb{M}, w \models \phi \to_2 \psi$ iff $\mathbb{M}, w \models \Box(\phi \supset \psi) \wedge \Box(A\phi \supset A\psi) \wedge A(\phi \to \psi)$

The following table compares the properties of indicative conditionals in L_{AC} with other existing theories. We use $\rhd_1$ to stand for Stalnaker conditionals [30,31] and use $\rhd_2$ to stand for Evidential Conditionals from Crupi and Iacona [7]. We use $\Rightarrow$ to stand for the strict interpretation of conditionals [18], that is to abbreviate $\Box(\phi \supset \psi)$.

As we can see from the table, L_{AC} satisfies several traditionally desired properties such as Restricted Identity, Modus Ponens, and Transitivity. This means that the logic is stronger than some other theories for indicative conditionals, which fail to satisfy principles like Modus Ponens [23] or Transitivity [7]. On the other hand, problematic principles such as Truth Consequent, False Antecedent, Linearity, Monotonicity, and And-to-If are invalid in L_{AC}, and counterexamples to these principles can be easily found in practical situations. The invalidity of these principles has also been analyzed and supported by Crupi & Iacona [7] and Berto & Özgün [4]. In addition, our theory can also get the desired invalidity of Necessary Consequent, Impossible Antecedent and Necessary And-to-If, which are seldom taking into account in many recent theories for conditionals.

Principles	$\supset$	$\rightarrow$	$\rightarrow_1$	$\rightarrow_2$	$\Rightarrow$	$\rhd_1$	$\rhd_2$
Modus Ponens	✓	✓	✓	✓	✓	✓	✓
Restricted Identity	✓	✓	✓	✓	✓	✓	✓
Transitivity	✓	✓	✓	✓	✓	×	×
Necessary Consequent	✓	×	×	×	✓	✓	✓
Impossible Antecedent	✓	×	×	×	✓	✓	✓
True Consequent	✓	×	×	×	×	×	×
False Antecedent	✓	×	×	×	×	×	×
Linearity	✓	×	×	×	×	×	×
Monotonicity	✓	×	×	×	✓	×	×
And-to-If	✓	×	×	×	×	✓	×
Necessary And-to-If	✓	×	×	×	✓	✓	✓
Contraposition	✓	×	✓	×	✓	×	✓

It is also noticeable that our theory works well together with the traditional method of material implication. Unlike suppositional theories of indicative conditionals, our theory does not rely on hypothetical belief and does not reject the truth-functionality of conditionals. In fact, the indicative conditionals in our system can be seen as material conditionals with relevance requirements. The conditionals satisfying these requirements are considered and valued just like material conditionals. If we imagine a maximal awareness model in which the agent is aware of everything in every world, indicative conditionals would be trivialized and reduced to material conditionals. This allows us to maintain the philosophical and systematic advantages of material condition theories within our framework [32].

Furthermore, nested conditionals can be clearly defined and easily handled in L_{AC}, whereas they pose difficulties in many other theories. For suppositional conditional theories generated directly from the Ramsey Test, nested conditionals require multiple levels of hypothesizing, which is highly complex and difficult [20]. In particular, probabilistic approaches following Adam's Thesis [1] do not account for nested conditionals at all. For many of those theories for conditionals based on modality, nested conditionals lack clear intuition due to the nested modal relations they involve. In contrast, the semantic values of nested conditionals are defined clearly and directly in L_{AC}. To check a nested conditional we only check a nested material conditional and check the relevance requirements between all its components.

References

1. Adams, E.W.: Probability and the logic of conditionals. In: Studies in Logic and the Foundations of Mathematics, vol. 43, pp. 265–316. Elsevier (1966)
2. Barnett, D.: Zif is if. Mind **115**(459), 519–566 (2006)

3. Bennett, J.: A Philosophical Guide to Conditionals. Clarendon Press, Oxford (2003)
4. Berto, F., Özgün, A.: Indicative conditionals: probabilities and relevance. Philos. Stud. **178**(11), 3697–3730 (2021)
5. Brandom, R.: Articulating Reasons: An Introduction to Inferentialism. Harvard University Press, Cambridge (2009)
6. Cosijn, E., Ingwersen, P.: Dimensions of relevance. Inf. Proc. Manage. **36**(4), 533–550 (2000)
7. Crupi, V., Iacona, A.: The evidential conditional. Erkenntnis **87**(6), 2897–2921 (2022)
8. Douven, I.: The Epistemology of Indicative Conditionals: Formal and Empirical Approaches. Cambridge University Press, Cambridge (2015)
9. Douven, I., Elqayam, S., Singmann, H., Wijnbergen-Huitink, J.V.: Conditionals and inferential connections: toward a new semantics. Think. Reason. **26**(3), 311–351 (2020)
10. Evans, J.S.B., Over, D.E., et al.: If: Supposition, Pragmatics, and Dual Processes. Oxford University Press, USA (2004)
11. Fagin, R., Halpern, J.Y.: Belief, awareness, and limited reasoning. Artif. Intell. **34**(1), 39–76 (1987)
12. Fagin, R., Halpern, J.Y., Moses, Y., Vardi, M.: Reasoning about Knowledge. MIT press, Cambridge (2004)
13. Fine, K.: Analytic implication. Notre Dame J. Formal Logic **27**(2), 169–179 (1986)
14. Halpern, J.Y.: Alternative semantics for unawareness. Games Econom. Behav. **37**(2), 321–339 (2001)
15. Halpern, J.Y., Rêgo, L.C.: Interactive unawareness revisited. Games Econom. Behav. **62**(1), 232–262 (2008)
16. Hjørland, B.: The foundation of the concept of relevance. J. Am. Soc. Inform. Sci. Technol. **61**(2), 217–237 (2010)
17. Hoaglund, J.: The logic of'If-Then'Propositions. Informal Logic **8**(3) (1986)
18. Iacona, A., et al.: Indicative conditionals as strict conditionals. Argumenta **4**, 177–192 (2018)
19. Krzyżanowska, K., Collins, P.J., Hahn, U.: Between a conditional's antecedent and its consequent: discourse coherence vs probabilistic relevance. Cognition **164**, 199–205 (2017)
20. Lindström, S., Rabinowicz, W.: The Ramsey test revisited (1995)
21. Lycan, W.G.: Real Conditionals. Clarendon Press, Oxford (2001)
22. Mackie, J.L.: Truth, Probability and Paradox: Studies in Philosophical Logic. Oxford University Press on Demand, Oxford (1973)
23. McGee, V.: A counterexample to modus ponens. J. Philos. **82**(9), 462–471 (1985)
24. Moore, G.E.: External and internal relations. In: Proceedings of the Aristotelian Society, vol. 20, pp. 40–62. JSTOR (1919)
25. Oaksford, M., Chater, N.: A rational analysis of the selection task as optimal data selection. Psychol. Rev. **101**(4), 608 (1994)
26. Özgün, A., Berto, F.: Dynamic hyperintensional belief revision. Rev. Symbolic Logic **14**(3), 766–811 (2021)
27. Ramsey, F.P.: General propositions and causality (1929)
28. Restall, G.: Four-valued semantics for relevant logics (and some of their rivals). J. Philos. Log. **24**, 139–160 (1995)
29. Schipper, B.C.: Awareness. SSRN 2401352 (2014)

30. Stalnaker, R.C.: A theory of conditionals. In: IFS: Conditionals, Belief, Decision, Chance and Time, vol. 15, pp. 41–55. Springer, Dordrecht (1968). https://doi.org/10.1007/978-94-009-9117-0_2
31. Stalnaker, R.C., Thomason, R.H.: A semantic analysis of conditional logic 1. Theoria **36**(1), 23–42 (1970)
32. Williamson, T.: Suppose and Tell: The Semantics and Heuristics of Conditionals. Oxford University Press, Oxford (2020)

A Logical Description of Priority Separable Games

Ramit Das[1], R. Ramanujam[2], and Sunil Simon[3(✉)]

[1] Institute of Mathematical Sciences and Homi Bhabha National Institute, Chennai, India
ramitd@imsc.res.in
[2] Azim Premji University, Bengaluru, India
jam@imsc.res.in
[3] Department of CSE, IIT Kanpur, Kanpur, India
simon@cse.iitk.ac.in

Abstract. When we reason about strategic games, implicitly we need to reason about arbitrary strategy profiles and how players can improve from each profile. This structure is exponential in the number of players. Hence it is natural to look for subclasses of succinct games for which we can reason directly by interpreting formulas on the (succinct) game description rather than on the associated improvement structure. Priority separable games are one of such subclasses: payoffs are specified for pairwise interactions, and from these, payoffs are computed for strategy profiles. We show that equilibria in such games can be described in Monadic Least Fixed Point Logic (MLFP). We then extend the description to games over arbitrarily many players, but using the monadic least fixed point extension of existential second order logic.

1 Introduction

Finite strategic games are a well-studied formalism used to analyse strategic behaviour of rational agents. A strategic game is specified by a finite set of players along with a finite set of strategies and a payoff function for each player. Players choose strategies simultaneously and for each player, the corresponding payoff function specifies the utility for the player given the profile of choices. In terms of structural and computational analysis of games, a major drawback of strategic games is that the representation is not compact. The two main parameters in the representation are the number of players and the strategies available for each player. An explicit representation of the payoff functions is exponential in the number of players. Identifying subclasses of strategic game with compact representation, is thus an important first step towards analysing the structural and computational properties of the game model.

There are two possible approaches which are commonly adopted to achieve concise representation in games. First is to retain quantitative payoffs and impose restrictions on the payoff functions. The second is to use an appropriate logical formalism to describe payoffs in a qualitative manner.

N. Alechina et al. (Eds.): LORI 2023, LNCS 14329, pp. 31–46, 2023.
https://doi.org/10.1007/978-3-031-45558-2_3

In the classical approach with quantitative payoffs, it is possible to achieve compact representation with a careful analysis of the underlying dependency structure in the payoff functions. Such an approach, which restricts the dependency of payoff functions to a "small" number of other agents is adopted in *graphical games* [26]. Another approach is to explicitly impose restrictions on the payoff functions. A simple constraint is to insist that the payoff functions are *pairwise separable*. This results in the well-studied class of games with a compact representation, called *polymatrix games* [10,11,25,35]. In the specific context of coalition formation games, the restriction to pairwise separable payoffs results in the well studied game model called *additively separable hedonic games* [5,23].

The second approach is to view payoffs as qualitative outcomes described using some logical formalism. Boolean games [19] is a well studied model adopting this approach. In Boolean games, each player controls a disjoint subset of atomic propositions and the payoffs are represented using Boolean formulas over the union of these propositions. Though originally defined as a qualitative version of two player zero-sum games, the model has been extended to reason about multi-player games [9,20]. Epistemic Boolean games, where payoffs are specified as epistemic formulas, were studied in [2,21].

In this paper we propose a subclass of strategic games that combine both approaches. In our model, the payoffs are qualitative (but not necessarily Boolean). The payoff functions are restricted to be pairwise separable. This results in a game model which is concise while at the same time being able to represent genuine multi-player games with non-zero sum objective. We show that these priority separable games need not always have a pure Nash equilibrium. Then an immediate question is whether there is an efficient procedure to check if a Nash equilibrium exists in this class of games. We show that checking for the existence of a Nash equilibrium is NP-complete. Nash's Theorem states that the mixed extension of every finite strategic game has a Nash equilibrium (in mixed strategies). It is known that computing such a mixed strategy Nash equilibrium is PPAD-hard [15]. In this paper, we restrict our study to pure strategies and pure Nash equilibrium.

We express the existence of pure Nash equilibria in priority separable games in a logical language: the *monadic least fixed-point logic* ([33]). This is an extension of first order logic with monadic least fixed-point operators. In this, we follow the spirit of descriptive complexity [24], where extensions of first order logics describe complexity classes. The major question of interest in such investigation is to ask what logical resources are needed to describe the property (which is Nash equilibrium in this case). We would like minimal use of the logical resources needed. Since equilibrium computation involves iterative exploration of the strategy space, by considering every possible player improvement and player response to it, a least fixed-point operator is natural to use in this context. However, we show that it suffices to use a *monadic* least fixed-point operator, where the operator is applied only on sets (rather than on arbitrary relations).

A natural question that arises in the logical study is games over unboundedly many players. In game theory, we typically specify every game with the number of players playing in it. This seems reasonable for logical descriptions as well: we can ask, how many players are *forced* by the structure present in the reasoning? A formula can have different game models, with different numbers of players. With the addition of binary second order variables to the logic (and monadic least fixed-point operators), we describe Nash equilibrium in the logic. There are complexity theoretic implications in this but we do not take it up for study in this paper.

Various logical formalisms have been used in the literature to reason about games and strategies. Action indexed modal logics have often been used to analyse finite extensive form games where the game representation is interpreted as models of the logical language [6–8]. A dynamic logic framework can then be used to describe games and strategies in a compositional manner [18,30,31] and encode existence of equilibrium strategies [20]. The work in [13] employs modal logics similarly to study large games. Alternating temporal logic (ATL) [3] and its variants [12,22,36] constitute a popular framework to reason about strategic ability in games. Strategy Logic ([12]) is of specific interest in its explicit use of strategy quantifiers, and hence the existence of solution concepts like Nash equilibrium can be expressed in it. In the context of two-player turn-based sequential games on graphs Strategy Logic provides a logical mechanism for description of a variety of equilibria. [32] provides another logical treatment of explicit reasoning about structured strategies.

Our approach here is different from these, both in the structure of the games studied and in the use of logic for checking properties as in descriptive complexity theory. The models of the logic are strategy spaces: every node is a strategy profile, and edges denote deviations by players. Iteration of such deviations until no more deviation is possible suggests the use of fixed-point operators. These are one-shot strategic form games, as distinct from models of logics like ATL, Strategy Logic and the logic of structured strategies in [32] where the games are turn based and of infinite duration. Moreover, while Strategy Logic discusses two-player games, we talk of multi-player games which are determined by pairwise interactions among players. The work presented here follows [14] but the emphasis there was on improvement graph dynamics, whereas we reason directly with game descriptions here.

2 Our Model

2.1 Background

Let $N = \{1, \ldots, n\}$ be the set of players. A *strategic game* defined as $\mathcal{G} = (N, \{S_i\}_{i \in N}, \{p_i\}_{i \in N})$ consists of the set of players N, and for each player i, a set S_i of strategies along with a payoff function $p_i : S_1 \times \cdots \times S_n \to \mathbb{R}$. A strategy profile is a tuple of strategies, $s = (s_1, \ldots, s_n)$ where for all players i, $s_i \in S_i$. Given a strategy profile s and a player i, let $s_{-i} = (s_1, \ldots, s_{i-1}, s_{i+1}, \ldots, s_n)$.

Thus $s = (s_i, s_{-i})$. Let $S = S_1 \times \ldots \times S_n$ denote the set of all strategy profiles and $S_{-i} = S_1 \times \ldots S_{i-1} \times S_{i+1} \times \ldots \times S_n$.

We say that the strategy $s_i \in S_i$ of player i is a *best response* to $s_{-i} \in S_{-i}$ if for all $s'_i \in S_i$, $(s'_i, s_{-i}) \preceq_i s$. A strategy profile s is a *Nash equilibrium* if for all $i \in N$, s_i is a best response to s_{-i}. Existence of Nash equilibrium and computation of an equilibrium profile (when it exists) are important questions in the context of strategic form games. While strategic games are well-studied as a model for games, it has the drawback that the representation is not concise. An explicit representation of the payoff functions is exponential in the number of players. To analyse the computational properties of games, it is important to identify subclasses of strategic form games which have a compact representation. *Polymatrix games* [25] form such a subclass, where the payoff functions are restricted to be pairwise separable. Formally, a *polymatrix game* is a strategic game $\mathcal{G} = (N, \{S_i\}_{i \in N}, \{p_i\}_{i \in N})$ where for all players i and for all $j \neq i$, there exists a partial payoff function $p_{i,j}$ such that for any strategy profile s, $p_i(s) = \Sigma_{j \neq i}\, p_{i,j}(s_i, s_j)$. It can be observed that polymatix games have compact representation, polynomial in $|N|$ and $\max_{i \in N} |S_i|$.

It is often useful to explicitly specify the dependency of the pairwise separable payoff functions in terms of a neighbourhood graph. Let $G = (N, E)$ be a directed graph (without self loops) over the set of players N and for each $i \in N$, let $R(i) = \{\ j \mid (j, i) \in E\ \}$ be the neighbourhood of i in G. For the players not in the neighbourhood of a certain player, say i, we define the partial payoff function values on those instances as 0. That is, for all strategy profiles s, for all $i \in N$, whenever $j \notin R(i)$ then $p_{i,j}(s_i, s_j) = 0$.

2.2 Priority Separable Games

Since we are interested in the logical study of games, we define a qualitative subclass of polymatrix games called *priority separable games* as follows.

Let N be a finite set of players and for each $i \in N$, let S_i be a finite set of strategies for player i. Let Ω be a finite set of outcomes and for all $i \in N$, let $\ll_i \subseteq \Omega \times \Omega$ be a strict total ordering over the outcome set.

We explicitly model the dependency on payoffs using a graphical structure. Let $G = (N, E)$ be a directed graph (without self loops) and $R(i)$ be the neighbourhood of i in G as defined earlier. We associate a linear priority ordering within the neighbourhood for each node $i \in N$ and denote this by the relation $\triangleright_i \subseteq R(i) \times R(i)$

For $i, j \in N$, let $p_{i,j} : S_i \times S_j \to \Omega$ be a partial payoff function. Given a strategy profile s, the payoff for player $i \in N$ is then defined as the tuple $p_i(s) = (p_{i,j}(s_i, s_j))_{j \in R(i)}$.

A *priority separable game* is defined as the tuple

$$\mathcal{G} = (G, (S_i)_{i \in N}, \Omega, (\ll_i)_{i \in N}, (\triangleright_i)_{i \in N}, (p_{i,j})_{i,j \in N}).$$

Note that in a priority separable game $\mathcal{G}$, the number of payoff entries that need to be specified in $\mathcal{G}$ is bounded by $2 \cdot \max_{i \in N} |S_i|^2 \cdot |N|^2$. Thus $\mathcal{G}$ has a compact representation that is polynomial in both $|N|$ and $\max_{i \in N} |S_i|$.

Given a strategy profile s, let $p_i^*(s)$ denote the reordering of the tuple $p_i(s)$ in decreasing order of the priority of neighbours of i. That is, if $R(i) = \{i_1, \dots, i_k\}$ and $i_1 \rhd_i i_2 \rhd_i \cdots \rhd_i i_k$, then for $j \in \{1, \dots, k\}$, $(p_i^*(s))_j = p_{i,i_j}(s_i, s_{i_j})$. In order to analyse the strategic aspect of the game, we need to define how players compare between strategy profiles. For $i \in N$, we define the relation $\preceq_i \subseteq S \times S$ as follows: $s \preceq_i s'$ if $p_i^*(s) \preceq^{lex} p_i^*(s')$ where $\preceq^{lex}$ denotes the lexicographic ordering.

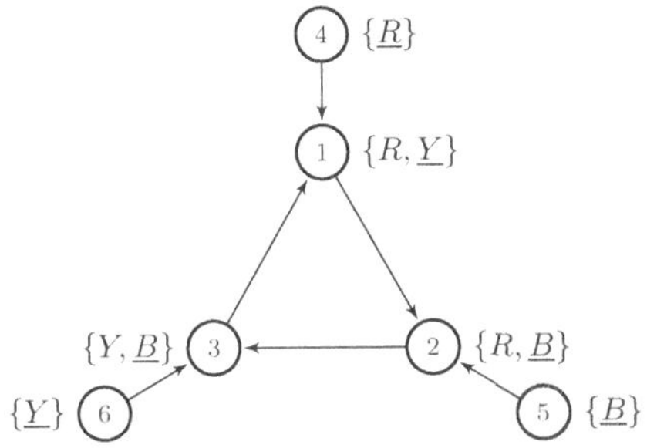

Fig. 1. A priority separable game

Example 1. Consider the game where $N = \{1, \dots, 6\}$ and the graph G is as given in Fig. 1. For $i \in N$ the set of strategies S_i is specified in Fig. 1 as a label next to each node in G. Let $\Omega = \{0, 1\}$ with $0 \ll_i 1$ for all $i \in N$. For $i, j \in N$, let $p_{i,j} = 1$ if $s_i = s_j$ and $p_{i,j} = 0$ if $s_i \neq s_j$. Let $3 \rhd_1 4$, $1 \rhd_2 5$ and $2 \rhd_3 6$. For $j \in \{4, 5, 6\}$, $|S_j| = 1$ and $R(j) = \emptyset$. Consider the strategy profile $s = (Y, B, B, R, B, Y)$ which is denoted with an underline in Fig. 1. Note that in s player 1 is not playing its best response and has a profitable deviation to R. Therefore s is not a Nash Equilibrium.

Some Classes of Priority Separable Games. Two player zero-sum games form a well studied subclass of strategic games that has compact representation and good computational properties. For instance, a (mixed) Nash equilibrium in two player zero-sum games can be computed in polynomial time. It can be observed that every two player game is a priority separable game. The restriction to separable payoff functions extends the underlying idea of two player interaction to multi-player games while retaining the attractive property of having a concise representation. Priority separable games form a subclass of polymatrix games with qualitative payoffs, making it an ideal model for logical analysis.

Note that it is not the case that all polymatrix games can be translated into a priority separable game which preserves the set of Nash equilibria. This is illustrated in the example given below.

Example 2. Consider the polymatrix game $\mathcal{H}$ defined as follows. The player set $N = \{1, 2, 3, 4\}$ and the graph G is as given in Fig. 2. For $i \in N$, S_i is specified in Fig. 2 as a label next to each node in G. Note that for all $i \in \{1, 2, 3\}$, $R(i) = \emptyset$. Consider the partial payoff functions given below.

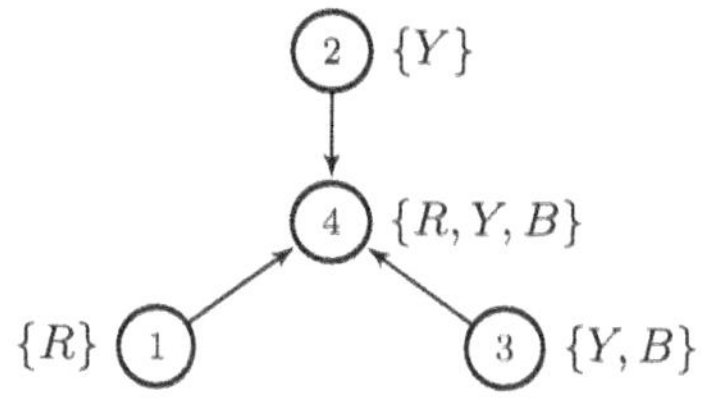

Fig. 2. A priority separable game

- $p_{4,1}(s_4, s_1) = 3$ if $s_1 = s_4$ and 0 otherwise.
- $p_{4,2}(s_4, s_2) = 2$ if $s_2 = s_4$ and 0 otherwise.
- $p_{4,3}(s_4, s_3) = 2$ if $s_3 = s_4$ and 0 otherwise.

Now consider the priority separable game, $\mathcal{G}$, defined over the same set of players N, the same graph G and the same partial payoff functions $p_{i,j}$ where $\Omega = \{2, 3\}$. For all $i \in N$, let $2 \ll_i 3$. Let $\rhd_4$ be chosen arbitrarily. Let $NE(\mathcal{H})$ and $NE(\mathcal{G})$ denote the set of Nash equlibria in $\mathcal{H}$ and $\mathcal{G}$ respectively.

Below we list all possible orderings for $\rhd_4$ and argue that in each case $NE(\mathcal{H}) \neq NE(\mathcal{G})$.

- $1 \rhd_4 2 \rhd_4 3$: The strategy profile $(R, Y, Y, R) \in NE(\mathcal{G})$ but $(R, Y, Y, R) \notin NE(\mathcal{H})$.
- $2 \rhd_4 1 \rhd_4 3$ or $2 \rhd_4 3 \rhd_4 1$: The strategy profile $(R, Y, B, Y) \in NE(\mathcal{G})$ but $(R, Y, B, Y) \notin NE(\mathcal{H})$.
- $3 \rhd_4 2 \rhd_4 1$ or $3 \rhd_4 1 \rhd_4 1$: The strategy profile $(R, Y, B, B) \in NE(\mathcal{G})$ but $(R, Y, B, B) \notin NE(\mathcal{H})$.

There are various interesting classes of polymatrix games which can be viewed as priority separable games. For instance, consider a polymatrix game over a graph G where for every $i \in N$, we can define an ordering $>_i \subseteq R(i) \times R(i)$ such that for all $s \in S$ and for all $j \in R(i)$, $p_{i,j}(s) > \Sigma_{k \in R(i): j >_i k}\, p_{i,k}(s)$. Such a game can be converted into a priority separable game which is strategically equivalent by defining $\rhd_i$ as $>_i$ for all $i \in N$ and taking $\ll_i$ to be the natural ordering over numbers. Note that priority separable games allows $\ll_i$ to be different for each player $i \in N$.

A Resource Allocation Model. The restriction to priority separable payoffs also arises naturally in other domains like resource allocation. A well-studied model for allocation of indivisible items is the Shapley-Scarf housing market [34] defined as follows. Let $N = \{1, \dots, n\}$ be a finite set of agents and $A = \{a_1, \dots, a_n\}$ be a finite set of indivisible items where $|N| = |A|$. An allocation is a bijection $\pi : N \rightarrow A$. In the most commonly studied setting, each agent i has a preference ordering over its allocation $\pi(i)$ and is independent of the allocation of the other agents. However, in many practical instances, agents preferences could depend on externalities like the allocation of other agents. This is particularly relevant in the housing market where an agent's preference for a house could depend on identity of other agents in its immediate neighbourhood. This is also a natural criterion in the allocation of office space where individuals might prefer to be located close to their group members. Priority separable externalities can capture many of these situations.

If agents are allowed to exchange items with each other, stability of allocation is a very natural solution concept to study. Core stable outcomes are defined as allocations in which no group of agents have an incentive to exchange their items as part of an internal redistribution within the coalition. For the housing market without externalities, a simple and efficient procedure often termed as Gale's Top Trading Cycle, can compute a stable allocation that is core stable [34]. In the presence of pairwise separable externalities with quantitative payoffs, the computational properties of the model are studied in [17,28].

Existence of Nash Equilibrium. A natural question is whether priority separable games always have a pure Nash equilibrium. Below we show that the class of priority separable games need not always have a pure Nash equilibrium using an example which is similar to the one given in [4] for polymatrix games.

Example 3. Consider the game given in example 1 along with the neighbourhood graph given in Fig. 1. For players $i \in \{4, 5, 6\}$, $R_i = \emptyset$ and since $|S_i| = 1$, for all $s \in S$, s_i is a best response to s_{-i}. Thus in each strategy profile s only the choices made by players 1, 2 and 3 are relevant. Below we enumerate all such strategy profiles and underline a strategy which is not a best response for each strategy profile. It then follows that this game does not have a Nash equilibrium. $(R, R, \underline{B})$, $(\underline{R}, R, Y)$, $(R, \underline{B}, B)$, $(R, \underline{B}, Y)$, $(\underline{Y}, R, B)$, $(Y, \underline{R}, Y)$, $(\underline{Y}, B, B)$, $(Y, B, \underline{Y})$.

3 Computing Nash Equilibria in Priority Separable Games

Given that priority separable games need not always have a Nash equilibrium, an immediate question is whether there is an efficient procedure to check if a Nash equilibrium exists in this class of games. We show that checking for the existence of a Nash equilibrium is NP-complete. While the upper bound is straightforward, to show NP-hardness we give a reduction from 3-SAT using an argument similar to the one in [4].

Theorem 1. *Given a priority separable game $\mathcal{G}$, deciding if $\mathcal{G}$ has a Nash equilibrium is NP-complete.*

Proof. Given a priority separable game $\mathcal{G}$ and a strategy profile s, we can verify if s is a Nash equilibrium in $\mathcal{G}$ in time polynomial in $|N|$ and $\max_{i \in N} |S_i|$. It follows that the problem is in NP. We show NP-hardness by giving a reduction from 3-SAT.

Consider an instance of 3-SAT given by the formula $\varphi = (a_1 \vee b_1 \vee c_1) \wedge (a_2 \vee b_2 \vee c_2) \wedge \ldots \wedge (a_k \vee b_k \vee c_k)$ with k clauses and m propositional variables $x_1, \ldots, x_m$. For $j \in \{1, \ldots, k\}$, a_j, b_j and c_j are literals of the form x_ℓ or $\neg x_\ell$ for some $\ell \in \{1,\ldots,m\}$. We construct in poly-time a priority separable game $\mathcal{G}_\varphi$ with the neighbourhood graph structure $G = (N, E)$ such that $\mathcal{G}_\varphi$ has a Nash equilibrium iff φ is satisfiable.

For every propositional variable x_ℓ where $\ell \in \{1,\ldots,m\}$, we add a player X_ℓ in $\mathcal{G}_\varphi$ with $S_{X_\ell} = \{\top, \bot\}$. With each clause $a_j \vee b_j \vee c_j$ for $j \in \{1,\ldots,k\}$, we associate 9 players whose neighbourhood is specified by the graph given in Fig. 3. The strategy set for each such node (or player) in the graph is specified as a label next to the node. We use x, y, z as variables where $x, y, z \in \{\top, \bot\}$ whose values are specified as part of the reduction. We denote this graph by $F_j(x, y, z)$ indicating that x, y and z are parameters.

For a literal d, let $\lambda(d) = \top$ if d is a positive literal and $\lambda(d) = \bot$ if d is a negative literal. For each clause with literals a_j, b_j and c_j, which is of the form x_ℓ or $\neg x_\ell$, we add to $\mathcal{G}_\varphi$ the subgraph $F_j(\lambda(a_j), \lambda(b_j), \lambda(c_j))$ and an edge from X_ℓ to the node A_j, B_j or C_j. Let $\Omega = \{0, 1\}$ with $0 \ll_i 1$ for all $i \in N$. For all $i, i' \in N$, we define $p_{i,i'} = 1$ if $s_i = s_{i'}$ and $p_{i,i'} = 0$ if $s_i \neq s_{i'}$. For each subgraph $F_j(x, y, z)$ corresponding to the clause $(a_j \vee b_j \vee c_j)$ and nodes A_j, B_j and C_j let $X_{\ell[A_j]}, X_{\ell[B_j]}, X_{\ell[C_j]}$ denote the nodes such that $(X_{\ell[A_j]}, A_j) \in E$, $(X_{\ell[B_j]}, B_j) \in E$ and $(X_{\ell[C_j]}, C_j) \in E$ respectively for $\ell[A_j], \ell[B_j], \ell[C_j] \in \{1,\ldots,m\}$. We specify the priority ordering for all players i with $|R(i)| > 1$ as follows. For each subgraph $F_j(x, y, z)$ we have,

- $X_{\ell[A_j]} \rhd_{A_j} 6_j \rhd_{A_j} 7_j \rhd_{A_j} C_j$.
- $X_{\ell[B_j]} \rhd_{B_j} 4_j \rhd_{B_j} 8_j \rhd_{B_j} A_j$.
- $X_{\ell[C_j]} \rhd_{C_j} 5_j \rhd_{C_j} 9_j \rhd_{C_j} B_j$.

The crucial observation used in the reduction is the following. Consider the subgraph H_j induced by the nodes in $F_j(x, y, z)$ for $j \in \{1,\ldots,k\}$ along the nodes $X_{\ell[A_j]}, X_{\ell[B_j]}, X_{\ell[C_j]}$. Consider the priority separable game $\mathcal{G}(H_j)$ induced by nodes in H_j and the neighbourhood structure specified by H_j. Observe that a strategy profile t in $\mathcal{G}(H_j)$ is a Nash equilibrium iff at least one of the following conditions hold: $t_{A_j} = t_{X_{\ell[A_j]}}$ or $t_{B_j} = t_{X_{\ell[B_j]}}$ or $t_{B_j} = t_{X_{\ell[B_j]}}$. Using this observation, we can argue that $\mathcal{G}_\varphi$ has a Nash equilibrium iff φ is satisfiable.

Suppose s is a Nash equilibrium in $\mathcal{G}_\varphi$. Consider the valuation function $v_s : \{x_1,\ldots,x_m\} \to \{\top, \bot\}$ defined as follows: $x_\ell = s_{X_\ell}$. From the observation above, it follows that for every $F_j(x, y, z)$ for $j \in \{1,\ldots,k\}$ at least one of the following conditions hold: $s_{A_j} = s_{X_{\ell[A_j]}}$ or $s_{B_j} = s_{X_{\ell[B_j]}}$ or $s_{B_j} = s_{X_{\ell[B_j]}}$. Assume without loss of generality that $s_{A_j} = s_{X_{\ell[A_j]}}$. By the definition of $\mathcal{G}_\varphi$, we have $S_{A_j} \cap S_{X_{\ell[A_j]}} = \{\lambda(a_j)\}$. By the definition of v_s we have $v_s(x_{\ell[A_j]}) = \lambda(a_j)$. This implies that $v_s \models a_j$ and therefore $v_s \models a_j \vee b_j \vee c_j$. Since this holds for all clauses, it follows that $v_s \models \varphi$.

Conversely, suppose φ is satisfiable and let $v \models \varphi$ for some valuation $v : \{x_1,\ldots,x_m\} \to \{\top, \bot\}$. Consider the partially defined strategy profile s^v where $s^v_{X_\ell} = v(x_\ell)$ for all $\ell \in \{1,\ldots,m\}$. Since $v \models \varphi$, for all clauses $a_j \vee b_j \vee c_j$, for $j \in \{1,\ldots,k\}$ we have $v \models a_j$ or $v \models b_j$ or $v \models c_j$. Without loss of generality suppose $v \models a_j$. By definition of $\mathcal{G}_\varphi$ we have $S_{A_j} \cap S_{X_{\ell[A_j]}} = \{\lambda(a_j)\}$. Therefore, the unique best response for node A_j in the game $\mathcal{G}_\varphi$ is the strategy $\lambda(a_j)$. This holds for all clauses and therefore, it is possible to extend s^v to a strategy profile which is a Nash equilibrium in $\mathcal{G}_\varphi$.

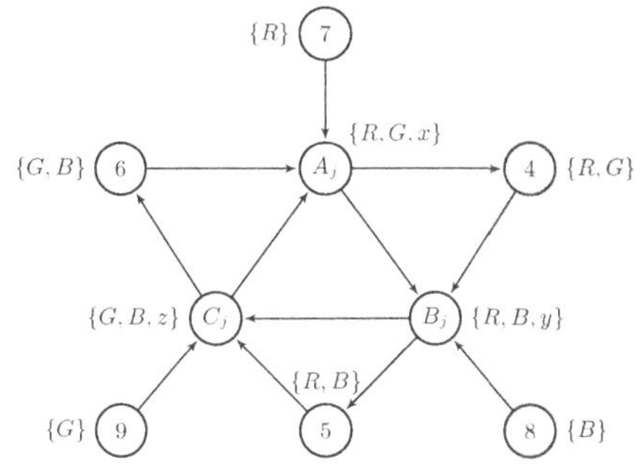

Fig. 3. Gadget $F_j(x, y, z)$

4 Monadic Least Fixed Point Logic

We now present Monadic least fixed point logic (*MLFP*) [33], the logical language we will use to reason about separable games. As fist-order logic cannot express properties like the transitive closure of a relation, its extension with the least fixed point operator, termed *FO*(*LFP*) is used to describe path properties on graphs. *MLFP* is a monadic restriction of *FO*(*LFP*) where the fixed-point operator can be applied only to unary relation variables. The model checking problem for *MLFP* over finite relational structures can be solved in time polynomial in the size of the model, for a fixed formula (in the sense of data complexity ([16])). The logic is also expressive enough to describe various interesting properties of games on finite graphs, as it can describe the transitive closure of a binary relation.

4.1 Syntax

Let $V = \{x_0, x_1, \ldots\}$ be a countable set of first-order variables, and $SV = \{S_0, S_1, \ldots\}$ be a countable set of second-order variables. These sets are disjoint. A first-order vocabulary $\sigma = \{R_0^{k_0}, R_1^{k_1}, \ldots\}$ is a countable set of relation symbols R_i of arity $k_i > 0$.

The set of all *MLFP* formulas is defined inductively as follows:

$$\alpha \in \Phi_{MLFP} := R(x_1, \ldots, x_k) \mid x = y \mid S(x) \mid \neg\alpha \mid \alpha \wedge \alpha \mid \exists x\alpha \mid [\mathbf{lfp}_{S,x}\alpha](u)$$

where R is of arity k, $x_1, \ldots, x_k, x \in V$, $S \in SV$, u does not occur free in α, and all occurrences of S in α are *positive*.

The notion of free occurrence of a variable in a formula is standard. We say that an occurrence of a relation symbol or a second order variable in a formula is *positive* if it occurs under the scope of an even number of negations. (Otherwise we call the occurrence negative.) In the formula, $\phi(R) \stackrel{\text{def}}{=} \neg\forall x\forall y\neg(R(y)\wedge\neg R(x))$ the first occurrence of R is positive and the second is negative. This restriction is needed to provide semantics, where positivity ensures monotonicity of the associated operator and hence guaranteeing existence of a least fixed-point.

Note that in a formula $[\mathbf{lfp}_{S,x}\alpha](u)$ there can be free variables other than x (these are often called parameters).

4.2 Semantics

A σ-structure is a pair $\mathcal{D} = (A, \iota)$ where the non-empty domain of the structure is A, ι is an interpretation such that a relation symbol R of arity k is interpreted as a k-ary relation over A. A model is a tuple $M = (A, \iota, \rho_1, \rho_2)$ where (A, ι) is a σ-structure, and ρ_1, ρ_2 are interpretations of first order and second order variables respectively. For a first order variable x, $\rho_1(x) \in A$ and for a second order variable S, $\rho_2(S) \subseteq A$.

The notion that a formula α holds in a model M is defined inductively as follows:

- $M \models R(x_1, \ldots, x_k)$ iff the tuple $(\rho_1(x_1), \ldots, \rho_k(x_k)) \in \iota(R)$.
- $M \models x = y$ iff $\rho_1(x) = \rho_1(y)$.
- $M \models S(x)$ iff $\rho_1(x) \in \rho_2(S)$.
- $M \models \neg\alpha$ iff $M \not\models \alpha$.
- $M \models \alpha \wedge \beta$ iff $M \models \alpha$ and $M \models \beta$.
- $M \models \exists x.\alpha$ iff for some $a \in A$, $M[x \to a] \models \alpha$.
- $M \models [\mathbf{lfp}_{S,x}\alpha](u)$ iff $\rho_1(u) \in \mathrm{lfp}(f_\alpha)$ (where the map f_α is defined below).

Above, $M[x \to a]$ is the model variant $(A, \iota, \rho_1', \rho_2)$ where $\rho_1'(y) = \rho_1(y)$ for $y \neq x$ and $\rho_1'(x) = a$. Similarly, define $M[x \to a, S \to B]$ is the model variant $(A, \iota, \rho_1', \rho_2')$ where $\rho_1'(y) = \rho_1(y)$ for $y \neq x$ and $\rho_1'(x) = a$, and $\rho_2'(S') = \rho_2(S')$ for $S' \neq S$ and $\rho_2'(S) = B$.

For any formula with x and S occurring free in β, $f_\beta : \wp(A) \mapsto \wp(A)$ is defined by: $f_\beta(B) = \{a \in A \mid M[x \to a, S \to B] \models \beta\}$. The map f_β is an operator on the powerset of elements on the structure ordered by inclusion. The positivity restriction ensures that the operator is monotone and hence has a least fixed-point due to the Knaster-Tarski Theorem [27].

5 Expressing Nash Equilibria in MLFP

We now consider priority separable games as models for the logic. This requires some interpreted relations, and accordingly in the syntax we use special atomic formulas. We will use many-sorted domains, with elements of different types: players, strategies, outcomes etc. Types will be specified by formulas: $P(x)$ would denote that the domain element denoted by x is a player, whereas $O(x)$ would similarly denote x as an outcome, and $T(x)$ denotes x as a strategy and so on.

Towards this, we fix $n > 0$ and consider only models from n-player games. Let $N = \{1, \dots, n\}$. The *game vocabulary* is a tuple

$$\sigma_{\mathcal{G}} = (P^1, T^1, O^1, E^2, u^3, (S_i^1)_{i \in N}, \mathcal{S}^{n+2}, \mathcal{O}^{n+2}, \ll^3, \rhd^2),$$

where the superscripts represent the arities of the relational symbols. In $\sigma_{\mathcal{G}}$, P is intended to denote players, T for strategies, O for outcomes, E for edge relation on players, u for the pairwise partial payoffs, S_i for strategies of player i, $\mathcal{S}$ for relating players and their strategies, $\mathcal{O}$ for relating players and their outcomes, $\ll$ for the preference on outcomes of players, and $\rhd$ for the priority relation on players. In addition, we identify a subset of first order and second order *player variables* V_P and SV_P, respectively. We use p, q, p', q' etc. to distinguish player variables and Q, Q' etc. to designate player set variables.

Correspondingly, $\sigma_{\mathcal{G}}$ structures are interpreted as priority separable games. We abuse notation and use the same symbols for the interpreted relations as well. Thus $P \subseteq N$, $T, O \subseteq D$ (the domain that consisted of S_i and Ω) such that $T \cap O = \emptyset$, $E \subseteq (N \times N)$, $u \subseteq (N \times N \times O)$, and for each $i \in N$, $S_i \subseteq T$, $\mathcal{O} \subseteq (N \times O)$, $\ll \subseteq (N \times O \times O)$ and $\rhd \subseteq (N \times N \times N)$. For variable assignments, we ensure that player variables map to N and player set variables map to subsets of N. However we use x, y etc. to denote variables of all types.

We use the abbreviation $\boldsymbol{x}$ to denote the tuple of variables $(x_1, \dots, x_n)$. The abbrevation $St(\boldsymbol{x})$ will be used for $\bigwedge_{j \in N} S_j(x_j)$, to denote strategy profiles. $Out(\boldsymbol{y})$ for $\bigwedge_{j \in N} O(y_j)$ for tuples of outcomes. We use $\mathcal{S}(\boldsymbol{x}, p, h)$ to denote that the player p chooses strategy h in profile $\boldsymbol{x}$, and $\mathcal{O}(\boldsymbol{a}, p, a)$ to denote that the player p gets outcome a in the vector of outcomes $\boldsymbol{a}$.

Our goal is to describe Nash equilibrium. A profile $\boldsymbol{x}$ is a Nash equilibrium if there is no other profile $\boldsymbol{y}$ such that $\boldsymbol{y}$ is an improvement over $\boldsymbol{x}$ for some player $i \in N$. Thus the formula we are looking for is $Imp(p, \boldsymbol{x}, \boldsymbol{y})$ to denote such an improvement for player p.

There are several steps involved in specifying a player i improvement, some routine, some tricky. For instance we need to specify that for other players, the strategy choices remain the same, which is routine. In each profile, from the pairwise interactions and player priorities we need to compute the outcomes for all players, which is tricky. Indeed this is where we use the least fixed-point operator. We now build the improvement formula, step-by-step.

We first describe formulas relating to the graph structure on players, their priorities and preferences over outcomes.

- p is the player in the neighbourhood of q with the highest priority:

$$\text{first}(p, q) = E(p, q) \wedge (\forall q'(E(q, q') \implies \rhd(p, q', q))$$

- In the priority preference for player p, player q comes immediately before q':

$$\text{nxt}(p,q,q') = E(q,p) \wedge E(q',p) \wedge \triangleright(p,q,q') \wedge \forall p'((E(p',p) \wedge \triangleright(p,q,p') \wedge \triangleright(p,p,q')) \implies (q = p' \vee p' = q'))$$

The following formulas relate strategy profiles and outcomes.

- Player 1-step: this relates two strategy profiles that differ only in the strategy of player p:

$$\text{1-step}(p,\boldsymbol{x},\boldsymbol{y}) = St(\boldsymbol{x}) \wedge St(\boldsymbol{y}) \wedge (\mathcal{S}(\boldsymbol{x},p,g) \wedge \mathcal{S}(\boldsymbol{y},p,h)) \implies g \neq h \wedge \forall q(q \neq p \wedge \mathcal{S}(\boldsymbol{x},q,g) \wedge \mathcal{S}(\boldsymbol{y},q,h)) \implies g = h$$

- The outcome vector $\boldsymbol{y}$ is consistent with the pairwise utility relation u and the tuple of strategies $\boldsymbol{x}$:

$$\text{Con}(\boldsymbol{x},\boldsymbol{y}) = St(\boldsymbol{x}) \wedge Out(\boldsymbol{y}) \wedge \bigwedge_{i \in N} \bigwedge_{j \neq i} u(x_i, x_j, y_j)$$

- Fix $i, j \in N$. We want to specify that $p_{i,j}(\boldsymbol{x}) \lll_i p_{i,j}(\boldsymbol{y})$:

$$\psi_1(p,q,\boldsymbol{x},\boldsymbol{a},\boldsymbol{y},\boldsymbol{b}) = \text{Con}(\boldsymbol{x},\boldsymbol{a}) \wedge \text{Con}(\boldsymbol{y},\boldsymbol{b}) \wedge (\mathcal{O}(\boldsymbol{a},q,a) \wedge \mathcal{O}(\boldsymbol{b},q,b)) \implies \lll (p,a,b)$$

- For a player p the outcomes are equal in two outcome tuples:

$$\psi_2(p,\boldsymbol{a},\boldsymbol{b}) = Out(\boldsymbol{a}) \wedge Out(\boldsymbol{b}) \wedge (\mathcal{O}(\boldsymbol{a},p,a) \wedge \mathcal{O}(\boldsymbol{b},p,b)) \implies a = b$$

Given two outcome vectors $\boldsymbol{a}$ and $\boldsymbol{b}$ we can now compare the outcome for player p using these formulas. Following this we can then write the formula $\text{Dev}(p,\boldsymbol{x},\boldsymbol{a},\boldsymbol{y},\boldsymbol{b})$ that states that $\boldsymbol{y}$ with outcome $\boldsymbol{b}$ is an improvement for player p from $\boldsymbol{x}$ with outcome $\boldsymbol{a}$ by a 1-step deviation.

$$\text{Dev}(p,\boldsymbol{x},\boldsymbol{a},\boldsymbol{y},\boldsymbol{b}) = \text{Con}(\boldsymbol{x},\boldsymbol{a}) \wedge \text{Con}(\boldsymbol{y},\boldsymbol{b}) \wedge \text{1-step}(p,\boldsymbol{x},\boldsymbol{y}) \wedge \exists v.([\mathbf{lfp}_{M,w}\alpha](p,v,\boldsymbol{x},\boldsymbol{a},\boldsymbol{y},\boldsymbol{b}) \wedge \psi_1(p,v,\boldsymbol{x},\boldsymbol{a},\boldsymbol{y},\boldsymbol{b})))$$

$$\alpha(M,p,w,\boldsymbol{x},\boldsymbol{a},\boldsymbol{y},\boldsymbol{b}) = \text{first}(p,w) \vee \exists v\big[M(v) \wedge \text{nxt}(p,w,v) \wedge \psi_2(v,\boldsymbol{a},\boldsymbol{b})\big]$$

The least fixed point formula helps in checking whether for player i has a lexicographic better payoff from strategy profile $\boldsymbol{x}$ to $\boldsymbol{y}$ by checking against $p_i^*(\boldsymbol{x}) \preceq^{lex} p_i^*(\boldsymbol{y})$. The formula α does the iteration for the lexicographic checking. As long as the partial payoffs are equal (which is why ψ_2 features there), it keeps accumulating the players from the neighbourhood set of p. The **lfp** computation on the operator due to the formula α, scans across the payoffs $p_i^*(\boldsymbol{x})$ and $p_i^*(\boldsymbol{y})$. When there is a mismatch in the payoffs the iteration halts or it accumulates all the vertices, which would mean that the payoffs are same.

Finally, in the Dev formula we have $\exists v \psi_1$ which checks if there is a player in the neighbourhood of i for whom there is a lexicographically greater outcome, and along with the other conditions set we are able to express that $\boldsymbol{x} \preceq_p \boldsymbol{y}$. Now, the existence of Nash equilibrium can be characterised using the following formula:

- $\text{Imp}(p, \boldsymbol{x}, \boldsymbol{y}) = (\exists \boldsymbol{b} \exists \boldsymbol{a} \wedge \ \text{Dev}(p, \boldsymbol{x}, \boldsymbol{a}, \boldsymbol{y}, \boldsymbol{b}))$.
- $\text{NE}(\boldsymbol{x}) = \forall p \forall \boldsymbol{y} \neg \text{Imp}(p, \boldsymbol{x}, \boldsymbol{y})$.
- $\mathcal{G} \models \exists \boldsymbol{x} \ \text{NE}(\boldsymbol{x})$

6 Unboundedly Many Players

In game theory, we typically specify every game with the number of players playing in it, and there is no significant difference between analysing equilibria in 23-player games and in 29-player games. In this sense, it is rather unnatural to fix n, the number of players for the universe of games, parameterise the logic by n and consider only models based on n-player games. (But this is customary in modal logics over n-agent systems: for instance, we use epistemic modalities indexed by each of the n agents.) However, if every model specifies its own number of players, the number is potentially unbounded, and the syntax must allow for infinitely many players. For reasoning about games, there is a more tricky representational issue: strategy profiles are no longer n-tuples but maps from a finite subset of player names to the set of strategies.

On the other hand, it should be noted that several of these representational issues have already been faced in our *MLFP* specification of Nash equilibria in priority separable games. Hence, with some extra logical effort, we can consider models with unboundedly many players. In this section, we sketch the development of such a logic.

The important extension we need to make to the syntax of the logic is the addition of *a set of binary second order variables* to the logic. Let $(\mathbf{X}_i)_{i \in \mathbb{N}}$ be a set of binary second order variables. We use X, Y, X', Y' etc. to denote such binary second order variables. These variables are intended to take strategy profiles as values; semantically, they are interpreted as $\mathbf{X} \subseteq (N \times \cup_i S_i)$. Further we will have another set of binary second order variables, $(\mathbf{U}_i)_{i \in \mathbb{N}}$, whose members are denoted by U, V, U', V'etc. These are interpreted as $\mathbf{U} \subseteq (N \times O)$, associating players with outcomes.

Since we have three different types of domain elements, we represent them by three different types of variables - p, u, v, w for player variables, o, a, b - for outcome variables and g, h - for strategy variables.

The set of all formulas of the extended language is defined as before: the only new addition are atomic formulas of the form $X(p, g)$ and $U(p, a)$ where X, U are binary second order variable, p is a player variable, g is a strategy variable and a an outcome variable. We need to specify the relation value taken by a second order variable X or U is indeed a function. This is specified by the formulas:

$$\text{func}(\mathbf{X}) = \forall \mathbf{p} \forall \mathbf{g} \forall \mathbf{h} (\mathbf{P}(\mathbf{p}) \wedge \mathbf{T}(\mathbf{g}) \wedge \mathbf{T}(\mathbf{h}) \wedge ((\mathbf{X}(\mathbf{p}, \mathbf{g}) \wedge \mathbf{X}(\mathbf{p}, \mathbf{h})) \implies \mathbf{g} = \mathbf{h})$$
$$\text{func}(\mathbf{U}) = \forall \mathbf{p} \forall \mathbf{a} \forall \mathbf{b} (\mathbf{P}(\mathbf{p}) \wedge \mathbf{O}(\mathbf{a}) \wedge \mathbf{O}(\mathbf{b}) \wedge ((\mathbf{U}(\mathbf{p}, \mathbf{a}) \wedge \mathbf{U}(\mathbf{p}, \mathbf{b})) \implies \mathbf{a} = \mathbf{b})$$

Now note that all the formulas from the previous section can be translated to formulas in the extended logic, with $\boldsymbol{x}$ being systematically replaced by $\mathbf{X}$ and with $\boldsymbol{a}$ being systematically replaced by $\mathbf{U}$, respectively. For instance:

$$\psi_1(p, q, \mathbf{X}, \mathbf{U}, \mathbf{Y}, \mathbf{V}) =$$
$$\text{Con}(\mathbf{X}, \mathbf{U}) \wedge \text{Con}(\mathbf{Y}, \mathbf{V}) \wedge (\mathcal{O}(U, q, a) \wedge \mathcal{O}(V, q, b)) \implies \ll (p, a, b)$$

specifies outcome preference on partial payoffs. Now, existence of Nash equilibrium can be characterised using the following formula:

- $\text{Imp}(\mathbf{X},\mathbf{Y}) = \exists p(P(p) \wedge \text{Dev}(p, \mathbf{X}, \mathbf{Y}))$
- $\text{NE}(\mathbf{X}) = \forall \mathbf{Y} \neg \text{Imp}(\mathbf{X}, \mathbf{Y})$
- $\mathcal{G} \models \exists\ \mathbf{X}\ \text{NE}(\mathbf{X})$

7 Discussion

In this paper we have studied Nash equilibria in priority separable games and their description in monadic least fixed point logic. Proceeding further, we would like to delineate bounds on the use of logical resources for game theoretic reasoning. For instance, one natural question is the characterization of equilibrium dynamics definable with at most one second order (fixed-point) variable. Moreover, delineating the precise complexity of the logic over games with unboundedly many players requires further work. In this context, it would be especially interesting to explore the framework of parameterized verification ([1]).

Strategy Logic ([12]) is a natural logical framework for description of Nash equilibria. It would be interesting to explore reasoning in such logics over subclasses of polymatrix games, especially in terms of the impact on the model checking problem ([29]).

An important direction is the study of infinite strategy spaces. Clearly the model checking algorithm needs a finite presentation of the input but this is possible and it is then interesting to explore convergence of fixed-point iterations.

Acknowledgements. We thank the anonymous reviewers for their comments which were very helpful in improving the presentation. The first author was partially supported by the Research-I foundation, IIT Kanpur. The third author was partially supported by the grant CRG/2022/006140.

References

1. Abdulla, P.A., Delzanno, G.: Parameterized verification. Int. J. Softw. Tools Technol. Transfer **18**(5), 469–473 (2016). https://doi.org/10.1007/s10009-016-0424-3
2. Ågotnes, T., Harrenstein, P., van der Hoek, W., Wooldridge, M.: Boolean games with epistemic goals. In: Grossi, D., Roy, O., Huang, H. (eds.) LORI 2013. LNCS, vol. 8196, pp. 1–14. Springer, Heidelberg (2013). https://doi.org/10.1007/978-3-642-40948-6_1
3. Alur, R., Henzinger, T.A., Kupferman, O.: Alternating-time temporal logic. J. ACM **49**, 672–713 (2002). https://doi.org/10.1145/585265.585270
4. Apt, K., Simon, S., Wojtczak, D.: Coordination games on directed graphs. In: Proceedings of the 15th International Conference on Theoretical Aspects of Rationality and Knowledge (2015). https://doi.org/10.4204/EPTCS.215.6
5. Aziz, H., Savani, R.: Hedonic games. Handbook of Computational Social Choice, chap. 15, pp. 356–376. Cambridge University Press, Cambridge (2016)
6. Benthem, J.: Games in dynamic epistemic logic. Bull. Econ. Res. **53**(4), 219–248 (2001). https://doi.org/10.1111/1467-8586.00133

7. Benthem, J.: Extensive games as process models. J. Logic Lang. Inform. **11**, 289–313 (2002). https://doi.org/10.1023/A:1015534111901
8. Bonanno, G.: Branching time logic, perfect information games and backward induction. Games Econom. Behav. **36**(1), 57–73 (2001). https://doi.org/10.1006/game.1999.0812
9. Bonzon, E., Lagasquie-Schiex, M., Lang, J., Zanuttini, B.: Boolean games revisited. In: Brewka, G., Coradeschi, S., Perini, A., Traverso, P. (eds.) Proceedings of the 17th ECAI. Frontiers in Artificial Intelligence and Applications, vol. 141, pp. 265–269. IOS Press (2006)
10. Cai, Y., Candogan, O., Daskalakis, C., Papadimitriou, C.: Zero-sum polymatrix games: a generalization of minmax. Math. Oper. Res. **41**(2), 648–655 (2016). https://doi.org/10.1287/moor.2015.0745
11. Cai, Y., Daskalakis, C.: On minmax theorems for multiplayer games. In: Proceedings of the SODA 2011, pp. 217–234. SIAM (2011)
12. Chatterjee, K., Henzinger, T., Piterman, N.: Strategy logic. Inf. Comput. **208**(6), 677–693 (2010). https://doi.org/10.1016/j.ic.2009.07.004
13. Das, R., Padmanabha, A., Ramanujam, R.: Reasoning in large games with unboundedly many players. In: Ghosh, S., Icard, T. (eds.) LORI 2021. LNCS, vol. 13039, pp. 41–57. Springer, Cham (2021). https://doi.org/10.1007/978-3-030-88708-7_4
14. Das, R., Ramanujam, R., Simon, S.: Reasoning about social choice and games in monadic fixed-point logic. In: Moss, L.S. (ed.) Proceedings Seventeenth Conference on Theoretical Aspects of Rationality and Knowledge, TARK 2019, Toulouse, France, 17–19 July 2019. EPTCS, vol. 297, pp. 106–120 (2019). https://doi.org/10.4204/EPTCS.297.8
15. Daskalakis, C., Goldberg, P.W., Papadimitriou, C.H.: The complexity of computing a Nash equilibrium. SIAM J. Comput. **39**(1), 195–259 (2009). https://doi.org/10.1137/070699652
16. Ebbinghaus, Heinz-Dieter., Flum, Jörg.: Finite Model Theory. SMM, Springer, Heidelberg (1995). https://doi.org/10.1007/3-540-28788-4
17. Ghodsi, M., Saleh, H., Seddighin, M.: Fair allocation of indivisible items with externalities. CoRR abs/1805.06191 (2018). http://arxiv.org/abs/1805.06191
18. Goranko, V.: The basic algebra of game equivalences. Stud. Logica. **75**(2), 221–238 (2003). https://doi.org/10.1023/A:1027311011342
19. Harrenstein, B., van der Hoek, W., Meyer, J.J., Witteveen, C.: Boolean games. In: van Benthem, J. (ed.) Proceedings of the 8th TARK, pp. 287–298. Morgan Kaufmann, San Francisco (2001)
20. Harrenstein, P., Hoek, W., Meyer, J., Witteven, C.: A modal characterization of Nash equilibrium. Fund. Inform. **57**(2–4), 281–321 (2003)
21. Herzig, A., Lorini, E., Maffre, F., Schwarzentruber, F.: Epistemic Boolean games based on a logic of visibility and control. In: Kambhampati, S. (ed.) Proceedings of the 25th IJCAI, pp. 1116–1122. IJCAI/AAAI Press (2016)
22. Hoek, W., Jamroga, W., Wooldridge, M.: A logic for strategic reasoning. In: Proceedings of the Fourth International Joint Conference on Autonomous Agents and Multi-Agent Systems, pp. 157–164 (2005). https://doi.org/10.1145/1082473.1082497
23. Igarashi, A., Elkind, E.: Hedonic games with graph-restricted communication. In: Jonker, C.M., Marsella, S., Thangarajah, J., Tuyls, K. (eds.) Proceedings of the 2016 International Conference on Autonomous Agents & Multiagent Systems, Singapore, 9–13 May 2016, pp. 242–250. ACM (2016)

24. Immerman, N.: Descriptive Complexity. Springer, Berlin (2012)
25. Janovskaya, E.: Equilibrium points in polymatrix games. Litovskii Matematicheskii Sbornik **8**, 381–384 (1968)
26. Kearns, M.J., Littman, M.L., Singh, S.: Graphical models for game theory. In: Breese, J.S., Koller, D. (eds.) UAI 2001: Proceedings of the 17th Conference in Uncertainty in Artificial Intelligence, University of Washington, Seattle, Washington, USA, 2–5 August 2001, pp. 253–260. Morgan Kaufmann (2001)
27. Libkin, L.: Elements of Finite Model Theory, vol. 41. Springer, Heidelberg (2004)
28. Massand, S., Simon, S.: Graphical one-sided markets. In: Proceedings of the Twenty-Eighth International Joint Conference on Artificial Intelligence, IJCAI-19, pp. 492–498. International Joint Conferences on Artificial Intelligence Organization (2019). https://doi.org/10.24963/ijcai.2019/70
29. Mogavero, F., Murano, A., Perelli, G., Vardi, M.: Reasoning about strategies: on the model-checking problem. ACM Trans. Comput. Logic **15**(4), 1–47 (2014). https://doi.org/10.1145/2631917
30. Parikh, R.: The logic of games and its applications. Ann. Discrete Math. **24**, 111–140 (1985). https://doi.org/10.1016/S0304-0208(08)73078-0
31. Ramanujam, R., Simon, S.: Dynamic logic on games with structured strategies. In: Proceedings of the 11th International Conference on Principles of Knowledge Representation and Reasoning (KR-08), pp. 49–58. AAAI Press (2008)
32. Ramanujam, R., Simon, S.E.: Structured strategies in games on graphs. In: Flum, J., Grädel, E., Wilke, T. (eds.) Logic and Automata: History and Perspectives [in Honor of Wolfgang Thomas]. Texts in Logic and Games, vol. 2, pp. 553–574. Amsterdam University Press (2008)
33. Schweikardt, N.: On the expressive power of monadic least fixed point logic. Theoret. Comput. Sci. **350**, 325–344 (2006). https://doi.org/10.1016/j.tcs.2005.10.025
34. Shapley, L.S., Scarf, H.: On cores and indivisibility. J. Math. Econ. **1**(1), 23–37 (1974). https://doi.org/10.1016/0304-4068(74)90033-0
35. Simon, S., Wojtczak, D.: Constrained pure Nash equilibria in polymatrix games. In: Proceedings of the AAAI Conference on Artificial Intelligence, vol. 31 (2017)
36. Walther, D., Hoek, W., Wooldridge, M.: Alternating-time temporal logic with explicit strategies. In: Proceedings of the 11th Conference on Theoretical Aspects of Rationality and Knowledge (TARK-2007), pp. 269–278 (2007). https://doi.org/10.1145/1324249.1324285

Modal Logics with Non-rigid Propositional Designators

Yifeng Ding(✉)

Peking University, Beijing, China
yf.ding@pku.edu.cn

Abstract. In most modal logics, atomic propositional symbols are directly representing the meaning of sentences (such as sets of possible worlds). In other words, they use only rigid propositional designators. This means they are not able to handle uncertainty in meaning directly at the sentential level. In this paper, we offer a modal language involving non-rigid propositional designators which can also carefully distinguish *de re* and *de dicto* use of these designators. Then, we axiomatize the logics in this language with respect to all Kripke models with multiple modalities and with respect to S5 Kripke models with a single modality.

Keywords: Modal Logic · Epistemic Logic · Non-rigid Designator · Ambiguity · Propositional Quantifier

1 Introduction

We frequently fail to grasp the meaning of sentences. People who learned English only from textbooks may not get a certain contextual meaning of "this is sick!", and anyone who is not well versed in set theory is unlikely to fully grasp even the literal meaning of "the Ultimate-L conjecture". We also intentionally hide the meaning of symbols by designing secret interpretations of symbols to communicate private information in public: cryptographic protocols are essentially doing this, and the same string of zeros and ones can mean different things when decoded by different keys.

The famous Frege's puzzle can also be understood in this way. To people unfamiliar with the fact that "Lewis Carroll" is the pen name of Charles Dodgson who is also a logician and responsible for Dodgson's method in voting theory, "Lewis Carroll authored *Alice in Wonderland*" and "Charles Dodgson authored *Alice in Wonderland*" express different propositions. Indeed, they are likely to believe that the first sentence is true while the second sentence is false. However, given that Lewis Carroll is actually Charles Dodgson, "Lewis Carroll authored

This work is supported by NSSF 22CZX066. The author also thanks the anonymous referees and the audience of the 2023 Beijing International Summer Workshop on Formal Philosophy for their helpful comments and suggestions.

N. Alechina et al. (Eds.): LORI 2023, LNCS 14329, pp. 47–62, 2023.
https://doi.org/10.1007/978-3-031-45558-2_4

Alice in Wonderland" and "Charles Dodgson authored *Alice in Wonderland*" in fact express the same proposition.

In epistemic logic in its basic form, this ubiquitous phenomenon of uncertainty in meaning is not modeled at all. A propositional symbol p is meant to directly designate a proposition (a set of possible worlds) much like in first-order (modal) logic an individual variable x is meant to directly designate an object in the domain, and one can never be uncertain about what p means but only what's p's truth value since p is already 'interpreted'.

To our best knowledge, attempts to model uncertainty in meaning in the modal logic and possible world semantics paradigm are scarce. One notable work is [19] where different agents may interpret the same propositional symbol differently. In the usual setting of possible world semantics with multiple agents in Agt, this can be understood as taking a model to be $(W, \{R_i\}_{i\in\mathsf{Agt}}, \{V_i\}_{i\in\mathsf{Agt}})$ where R_i is the accessibility relation for i (we write the corresponding modal operator as 'B_i', 'B' for 'Belief') and for each $i \in \mathsf{Agt}$, V_i is a valuation function assigning to each propositional symbol p a set in $\wp(W)$. Then, B_ip is true at a world w iff $V_i(p) \subseteq R_i(w)$; that is B_ip says that i believes the proposition she takes p to mean. More generally, $B_iB_jB_kp$ means that i believes that j believes that k believes that p as interpreted by k. In other words, an occurrence of p is always interpreted by the last agent i whose belief operator scopes over that occurrence. This restriction is lifted in [18], where we can form propositional symbols p_i indexed by agent i so that p_i is always interpreted by V_i. This in a sense means that if the only uncertainty to the meaning of a propositional symbol p is how different agents may interpret it differently but unambiguously, the standard epistemic logic can simulate this by using more propositional symbols.

Another important relevant work is [17]. There, the meaning of a propositional symbol p is not merely determined by the set of possible worlds assigned to it by the valuation function V, but fundamentally by a syntactic definition $\mathsf{DEF}_w(p)$ of it using other propositional symbols, and the definition could vary from worlds to worlds. Of course, the definitions and the valuation must cohere. Then, while an agent still knows what is the proposition assigned to p by the valuation function V, the agent may not know the *definition* of p, and further the proposition expressed by the definition of p.

In this paper, we take perhaps the most straightforward way to allow uncertainty in meaning: we simply let propositional symbols be *non-rigid* designators of sets of possible worlds. In other words, we let the valuation function be world relative. This approach has been taken up in [21] to formalize definite descriptions of propositions and the Brandenburger-Keisler paradox. The paradox involves sentences such as:

(A1) Ann believes that the strangest proposition that Bob believes is that neutrinos travel at twice the speed of light.
(A2) Ann believes that the strangest proposition that Bob believes is true.

In [21], the first sentence is formalized as $B_a(\gamma \,\mathsf{is}\, \varphi)$, and the second sentence's *de dicto* and *de re* readings are formalized as $B_a^{re}\mathsf{T}(\gamma)$ and $B_a^{dicto}\mathsf{T}(\gamma)$, respectively, where γ is a definite description (non-rigid designator) for the strangest proposition that Bob believes. We find the formalism slightly cumbersome and not fully general. Taking inspiration from concept abstraction used in first-order intentional modal logic [11,13] and assignment operators used in [6,17,25,30], we relabel the syntactic category of propositional variables x which are rigid designators and use $[p/x]\varphi$ to mean "letting x be the proposition expressed by p, φ". Since the propositional variables x are only playing the role p used to play, we are only extending the basic language of modal logic by the binders $[p/x]$. With this minimal perturbation, we can already easily distinguish

- $[p/x]B_iB_jx$: letting x be the proposition p actually means, i believes that j believes that x is true;
- $B_i[p/x]B_jx$: i believes that, with x being p's meaning, j believes that x;
- $B_iB_j[p/x]x$: i believes that j believes that p is true.

The Ann and Bob sentence above can also be formalized with the help of a necessity modality $\Box$ that quantifies over all possible worlds, in which case when $\Box(x \leftrightarrow y)$ is true, x and y denote the same proposition.

- $B_a[p/x] \Box (x \leftrightarrow y)$ formalizes (A1) where y directly denotes the proposition expressed by 'neutrinos travel at twice the speed of light'.
- $[p/x]B_ax$ formalizes the *de re* reading of (A2).
- $B_a[p/x]x$ formalizes the *de dicto* reading of (A2).

The semantic type of functions from worlds to sets of worlds appears in various kinds of higher-order modal logics [12,15,27]. Indeed, there is a way to embed our language in the higher-order intentional language presented in [15]. Objects of the said type also bear the name 'two-dimensional content' and are used in for example [3,4,23,24,28]. The semantic function of the operator $[p/x]$ can also be understood as 'rigidifying' the non-rigid designator p. From this perspective, our work is related to generalized versions of hybrid logic [2]. Further discussion of relations to higher-order and hybrid modal logics are included in Sect. 2 after we formally introduce our minimalist language and its semantics.

Our main technical contributions are two axiomatization results, one with respect to all multiagent models, and one with respect to single-agent models where the accessibility relation is the universal relation (single-agent epistemic models). Axiomatization in our setting poses an interesting challenge that echos with the following 'paradox' on a Cantorian level: one cannot be completely ignorant of the meaning of p in the possible world framework, because there are always more possible meanings of p (sets of possible worlds) than there are possible worlds, but for different meanings X of p, we need different possible worlds to model the possibility that the agent takes X to be the meaning of p. In completeness proofs with assignment operators, one typically extends language so that in each maximally consistent set (MCS), each non-rigid designator has a witness. Let p be such a non-rigid designator. Now different MCSs should have different witnesses for p, as otherwise they are forced to take p to mean the same thing. Put in another way, we in principle need fresh witnesses for each MCS to maintain consistency when adding those witnesses. But then, we are back in

Cantor's trap: no matter how we extend our language, there will always be more MCSs than there are variables. We will bypass this difficulty using step-by-step constructions.

The rest of the paper is organized as follows: in Sect. 2, we formally introduce the language and the semantics. We will also comment on the undecidability of the set of validities for the class of universal models (single-agent S5 case) and discuss how our language compares to higher-order and hybrid modal logics. Section 3 deals with the class of all models, i.e., the multi-agent K case, and in Sect. 4, we consider the class of universal models, i.e., single-agent S5 case. Finally, we conclude in Sect. 5 with possible future research directions.

2 Formal Language and Semantics

Definition 1. *We fix a countably infinite set* Prop *of propositional names, a countably infinite set* Var *of propositional variables, and a non-empty set* Agt *of unary modal operators. Then, define language $\mathcal{L}$ by the following grammar:*

$$\mathcal{L} \ni \varphi ::= x \mid \neg\varphi \mid (\varphi \wedge \varphi) \mid \Box\varphi \mid [p/x]\varphi$$

where $x \in$ Var, $p \in$ Prop, and $\Box \in$ Agt. The usual abbreviations apply. Also, we treat $[p/x]$ as a quantifier that binds the variable x. Thus the usual notions of free and bound variables, free for substitution (substitutability), and so on apply as well. $\varphi[y/x]$ is the result of replacing all free occurrences of x in φ by y. We will usually accompany this notation with a substitutability requirement.

Here symbols in Prop are non-rigid propositional designators while symbols in Var are rigid propositional designators. Syntactically we do not allow for $p \in$ Prop to appear as an atomic formula since for example, $B_iB_jB_kp$ is ambiguous. Of course, we could write $B_iB_jB_k[p/x]x$ when that is the intended expression.

Definition 2. *A* Kripke model with non-rigid propositional designators *('model' for short) is a tuple $(W, \{P_p\}_{p\in\mathsf{Prop}}, \{R_\Box\}_{\Box\in\mathsf{Agt}})$ where*

- *W is a non-empty set, intuitively the set of possible worlds;*
- *for each $p \in$ Prop, P_p is a function from W to $\wp(W)$, with $P_p(w)$ understood as the proposition p designates at w;*
- *for each $\Box \in$ Agt, $R_\Box \subseteq W^2$, the accessibility relation for $\Box$.*

Given a model $\mathcal{M} = (W, \{P_p\}_{p\in\mathsf{Prop}}, \{R_\Box\}_{\Box\in\mathsf{Agt}})$, an assignment *$\sigma$ for $\mathcal{M}$ is a function from* Var *to $\wp(W)$. Truth in a model $\mathcal{M} = (W, \{P_p\}_{p\in\mathsf{Prop}}, \{R_\Box\}_{\Box\in\mathsf{Agt}})$ is defined recursively relative to worlds and assignments as follows:*

$$\begin{aligned}
\mathcal{M}, w, \sigma \vDash x &\iff w \in \sigma(x)\\
\mathcal{M}, w, \sigma \vDash \neg\varphi &\iff \mathcal{M}, w, \sigma \nvDash \varphi\\
\mathcal{M}, w, \sigma \vDash (\varphi \wedge \psi) &\iff \mathcal{M}, w, \sigma \vDash \varphi \text{ and } \mathcal{M}, w, \sigma \vDash \psi\\
\mathcal{M}, w, \sigma \vDash \Box_i\varphi &\iff \forall v \in W, wR_\Box v \Rightarrow \mathcal{M}, v, \sigma \vDash \varphi\\
\mathcal{M}, w, \sigma \vDash [p/x]\varphi &\iff \mathcal{M}, w, \sigma[P_p(w)/x] \vDash \varphi.
\end{aligned}$$

Here $\sigma[P_p(w)/x]$ is the function that is identical to σ except that $\sigma[P_p(w)/x](x) = P_p(w)$. This '$f[a/x]$' notation is used for all functions. A formula φ is valid on a model $\mathcal{M}$ if it is true at all worlds relative to all assignments (written $\mathcal{M} \vDash \varphi$). φ is valid on a class $\mathcal{K}$ of models if it is valid on all models in the class $\mathcal{K}$.

The analogue of the substitution lemma in first-order logic holds as well.

Lemma 1. *For any model $\mathcal{M} = (W, \{P_p\}_{p \in \mathsf{Prop}}, \{R_\Box\}_{\Box \in \mathsf{Agt}})$, $w \in W$, assignment σ for $\mathcal{M}$, formula $\varphi \in \mathcal{L}$, $x, y \in \mathsf{Var}$, if y is substitutable for x in φ, then $\mathcal{M}, w, \sigma \vDash \varphi[y/x]$ iff $\mathcal{M}, w, \sigma[\sigma(y)/x] \vDash \varphi$.*

We will also be interested in the case when Agt is a singleton $\{\Box\}$, and the relation $R_\Box$ is the universal relation. Since the universal relation is uniquely determined by the set of possible worlds, we will simply dispense with it.

Definition 3. *A* universal Kripke model with non-rigid propositional designators *('universal model' for short) is a tuple $(W, \{P_p\}_{p \in \mathsf{Prop}})$ where W is a non-empty set and for each $p \in \mathsf{Prop}$, $P_p : W \to \wp(W)$. When $\mathsf{Agt} = \{\Box\}$, we interpreted $\mathcal{L}$ on universal models $\mathcal{M} = (W, \{P_p\}_{p \in \mathsf{Prop}})$ just like in Definition 2 except that $\mathcal{M}, w, \sigma \vDash \Box\varphi$ iff forall $v \in W$, $\mathcal{M}, v, \sigma \vDash \varphi$.*

These models can be used to model an S5 agent, for which the $R_\Box$ relation is an equivalence relation since truth in $\mathcal{L}$ is preserved under generated submodel. Due to lack of space, we will not define and prove this formally, but in fact, more generally, $\mathcal{L}$ can be translated into the guarded fragment, though not the two-variable fragment. Now, on universal models, the 'guard' does not really do anything, and indeed, for the class of universal models, its set of validities is undecidable. For a starter, note that:

Proposition 1. *$\Box[p/x]\Diamond[p/y](\Box(x \to y) \wedge \Diamond(y \wedge \neg x))$ is satisfiable by a universal model, and all such models are infinite.*

The idea is that $\Box[p/x] \Diamond [p/y](\Box(x \to y) \wedge \Diamond(y \wedge \neg x))$ entails there must be an infinite strictly ascending chain of sets of possible worlds. The 'paradox' mentioned in the introduction is also formalizable as $\Diamond[p/y] \Box (x \leftrightarrow y)$, and indeed no universal model can validate this formula.

Again, due to lack of space, we will not formally prove undecidability, but the idea is to use the formula $\Box[p/x](x \wedge \Box(x \to [p/y] \Box (y \leftrightarrow x)))$ so that we can use $\Box[p/x]$ to simulate the first-order quantifier $\forall x$ and use another $q \in \mathsf{Prop}$ to simulate a binary relation R so that $R(x, y)$ translates to $\Diamond(x \wedge [q/z] \Diamond (y \wedge z))$. Then we can translate first-order logic with a binary relation into $\mathcal{L}$.

Now we briefly comment on how our language and semantics compare to the semantics of higher-order modal logics and hybrid logics. First, we consider the influential system **IL** (Intentional Logic) presented in [15]. As a higher-order logic, we first need to define the types of its language. To simplify the presentation, we omit the basic type e for individuals (tables and chairs) and the complex types using it. Thus, the basic type names are s and t where s

names the type for possible worlds and t names the type for truth values. All types can be generated by the following BNF grammar

$$\mathsf{Type} \ni \alpha ::= t \mid (\alpha \to \alpha) \mid (s \to \alpha).$$

Note that s is not by itself a type. When parentheses are omitted, we assume right-association, e.t. $s \to s \to t$ means $(s \to (s \to t))$. For each $\alpha \in \mathsf{Type}$ we assume that there are countably infinitely many constants c and variables x of the type α (when we highlight their type, we write c_α and x_α). Then, the set T_α of the terms of type α are defined inductively by the following clauses:

- Constants and variables of type α are in T_α.
- If $A \in \mathsf{T}_{\alpha\to\beta}$ and $B \in \mathsf{T}_\alpha$, then $(AB) \in \mathsf{T}_\beta$.
- If $A \in \mathsf{T}_\beta$ and x is a variable of type α, then $(\lambda x.A) \in \mathsf{T}_{\alpha\to\beta}$.
- If $A, B \in \mathsf{T}_\alpha$, then $(A = B) \in \mathsf{T}_t$.
- If $A \in \mathsf{T}_\alpha$, then $(\hat{}A) \in \mathsf{T}_{s\to\alpha}$.
- If $A \in \mathsf{T}_{s\to\alpha}$, then $(\check{}A) \in \mathsf{T}_\alpha$.

Again, we write A_α to highlight that A is of type α and assume left-association when parentheses are omitted. Truth-functional operators such as $\neg \in \mathsf{T}_{t\to t}$ and $\wedge \in \mathsf{T}_{t\to t\to t}$ are not included as they can be defined by lambda terms, and the meaning of $\hat{}$ and $\check{}$ will become clear below.

Semantically, an object of type t is a truth value while an object of type s is understood as a possible world, and an object of type $\alpha \to \beta$ is a function from objects of type α to objects of type β. Each term A of type α extensionally denotes an object of type α and intentionally denotes an object of type $s \to \alpha$, namely a function from possible worlds to objects of type α. Thus, for a set-theoretical formal semantics, given a non-empty set W for possible worlds, we define the full domain D^W_α for each type α recursively by $D^W_t = \{0, 1\}$, $D^W_s = W$, and $D^W_{\alpha\to\beta} = (D^W_\beta)^{D^W_\alpha}$, the set of all functions from D^W_α to D^W_β (here we allow α to be s). Then, a standard model for **IL** is a pair (W, I) where W is a non-empty set (of possible worlds) and I is a function that maps, for all type α, the constants c of type α to $I(c) \in D^W_{s\to\alpha}$, which we take as the intention of c in this model. An assignment σ for a model (W, I) is a function that maps each variable x of its type α to $\sigma(x) \in D^W_\alpha$. Then the denotation $|A|^{W,I,w,\sigma}$ of terms A at world w relative to assignment σ in model (W, I) is defined recursively:

- $|c|^{W,I,w,\sigma} = I(c)(w)$ and $|x|^{W,I,w,\sigma} = \sigma(x)$.
- $|AB|^{W,I,w,\sigma} = |A|^{W,I,w,\sigma}(|B|^{W,I,w,\sigma})$.
- $|\lambda x_\alpha.A_\beta|^{W,I,w,\sigma} = \{(a, |A_\beta|^{W,I,w,\sigma[a/x_\alpha]}) \mid a \in D^W_\alpha\}$.
- $|A = B|^{W,I,w,\sigma} = 1$ if $|A|^{W,I,w,\sigma} = |B|^{W,I,w,\sigma}$, and is 0 otherwise.
- $|\hat{}A|^{W,I,w,\sigma} = \{(v, |A|^{W,I,v,\sigma}) \mid v \in W\}$.
- $|\check{}A_{s\to\alpha}|^{W,I,w,\sigma} = |A_{s\to\alpha}|^{W,I,w,\sigma}(w)$.

At any world w, the idea of $\hat{}A$ is to obtain the intention of A, the total function from worlds to A's denotation at those worlds, as its denotation. Conversely, when we have a term $A_{s\to\alpha}$, its denotation is already 'intentionally of type α', and the idea of $\check{}A_{s\to\alpha}$ is to get the extension of A's denotation.

Now let us try to translate our language $\mathcal{L}$ into the language of **IL** with its standard semantics. If we were working with the basic language of propositional modal logic, then it would be a natural choice to regard $\varphi \in \mathcal{L}$ as terms of type t. This would require us to take atomic propositional symbols p as constants of type t so that they could have non-constant intentions, and take modal operators $\Box$ as terms of type $(s \to t) \to t$ as they operate on the intention of formulas. Then $\Box\varphi$ should be translated as $\Box(\hat{}\varphi)$. However, we cannot treat propositional variables x as constants of type t since syntactically we must be able to bind them. So they must be a variable of some type. The natural choice we have then is variables of type $s \to t$, since the natural translation for propositional names $p \in \mathsf{Prop}$ are constants of type $s \to t$, and $[p/x]\varphi$ can be understood as $(\lambda x.\varphi)p$. But it is also natural to take formulas as terms of type t, which coheres well with the truth-functional operators, so we must deal with the fact that in $\mathcal{L}$, propositional variables are also formulas. The solution is simple: always use $\check{}x$.

More formally, define a translation T on $\mathcal{L}$:

- $T(x) = (\check{}x_{s\to t})$ where $x_{s\to t}$ is a variable corresponding to x.
- $T(\neg\varphi) = \neg T(\varphi)$, and $T(\varphi \wedge \psi) = (\wedge T(\varphi))T(\psi)$.
- $T(\Box\varphi) = \Box_{(s\to t)\to t}(\hat{}T(\varphi))$ where $\Box_{(s\to t)\to t}$ is a constant corresponding to $\Box$.
- $T([p/x]\varphi) = (\lambda x_{s\to t}.T(\varphi))p_{s\to t}$ where $p_{s\to t}$ is a constant corresponding to p.

Then it is not hard to check that, with the obvious way to expand a model and assignment for $\mathcal{L}$ into a standard model and assignment for **IL**, φ and $T(\varphi)$ are true at precisely the same worlds.

For the hybrid way to understand $\mathcal{L}$, consider the following variation $\mathcal{L}@$ of $\mathcal{L}$ where instead of a set Var of propositional variables, we use a set Nom of nominal variables. Then $\mathcal{L}@$ is defined by the grammar

$$\varphi ::= (p@i) \mid \neg\varphi \mid (\varphi \wedge \varphi) \mid \Box\varphi \mid \downarrow i\varphi$$

where $p \in \mathsf{Prop}$, $i \in \mathsf{Nom}$, and $\Box \in \mathsf{Agt}$. Given a model $\mathcal{M} = (W, \{P_p\}_{p\in\mathsf{Prop}}, \{R_\Box\}_{\Box\in\mathsf{Agt}})$ and a nominal assignment $\nu : \mathsf{Nom} \to W$, we define the semantics by $\mathcal{M}, w, \nu \vDash (p@i)$ iff $w \in P_p(\nu(i))$ and $\mathcal{M}, w, \nu \vDash \downarrow i\varphi$ iff $\mathcal{M}, w, \nu[w/i] \vDash \varphi$. Then it is also not hard to see that $[p/x]\varphi$ can be understood as $\downarrow i_x\varphi[(p@i_x)/x]$ where i_x is a nominal variable corresponding to the variable x and $\varphi[(p@i_x)/x]$ is the result of replacing free occurrences of x in φ with $(p@i)$. Thus, a truth-preserving translation from sentences (formulas without free propositional variables) in $\mathcal{L}$ to sentences (formulas without free nominal variables) in $\mathcal{L}@$ can be defined.

3 Axiomatization for Multi-agent K

In this section, we deal with the class of all models. Our completeness proof requires adding new variables to the language. Thus let us fix a set Var^+ that is a superset of Var and $\mathsf{Var}^+ \setminus \mathsf{Var}$ is also countably infinite. Then by $[\mathsf{Var}, \mathsf{Var}^+]$ we mean the set $\{X \mid \mathsf{Var} \subseteq X \subseteq \mathsf{Var}^+\}$.

Definition 4. *For any* $X \in [\mathsf{Var}, \mathsf{Var}^+]$*, define* $\mathcal{L}(X)$ *by the following grammar:*

$$\mathcal{L}(X) \ni \varphi ::= x \mid \neg\varphi \mid (\varphi \wedge \varphi) \mid \Box\varphi \mid [p/x]\varphi$$

where $x \in X$*,* $p \in \mathsf{Prop}$*, and* $\Box \in \mathsf{Agt}$*. Obviously* $\mathcal{L} = \mathcal{L}(\mathsf{Var})$*.*

Now we define the logic NPK (non-rigid propositional K).

Definition 5. *For any* $X \in [\mathsf{Var}, \mathsf{Var}^+]$*, let* $\mathsf{NPK}(X)$ *be the set of formulas in* $\mathcal{L}(X)$ *axiomatized by the following axioms and rules:*

- *(PL) All instances of propositional tautologies in* $\mathcal{L}(X)$
- *(K)* $\Box(\varphi \to \psi) \to (\Box\varphi \to \Box\psi)$
- *(Comm)* $[p/x](\varphi \to \psi) \to ([p/x]\varphi \to [p/x]\psi)$ *and* $[p/x]\neg\varphi \leftrightarrow \neg[p/x]\varphi$
- *(Triv)* $\varphi \leftrightarrow [p/x]\varphi$ *where* x *does not occur free in* φ
- *(Sub)* $[p/y]([p/x]\varphi \leftrightarrow \varphi[y/x])$ *whenever* y *is substitutable for* x *in* φ
- *(Perm)* $[p/x][q/y]\varphi \leftrightarrow [q/y][p/x]\varphi$ *where* x *and* y *are distinct variables*
- *(MP) from* φ *and* $\varphi \to \psi$ *derive* ψ
- *(Nec) from* φ *derive* $\Box\varphi$ *for every* $\Box \in \mathsf{Agt}$
- *(Inst) from* φ *derive* $[p/x]\varphi$

As usual, we write $\Gamma \vdash_{\mathsf{NPK}(X)} \varphi$ *to mean that* $\Gamma \cup \{\varphi\} \subseteq \mathcal{L}(X)$ *and there is a finite conjunction* γ *of formulas in* Γ *such that* $\gamma \to \varphi$ *is in* $\mathsf{NPK}(X)$*. By* NPK *we mean* $\mathsf{NPK}(\mathsf{Var})$*.*

The soundness of these axioms and rules is easy to check, where (Sub) is the syntactic version of the substitution lemma. We collect some basic facts about the logic in the following lemma:

Lemma 2. *Let* $X, Y \in [\mathsf{Var}, \mathsf{Var}^+]$ *such that* $X \subseteq Y$*.*

- $\mathsf{NPK}(X)$ *proves equivalence under renaming of bound variables.*
- *If* $\Gamma \subseteq \mathcal{L}(X)$ *is consistent in* $\mathsf{NPK}(X)$ *(that is,* $\Gamma \nvdash_{\mathsf{NPK}(X)} \bot$*), then there is a maximally consistent set (MCS for short)* Δ *w.r.t.* $\mathsf{NPK}(X)$ *extending* Γ*. We use choice to fix such a set uniformly as* $Ext_{\mathsf{NPK}(X)}(\Gamma)$*.*
- *For any* $\Gamma \subseteq \mathcal{L}(X)$ *and any* $\Box \in \mathsf{Agt}$*, define* $\Box^{-1}\Gamma = \{\varphi \mid \Box\varphi \in \Gamma\}$*. Then if* Γ *is consistent in* $\mathsf{NPK}(X)$*, for any formula* $\neg\Box\varphi \in \Gamma$*,* $\{\neg\varphi\} \cup \Box^{-1}\Gamma$ *is consistent.*
- $\mathsf{NPK}(Y)$ *is conservative over* $\mathsf{NPK}(X)$*:* $\mathsf{NPK}(Y) \cap \mathcal{L}(X) = \mathsf{NPK}(X)$*.*

The first three points are standard exercises. For the last point, note that any proof in $\mathsf{NPK}(Y)$ uses only finitely many variables. Thus we can always find unused variables in X and uniformly replace variables in $Y \setminus X$ used in the proof by these new variables in X.

Definition 6. *A* witness assignment *is an injective function* v *from* Prop *to* Var^+*. We often write* v_p *for* $v(p)$*. For any* $X \in [\mathsf{Var}, \mathsf{Var}^+]$*, a witness assignment* v *is* fresh for X *if* $ran(v) \cap X = \varnothing$*. For any witness assignment* v *and* $X \in [\mathsf{Var}, \mathsf{Var}^+]$*, define* $\mathrm{WF}(v, X)$ *(witnessing formulas in* $\mathcal{L}(X)$ *using* v*) to be*

$$\{[p/x]\alpha \leftrightarrow \alpha[v_p/x] \mid [p/x]\alpha \in \mathcal{L}(X), v_p \text{ is substitutable for } x \text{ in } \alpha\}.$$

We also write $X + v$ *for* $X \cup ran(v)$*.*

Lemma 3. *For any* $X \in [\mathsf{Var}, \mathsf{Var}^+]$, Γ *a MCS in* $\mathsf{NPK}(X)$, *and* v *a witness assignment fresh for* X, *the set* $\Gamma' = \Gamma \cup \mathrm{WF}(v, X+v)$ *is consistent in* $\mathsf{NPK}(X+v)$ *and has exactly one MCS extension in* $\mathsf{NPK}(X+v)$. *We denote this extension by* $\Gamma + v$.

Proof. In this proof, we write $\vdash$ for $\vdash_{\mathsf{NPK}(X)}$ and $\vdash_{+v}$ for $\vdash_{\mathsf{NPK}(X+v)}$. First, we show consistency. Suppose not, then we have a finite $\{\alpha_1, \ldots, \alpha_n\} \subseteq \Gamma$ and a finite $\{[p_i/x_i]\beta_i \leftrightarrow \beta_i[v_{p_i}/x_i]\}_{i=1}^m \subseteq \mathrm{WF}(v, X+v)$ with

$$\vdash_{+v} \bigwedge_i \alpha_i \to \neg \bigwedge_i ([p_i/x_i]\beta_i \leftrightarrow \beta_i[v_{p_i}/x_i]). \tag{1}$$

Now we have the following derivable formulas:

$$\vdash_{+v} \bigwedge \alpha_i \to \neg \bigwedge_i [p_1/v_{p_1}] \ldots [p_m/v_{p_m}]([p_i/x_i]\beta_i \leftrightarrow \beta_i[v_{p_i}/x_i]). \tag{2}$$

$$\vdash_{+v} [p_i/v_{p_i}]([p_i/x_i]\beta_i \leftrightarrow \beta_i[v_{p_i}/x_i]). \tag{3}$$

$$\vdash_{+v} [p_1/v_{p_1}] \ldots [p_m/v_{p_m}]([p_i/x_i]\beta_i \leftrightarrow \beta_i[v_{p_i}/x_i]). \tag{4}$$

$$\vdash_{+v} \bigwedge \alpha_i \to \bigwedge_i [p_1/v_{p_1}] \ldots [p_m/v_{p_m}]([p_i/x_i]\beta_i \leftrightarrow \beta_i[v_{p_i}/x_i]). \tag{5}$$

(2) is obtained from (1) by repeated use of (Inst), (Comm), and (Triv). (3) are simply instances of (Sub). (4) are obtained from (3) by (Inst) and (Perm). (5) is simply combining (4) for all i and add an antecedent. Thus, Γ is inconsistent in $\mathsf{NPK}(X+v)$. By the conservativity of $\mathsf{NPK}(X+v)$ over $\mathsf{NPK}(X)$, Γ is inconsistent in $\mathsf{NPK}(X)$, contradicting the assumption.

Now we show that Γ' has at most one maximally consistent extension in $\mathsf{NPK}(X+v)$. For this, it is enough to show that for any $\varphi \in \mathcal{L}(X+v)$, if $\Gamma' \nvdash_{+v} \varphi$, then $\Gamma' \vdash_{+v} \neg\varphi$. So suppose $\Gamma' \nvdash_{+v} \varphi$. Let ψ be the result of renaming bound variables in φ so that all bound variables are in X. Then $\Gamma' \nvdash_{+v} \psi$ as $\mathsf{NPK}(X+v)$ proves equivalence under such renamings (and renamings are reversible by renamings again). Now list the free variables of ψ in $ran(v)$ as $v_{p_1}, v_{p_2}, \ldots, v_{p_l}$ and pick distinct variables $x_1, x_2, \ldots, x_l$ in X that does not appear in ψ. Then inductively define the formulas $\alpha_0 = \psi$, $\alpha_{i+1} = [p_{i+1}/x_{i+1}](\alpha_i[x_{i+1}/v_{p_{i+1}}])$. Then α_l is in fact $[p_l/x_l] \ldots [p_1/x_1](\psi[x_1/v_{p_1}] \ldots [x_l/v_{p_l}])$ and moreover, for each $i = 0 \ldots l-1$, $\alpha_{i+1} \leftrightarrow \alpha_i$ is in Γ', since $\alpha_i[x_{i+1}/v_{p_{i+1}}][v_{p_{i+1}}/x_{i+1}]$ is identical to α_i and hence $\alpha_{i+1} \leftrightarrow \alpha_i$ is in the form of $[p/x]\beta \leftrightarrow \beta[v_p/x]$. Thus $\Gamma' \nvdash_{+v} \alpha_l$. But now $\alpha_l \in \mathcal{L}(X)$, so by the maximality of Γ, $\Gamma' \vdash_{v+} \neg\alpha_l$. By (Comm),

$$\Gamma' \vdash_{v+} [p_l/x_l] \ldots [p_1/x_1](\neg\psi[x_1/v_{p_1}] \ldots [x_l/v_{p_l}])$$

Then by using formulas in $\mathrm{WF}(v, X)$, we see that $\Gamma' \vdash_{+v} \neg\psi$.

Let Γ be a maximally consistent set for NPK. To prepare for the model building for Γ, first pick for each $i \in \mathbb{N}$ a witness assignment v^i such that for any $i \neq j$, $ran(v^i) \cap ran(v^j) = \varnothing$ and $ran(v^i) \cap \mathsf{Var} = \varnothing$. Also, let $\mathsf{Var}_0 = \mathsf{Var} + v^0$ and $\mathsf{Var}_{i+1} = \mathsf{Var}_i + v^{i+1}$. Then let $\mathsf{Var}_\omega = \bigcup_{i \in \mathbb{N}} \mathsf{Var}_i$.

Now we construct a tree model for Γ in stages. Each node of the tree is of the form (s, Δ) where s is a sequence of modal operators in Agt and Δ is a MCS in $\mathcal{L}(\mathsf{Var}_{len(s)})$ such that $WF(v^{len(s)}, \mathsf{Var}_{len(s)}) \subseteq \Delta$. $len(s)$ is the length of s.

At stage 0, the tree is $T_0 = \{(\epsilon, \Gamma + v^0))\}$ where ϵ is the empty sequence. Then inductively, we define T_{i+1} as the result of adding to T_i for each leaf node (s, Δ) (it is a leaf in the sense that $len(s) = i$), for each $\Box \in \mathsf{Agt}$, and for each formula in Δ of the form $\neg\Box\varphi$, the pair $(s+\Box, Ext_{\mathsf{NPK}(\mathsf{Var}_i)}(\{\neg\varphi\} \cup \Box^{-1}(\Delta)) + v^{i+1})$. Here $s+\Box$ is the sequence that extends s by $\Box$. Finally, set $\mathcal{M}^{\mathsf{NPK}} = (T, \{P_p\}_{p \in \mathsf{Prop}}, \{R_\Box\}_{\Box \in \mathsf{Agt}})$ where:

- $T = \bigcup_{i \in \mathbb{N}} T_i$;
- $(s_1, \Delta_1) R_\Box (s_2, \Delta_2)$ iff $s_2 = s_1 + \Box$
- $P_p((s, \Delta)) = \{(s', \Delta') \in T \mid v_p^{len(s)} \in \Delta'\}$.

Definition 7. *A formal assignment g for $\mathcal{M}^{\mathsf{NPK}}$ is a function from Var to Var_ω such that $g(x)$ is either x itself or is in $\mathsf{Var}_\omega \setminus \mathsf{Var}$. We extend g so that for any $\varphi \in \mathcal{L}$, $g(\varphi) = \varphi[g(x)/x]$. Note that $g(x)$ is always substitutable for x in φ. For each formal assignment g for $\mathcal{M}^{\mathsf{NPK}}$, define assignment $\overline{g}$ by*

$$\overline{g}(x) = \{(s, \Delta) \in T \mid g(x) \in \Delta\}.$$

Also, for each $(s, \Delta) \in T$, we say that a formal assignment g is admissible *for (s, Δ) if $ran(g) \subseteq \mathsf{Var}_{len(s)}$.*

Lemma 4. *For any formula $\varphi \in \mathcal{L}$, any formal assignment g for $\mathcal{M}^{\mathsf{NPK}}$, and any $(s, \Delta) \in T$, if g is admissible for (s, Δ), then $\mathcal{M}^{\mathsf{NPK}}, (s, \Delta), \overline{g} \models \varphi$ iff $g(\varphi) \in \Delta$.*

Proof. For the base case, note that for any $x \in \mathsf{Var}$, trivially by definition,

$$\mathcal{M}^{\mathsf{NPK}}, (s, \Delta), \overline{g} \models x \Leftrightarrow (s, \Delta) \in \overline{g}(x) \Leftrightarrow g(x) \in \Delta.$$

For the Boolean cases, we only need to note that $g(\neg\alpha) = \neg g(\alpha)$ and $g((\alpha \wedge \beta)) = (g(\alpha) \wedge g(\beta))$ and that Δ is maximally consistent.

For one direction of the modal cases, suppose $g(\Box\varphi) \in \Delta$. Then $\Box g(\varphi) \in \Delta$. By the construction of T, for any $(s+\Box, \Delta') \in T$, $g(\varphi) \in \Delta'$. By Induction Hypothesis (IH), and noting that since g is admissible for (s, Δ), g must also be admissible for $(s+\Box, \Delta')$, $\mathcal{M}^{\mathsf{NPK}}, (s+\Box, \Delta'), \overline{g} \models \varphi$. By the definition of $R_\Box$, $\mathcal{M}^{\mathsf{NPK}}, (s, \Delta), \overline{g} \models \Box\varphi$.

For the other direction of the modal cases, suppose $g(\Box\varphi) \notin \Delta$. Since g is admissible for (s, Δ), $g(\Box\varphi) \in \mathcal{L}(\mathsf{Var}_{len(s)})$. Since Δ is a MCS of $\mathsf{NPK}(\mathsf{Var}_{len(s)})$, $\neg\Box g(\varphi) \in \Delta$. By the construction of T, there is $(s+\Box, \Delta') \in T$ such that $\neg g(\varphi) \in \Delta'$, and then $g(\varphi) \notin \Delta'$. By IH, $\mathcal{M}^{\mathsf{NPK}}, (s+\Box, \Delta'), \overline{g} \not\models \varphi$, and thus $\mathcal{M}^{\mathsf{NPK}}, (s, \Delta), \overline{g} \not\models \Box\varphi$.

Finally, for the assignment operator case, consider any formula $[p/x]\varphi \in \mathcal{L}$.

$$\mathcal{M}^{\mathsf{NPK}}, (s, \Delta), \overline{g} \models [p/x]\varphi \Leftrightarrow \mathcal{M}^{\mathsf{NPK}}, (s, \Delta), \overline{g[v_p^{len(s)}/x]} \models \varphi \Leftrightarrow g[v_p^{len(s)}/x](\varphi) \in \Delta.$$

The first equivalence is due to our definition of P_p, $\overline{g}[P_p(s, \Delta)/x] = \overline{g[v_p^{len(s)}/x]}$, and the second is by IH. Observe that the formula $g([p/x]\varphi) \leftrightarrow g[v_p^{len(s)}/x](\varphi)$ is precisely of the form $[p/x]\beta \leftrightarrow \beta[v_p^{len(s)}/x] \in \mathrm{WF}(v^{len(s)}, \mathsf{Var}_{len(s)})$, where β is the result of replacing each free variable $y \neq x$ in φ by $g(y)$. Since g is admissible for Δ so that $\beta \in \mathcal{L}(\mathsf{Var}_{len(s)})$ and by construction Δ is of the form $\Xi + v^{len(s)}$ where Ξ is maximally consistent, the formula $g([p/x]\varphi) \leftrightarrow g[v_p^{len(s)}/x](\varphi)$ is in Δ. Thus $g[v_p^{len(s)}/x]\varphi \in \Delta$ iff $g([p/x]\varphi) \in \Delta$.

Given the last truth lemma, $\mathcal{M}^{\mathsf{NPK}}, (\epsilon, \Gamma + v^0), \overline{id}$ satisfies Γ, where id is the identity function from Var to Var. Thus,

Theorem 1. *NPK is sound and strongly complete with respect to the class of all Kripke models with non-rigid propositional designators.*

4 Axiomatization for Single Agent S5

In this section, we deal with the case where $\mathsf{Agt} = \{\Box\}$ and models are universal. To facilitate describing a special axiom for S5, where $\vec{p} = (p_1, \ldots, p_n)$ is a finite sequence from Prop of length $n \in \mathbb{N}^+$, and $\vec{x} = (x_1, \ldots, x_n)$ is a finite sequence from Var^+ of equal length n, by $[\vec{p}/\vec{x}]$ we mean the stack of assignment operators $[p_1/x_1] \cdots [p_n/x_n]$. Also, when v is an injective function from Prop to Var^+, by $v_{\vec{p}}$ we mean the sequence $(v_{p_1}, \ldots, v_{p_n})$. Thus $[\vec{p}/v_{\vec{p}}]$ is $[p_1/v_{p_1}] \cdots [p_n/v_{p_n}]$.

Definition 8. *For any $X \in [\mathsf{Var}, \mathsf{Var}^+]$, let $\mathsf{NPS5}(X)$ be the set of formulas in $\mathcal{L}(X)$ axiomatized by all the axioms and rules defining $\mathsf{NPK}(X)$ and also:*

- *All instances of the usual S5 axioms.*
- *(SymSub) $[\vec{p}/v_{\vec{p}}](\gamma \to \Box[\vec{q}/\vec{z}] \Diamond (\gamma \wedge \bigwedge_{i=1}^{m}([p_i/x_i]\varphi_i \leftrightarrow \varphi_i[v_{p_i}/x_i])))$ where $\vec{p} = (p_1, \ldots, p_n)$ is from Prop, v is an injection from Prop to X, $\vec{q}$ and $\vec{z}$ are sequences of equal length from Prop and X respectively, variables in $\vec{z}$ does not occur in γ or $v_{\vec{p}}$, and v_{p_i} is substitutable for x_i in φ_i.*

(SymSub) says something stronger than (Sub): under the assignment $[p/y]$, even if some other variable z in φ is bound by the value of p at some other world, still $[p/x]\varphi \leftrightarrow \varphi[y/x]$ at this world. We can return to 'this world' by $\Box\Diamond$ since the underlying accessibility relation is universal (and hence symmetric). The extra formula γ further solidifies that we are returning to 'this world'. Then, the soundness of these axioms and rules over universal models is not hard to check.

The analogue of Lemma 2 and Lemma 3 holds also for $\mathsf{NPS5}$ since they only use the NPK part, and for the lack of space we do not repeat then here. The following technical lemma shows the use of (SymSub).

Lemma 5. *Suppose $X \in [\mathsf{Var}, \mathsf{Var}^+]$, v^1 is a witness assignment such that $ran(v^1) \subseteq X$, v^2 is a witness assignment fresh for X, Γ_1 and Γ_2 are both MCSs in $\mathsf{NPS5}(X)$ such that $\mathrm{WF}(v^1, X) \subseteq \Gamma_1$ and $\Box^{-1}(\Gamma_1) \subseteq \Gamma_2$, and finally $\Delta_2 = \Gamma_2 + v^2$. Then $\Gamma_1 \cup \mathrm{WF}(v^1, X + v^2) \cup \Box^{-1}\Delta_2$ is consistent in $\mathsf{NPS5}(X + v^2)$.*

Proof. We write $\vdash$ for $\vdash_{\mathsf{NPS5}(X)}$ and $\vdash_{+v^2}$ for $\vdash_{\mathsf{NPS5}(X+v^2)}$. Suppose toward a contradiction that $\Gamma_1 \cup \mathrm{WF}(v^1, X + v^2) \cup \Box^{-1}\Delta_2$ is inconsistent. Since Γ_1 and $\Box^{-1}\Delta_2$ are closed under conjunctions, there are $\gamma \in \Gamma_1$, $\delta \in \Box^{-1}\Delta_2$, and formulas $[p_i/x_i]\varphi_i \leftrightarrow \varphi_i[v^1_{p_i}/x_i]$ $(i = 1 \ldots m)$ from $\mathrm{WF}(v^1, X + v^2)$ such that

$$\gamma, \bigwedge_{i=1}^{m}([p_i/x_i]\varphi_i \leftrightarrow \varphi_i[v^1_{p_i}/x_i]), \delta \vdash_{+v^2} \bot.$$

Since NPS5 proves equivalence under renaming bound variables, without loss of generality, we can assume that all the bound variables in all the φ_i appear in X. By Boolean and normal modal reasoning, we have

$$\delta \vdash_{+v^2} \gamma \to \neg \bigwedge_{i=1}^{m}([p_i/x_i]\varphi_i \leftrightarrow \varphi_i[v^1_{p_i}/x_i]),$$

$$\Box\delta \vdash_{+v^2} \Box(\gamma \to \neg \bigwedge_{i=1}^{m}([p_i/x_i]\varphi_i \leftrightarrow \varphi_i[v^1_{p_i}/x_i])),$$

$$\Box\delta \vdash_{+v^2} \neg \Diamond (\gamma \wedge \bigwedge_{i=1}^{m}([p_i/x_i]\varphi_i \leftrightarrow \varphi_i[v^1_{p_i}/x_i])).$$

Since $\delta \in \Box^{-1}\Delta_2$, this means that $\neg \Diamond (\gamma \wedge \bigwedge_{i=1}^{m}([p_i/x_i]\varphi_i \leftrightarrow \varphi_i[v^1_{p_i}/x_i]))$ is also in Δ_2. We will show that

$$\Diamond(\gamma \wedge \bigwedge_{i=1}^{m}([p_i/x_i]\varphi_i \leftrightarrow \varphi_i[v^1_{p_i}/x_i])) \qquad (\beta)$$

is also in Δ_2, rendering Δ_2 inconsistent. Since Γ_2 is consistent, by Lemma 3, Δ_2 should also be consistent, a contradiction.

Enumerate the set $\{p \in \mathsf{Prop} \mid v^2_p \text{ occurs in } \beta\}$ as $\vec{q} = (q_1, \dots, q_l)$. Then pick fresh variables $\vec{z} = (z_1, \dots, z_l)$ from X and let $\psi_i = \varphi_i[z_1/v^2_{q_1}] \cdots [z_l/v^2_{q_l}]$ for each $i = 1 \dots m$. Note that since γ is from Γ_1, no variables in $ran(v^2)$ occurs in γ. Now consider the formula

$$[\vec{p}/v^1_{\vec{p}}](\gamma \to \Box[\vec{q}/\vec{z}] \Diamond (\gamma \wedge \bigwedge_{i=1}^{m}([p_i/x_i]\psi_i \leftrightarrow \psi_i[v^1_{p_i}/x_i]))).$$

Note that this is in the form of the axiom (SymSub) and is in $\mathcal{L}(X)$. Thus it is in Γ_1. But since $\mathrm{WF}(v^1, X) \subseteq \Gamma_1$, $\gamma \to \Box[\vec{q}/\vec{z}] \Diamond (\gamma \wedge \bigwedge_{i=1}^{m}([p_i/x_i]\psi_i \leftrightarrow \psi_i[v^1_{p_i}/x_i]))$ is in Γ_1. Since $\gamma \in \Gamma_1$, $\Box[\vec{q}/\vec{z}] \Diamond (\gamma \wedge \bigwedge_{i=1}^{m}([p_i/x_i]\psi_i \leftrightarrow \psi_i[v^1_{p_i}/x_i]))$ is also in Γ_1. Since $\Box^{-1}\Gamma_1 \subseteq \Gamma_2$,

$$[\vec{q}/\vec{z}] \Diamond (\gamma \wedge \bigwedge_{i=1}^{m}([p_i/x_i]\psi_i \leftrightarrow \psi_i[v^1_{p_i}/x_i])) \qquad (\alpha)$$

is in Γ_2, and hence is in $\Delta_2 = \Gamma_2 + v^2$. But observe that β can be obtained by iteratively removing $[q_i/z_i]$ and instantiate z_i with $v^2_{q_i}$ (reversing the process of constructing ψ_i from φ_i). Since $\Delta_2 = \Gamma_2 + v^2$, $\mathrm{WF}(v^2, X + v^2) \subseteq \Delta_2$. So Δ_2 proves $\alpha \leftrightarrow \beta$, and hence β is in Δ_2.

Let Γ be a maximally consistent set for NPS5. To build a universal model for Γ, pick fresh witness assignments $\{v^i \mid i \in \mathbb{N}\}$ and corresponding variable sets Var_i and Var_ω as before. It is useful to note that $\mathcal{L}(\mathsf{Var}_\omega) = \bigcup_{i \in \mathbb{N}} \mathcal{L}(\mathsf{Var}_i)$ since each formula is finite and uses only finitely many variables in Var_ω. We fix an enumeration of $(\neg \Box \chi_1, \neg \Box \chi_2, \dots)$ of all formulas in $\mathcal{L}(\mathsf{Var}_\omega)$ of the form $\neg \Box \varphi$.

Now we build a model for Γ in stages, where at stage i, we build a sequence $\Sigma^i = (\Sigma^i_0, \ldots, \Sigma^i_i)$ of MCSs in $\mathsf{NPS5}(\mathsf{Var}_i)$, a set $\Pi^i \subseteq \mathcal{L}(\mathsf{Var}_i)$, and a formula $\neg\Box\theta_i$ (here $i > 0$) with set $H^i = \{\neg\Box\theta_1, \ldots, \neg\Box\theta_i\}$ (with $H^0 = \varnothing$) such that:

- for each $j = 0 \ldots i$, $WF(v^j, \mathsf{Var}_i) \subseteq \Sigma^i_j$;
- for each $j = 0 \ldots i$, $\Box^{-1}\Sigma^i_j = \Pi^i$;
- for each $j = 1 \ldots i$, $\neg\theta_j \in \Sigma^i_j$ and $\neg\Box\theta_j$ is the first formula in the sequence $(\neg\Box\chi_1, \neg\Box\chi_2, \ldots)$ that appears in $\Pi^{j-1} \setminus H^{j-1}$.

Intuitively, Π^i is the 'modal theory' of the model at stage i, and H^i is the set of $\neg\Box$ formulas processed before and at stage i.

We start the process with $\Sigma^0 = (\Sigma^0_0)$ where $\Sigma^0_0 = \gamma + v^0$ and $\Pi^0 = \Box^{-1}\Sigma^0_0$. Then, inductively for each $i \in \mathbb{N}$, we define Σ^{i+1} and Π^{i+1} as follows:

- Let $\neg\Box\theta_{i+1}$ be the first in $(\neg\Box\chi_1, \neg\Box\chi_2, \ldots)$ that is in $\Pi^i \setminus H^i$. There must be one since H^i is finite while using redundant conjuncts, there are infinitely many formulas of the form $\neg\Box\varphi$ in Π^i.
- Since $\Pi^i = \Box^{-1}\Sigma^i_i$, by (S5), $\Pi^i \cup \{\neg\theta_{i+1}\}$ is consistent in $\mathsf{NPS5}(\mathsf{Var}_i)$. Let $\Sigma^{i+1}_{i+1} = Ext_{\mathsf{NPS5}(\mathsf{Var}_i)}(\Pi^i \cup \{\neg\theta_{i+1}\}) + v^{i+1}$. Then let $\Pi^{i+1} = \Box^{-1}\Sigma^{i+1}_{i+1}$.
- For each $j{=}0 \ldots i$, by construction $WF(v^j, \mathsf{Var}_i) \subseteq \Sigma^i_j$. Also, since $\Box^{-1}(\Sigma^i_j) = \Pi^i$, $\Box^{-1}(\Sigma^i_j) \subseteq Ext_{\mathsf{NPS5}(\mathsf{Var}_i)}(\Pi^i \cup \{\neg\theta_{i+1}\})$. Moreover, $\mathsf{Var}_{i+1} = \mathsf{Var}_i + v^{i+1}$. Thus, Lemma 5 applies, and $\Sigma^i_j \cup WF(v^j, \mathsf{Var}_{i+1}) \cup \Pi^{i+1}$ is consistent. We let $\Sigma^{i+1}_j = Ext_{\mathsf{NPS5}(\mathsf{Var}_{i+1})}(\Sigma^i_j \cup WF(v^j, \mathsf{Var}_{i+1}) \cup \Pi^{i+1})$. By (S5), $\Box^{-1}\Sigma^{i+1}_j = \Pi^{i+1}$.

Now we combine the sequences into a single model. For each $i \in \mathbb{N}$, let $\Delta_i = \bigcup_{j \geq i} \Sigma^j_i$. We also set $\Pi = \bigcup_{i \in \mathbb{N}} \Pi^i$ and $H = \bigcup_{i \geq 1} H^i$, which is $\{\neg\Box\theta_1, \neg\Box\theta_2, \ldots\}$.

Lemma 6. *For each $i \in \mathbb{N}$: Δ_i is a MCS of $\mathsf{NPS5}(\mathsf{Var}_\omega)$, $WF(v^i, \mathsf{Var}_\omega) \subseteq \Delta_i$, and $\Box^{-1}\Delta_i = \Pi$. Moreover, for any formula of the form $\neg\Box\varphi \in \Pi$, there is $i \in \mathbb{N}$ such that $\neg\varphi \in \Delta_i$.*

Proof. Since each Σ^j_i is a MCS of $\mathsf{NPS5}(\mathsf{Var}_j)$ and $(\Sigma^j_i)_{j \geq i}$ is also an ascending sequence, Δ_i is a MCS of $\mathsf{NPS5}(\mathsf{Var}_\omega)$. Also, for each $j \geq i$, $WF(v^i, \mathsf{Var}_j) \subseteq \Sigma^j_i \subseteq \Delta_i$. This means $WF(v^i, \mathsf{Var}_\omega) \subseteq \Delta_i$. The proof for $\Box^{-1}\Delta_i = \Pi$ is also not hard.

Now take any formula $\neg\Box\varphi \in \Pi$. Let i be the smallest such that $\neg\Box\varphi \in \Pi^i$, and also let j be such that $\neg\Box\varphi = \neg\Box\chi_j$. Then by construction, $\neg\Box\varphi$ must be in H^{i+j} since at every stage after i, a formula $\neg\Box\chi_k$ before $\neg\Box\chi_j$ must be processed if $\neg\Box\chi_j$ is not processed at that stage. This means there $k \leq i + j$ such that $\neg\varphi \in \Sigma^k_k \subseteq \Delta_k$.

Given the above lemma, we define $\mathcal{M}^{\mathsf{NPS5}} = (D, \{P_p\}_{p \in \mathsf{Prop}})$ where $D = \{\Delta_i \mid i \in \mathbb{N}\}$ and for any $\Delta_i \in D$, $P_p(\Delta_i) = \{\Delta_j \in D \mid v^i_p \in \Delta_j\}$. Then, similar to Definition 7, a formal assignment g for $\mathcal{M}^{\mathsf{NPS5}}$ is a function from Var to Var_ω such that $g(x)$ is either x itself or is not in Var. Then for any $\varphi \in \mathcal{L}$, $g(\varphi) = \varphi[g(x)/x]$. Further, define the corresponding assignment $\overline{g}$ for $\mathcal{M}^{\mathsf{NPS5}}$ by $\overline{g}(x) = \{\Delta_j \in D \mid g(x) \in \Delta_j\}$. The concept of admissibility is not needed here.

Lemma 7. *For any formula $\varphi \in \mathcal{L}$, any formal assignment g for $\mathcal{M}^{\mathsf{NPS5}}$, and any $\Delta_i \in D$, $\mathcal{M}^{\mathsf{NPS5}}, \Delta_i, \overline{g} \vDash \varphi$ iff $g(\varphi) \in \Delta_i$.*

Proof. The base case and the Boolean cases are again easy. For the modal case, if $g(\Box\varphi) = \Box g(\varphi) \in \Delta_i$, then by S5 logic, $\Box\Box g(\varphi) \in \Delta_i$. Then $\Box g(\varphi) \in \Pi$ and $g(\varphi) \in \Delta_j$ for any $\Delta_j \in D$ by Lemma 6. By IH, for any $\Delta_j \in D$, $\mathcal{M}^{\mathsf{NPS5}}, \Delta_j, \overline{g} \vDash \varphi$. Then $\mathcal{M}^{\mathsf{NPS5}}, \Delta_i, \overline{g} \vDash \Box\varphi$.

If $g(\Box\varphi) = \Box g(\varphi) \notin \Delta_i$, then by maximality, $\neg \Box g(\varphi) \in \Delta_i$. By S5 logic, $\Box \neg \Box g(\varphi) \in \Delta_i$, and $\neg \Box g(\varphi) \in \Pi$. By Lemma 6, there is $\Delta_j \in D$ such that $\neg g(\varphi) \in \Delta_j$. By consistency and IH, $\mathcal{M}^{\mathsf{NPS5}}, \Delta_j, \overline{g} \nvDash \varphi$. Then $\mathcal{M}^{\mathsf{NPS5}}, \Delta_i, \overline{g} \nvDash \Box \varphi$.

Finally, for the assignment operator case, consider any formula $[p/x]\varphi \in \mathcal{L}$. Now because $\overline{g}[P_p(\Delta_i)/x] = \overline{g[v_p^i/x]}$ and IH,

$$\mathcal{M}^{\mathsf{NPS5}}, \Delta_i, \overline{g} \vDash [p/x]\varphi \Leftrightarrow \mathcal{M}^{\mathsf{NPS5}}, \Delta_i, \overline{g[v_p^i/x]} \vDash \varphi \Leftrightarrow g[v_p^i/x](\varphi) \in \Delta_i.$$

Then, noting that $g([p/x]\varphi) \leftrightarrow g[v_p^i/x](\varphi)$ is in $\mathrm{WF}(v^i, \mathsf{Var}_\omega) \subseteq \Delta_i$, $\mathcal{M}^{\mathsf{NPS5}}, \Delta_i, \overline{g} \vDash [p/x]\varphi \Leftrightarrow g([p/x]\varphi) \in \Delta_i$.

By the above truth lemma, we have

Theorem 2. NPS5 *is sound and strongly complete with respect to the class of all universal Kripke models with non-rigid propositional designators.*

5 Conclusion

We have only scratched the surface of the formalism proposed in this paper. The immediate next step is to consider the logic of multi-agent epistemic models, be it with equivalence relations, transitive relations, or some other special relations of interest, since only then can we start talking about uncertainty in meaning in a multi-agent setting, and consider the information dynamics on the meaning of sentences. Models with a universal modality are also very important since the universal modality can help us express equality between propositions. We believe that by combining our two constructions in this paper, axiomatizations can be obtained in most of the cases.

Once we start working in a multi-agent epistemic setting, the ideas in [18,19] and [17] are worth incorporating. When talking about how different people may interpret p differently, an obvious drawback of our semantics is that there is always a ground truth of what p actually means. But in many situations, the meaning of p may be completely relative (before a convention is reached). In that case, the best we can do is to have versions p_i of p for each agent i, representing how agent i interprets p, just like in [18]. But importantly, and differently from [18], each p_i is still non-rigid, since an agent i may well be uncertain how j interprets p, i.e., what p_j means. The issue of definition in [17] can also be discussed in our framework. For example, when $\Box$ is the universal modality, $\Box[p/x][q/y][r/z]\Box(x \leftrightarrow (y \wedge z))$ seems to say that p is defined by $q \wedge r$. Of course, one may take this as only saying that p and $q \wedge r$ are necessarily equal in the proposition expressed, and definitions are more hyperintentional than that. But our framework is already hyperintentional in a sense: if we take functions from worlds to truth values, namely sets of possible worlds, as intentional, then functions from worlds to sets of possible worlds seem deserving of the description 'hyperintentional'. A discussion of how our framework relates to other hyperintentional frameworks such as [23,24,26,29] is needed here. Another important addition to consider is information dynamics. Since p is now non-rigid, updating with p relates to externalism in epistemology [5,9,16].

The extra axiom (SymSub) may look unseemly to many. We see two possible ways to eliminate it. The first is through nominals [1]: then we believe the axiom can be replaced by $[p/y](i \to \Box[q/z]\,\Box\,(i \to ([p/x]\varphi \leftrightarrow \varphi[y/x])))$. If we use $(p@i)$ and $\downarrow i$ as in $\mathcal{L}$@ mentioned in Sect. 2, then as the assignment operator can be eliminated, a simple axiomatic system may be obtained. Another way is by introducing propositional quantifiers $\forall x$ binding propositional variables $x \in \mathsf{Var}$ [10] since what we really want is $[p/y]\forall z(\varphi \leftrightarrow \varphi[y/x])$ where z could range over propositions denoted by some q at other worlds. A Barcan formula $\forall x \Box \varphi \longrightarrow \Box \forall x \varphi$ and an instantiation axiom $\forall x\varphi \to [p/x]\varphi$ intuitively correspond to the minimal requirement on the range of propositional variables. Note that if we insist that the semantics of $\forall x$ considers all sets of possible worlds, we will immediately run into non-axiomatizability even in single-agent cases, unlike in situations without non-rigid propositional designators and assignment operators [7,8,20], since those non-rigid designators can simulate arbitrary modal operators, and results such as [14,22] would apply. But without this 'full domain' requirement, we believe axiomatizations are within reach. Generalizing to an algebraic setting that can avoid assuming that there are possible worlds (world propositions) may also be interesting. Here it may be useful to interpret $[p/x]\varphi$ as $\forall x([p]x \to \varphi)$ where for each $p \in \mathsf{Prop}$, $[p]$ is a unary modality so that $[p]x$ means 'what p means is x'. This essentially goes back to the expression $(\gamma\ \mathsf{is}\ \varphi)$ used in [21]. An assumption we have made throughout the paper is that there is always a unique proposition meant by p. This can be expressed by $\exists x([p]x \land \forall y([p]y \to (x = y)))$ using equality between propositions. It remains to be seen what axiomatizability results follow from this setting.

References

1. Areces, C., ten Cate, B.: 14 hybrid logics. In: Studies in Logic and Practical Reasoning, vol. 3, pp. 821–868. Elsevier (2007)
2. Blackburn, P., Martins, M., Manzano, M., Huertas, A.: Rigid first-order hybrid logic. In: Iemhoff, R., Moortgat, M., de Queiroz, R. (eds.) WoLLIC 2019. LNCS, vol. 11541, pp. 53–69. Springer, Heidelberg (2019). https://doi.org/10.1007/978-3-662-59533-6_4
3. Blumberg, K.: Counterfactual attitudes and the relational analysis. Mind **127**(506), 521–546 (2018)
4. Blumberg, K.: Wishing, decision theory, and two-dimensional content. J. Philos. **120**(2), 61–93 (2023)
5. Cohen, M.: Opaque updates. J. Philos. Log. **50**(3), 447–470 (2021)

6. Cohen, M., Tang, W., Wang, Y.: De re updates. In: Halpern, J.Y., Perea, A. (eds.) Proceedings Eighteenth Conference on Theoretical Aspects of Rationality and Knowledge, TARK 2021. EPTCS, vol. 335, pp. 103–117 (2021)
7. Ding, Y.: On the logics with propositional quantifiers extending S5Π. In: Bezhanishvili, G., D'Agostino, G., Metcalfe, G., Studer, T. (eds.) Advances in Modal Logic 12, pp. 219–235. College Publications (2018)
8. Ding, Y.: On the logic of belief and propositional quantification. J. Philos. Log. **50**(5), 1143–1198 (2021)
9. Dorst, K.: Evidence: a guide for the uncertain. Philos. Phenomenol. Res. **100**(3), 586–632 (2020)
10. Fine, K.: Propositional quantifiers in modal logic. Theoria **36**(3), 336–346 (1970)
11. Fitting, M.: Modal logics between propositional and first-order. J. Log. Comput. **12**(6), 1017–1026 (2002)
12. Fitting, M.: Types, Tableaus, and Gödel's God, vol. 12. Springer, Science, Dordrecht (2002)
13. Fitting, M.: First-order intensional logic. Ann. Pure Appl. Logic **127**(1–3), 171–193 (2004)
14. Fritz, P.: Axiomatizability of propositionally quantified modal logics on relational frames. J. Symbolic Logic, 1–38 (2022)
15. Gallin, D.: Intensional and Higher-Order Modal Logic. Elsevier, Amesterdam (2016)
16. Gallow, J.D.: Updating for externalists. Noûs **55**(3), 487–516 (2021)
17. Gattinger, M., Wang, Y.: How to agree without understanding each other: public announcement logic with boolean definitions. In: Electronic Proceedings in Theoretical Computer Science in Proceedings TARK 2019, pp. 297, 206–220 (2019)
18. Halpern, J.Y., Kets, W.: A logic for reasoning about ambiguity. Artif. Intell. **209**, 1–10 (2014)
19. Halpern, J.Y., Kets, W.: Ambiguous language and common priors. Games Econom. Behav. **90**, 171–180 (2015)
20. Holliday, W.: A note on algebraic semantics for S5 with propositional quantifiers. Notre Dame J. Formal Logic **60**(2), 311–332 (2019)
21. Holliday, W., Pacuit, E.: Beliefs, propositions, and definite descriptions (2016)
22. Kaminski, M., Tiomkin, M.: The expressive power of second-order propositional modal logic. Notre Dame J. Formal Logic **37**(1), 35–43 (1996)
23. Kocurek, A.W.: The logic of hyperlogic. Part A: foundations. Rev. Symbolic Logic, 1–28 (2022)
24. Kocurek, A.W.: The logic of hyperlogic. Part B: extensions and restrictions. Rev. Symbolic Logic, 1–28 (2022)
25. Kooi, B.: Dynamic term-modal logic. In: van Benthem, J., Ju, S., Veltman, F. (eds.) A meeting of the minds. In: Proceedings of the Workshop on Logic, Rationality and Interaction, Beijing, 2007, pp. 173–185. College Publications (2008)
26. Leitgeb, H.: Hype: a system of hyperintensional logic (with an application to semantic paradoxes). J. Philos. Log. **48**(2), 305–405 (2019)
27. Muskens, R.: 10 higher order modal logic. In: Studies in Logic and Practical Reasoning, vol. 3, pp. 621–653. Elsevier (2007)
28. Ninan, D.: Imagination, content, and the self. Ph.D. thesis, Massachusetts Institute of Technology (2008)
29. Sedlár, I.: Hyperintensional logics for everyone. Synthese **198**(2), 933–956 (2021)
30. Wang, Y., Wei, Y., Seligman, J.: Quantifier-free epistemic term-modal logic with assignment operator. Ann. Pure Appl. Log. **173**(3), 103071 (2022)

An Arrow-Based Dynamic Logic of Normative Systems and Its Decidability

Hans van Ditmarsch[1], Louwe Kuijer[2], and Mo Liu[3](✉)

[1] University of Toulouse, CNRS, IRIT, Toulouse, France
hans.van-ditmarsch@irit.fr
[2] University of Liverpool, Liverpool, UK
louwe.kuijer@liverpool.ac.uk
[3] Sun Yet-sen University, Guangzhou, China
mo.liu26@mail.sysu.edu.cn

Abstract. Normative arrow update logic (NAUL) is a logic that combines normative temporal logic (NTL) and arrow update logic (AUL). In NAUL, norms are interpreted as arrow updates on labeled transition systems with a CTL-like logic. We show that the satisfiability problem of NAUL is decidable with a tableau method and it is in EXPSPACE.

Keywords: normative system · arrow update logic · tableau method

1 Introduction

Deontic logic is the study of rules, norms, obligations and permissions, through logical means [5,7,10,13,17], and this has also been extensively investigated in dynamic modal logics [6,11,12,16,20]. In the field of deontic logic, there is a sub-field that studies rules or norms by comparing the situation where a rule is not in effect, or not being followed, to the situation where the rule/norm is obeyed. There is no universally accepted name for this sub-field, but "social laws" [9,18,19] and "normative systems" [1,3] are often used. We will use the term *normative systems*, and refer to the behavioural restrictions under consideration as *norms*.

A logic of normative systems is concerned with what things agents are capable of doing, and what they are allowed to do if a norm is enacted. It therefore requires a model of agency at its core. Any model of agency will do, but the most commonly used choices are labeled transition systems with a CTL-like logic of agency [8] and outcome function transition systems with ATL-like logic [4]. Here, we will follow the CTL-style approach of *normative temporal logic* [2]. This means that a model is a labeled transition system, i.e., it contains a set S of states and a set $\{R(a) \mid a \in \mathcal{A}\}$ of accessibility relations, where $R(a) \subseteq S \times S$. A transition $(s_1, s_2) \in R(a)$, is an action or an agent that changes the state of the world from s_1 to s_2.

N. Alechina et al. (Eds.): LORI 2023, LNCS 14329, pp. 63–76, 2023.
https://doi.org/10.1007/978-3-031-45558-2_5

In order to choose a course of action, we need to decide whether we should adopt a norm and then check if an action is allowed by the norm. Whether an action a is allowed may depend on a logical condition φ before the action takes place, so on the situation in s_1, and also may depend on a logical condition ψ after the action took place, so on a condition satisfied in s_2. We refer to s_1 as the *source* of the action, to φ as a *source condition*, to s_2 as the *target*, and to ψ as a *target condition*. For norms with both source and target conditions one cannot reduce multiple source conditions to one (for example by taking the disjunction), nor multiple target condition to one. A norm in our formalism will be therefore represented by a list of clauses, each with a source condition and a target condition. This is as in *arrow update logic* [14,21]. The arrow eliminating updates in arrow update logic now correspond to adherence to norms.

We will also introduce more complex ways to describe norms, so we will refer to such a list of clauses as an *atomic norm*. We distinguish four ways to combine norms. If N_1 and N_2 are norms, then

- $-N_1$ is the negation of N_1, and allows exactly those actions that are disallowed by N_1,
- $N_1 + N_2$ is the additive combination of N_1 and N_2, and allows exactly those actions that are allowed by N_1 or N_2,
- $N_1 \times N_2$ is the multiplicative combination of N_1 and N_2, and allows exactly those actions that are allowed by both N_1 and N_2.
- $N_1 \circ N_2$ is the sequential composition of N_1 and N_2, and allows exactly those actions that are allowed by N_2 in the transition system restricted to those actions that are allowed by N_1.

We further distinguish *static* from *dynamic* applications of norms. A liveness condition such as "if the norm N is obeyed, then φ is guaranteed to be true at every time in the future" can be formalized in two ways, which we denote $[N]G\varphi$ (dynamic) and $G_N\varphi$ (static). The difference lies in whether the norm N is assumed to hold during the evaluation of φ: when evaluating $[N]G\varphi$, everything inside the scope of $[N]$ is considered in the transition system restricted to the actions allowed by N. When evaluating $G_N\varphi$, on the other hand, the "forever in the future" operator G is evaluated in the system restricted to N-allowed actions, but φ is evaluated in the non-restricted system.

The dynamic operator $[N]$ can be expressed using only the static operators, and the combined norms can be expressed using only atomic norms. They do not affect the expressivity. However, the combined and dynamic norms affect the succinctness of the language, and thus the complexity of decision problems. The logic will be called NAUL, Normative Arrow Update Logic. A preliminary investigation of NAUL was presented in [15], with a focus on expressivity (NAUL is strictly more expressive than CTL and AUL*) and complexity of model checking (in PTIME). Here we will investigate the complexity of the satisfiability problem, by introducing a tableaux method for deciding satisfiability. First, however, we will formally define the syntax and semantics of NAUL, and discuss an example of its application. We will now formally define its syntax and semantics and then investigate the complexity of satisfiability with a tableau method.

2 Language and Semantics

Let $\mathcal{A}$ be a finite set of agents and $\mathcal{P}$ a countably infinite set of propositional variables.

Definition 1. *The formulas of $\mathcal{L}_{NAUL}$ are given by*

$$\varphi ::= p \mid \neg\varphi \mid \varphi \vee \varphi \mid [N]\varphi \mid \Box_N\varphi \mid G_N\varphi \mid F_N\varphi$$
$$\mathcal{N} ::= (\varphi, B, \varphi) \mid \overline{(\varphi, B, \varphi)} \mid \mathcal{N}, (\varphi, B, \varphi) \mid \mathcal{N}, \overline{(\varphi, B, \varphi)}$$
$$N ::= \mathcal{N} \mid -N \mid N + N \mid N \times N \mid N \circ N$$

where $p, \in \mathcal{P}$ and $B \subseteq \mathcal{A}$.

Remark 1. In NAUL we use only three temporal operators: $\Box_N$, G_N and F_N. These temporal operators include an implicit universal quantification over all paths, so we could have denoted them in a more CTL-like fashion as AX_N, AG_N and AF_N. Operators corresponding to the other temporal operators from CTL can be defined in NAUL. For example, $E(\varphi_1 U_N \varphi_2)$ can be defined as $\neg G_{(\varphi_1, \mathcal{A}, \top) \times N} \neg\varphi_2$.

In NAUL, the set of subformulas ($SubF$) or subnorms ($SubN$) of a formula φ (or a norm N) includes all formulas or norms *occurring* in φ (or N).

Strictly speaking a norm of type $\mathcal{N}$ is a list of clauses, but we abuse notation by identifying it with the set of its clauses. Additionally, we use a number of abbreviations. We refer to norms of type $\mathcal{N}$ as *atomic norms* and norms of type N simply as *norms*. Note that every atomic norm is also a norm.

Definition 2. *We use $\wedge, \rightarrow, \leftrightarrow, \bigwedge, \bigvee$ and $\Diamond_N$ in the usual way as abbreviations. Furthermore, we use $\hat{G}_N$ and $\hat{F}_N$ as abbreviations for $\neg G_N \neg$ and $\neg F_N \neg$. We write $\Box_B$ for $\Box_{(\top, B, \top)}$, G_B for $G_{(\top, B, \top)}$ and F_B for $F_{(\top, B, \top)}$. Finally, we use $\Box, G$ and F for $\Box_{\mathcal{A}}$, $G_{\mathcal{A}}$ and $F_{\mathcal{A}}$.*

Definition 3. *A* model $\mathcal{M}$ *is a triple $\mathcal{M} = (S, R, v)$ where S is a set of states, $R : \mathcal{A} \rightarrow 2^{S \times S}$ maps each agent to an accessibility relation on S, and $v : \mathcal{P} \rightarrow 2^S$ is a valuation. A* pointed model *is a pair $(\mathcal{M}, s)$ where $\mathcal{M} = (S, R, v)$ is a model and $s \in S$.*

A pair $(s_1, s_2) \in R(a)$ is also called *transition* in $\mathcal{M}$. It is denoted $s_1 \xmapsto{a_1} s_2$. A *path* in $\mathcal{M}$ is a (possibly infinite) sequence $s_1 \xmapsto{a_1} s_2, s_2 \xmapsto{a_2} s_3, \cdots$ of transitions in $\mathcal{M}$. A path P' *extends* a path P if P is an initial segment of P'. The semantics of $\mathcal{L}_{NAUL}$ are given by the following two interdependent definitions.

Definition 4. *Let $\mathcal{M} = (S, R, v)$ be a relational model and N a norm. A transition $s_1 \xmapsto{a} s_2$* satisfies N *in $\mathcal{M}$ if one of the following holds:*

1. *N is an atomic norm, there is a positive clause $(\varphi, B, \psi) \in N$ such that $\mathcal{M}, s_1 \models \varphi$, $a \in B$ and $\mathcal{M}, s_2 \models \psi$. Furthermore, there is no negative clause $\overline{(\varphi, B, \psi)} \in N$ such that $\mathcal{M}, s_1 \models \varphi$, $a \in B$ and $\mathcal{M}, s_2 \models \psi$,*

2. N is of the form $-N_1$ and $s_1 \overset{a}{\longmapsto} s_2$ does not satisfy N_1,
3. N is of the form $N_1 + N_2$ and $s_1 \overset{a}{\longmapsto} s_2$ satisfies N_1 or N_2 in $\mathcal{M}$,
4. N is of the form $N_1 \times N_2$ and $s_1 \overset{a}{\longmapsto} s_2$ satisfies N_1 and N_2 in $\mathcal{M}$,
5. N is of the form $N_1 \circ N_2$, $s_1 \overset{a}{\longmapsto} s_2$ satisfies N_1 in $\mathcal{M}$ and the transition $s_1 \overset{a}{\longmapsto} s_2$ satisfies N_2 in $\mathcal{M} * N_1$.

A path $s_1 \overset{a_1}{\longmapsto} s_2 \overset{a_2}{\longmapsto} s_3 \cdots$ is an N-path in *$\mathcal{M}$ if every transition $s_i \overset{a_i}{\longmapsto} s_{i+1}$ in the path satisfies N in $\mathcal{M}$. An N-path is* full in *$\mathcal{M}$ if there is no N-path in $\mathcal{M}$ that extends it.*

When the model $\mathcal{M}$ is clear from context, we say simply that a transition satisfies N or that a path is an N-path.

Definition 5. *Let $\mathcal{M} = (S, R, v)$ be a transition system and $s \in S$. The relation $\models$ is given as follows.*

$$
\begin{array}{ll}
\mathcal{M}, s \models p & \Leftrightarrow s \in v(p) \text{ for } p \in \mathcal{P} \\
\mathcal{M}, s \models \neg\varphi & \Leftrightarrow \mathcal{M}, s \not\models \varphi \\
\mathcal{M}, s \models \varphi_1 \vee \varphi_2 & \Leftrightarrow \mathcal{M}, s \models \varphi_1 \text{ or } \mathcal{M}, s \models \varphi_2 \\
\mathcal{M}, s \models \Box_N \varphi & \Leftrightarrow \mathcal{M}, s' \models \varphi \text{ for every transition } s \longmapsto s' \text{ that satisfies } N \\
\mathcal{M}, s \models G_N \varphi & \Leftrightarrow \text{for every } N\text{-path } P \text{ starting in } s \text{ and} \\
 & \quad\ \text{every } s' \in P \text{ we have } \mathcal{M}, s' \models \varphi \\
\mathcal{M}, s \models F_N \varphi & \Leftrightarrow \text{for every full } N\text{-path } P \text{ starting in } s \text{ there is} \\
 & \quad\ \text{some } s' \in P \text{ such that } \mathcal{M}, s' \models \varphi \\
\mathcal{M}, s \models [N]\varphi & \Leftrightarrow \mathcal{M} * N, s \models \varphi
\end{array}
$$

where $\mathcal{M} * N = (S, R * N, v)$ and, for every $a \in \mathcal{A}$,

$$R * N(a) = \{(s, s') \in R(a) \mid s \overset{a}{\longmapsto} s' \text{ satisfies } N\}.$$

Recall that the single state s is a degenerate path with no transitions. So every transition in s satisfies every norm N, so it is an N-path. As a result, $\mathcal{M}, s \models G_N \varphi$ implies $\mathcal{M}, s \models \varphi$.

3 Example: Self-driving Cars

We will give a simple example of NAUL. Suppose we have a racetrack where a number of self-driving cars operate. We want to equip cars with norms that will guarantee that they avoid

(a) collisions with each other and stationary objects;
(b) "deadlock" situations where no one can act.

Let *coll* be the proposition variable that represents "a collision happens". Note that situations where no one can act are represented by $\Box\bot$.

For (a), we create a norm N_c such that if no collision has occurred then it should prevent collisions for every point in the future. N_c is therefore successful if we have $\neg coll \rightarrow [N_c]G\neg coll$. The simplest way is to disallow any action, then

N_c is $(\bot, \mathcal{A}, \bot)$. However, we would like to let N_c allow at least one action to avoid deadlock. Thus we take $N_c := (\top, \mathcal{A}, \neg F\, coll)$. It is indeed successful as we have $\models \neg coll \to [N_c]G\neg coll$.

For (b), we interpret it as "there must be some available action that is not only possible but also allowed", and then we construct a N_d such than $[N_d]G\Diamond\top$ holds. we should take $N_d := (\top, \mathcal{A}, \neg F\Box\bot)$. This gives us $\models \neg F\Box\bot \to [N_d]G\Diamond\top$. In other words, as long as there is an infinite path the norm N_d forces agents to follow such a path.

For combining N_c and N_d, $N_c \times N_d$ allows agents to perform actions that result in a situation where movement, while possible, is disallowed because it will lead to a collision. The sequential combination solves this problem: the norm $N_c \circ N_d$ allows exactly those actions that lead to neither collisions nor situations where agents cannot or are not allowed to act. In other words, we have $\models \neg F(coll \vee \Box\bot) \to [N_c \circ N_d]G(\neg coll \wedge \Diamond\top)$.

The self-driving cars example is also useful for illustrating the difference between the static operators $\Box_N$, G_N, and F_N on the one hand, and the dynamic operator $[N]$ on the other. We have $\mathcal{M}, s \models G_N\varphi$ if φ holds after every sequence of actions that starts in s and is allowed by N. Importantly, during the evaluation of φ it is not assumed that everyone follows N. We have $\mathcal{M}, s \models [N]G\varphi$ if, under the assumption that all agents follow N permanently from now on, every sequence of actions leads to a φ state. In this case, during the evaluation of φ, we do assume that all agents follow N.

Sometimes we may require that N_c not only avoids collisions, but also situations where a single mistake could cause a collision. We cannot phrase this stronger success condition as $[N_c]\varphi$ for any φ. After all, the φ in $[N_c]\varphi$ is evaluated under the assumption that all agents follow the norm N_c—so no mistakes are made. This is where the static operator G_{N_c} is useful. Consider the formula $G_{N_c}(\neg coll \wedge \Box\neg coll)$. The $\Box$ in that formula is not evaluated under the assumption that the agents follow N_c, so $G_{N_c}(\neg coll \wedge \Box\neg coll)$ holds exactly if every sequence of actions allowed by N_c leads to a state where there is no collision and no single action can cause a collision.

4 Satisfiability Problem

In this section, we present a tableau method to show that the satisfiability problem of NAUL is decidable. We will use *negation normal form* (NNF) of formulas or norms. An NNF formula only has negation on literals. An NNF norm only has negations on atomic norms instead of clauses.

Definition 6 (Negation normal form (NNF)). *Given a set of variables* $\mathbf{P}$ *and a finite set of agents* $\mathcal{A}$.

$$\varphi ::= p \mid \neg p \mid \varphi\wedge\varphi \mid \varphi\vee\varphi \mid \Box_N\varphi \mid \Diamond_N\varphi \mid G_N\varphi \mid \hat{G}_N\varphi \mid F_N\varphi \mid \hat{F}_N\varphi \mid [N]\varphi \mid \langle N\rangle\varphi$$

$$\mathcal{N} ::= (\varphi, a, \varphi) \mid \mathcal{N}, (\varphi, a, \varphi)$$

$$N ::= \mathcal{N} \mid \overline{\mathcal{N}} \mid N + N \mid N \times N \mid N \circ N$$

where $p \in \mathbf{P}$, $a \in \mathcal{A}$.

Definition 7. *The satisfiability problem for NAUL is defined as follows:*

- ***Input:*** *an NNF formula* φ.
- ***Output:*** *YES if and only if there is a model* $(\mathcal{M}, s)$ *such that* $\mathcal{M}, s \models \varphi$.

Theorem 1. *Every NAUL-formula or norm can be transformed to an equivalent formula or norm in NNF.*

Proof. For NAUL-formulas, it can be shown easily by an induction. As for atomic norms, since the order of clauses in an atomic norm does not matter, given an atomic NAUL-norm $\mathcal{N}$, and $\mathcal{N}^+$ as all positive clauses, $\mathcal{N}^-$ as all negative clauses of $\mathcal{N}$, clearly $\mathcal{N}$ is equivalent to $\mathcal{N}^+ + \mathcal{N}^-$ which is an NNF norm. As for the negations of combined norms, we have the following transformations:

- $\overline{\overline{N}} = N$
- $\overline{N_1 + N_2} = \overline{N_1} \times \overline{N_2}$
- $\overline{N_1 \times N_2} = \overline{N_1} + \overline{N_2}$
- $\overline{N_1 \circ N_2} = \overline{N_1} + N_1 \circ \overline{N_2}$

Given an NAUL-formula φ or NAUL-norm N, the time of transforming it into an NNF formula φ' or NNF norm N' and the size of ψ or N' is polynomial in the size of φ or N.

4.1 Tableau Method

We introduce some concepts related to tableau method.

Definition 8 (Term). *There are two types of terms:*

F-term $\langle s; \lambda; \varphi \rangle$ *where* $s \in S$, λ *is a sequence of norms,* φ *is a formula. It means the model has been updated by* λ *and* φ *is true on* s.
N-term $\langle s_1 \overset{a}{\mapsto} s_2; \lambda; \eta \rangle$ *where* $s_1, s_2 \in S$, λ, η *are sequences of norms. It means the transition* $s_1 \overset{a}{\mapsto} s_2$ *satisfies* η *successively after the model is updated by* λ.

Definition 9 (Tableau). *A tableau* T *is a structure* $T = (W, V, E, \pi)$ *where* W *is an infinite set of states, and* V *is a finite set of nodes,* E *is a binary relation on* V. *Given a set of terms* L, $\pi : V \to \mathbf{P}(L)$ *is a labelling map.*

Let $A, C_1, \cdots, C_n$ *be sets of terms. A tableau rule is represented as*

$$\frac{A}{C_1 \mid \cdots \mid C_n}$$

Above the line, A *is the antecedent; below the line, there are consequents. A tableau rule is applicable on a node if the node has terms as an instance of the antecedent. If there are multiple consequents after applying a rule, one needs to choose one of them.*

Definition 10 (Interpretability). *Given a model* $\mathcal{M} = (S, R, v)$, *it interprets (noted as* $\models_T$*) a set of terms* T *if any term in* T *satisfies:*

- $\mathcal{M} \models_T \langle s; \lambda; \varphi \rangle$ *if and only if* $\mathcal{M} * \lambda, s \models \varphi$.
- $\mathcal{M} \models_T \langle s_1 \overset{a}{\mapsto} s_2; \lambda; \eta \rangle$ *if and only if* $s_1 \overset{a}{\mapsto} s_2$ *satisfies* η *on* $\mathcal{M} * \lambda$.

A set of terms T *is interpretable if there exists a model* $\mathcal{M}$ *such that* $\mathcal{M}$ *interprets all terms in* T.

Definition 11. *Given a tableau* T*, we define an order* $\prec$ *on all terms of* T *as*

- $\langle s; \lambda; \varphi \rangle \prec \langle s; \lambda'; \psi \rangle$ *if* φ *is a subformula of* ψ.
- $\langle s \overset{a}{\mapsto} s'; \lambda; \eta \rangle \prec \langle s; \lambda'; \varphi \rangle$ *if* η *is a parameter of some operator in* φ.
- $\langle s; \lambda; \varphi \rangle \prec \langle s \overset{a}{\mapsto} s'; \lambda'; \eta \rangle$ *if* φ *is in some clause of* η.
- $\langle s \overset{a}{\mapsto} s'; \lambda; \eta \rangle \prec \langle s \overset{a}{\mapsto} s'; \lambda; \eta' \rangle$ *if* η *is a sub-norm of* η';
- $\langle s \overset{a}{\mapsto} s'; \lambda; N' \rangle \prec \langle s \overset{a}{\mapsto} s'; \lambda'; N \rangle$ *if* λ *is an initial segment of* λ'.

Now we define the tableau rules for NAUL. We omit terms which remain the same after applying a certain rule. Let ϵ be the norm $(\top, \mathcal{A}, \top)$ after which nothing is updated.

Definition 12 (Tableau rules).

$$(lit)\ \frac{\langle s; \lambda; p \rangle}{\langle s; \epsilon; p \rangle} \quad \frac{\langle s; \lambda; \neg p \rangle}{\langle s; \epsilon; \neg p \rangle} \quad (\wedge)\ \frac{\langle s; \lambda; \varphi \wedge \psi \rangle}{\langle s; \lambda; \varphi \rangle, \langle s; \lambda; \psi \rangle} \quad (\vee)\ \frac{\langle s; \lambda; \varphi \vee \psi \rangle}{\langle s; \lambda; \varphi \rangle \mid \langle s; \lambda; \psi \rangle}$$

$$(G)\ \frac{\langle s; \lambda; G_N \varphi \rangle}{\langle s; \lambda; \varphi \rangle, \langle s; \lambda; \Box_N G_N \varphi \rangle} \qquad (\hat{G})\frac{\langle s; \lambda; \hat{G}_N \varphi \rangle}{\langle s; \lambda; \varphi \rangle \mid \langle s; \lambda; \Diamond_N \hat{G}_N \varphi \rangle}$$

$$(F)\ \frac{\langle s; \lambda; F_N \varphi \rangle}{\langle s; \lambda; \varphi \rangle \mid \langle s; \lambda; \Diamond_N \top \rangle, \langle s; \lambda; \Box_N F_N \varphi \rangle}$$

$$(\hat{F})\ \frac{\langle s; \lambda; \hat{F}_N \varphi \rangle}{\langle s; \lambda; \varphi \rangle, \langle s; \lambda; \Box_N \bot \rangle \mid \langle s; \lambda; \varphi \rangle, \langle s; \lambda; \Diamond_N \hat{F}_N \varphi \rangle}$$

$$(\Diamond)\ \frac{\langle s; \lambda; \Diamond_N \varphi \rangle}{\langle s'; \lambda; \varphi \rangle, \langle s \overset{a_1}{\mapsto} s'; \lambda; N \rangle \mid \cdots \mid \langle s'; \lambda; \varphi \rangle, \langle s \overset{a_n}{\mapsto} s'; \lambda; N \rangle}$$

$$(\Box)\ \frac{\langle s; \lambda; \Box_N \varphi \rangle, \langle s \overset{a}{\mapsto} s'; \epsilon; \lambda \rangle}{\langle s'; \lambda; \varphi \rangle, \langle s \overset{a}{\mapsto} s'; \lambda; N \rangle \mid \langle s \overset{a}{\mapsto} s'; \lambda; -N' \rangle} \qquad (-N' \text{ is the NNF of } -N)$$

$$(Dyn)\ \frac{\langle s; \lambda; [N]\varphi \rangle}{\langle s; \lambda, N; \varphi \rangle} \quad \frac{\langle s; \lambda; \langle N \rangle \varphi \rangle}{\langle s; \lambda, N; \varphi \rangle}$$

$$(At)\ \frac{\langle s \overset{a_i}{\mapsto} s'; \lambda; \mathcal{N} \rangle}{\langle s; \lambda; \varphi_i \rangle, \langle s'; \lambda; \psi_i \rangle} \qquad \text{where } (\varphi_i, a_i, \psi_i) \in \mathcal{N}$$

$$(Neg)\ \frac{\langle s \overset{a}{\mapsto} s'; \lambda; -\mathcal{N} \rangle}{\langle s; \lambda; \bigwedge_{j \in K_1} \varphi'_j \rangle, \langle s'; \lambda; \bigwedge_{j \in K_2} \psi'_j \rangle \mid \cdots} \qquad (@)$$

$$(Add)\ \frac{\langle s \overset{a}{\mapsto} s'; \lambda; N_1 + N_2 \rangle}{\langle s \overset{a}{\mapsto} s'; \lambda; N_1) \rangle \mid \langle s \overset{a}{\mapsto} s'; \lambda; N_2) \rangle}$$

$$(Multi)\frac{\langle s \overset{a}{\mapsto} s'; \lambda; N_1 \times N_2 \rangle}{\langle s \overset{a}{\mapsto} s'; \lambda; N_1) \rangle, \langle s \overset{a}{\mapsto} s'; \lambda; N_2) \rangle}$$

$$(Seq)\ \frac{\langle s \overset{a}{\mapsto} s'; \lambda; N_1 \circ N_2\rangle}{\langle s \overset{a}{\mapsto} s'; \lambda, N_1; N_2\rangle} \qquad (DN)\ \frac{\langle s \overset{a}{\mapsto} s'; \lambda, N_1; N_2\rangle}{\langle s \overset{a}{\mapsto} s'; \lambda; N_1\rangle}$$

(@): rule Neg is branching over all $K_1, K_2 \subseteq [1, n]$ such that $K_1 \cup K_2 = \{i \mid (\varphi_i, a, \psi_i) \in \mathcal{N}\}$, and $K_1 \cap K_2 = \varnothing$, and φ'_j and ψ'_j are the NNF of resp. $\neg\varphi_j$ and $\neg\psi_j$ with $(\varphi_j, a, \psi_j) \in \mathcal{N}$.

(lit), ($\wedge$) and ($\vee$) are Boolean rules. $(G), (F), (\hat{G}), (\hat{F})$ handle temporal modalities. (G) says if we have $G_N\varphi$ at a word s, then we have φ as well as $\Box_N G_N\varphi$ at s. (F) says if we have $F_N\varphi$ at s, then either we have φ, or s has some N-successor ($\Diamond_N\top$ is true) and $\Box_N F_N\varphi$. $(\hat{G})$ says if we have $\hat{G}_N\varphi$ at s, then either we have φ or $\Diamond_N\hat{G}_N\varphi$ at s. $(\hat{F})$ says if we have $\hat{F}_N\varphi$ at s, then we have φ at s and either s has no N-successor or we have $\Diamond_N\hat{F}_N\varphi$ at s. ($\Diamond$) says if we have $\Diamond_N\varphi$ at s, then we can choose an agent $a \in \mathcal{A}$ to "assume" that there is a transition $s \overset{a}{\mapsto} s'$ satisfying N and we have φ at s'. Note that ($\Diamond$) is the only rule that generates new states and either a state can be actually generated will be examined later. ($\Box$) says if we have $\Box_N\varphi$ at s and transition $s \overset{a}{\mapsto} s'$ exists, then whether we have φ at s' and $s \overset{a}{\mapsto} s'$ satisfies N or $s \overset{a}{\mapsto} s'$ does not satisfy N. (Dyn) handles dynamic operators. It says if we have $[N]\varphi$ (or $\langle N\rangle\varphi$) at s updated by λ, then we have φ at s updated by λ then by N.

The other rules handle norms. (Atomic) says if we have atomic norm $\mathcal{N}$ for $s \overset{a_i}{\mapsto} s'$ where a_i occurs in some clause $(\varphi_i, a_i, \psi_i) \in \mathcal{N}$, then we have φ_i at s and ψ_i at s'. (Neg) says if we have $\overline{\mathcal{N}}$ for $s \overset{a}{\mapsto} s'$, then given $\{i \mid (\varphi_i, a, \psi_i) \in \mathcal{N}\}$ we choose some $K_1, K_2 \subseteq [1, n]$ such that $K_1 \cup K_2 = \{i \mid (\varphi_i, a, \psi_i) \in \mathcal{N}\}$ and $\bigwedge_{j\in K_1} \neg\varphi_j$ is at s and $\bigwedge_{j\in K_2} \neg\psi_j$ is at s'. As a result, none of the clauses in $\mathcal{N}$ will be satisfied by $s \overset{a}{\mapsto} s'$. (Add), (Multi) and (Seq) are standard with respect to Definition 4. (DN) says if $s \overset{a}{\mapsto} s'$ satisfies some norm N_2 after updating by λ, N_1, then it satisfies N_1 after updating by λ. A special case of (DN) is

$$(\text{DN*})\ \frac{\langle s \overset{a}{\mapsto} s'; \lambda; N\rangle}{\langle s \overset{a}{\mapsto} s'; \epsilon; \lambda\rangle}$$

(DN*) says if transition $s \overset{a}{\mapsto} s'$ is updated by λ, then it satisfies λ.

Besides above tableau rules, we also need principles to delete inconsistent states, to set an order of applying rules, and to avoid infinite consequents.

Definition 13 (Tableau principles). *Given an NNF formula φ, we start from the root with label $\langle s_0; \epsilon; \varphi\rangle$. We have following the principles of generating a tableau of φ:*

(Inc) If a node has inconsistent literals, then mark it as "deleted". If all consequents are marked deleted, then mark the antecedent as deleted. In particular, if one node has no consequent then mark it as deleted directly.

(Exh) We should apply rules to terms with respect to one state until no rule is applicable on that state. When no rule is applicable on a state s, we mark s as "exhausted". After that, we can apply rules to terms on its successors.

(Cyc) When a state s is marked as "exhausted', one needs to check if there is some exhausted ancestor s^* of s which has the same F-terms as s on some node t^*. If so, we should add the pair of nodes of s and s^* $(t, t^*) \in E$ and mark s as "exhausted" as well. If a state s is merged with some ancestor, then all successors of s are also marked as "exhausted", and we stop to explore any term with respect to these successors further. In addition, let $\sim \subseteq S \times S$ be an equivalence relation, and use $s^* \sim s$ to "merge" these two state to a reflexive state.

(Eve$\hat{G}$) If all consequences of an antecedent t are marked as deleted, then mark t as "deleted". If $\hat{G}_N\varphi$ is in some term of a node t with respect to a state s, and there is no reachable state from s such that φ occurs in some term, then mark t as "deleted".

(EveF) If $F_N\varphi$ is in some term of a node t with respect to a state s, and there exists a full branch from s on which φ does not occur in any term of state on that branch, then mark t as "deleted".

If there is no rule applicable any more, the procedure of generating the tableau terminate, and the tableau is complete. If the root of a complete tableau T is not marked as "deleted", then we call a path from the root to a leaf node on T an open branch. If a complete tableau has at least one open branch, then we call it an open tableau.

Proposition 1. *For any NNF-formula φ, the procedure of generate a tableau for φ will terminate.*

For F-terms, boolean connectives, modal operators $\Box$ and $\Diamond$ are eliminated by the corresponding rule. Temporal operators may be retained after tableau rules and keep generating new states. The (Cyc) principle helps to avoid infinite generation of new states by merging states with the same terms. For N-terms, the composite norm will be disassembled into the atomic norm and eventually reduced to F-terms.

4.2 Soundness and Completeness

Proposition 2 (Soundness). *Given an NNF-formula φ, if φ is satisfiable then there is an open tableau rooted at $(s_0; \epsilon; \varphi)$.*

Proof. We show all tableau rules preserve interpretability. If a tableau rule has multiple consequents, then as least one of them is interpretable.

- (lit) and $(\wedge)$ preserve interpretability obviously. For $(\vee)$, if the antecedent is interpretable, then so is one of its consequences.
- For the rules (G), (F), $(\hat{G})$, $(\hat{F})$ and $(Dynamic)$, it can be shown by semantics. We present the case of (F) as an example. Suppose $\mathcal{M} * \lambda, s \models F_N\varphi$. By semantics, for every full N-path P starting from s, there is some $s' \in P$ such that $\mathcal{M} * \lambda, s' \models \varphi$. Since s is in every N-path starting from s, it is sufficient if $\mathcal{M} * \lambda, s \models \varphi$. Otherwise, we have there is some N-successor s' of s such that $\mathcal{M} * \lambda, s' \models F_N\varphi$. In this case, $\mathcal{M} * \lambda, s \models \Diamond_N \top$ and $\mathcal{M} * \lambda, s \models \Diamond_N F_N\varphi$.

- ($\Diamond_N$): Suppose $\mathcal{M} * \lambda, s \models \Diamond_N \varphi$. By semantics, there is a transition $s \overset{a}{\mapsto} s'$ satisfying N and $\mathcal{M} * \lambda, s' \models \varphi$ for some $a \in \mathcal{A}$.
- ($\Box_N$): Suppose $\mathcal{M} * \lambda, s \models \Box_N \varphi$ and $s \overset{a}{\mapsto} s'$ satisfies λ on $\mathcal{M}$. If $s \overset{a}{\mapsto} s'$ satisfies N on $\mathcal{M} * \lambda$, then by semantics we have $\mathcal{M} * \lambda, s' \models \varphi$. If $s \overset{a}{\mapsto} s'$ does not satisfy N on $\mathcal{M} * \lambda$, then it satisfies $-N$ on $\mathcal{M} * \lambda$.
- (Atomic): Suppose $s \overset{a_i}{\mapsto} s'$ satisfies $\mathcal{N}$ on $\mathcal{M} * \lambda$. It follows that $\mathcal{M} * \lambda, s \models \varphi_i$ and $\mathcal{M} * \lambda, s' \models \psi_i$. Thus $\mathcal{M} \models_T \langle s; \lambda; \varphi_i \rangle, \langle s'; \lambda; \psi_i \rangle$.
- (Neg): Suppose $s \overset{a_i}{\mapsto} v$ satisfies $\overline{\mathcal{N}}$ on $\mathcal{M} * \lambda$. It follows that $s \overset{a}{\mapsto} v$ satisfies no clause with respect to a in $\mathcal{N}$. Thus let $K_1 = \{i \mid \mathcal{M} * \lambda, v \models \neg\varphi_i$ for any $(\varphi_i, a, \psi_i) \in \mathcal{N}\}$ and $K_2 = \{i \mid \mathcal{M} * \lambda, v \models \neg\psi_i$ for any $(\varphi_i, a, \psi_i) \in \mathcal{N}\}$. Therefore, we have $K_1 \cup K_2 = \{i \mid (\varphi_i, a_i, \psi_i) \in \mathcal{N}\}$, and $\mathcal{M} \models_T \langle v; \lambda; \bigwedge_{i \in K_1} \neg\varphi_i \rangle$ and $\mathcal{M} \models_T \langle v; \lambda; \bigwedge_{i \in K_2} \neg\psi_i \rangle$.
- (Add), (Multi), (Seq) is straightforward by the definition.
- (DN): Suppose $s \overset{a}{\mapsto} s'$ satisfies N_2 on $\mathcal{M} * \lambda * N_1$. By definition, if $s \overset{a}{\mapsto} s'$ is on $\mathcal{M} * \lambda * N_1$, then it satisfies N_1 on $\mathcal{M} * \lambda$ as well. Thus $\mathcal{M} \models_T \langle s \mapsto s'; \lambda; N_1 \rangle$.

Note that the trace-back links by (Cyc) only connect nodes with the same terms on the same state. Thus interpretability is preserved as well.

Suppose φ is satisfiable, then there is a pointed model $\mathcal{M}, s \models \varphi$. Let s be s_0, then there is an open tableau rooted at $\langle s_0; \epsilon; \varphi \rangle$.

Proposition 3 (Completeness). *Given an NNF-formula φ, if there is an open tableau rooted at $\langle s_0; \epsilon; \varphi \rangle$, then φ is satisfiable.*

Proof. Suppose there is an open tableau T rooted at $\langle s_0; \lambda; \varphi \rangle$. Let T^* be a full branch on T. We construct a model $\mathcal{M} = (S, R, v)$ where

- $S = \{[s] \mid \langle s; \lambda; \psi \rangle$ is in $T^*\}$
- $R = \{s \overset{a}{\mapsto} s' \mid \left\langle s \overset{a}{\mapsto} s'; \epsilon; \epsilon \right\rangle \in T^*\} \cup \{s \overset{a}{\mapsto} s \mid a \in \mathcal{A}, |[s]| > 1\}$
- $v(s) = \{p \mid \langle s; \epsilon; p \rangle \in T^*\}$

where $[s] = \{s' \in S \mid s \sim s'\}$.

We show the following claims:

1. if $\langle s; \lambda; \psi \rangle$ is in T^*, then $\mathcal{M} * \lambda, s \models \psi$.
2. if $\left\langle s_1 \overset{a}{\mapsto} s_2; \lambda; N \right\rangle$ is in T^*, then $s_1 \overset{a}{\mapsto} s_2$ is in $\mathcal{M} * \lambda$ and satisfies N on $\mathcal{M} * \lambda$.

Make an induction on all terms by the order $\prec$ in Definition 11 to show the above claims. For Claim 2,

- If $\left\langle s_1 \overset{a_i}{\mapsto} s_2; \lambda; \mathcal{N} \right\rangle \in T^*$ where $\mathcal{N} = (\varphi_1, a_1, \psi_1), \cdots, (\varphi_n, a_n, \psi_n)$, $i \in [1, n]$, then by (Atomic) rule $\langle s_1; \lambda; \varphi_i \rangle, \langle s; \lambda; \psi_i \rangle \in T^*$. Then by IH, we have $\mathcal{M} * \lambda, s_1 \models \varphi$, $\mathcal{M} * \lambda, s_2 \models \psi$. Thus $s_1 \overset{a_i}{\mapsto} s_2$ satisfies $\mathcal{N}$ on $\mathcal{M} * \lambda$.
- If $\left\langle s_1 \overset{a}{\mapsto} s_2; \lambda; \overline{\mathcal{N}} \right\rangle \in T^*$ where $\mathcal{N} = (\varphi_1, a_1, \psi_1), \cdots, (\varphi_n, a_n, \psi_n)$, $i \in [1, n]$, then by (Neg) rule, $\langle s_1; \lambda; \bigwedge_{i \in K_1} \neg\varphi_i \rangle \in T^*$ and $\langle s_2; \lambda; \bigwedge_{i \in K_2} \neg\psi_i \rangle \in T^*$ for some disjoint $K_1 \cup K_2 = \{i \mid (\varphi_i, a, \psi_i) \in \mathcal{N}\}$. Thus, no φ_i and ψ_i are satisfied simultaneously so that no clause in $\mathcal{N}$ with respect to a is satisfied. Therefore, $s_1 \overset{a}{\mapsto} s_2$ satisfies $\overline{\mathcal{N}}$.

- The cases of $N_1 + N_2$, $N_1 \times N_2$, and $N_1 \circ N_2$ are straightforward by IH.

For Claim 1,

- If $\langle s; \lambda; p\rangle \in T^*$, then by by the rule (lit), $\langle s; \epsilon; p\rangle \in T^*$. Thus $\mathcal{M}, s \vDash p$, and then $\mathcal{M} * \lambda, s \vDash p$. Similarly, if $\langle s; \lambda; \neg p\rangle \in T^*$, then $\mathcal{M} * \lambda, s \vDash \neg p$. The boolean cases, dynamic case and $\langle s; \lambda; \Diamond_N \psi\rangle \in T^*$ are straightforward by IH.
- If $\langle s; \lambda; \Box_N \psi\rangle \in T^*$, then for any $\left\langle s \overset{a}{\mapsto} s'; \epsilon; \lambda\right\rangle \in T^*$, by $(\Box_N)$ rule, $\langle s'; \lambda; \psi\rangle \in T^*$ or $\left\langle s \overset{a}{\mapsto} s'; \lambda; -N\right\rangle \in T^*$. If $\langle s'; \lambda; \psi\rangle \in T^*$, then by IH $\mathcal{M} * \lambda, s' \vDash \psi$; If $\left\langle s \overset{a}{\mapsto} s'; \lambda; -N\right\rangle \in T^*$, then by Claim 2, $s \overset{a}{\mapsto} s'$ satisfies $-N$, that is to say, $s \overset{a}{\mapsto} s'$ does not satisfy N. Thus by semantics, $\mathcal{M}, s \vDash \Box_N \varphi$.
- Suppose $\langle s; \lambda; G_N \psi\rangle \in T^*$. Let $P = s \overset{a_1}{\mapsto} s_1 \overset{a_2}{\mapsto} s_2 \cdots \overset{a_n}{\mapsto} s_{n+1}$ be any N-path starting from s. We show that $\langle s'; \lambda; \Box_N G_N \psi\rangle \in T^*$ and $\mathcal{M} * \lambda, s' \vDash \psi$ for any $s' \in P$ by induction on $n+1$. By (G) rule, $\langle s; \lambda; \psi\rangle \in T^*$ and $\langle s; \lambda; \Box_N G_N \psi\rangle \in T^*$. Since $\langle s; \lambda; \psi\rangle \in T^*$, by IH we have $\mathcal{M} * \lambda, s \vDash \psi$. Assume $\langle s_n; \lambda; \Box_N G_N \psi\rangle \in T^*$ and $\mathcal{M} * \lambda, s_n \vDash \psi$. By (G) rule again, we have $\langle s_{n+1}; \lambda; \Box_N G_N \psi\rangle \in T^*$. Since $s_n \overset{a_n}{\mapsto} s_{n+1} \in R$, we have $\left\langle s_n \overset{a_n}{\mapsto} s_{n+1}; \lambda; N\right\rangle \in T^*$. Then by $(\Box_N)$ rule, we have $\langle s_{n+1}; \lambda; \psi\rangle$. By IH, we have $\mathcal{M} * \lambda, s_{n+1} \vDash \psi$. Thus for every $s' \in P$, we have $\mathcal{M} * \lambda, s' \vDash \psi$. As P is arbitrary, by semantics $\mathcal{M} * \lambda, s \vDash G_N \psi$. For terms with $F_N \psi, \hat{G}_N \psi$ and $\hat{F}_N \psi$, it is routine by IH.

Theorem 2. *For any NNF formula φ, φ is satisfiable if and only if there is an open tableau rooted at (s_0, ϵ, φ).*

Therefore, the satisfiability problem of NAUL is decidable. We wil show its upper bound is in EXPSPACE.

Theorem 3. *The satisfiability problem of NAUL is in EXPSPACE.*

Proof. Let φ be an NNF formula, and T be an open tableau for φ. We show the following claims:

1. The depth of T is at most exponential.
2. The width of T is at most double exponential.
3. The procedure can be done in double exponential amount of time.

Note that tableau rules does not decompose formulas strictly, thus the sizes of formulas in the consequents may be larger than the sizes of formulas in the antecedents. However we can give an upper bound of how many terms a single open branch in T has.

The *agenda* $Ag(\varphi)$ of a formula φ is the smallest set containing ϵ, $SubF(\varphi)$ as well as $SubN(\varphi)$ and satisfying the following conditions:

- If $\psi \in Ag(\varphi)$, then $\neg\psi^* \in Ag(\varphi)$;
- If $G_N \psi \in Ag(\varphi)$, then $\Box_N G_N \psi \in Ag(\varphi)$;
- If $\hat{G}_N \psi \in Ag(\varphi)$, then $\Diamond_N \hat{G}_N \psi \in Ag(\varphi)$;
- If $F_N \psi \in Ag(\varphi)$, then $\Diamond_N \top, \Box_N F_N \psi \in Ag(\varphi)$;

- If $\hat{F}_N\psi \in Ag(\varphi)$, then $\Box_N\bot, \Diamond_N\hat{F}_N\psi \in Ag(\varphi)$;
- If $N \in Ag(\varphi)$, then $\overline{N}^* \in Ag(\varphi)$.

Clearly, the cardinality of $Ag(\varphi)$ is polynomial in $|\varphi|$.

Proof of 1: For any F-term $\langle s; \lambda; \psi \rangle$ or N-term $\left\langle s \overset{a}{\mapsto} s'; \lambda; N \right\rangle$ occurring in T when s is marked as exhausted, it can be shown that ψ, $N \in Ag(\varphi)$ and all elements of λ are in $Ag(\varphi)$ by examining every rule. Firstly, we could give an upper bound of states in one open branch. We have shown the formulas of all F-terms are in $Ag(\varphi)$. Since two exhausted states get merged if they have the same F-terms, we can get at most exponential many states in the size of φ. Secondly, we could give an upper bound of how many transitions are generated from one state. Note that the $(\Diamond_N)$ rule is the only rule that generates new transitions. The frequency that $(\Diamond_N)$ rule is applied is bounded by the size of φ, one state has at most polynomial many arrows in the size of φ.

Therefore, the upper bounds of the amount of F-terms and N-terms are both at most exponential in the size of φ. One open branch has at most exponential depth as well, as there are at most exponentially many exhausted states with the same F-terms. This is because if a state is merged with some ancestor, then we will stop exploring terms of it. Therefore, the frequency that each state can be merged is no more than the number of paths starting from it. Since each exhausted state has polynomial many arrows to other states, it can be merged at most exponentially many times. In short, the depth of one open branch is in at most exponential.

Proof of 2: Only rule that leads to exponentially many branches is (Neg). Given an atomic norm $\mathcal{N}$, $|\overline{\mathcal{N}}|$ is bounded by $|\varphi|$. The cardinality of branches is in $O(2^{|\varphi|})$. As there are at most exponentially many terms in one branch, the width of T is in $O(2^{|\varphi|^2})$, so at most double exponential in the size of φ.

Proof of 3: The algorithm contains: applying tableau rules, checking, marking and pruning the tableau by principles, transforming formulas with negation into NNF. For each branch, as there are at most exponentially many terms in the size of φ, all of the three procedures above can be done in an exponential amount of time. To be specific, applying rules contains searching suitable premises and executing. The input of searching is the power set of labels on some node, which is exponential in the size of φ and the executions of applying rules are no more than the amount of terms; the input of checking inconsistency and states with the same terms is exponential in the size of φ and can be done in exponential time; the frequency of transforming NNF formulas is at most exponential and each transformation can be done in polynomial time.

To sum up, as we can reuse the space for each open branch, the procedure is in EXPSPACE.

5 Conclusion

We have presented a logic named *normative arrow update logic* (NAUL). In NAUL, we can combine norms in three ways: additive, multiplicative and sequential. We can also distinguish static and dynamic ways to consider norms. We have shown that the satisfiability problem of NAUL is decidable via a tableau method and the complexity of this problem is in EXPSPACE. For the further research, firstly, we conjecture the satisfiability problem of NAUL is EXPSPACE-hard but have no proof yet. Secondly, we are interested in finding tractable fragments of NAUL. Lastly, it may be interesting to develop a variant of arbitrary arrow update logic (AAUL) [21] for normative systems. It would have quantifier over norms and express "there is some norm that guarantees φ".

Acknowledgement. We thank the LORI reviewers for very detailed and helpful comments.

References

1. Ågotnes, T., van der Hoek, W., Rodríguez-Aguilar, J. A., Sierra, C., Wooldridge, M.: On the logic of normative systems. In Proceedings of 20th IJCAI, pp. 1175–1180 (2007)
2. Ågotnes, T., van der Hoek, W., Rodríguez-Aguilar, J.A., Sierra, C., Wooldridge, M.: A temporal logic of normative systems. In: Makinson, D., Malinowski, J., Wansing, H. (eds.) Towards Mathematical Philosophy. TL, vol. 28, pp. 69–106. Springer, Dordrecht (2009). https://doi.org/10.1007/978-1-4020-9084-4_5
3. Ågotnes, T., van der Hoek, W., Wooldridge, M.: Robust normative systems and a logic of norm compliance. Logic J. IGPL **18**, 4–30 (2010)
4. Alur, R., Henzinger, T.A., Kupferman, O.: Alternating-time temporal logic. J. ACM **49**, 672–713 (2002)
5. Anderson, A.R., Moore, O.K.: The formal analysis of normative concepts. Am. Sociol. Rev. **22**, 9–17 (1957)
6. Bartha, P.: Conditional obligation, deontic paradoxes, and the logic of agency. Ann. Math. Artif. Intell. **9**(1–2), 1–23 (1993)
7. Chisholm, R.M.: Contrary-to-duty imperatives and deontic logic. Analysis **24**, 33–36 (1963)
8. Clarke, E.M., Emerson, E.A.: Design and synthesis of synchronization skeletons using branching-time temporal logic. In: Kozen, D. (ed.) Logic of Programs 1981. LNCS, vol. 131, pp. 52–71. Springer, Heidelberg (1981). https://doi.org/10.1007/bfb0025774
9. Fitoussi, D., Tennenholtz, M.: Choosing social laws for multi-agent systems: minimality and simplicity. Artif. Intell. **119**, 61–101 (2000)
10. Føllesdal, D., Hilpinen, R.: Deontic logic: an introduction. In: Hilpinen, R. (ed.) Deontic Logic: Introductory and Systematic Readings. Synthese Library, vol. 33, pp. 1–35. Springer, Dordrecht (1971). https://doi.org/10.1007/978-94-010-3146-2_1
11. Herzig, A., Lorini, E., Moisan, F., Troquard, N.: A dynamic logic of normative systems. In: Walsh, T., (ed.), Proceedings of the Twenty-second International Joint Conference on Artificial Intelligence (IJCAI11), pp. 228–233 (2011)
12. Horty, J.F.: Agency and Deontic Logic. Oxford University Press, Oxford (2001)
13. Jones, A.J.I., Sergot, M.: On the characterisation of law and computer systems: the normative systems perspective. In: Deontic Logic in Computer Science: Normative System Specification, pp. 275–307 (1993)
14. Kooi, B., Renne, B.: Arrow update logic. Rev. Symbolic Logic **4**(4), 536–559 (2011)

15. Kuijer, L.B.: An arrow-based dynamic logic of norms. In: 3rd International Workshop on Strategic Reasoning (SR 2015) (2015)
16. Meyer, J.C.: A different approach to deontic logic: deontic logic viewed as a variant of dynamic logic. Notre Dame J. Formal Logic **29**, 109–136 (1988)
17. Ross, A.: Imperatives and logic. Theoria **7**, 53–71 (1941)
18. Shoham, Y., Tennenholtz, M.: On the synthesis of useful social laws for artificial agent societies. In: Proceedings of the Tenth National Conference on Artificial Intelligence (AAAI 1992), pp. 276–281 (1992)
19. Shoham, Y., Tennenholtz, M.: On social laws for artificial agent societies: off-line design. Artif. Intell. **73**, 231–252 (1995)
20. van der Meyden, R.: The dynamic logic of permission. J. Log. Comput. **6**(3), 465–479 (1996)
21. van Ditmarsch, H., van der Hoek, W., Kooi, B., Kuijer, L.B.: Arbitrary arrow update logic. Artif. Intell. **242**, 80–106 (2017)

Connexivity Meets Church and Ackermann

Luis Estrada-González[1(✉)] and Miguel Ángel Trejo-Huerta[1,2(✉)]

[1] Institute for Philosophical Research, UNAM, Mexico City, Mexico
loisayaxsegrob@comunidad.unam.mx, miguel.trejo@filosoficas.unam.mx
[2] Graduate Program in Philosophy of Science, UNAM, Mexico City, Mexico

Abstract. Here we study two connexive logics based on one of the conditionals introduced by Church in [4] and on some negations defined through falsity constants in the sense of Ackermann in [1].

Keywords: Churchian conditional · connexive logic · falsity constant

1 Introduction

Consider the following connexive features[1], with **L** some logic and $>$ and N any conditional and negation, respectively:

$\models_{\mathbf{L}} N(A > NA)$ $\quad\models_{\mathbf{L}} (A > B) > N(A > NB)$ $\quad\not\models_{\mathbf{L}} (A > B) > (B > A)$
$\models_{\mathbf{L}} N(NA > A)$ $\quad\models_{\mathbf{L}} (A > NB) > N(A > B)$

These schemas are, from left to right and from top to bottom, Aristotle's Thesis and its Variant, Boethius' Thesis and its Variant, and the Non-Symmetry of Implication.

In [4], Church introduced the conditional[2]

$A \rightarrow_{ch} B$	$\{1\}$	$\{1,0\}$	$\{0\}$
$\{1\}$	$\{1\}$	$\{1,0\}$	$\{0\}$
$\{1,0\}$	$\{0\}$	$\{1,0\}$	$\{0\}$
$\{0\}$	$\{1,0\}$	$\{1,0\}$	$\{1,0\}$

that apparently can validate the connexive features above since $A \rightarrow_{ch} B$ is not just true when A is untrue, very much like Wansing's connexive conditionals[3], it would only take pairing $\rightarrow_{ch}$ with a suitable negation. Here we only consider negations defined as $A \rightarrow_{ch} f$, for some falsity constant f. We aim to make at least a small contribution on

Supported by the PAPIIT projects IG400422 and IA105923.

[1] See [10,12,25] for useful overviews on connexive logics.

[2] 1 and 0 stand for truth and falsity, respectively. The notation has been adjusted using the mechanical procedure of [14] to transform a many-valued semantics to a bivalent Dunn semantics where interpretations are sets of truth values.

[3] See for example [24].

N. Alechina et al. (Eds.): LORI 2023, LNCS 14329, pp. 77–85, 2023.
https://doi.org/10.1007/978-3-031-45558-2_6

the conceptual linkings between connexivity and constants, which is a yet understudied topic in connexive logic.[4]

The plan of the paper is as follows. In Sect. 2, we give the technicalities to define a weak implicational logic with negation in the sense of Church. In Sect. 3, we characterize the first connexive logic based on Church's conditional. The negation of that logic is not without problems, so in Sect. 4 we introduce the Ackermannian falsity constants that will allow us to obtain less controversial negations and present the second connexive logic. We prove that this second logic is weak implicational, although it does not have a negation in Church's sense. Finally, in Sect. 5 we prove that the chosen Churchian conditional is not connexively stable.

2 Technical Preliminaries

Our base formal language $\mathcal{L}$ consists of formulas built, in the usual way, from a set of propositional variables *Prop* with the binary connective $\rightarrow_{ch}$. We express an expansion of a language by indicating at subscripts the symbols added. Thus, $\mathcal{L}_{\{c\}}$ denotes that the set of connectives is $\{\rightarrow_{ch}, c\}$. The formulas on expanded languages are also defined as usual. In many cases below, the exact shape of a language, and therefore of a logic, will be left implicit and will be indicated by the *de facto* use of connectives. The first capital letters of the Latin alphabet, 'A', 'B', 'C', ... will serve as variables ranging over arbitrary formulas.

Definition 1 (See [22] **and** [4]**).** *A* weak implicational logic *over* $\mathcal{L}$*, or WIL for short, is a logic where the following schemas are valid*[5]*:*

$(A > B) > ((C > A) > (C > B))$ $\quad$ $(C > (C > B)) > (C > B)$
$(C > (A > B)) > (A > (C > B))$ $\quad$ $(A > A)$

closed under Detachment —$A, A > B \vdash B$*— and Uniform Substitution —If* $A \vdash B$ *then* $A[q/p] \vdash B[q/p]$.

Definition 2 ([4])**.** *A WIL* with negation, *or WILN for short, has a propositional constant* f *such that* $NA =_{def.} A > f$ *and that satisfies the following schemas*[6]*:*

1. $(A > NB) > (B > NA)$
2. $(A > B) > (NB > NA)$
3. $A > NNA$
4. $NNNA > NA$
5. $(A > NA) > NA$
6. $(A > B) > (A > (NB > NA))$
7. $A > (NB > N(A > B))$
8. $NN(A > B) > (A > NNB)$

[4] Still, remember that Church's conditional was introduced and employed in a proto-relevant context. The exact connections between relevance and connexivity are still in need of explanation. (See [18], [19, chapters 1–4], [2,7,9,13,26] for further discussion.) Telling whether Church's conditional can shed some light on that topic requires a separate work.

[5] These axiom schemas are sometimes known by the names they are given in combinatorial logic: *B*, *C*, *W* and *I*, respectively. Thus, another name for Church's weak implicational logic is **BCWI**. They are also known by the names they are given in the relevance tradition: *Prefixing*, *Permutation*, *Contraction* and *Identity*, respectively.

[6] The list may be a bit redundant, but we decided to leave it as Church himself presented it.

9. $(NNA > NNB) > (NN(NNA > NNB))$
10. $(NN(NN(NNC > (NN(NNA > NNB)))) > (NN(NN(NNC > NNA)) > (NN(NNC > NNB)))$
11. $(NNA > NNB) > (NB > NA)$
12. $NN(NN(NNNNNA > NNNNNB)) > (NN(NN(B > NNA)))$

Church may be demanding too much from a negation; we will return to this issue in Sect. 4.

3 Ackermannian Falsity Constants and Our First Connexive Logic

Above we used notions from Dunn semantics to present Church's conditional. Let us introduce Dunn semantics more formally now.

Definition 3. *A* Dunn model *(or just a* model*) for $\mathcal{L}$ and its expansions is a function $V : Prop \longrightarrow \{\{0\}, \{1\}, \{0, 1\}\}$ and is then extended to functions σ to cover all formulas according to some evaluation conditions.*

Church intended his f to be false in all interpretations and to entail every proposition. Thus, the truth and falsity conditions for the intended falsity constant f_i would be $1 \notin \sigma(f_i)$ and $0 \in \sigma(f_i)$ for any σ. Then, the intended negation, defined as Church wanted, i.e. as $\sim_i A =_{def.} A \rightarrow_{ch} f_i$, can be presented tabularly as follows:

A	$\{1\}$	$\{1,0\}$	$\{0\}$
$\sim_i A$	$\{0\}$	$\{0\}$	$\{1,0\}$

This does not deliver connexivity, though: $\sigma(\sim_i (A \rightarrow_{ch} \sim_i A)) = \{0\}$ if $\sigma(A) = \{0\}$. On the other hand, $\sigma((A \rightarrow_{ch} B) \rightarrow_{ch} \sim_i (A \rightarrow_{ch} \sim_i B)) = \{0\}$ if $\sigma(A) = \{1\}$ and $\sigma(B) = \{0\}$.

This is where Ackermann's [1] understanding of the falsity constant can become in handy. For him, a falsity constant is false in all interpretations, but it need not entail every other proposition.[7] Consider a language with a stock of expressions that are *Ackermannian falsity constants*, defined as follows:

Definition 4. *An* Ackermannian falsity constant *in Dunn semantics is an expression f_j of the language such that, for all interpretation σ, $0 \in \sigma(f_j)$.*

Let us make two comments about this definition. First, the definition captures the idea behind Ackermann's falsity constant but duly adapted to the semantics we are working with, and that will come in handy in what follows. Second, the definition over the semantics makes room for interpretations in which f_j is true as well. This means that, in Dunn semantics one can distinguish between expressions that are assigned the same value, be it 1 or 0, in all interpretations, and expressions that are assigned exactly one interpretation, be it $\{1\}$, $\{1, 0\}$ or $\{0\}$. This means that falsity constants might have more than one interpretation, they just need that the value 0 belongs to all of them.

[7] See [20, §5.34] for a comparison between Ackermann's and Church's constants.

Consider the language $\mathcal{L}_{\{f_8\}}$.[8] The logic **CWILN**$_8$ can be obtained by extending valuations V in a Dunn model to interpretations σ for all formulas in $\mathcal{L}_{\{f_8\}}$ according to the following conditions:

$\sigma(p) = V(p)$

$1 \in \sigma(f_8)$
$0 \in \sigma(f_8)$

$1 \in \sigma(A \rightarrow_{ch} B)$ iff either (case I): $1 \notin \sigma(A)$, or (case II) $1 \in \sigma(B)$ and if $0 \in \sigma(A)$ then $0 \in \sigma(B)$
$0 \in \sigma(A \rightarrow_{ch} B)$ iff $0 \in \sigma(A)$ or $0 \in \sigma(B)$
A negation $\sim_8 A$, can be defined as $A \rightarrow_{ch} f_8$. It follows, by the evaluation conditions of the conditional, that $\sigma(f_8) = \sigma(\sim_8 A)$. $\sim_8 A$ can be presented in a tabular way as follows:

A	$\{1\}$	$\{1,0\}$	$\{0\}$
$\sim_8 A$	$\{1,0\}$	$\{1,0\}$	$\{1,0\}$

Finally, let A be a formula and Γ be a set of formulas of $\mathcal{L}_{\{f\}}$. A is a *logical consequence* of Γ in **CWILN**$_8$, $\Gamma \vDash_{\text{CWILN8}} A$ iff, for every σ, $1 \in \sigma(A)$ if $1 \in \sigma(B)$ for every $B \in \Gamma$.

Let us prove now some facts about **CWILN**$_8$.

Theorem 1. *CWILN*$_8$ *is a WILN.*

Proof. An inspection of the truth tables will show that **CWILN**$_8$ meets **Definition 1**. To prove that **CWILN**$_8$ meets **Definition 2**, the following will be useful:

Remark 1. For every A and σ, $\sigma(\sim_8 A) = \{1,0\}$. Thus, for any B and σ, $1 \in \sigma(B \rightarrow_{ch} \sim_8 A)$.

This delivers schemas 3, 4, 5, 9, 10 and 12. The remaining schemas are either of the form $C \rightarrow_{ch} (B \rightarrow_{ch} \sim_8 A)$ —like 1, 2, 7, 8 and 11— or of the form $A \rightarrow_{ch} (B \rightarrow_{ch} (\sim_8 C \rightarrow_{ch} \sim_8 D))$, like 6. In the former case, $\sigma(B \rightarrow_{ch} \sim_8 A) = \{1,0\}$ for every σ and hence $1 \in \sigma(C \rightarrow_{ch} (B \rightarrow_{ch} \sim_8 A))$ for every σ. In the latter case, $\sigma(\sim_8 C \rightarrow_{ch} \sim_8 D) = \{1,0\}$ for any σ, and hence $1 \in \sigma(A \rightarrow_{ch} (B \rightarrow_{ch} (\sim_8 C \rightarrow_{ch} \sim_8 D)))$ for every σ.

Theorem 2. *CWILN*$_8$ *validates Aristotle's and Boethius' Theses.*

Proof. By **Remark 1**, $1 \in \sigma(\sim_8 (A \rightarrow_{ch} \sim_8 A))$ and $1 \in \sigma(\sim_8 (\sim_8 A \rightarrow_{ch} A))$. For Boethius' Theses, a similar reasoning applies: since $1 \in \sigma(\sim_8 A)$ for every σ and A, $1 \in \sigma(\sim_8 (A \rightarrow_{ch} \sim_8 B))$. Hence, $1 \in \sigma((A \rightarrow_{ch} B) \rightarrow_{ch} (\sim_8 (A \rightarrow_{ch} \sim_8 B)))$. The proof of the Variant is similar.

Corollary 1. *CWILN*$_8$ *is negation-inconsistent. By* ***Remark 3.1****,* $A \rightarrow_{ch} \sim_8 A$ *is valid; by* ***Theorem 2****,* $\sim_8 (A \rightarrow_{ch} \sim_8 A)$ *is valid as well. Consider also* $\sim_8 A$ *and* $\sim_8 \sim_8 A$.

[8] The reason for the subscript will become apparent in the next section. Meanwhile, please simply take 'f_8' as a(n ugly) symbol for a certain nullary connective.

It can be argued that f_8 is not a good falsity constant because, although it is false under all interpretations, it is also true under all interpretations. On the other hand, $\sim_8 A$ have exactly the same evaluations as f_8, and it would seem preferable that the fs define negations that are not equivalent to the fs themselves. Nevertheless, if f_8 is both a falsity and a truth constant then it is, in particular, a falsity constant. It is then a matter of fact that to be a falsity constant does not imply not to be a truth constant nor vice versa. Nonetheless, no constant like f_8 is indispensable to make sense of connexive logics based on the Churchian conditional, as we will show in the next section.

4 The Second Churchian Connexive Logic

One further definition is needed:

Definition 5. *A* strict Ackermannian falsity constant *in a Dunn semantics is an Ackermannian falsity constant that, in some interpretation, is not true.*

Accordingly, these are the strict Ackermannian falsity constants expressible in a semantics like the one used here:

f_1	f_2	f_3	f_4	f_5	f_6	f_7
$\{0\}$	$\{1,0\}$	$\{0\}$	$\{0\}$	$\{1,0\}$	$\{0\}$	$\{1,0\}$
$\{0\}$	$\{0\}$	$\{1,0\}$	$\{0\}$	$\{1,0\}$	$\{1,0\}$	$\{0\}$
$\{0\}$	$\{0\}$	$\{0\}$	$\{1,0\}$	$\{0\}$	$\{1,0\}$	$\{1,0\}$

It is worth recalling that, in Dunn semantics, that an expression receives the same value in all interpretations is not the same that it receives only one interpretation. Clearly, 0 belongs in each line of the table for the fs; hence, every strict Ackermannian falsity constant is constantly false and we avoid making them also truth constants by requiring that they are not true in some interpretation.[9]

The following negations are defined as Church (and Ackermann) wanted, that is, as $A \rightarrow_{ch} f_k$, with $1 \leq k \leq 7$:

A	$\sim_1 A$	$\sim_2 A$	$\sim_3 A$	$\sim_4 A$	$\sim_5 A$	$\sim_6 A$	$\sim_7 A$
$\{1\}$	$\{0\}$	$\{1,0\}$	$\{0\}$	$\{0\}$	$\{1,0\}$	$\{0\}$	$\{1,0\}$
$\{1,0\}$	$\{0\}$	$\{0\}$	$\{1,0\}$	$\{0\}$	$\{1,0\}$	$\{1,0\}$	$\{0\}$
$\{0\}$	$\{1,0\}$	$\{1,0\}$	$\{1,0\}$	$\{1,0\}$	$\{1,0\}$	$\{1,0\}$	$\{1,0\}$

Note that $\sim_4$, $\sim_6$ and $\sim_7$ are but f_4, f_6 and f_7, respectively, so no new connective was defined in those cases, and $\sim_5$ is again the negation we are trying to do without. For definiteness, we will restrict ourselves to the connectives with the following truth condition: $1 \in \sigma(cA)$ iff $0 \in \sigma(A)$. This leaves us only with $\sim_3$, and so we expand $\mathcal{L}$ with the strict falsity constant f_3.

The logic **CWILN**$_3$ can be defined on top of Dunn models by extending valuations V to interpretations σ to all formulas in $\mathcal{L}_{\{f_3\}}$ by the following conditions:

[9] Note that one of the interpretations is repeated in all the fs from f_2 to f_7. This means that the constants have only two possible interpretations; we put the three rows to ease the comparison and calculation with other formulas.

$\sigma(p) = V(p)$

$1 \in \sigma(f_3)$ iff $1 \in \sigma(A)$ and $0 \in \sigma(A)$, for some fixed A
$0 \in \sigma(f_3)$

and the evaluation conditions for $A \rightarrow_{ch} B$ are as for **CWILN₈**.[10] A negation, $\sim_3 A$, can be defined as $A \rightarrow_{ch} f_3$ and it can be presented in a tabular way as follows[11]:

A	$\{1\}$	$\{1,0\}$	$\{0\}$
$\sim_3 A$	$\{0\}$	$\{1,0\}$	$\{1,0\}$

Logical consequence for **CWILN₃** is defined, mutatis mutandis, just as for **CWILN₈**.

Theorem 3. *CWILN₃ is a WIL, but not a WILN.*

Proof. Schemas 1, 2 and 4 are not valid. $\sigma(A) = \{1\}$ and $\sigma(B) = \{1,0\}$ provide a countermodel for 1 and 2. For 4, consider the case in which $\sigma(A) = \{1\}$.

Thus, **CWILN₃** does not have a negation in the sense of Church.[12] But it may be said to have a negation in less demanding ways. We will not enter the discussion about the necessary properties of negation; see [5,6,11,17,23] for discussions about negations that might validate much less than the properties asked by Church.

Remark 2. For every A and B, $\sigma(\sim_3(A \rightarrow_{ch} \sim_3 B)) = \sigma(\sim_3(\sim_3 A \rightarrow_{ch} B)) = \{1,0\}$.

Corollary 2. *CWILN₃ validates Aristotle's and Boethius' Theses.*

Proof. For Aristotle's Thesis, put A instead of B in $\sim_3 (A \rightarrow_{ch} \sim_3 B))$; for the Variant, put B instead of A in $\sim_3 (\sim_3 A \rightarrow_{ch} B)$. Boethius' Thesis and its Variant hold since $1 \in \sigma(A \rightarrow_{ch} B)$ whenever $\sigma(B) = \{1,0\}$.

Corollary 3. *CWILN₃ is also negation-inconsistent. For example, the schemas* $\sim_3 A \rightarrow_{ch} ((\sim_3 A \rightarrow_{ch} B) \rightarrow_{ch} \sim_3 (A \rightarrow_{ch} B))$ *and its negation are both valid in* **CWILN₃**. *Consider also* $\sim_3 \sim_3 A$ *and* $\sim_3 \sim_3 \sim_3 A$.

[10] The reference to an A is needed to distinguish between these different falsity constants and to be able to calculate interpretations in combination with other formulas. This makes the falsity constants look like unary connectives. If the reader wants to think of them in that way, that is no problem at all, their constant character is not altered by that fact.

[11] This negation can be found in [21]. See [15] for further discussion.

[12] Nevertheless, all the other schemas are valid. Schemas 3, 8, 9, 10 and 12 are of the form $A > NNB$, and we know that for every σ and every A, $1 \in \sigma(\sim_3 \sim_3 A)$. Then, for every σ we get $1 \in \sigma(A \rightarrow_{ch} \sim_3 \sim_3 B)$. For the proof of schema 5, suppose $\sigma((A \rightarrow_{ch} \sim_3 A) \rightarrow_{ch} \sim_3 A) = \{0\}$. Again, three cases arise, namely, $\sigma(A \rightarrow_{ch} \sim_3 A) = \{1\}$ and $\sigma(\sim_3 A) = \{0\}$, or $\sigma(A \rightarrow_{ch} \sim_3 A) = \{1,0\}$ and $\sigma(\sim_3 A) = \{1\}$ or $\sigma(\sim_3 A) = \{0\}$. It is easy to see that the three cases require interpretations that cannot be obtained. On the benefit of brevity, we suggest the reader verify the validity of the remaining schema, 7.

5 Connexive Stability

Following [3], we now investigate whether Church's conditional is connexively stable.

Definition 6. *A* standard negation *is a unary connective N satisfying that* $\sigma(NA) = \{0\}$ *if* $\sigma(A) = \{1\}$ *and* $\sigma(NA) = \{1\}$ *if* $\sigma(A) = \{0\}$.

Definition 7. *A negation N is* paraconsistent *(in a logic* **L***) iff* $A, NA \not\models_L B$.

Definition 8. *The* type of standard paraconsistent negations *(TSPN) is the set of standard paraconsistent negations definable according to a set of admissible evaluations.*

Remark 3. If there are only three admissible interpretations, logical consequence is truth-preservation under all interpretations and N is a standard negation, there are only two standard paraconsistent negations, namely, de Morgan's and Sette's negation [21], presented tabularly as follows:

A	$\{1\}$	$\{1,0\}$	$\{0\}$
$\neg A$	$\{0\}$	$\{1\}$	$\{1\}$

A	$\{1\}$	$\{1,0\}$	$\{0\}$
$\sim A$	$\{0\}$	$\{1,0\}$	$\{1\}$

Definition 9. *A conditional is* connexively stable *with respect to TSPN in* **L** *iff it meets the connexive features of Sect. 1 together with each N_i in TSPN.*

Now, it can be easily proved that

Theorem 4. *The Churchian conditional is not connexively stable with respect to TSPN.*

Proof. With $\neg$, Boethius' Thesis is just false when $\sigma(A) = \{1\}$ and $\sigma(B) = \{1, 0\}$.

None of the negations defined in Sect. 4 are standard in the specified sense, although they are standard in a generalized sense:

Definition 10. *A* generalized standard negation *is a unary connective N satisfying that* $0 \in \sigma(NA)$ *if* $1 \in \sigma(A)$ *and* $1 \in \sigma(NA)$ *if* $0 \in \sigma(A)$.[13]

Group all these negations, together with de Morgan negation (but not Sette's!), to form the type of generalized standard negations, TGSN. It can be easily checked that,

Theorem 5. *The Churchian conditional is not connexively stable with respect to TGSN.*

Proof. Aristotle's Thesis is invalid with $\sim_2$. Consider the case when $\sigma(A) = \{0\}$.

[13] Moreover, not all of those negations are paraconsistent, since $\sim_1$ and $\sim_4$ are not paraconsistent.

6 Conclusions

We did exactly what we promised in the abstract and in the introduction. There are four salient topics left for further work: (1) Investigating Church's conditional with other negations, e.g. de Morgan negation; (2) evaluating further connexive schemas involving other connectives, like Aristotle's Second Thesis, $N((A > B) \otimes (NA > B))$, or Abelard's Principle, $N((A > B) \otimes (A > NB))$, which involve a conjunction $\otimes$; (3) studying the compatibility connectives, if any, associated to Church's conditional[14]; (4) presenting the proof-theoretic versions of all the above.

Acknowledgments. We thank Sandra D. Cuenca, Fernando Cano-Jorge, Elisángela Ramírez-Cámara, Christian Romero-Rodríguez and Manuel Eduardo Tapia-Navarro as well as to the anonymous referees for their valuable comments.

References

1. Ackermann, W.: Begründung einer strengen implikation. J. Symbolic Logic **2**(21), 113–128 (1956)
2. Cano-Jorge, F.: Logical relevance in paraconsistent connexive logics I, unpublished typescript
3. Cano-Jorge, F., Estrada-González, L.: Mortensen logics. In: Indrzejczak, A., Zawidzki, M. (eds.) 10th International Conference on Non-Classical Logic. Theory and Applications (NCL 2022), vol. 358, pp. 189–201. EPTCS (2022)
4. Church, A.: The weak theory of implication. In: Burge, T., Enderton, H. (eds.) The Collected Works of Alonzo Church, chap. 8, pp. 307–318. MIT-Press, Massachusetts and London (2019). Originally published in Kontrolliertes Denken: Untersuchungen zum Logikkalkül und zur Logik der Einzelwissenschaften (Festschrift für Wilhelm Britzelmayr), edited by Albert Menne, Alexander Wilhelmy and Helmut Angstl, rotaprint, Kommissions-Verlag Karl Alber, Munich, 1951, pp. 22–37
5. Dunn, J.M.: Star and Perp: two treatments of negation. Philos. Perspect. **7**, 331–357 (1993)
6. Dunn, J.M., Zhou, C.: Negation in the context of gaggle theory. Stud. Logica. **80**, 235–264 (2005)
7. Estrada-González, L., Tanús-Pimentel, C.L.: Variable sharing in connexive logic. J. Philos. Log. **50**(6), 1377–1388 (2021)
8. Estrada-González, L., Nicolás-Francisco, R.A.: Connexive negation. Stud. Logica. (2023). to appear
9. Francez, N.: Relevant connexive logic. Logic Log. Philos. **28**(3), 409–425 (2019)
10. Francez, N.: A View of Connexive Logics. College Publications, London (2021)
11. Marcos, J.: On negation: pure local rules. J. Appl. Log. **3**(1), 185–219 (2005)
12. McCall, S.: A history of connexivity. In: Gabbay, D.M., Pelletier, F.J., Woods, J. (eds.) Handbook of the History of Logic, vol. 11. Logic: A History of its Central Concepts, pp. 415–449. Elsevier, Amsterdam (2012)
13. Omori, H.: A simple connexive extension of the basic relevant logic BD. IFCoLog J. Logics Appl. **3**(3), 467–478 (2016)
14. Omori, H., Sano, K.: Generalizing functional completeness in Belnap-Dunn logic. Stud. Logica. **103**, 883–917 (2015)

[14] On that topic, see [16] and [8].

15. Omori, H., Wansing, H.: Varieties of negation and contra-classicality in view of Dunn semantics. In: Bimbó, K. (ed.) Relevance Logics and other Tools for Reasoning. Essays in Honour of J. Michael Dunn, pp. 309–337. College Publications, London (2022)
16. Pizzi, C.: Cotenability and the logic of consequential implication. Logic J. IGPL **12**(6), 561–579 (2004)
17. Ripley, D.: Negation in Natural Language. Ph.D. thesis, Departament of Philosophy. University of North Carolina at Chapel Hill (2009)
18. Routley, R.: Semantics for connexive logics. I. Stud. Logica. **37**(4), 393–412 (1978)
19. Routley, R., Meyer, R.K., Plumwood, V., Brady, R.: Relevant Logics and Their Rivals. Ridgeview Publishing Company, Ohio (1982)
20. Schechter, E.: Classical and Nonclassical Logics: An Introduction to the Mathematics of Propositions. Princeton University Press, Princeton (2005)
21. Sette, A.M.: On the propositional calculus P1. Math. Japonicae **18**, 173–180 (1974)
22. Shaw-Kwei, M.: The deduction theorems and two new logical systems. Methodos **2**(5), 56–75 (1950)
23. Shramko, Y.: Dual intuitionistic logic and a variety of negations: the logic of scientific research. Stud. Logica. **18**, 347–367 (2005)
24. Wansing, H.: Connexive modal logic. In: Schmidt, R., Pratt-Hartmann, I., Reynolds, M., Wansing, H. (eds.) Advances in Modal Logic, vol. 5, pp. 367–383. College Publications, London (2005)
25. Wansing, H.: Connexive logic. In: Zalta, E.N. (ed.) The Stanford Encyclopedia of Philosophy. CSLI Stanford, Spring 2020 edn. (2020)
26. Weiss, Y.: Semantics for pure theories of connexive implication. Rev. Symbolic Logic **15**(3), 591–606 (2022)

Unknown Truths and Unknowable Truths

Jie Fan[1,2](✉)

[1] School of Humanities, University of Chinese Academy of Sciences, Beijing, China
[2] Institute of Philosophy, Chinese Academy of Sciences, Beijing, China
jiefan@ucas.ac.cn
https://teacher.ucas.ac.cn/~jiefan?language=en

Abstract. Notions of unknown truths and unknowable truths are important in formal epistemology, which are related to each other in e.g. Fitch's paradox of knowability. Although there have been some logical research on the notion of unknown truths and some philosophical discussion on the two notions, there seems to be no logical research on unknowable truths. In this paper, we propose a logic of unknowable truths, investigate the logical properties of unknown truths and unknowable truths, which includes the similarities of the two notions and the relationship between the two notions, and axiomatize this logic.

Keywords: unknown truths · unknowable truths · axiomatization · Fitch's paradox of knowability

1 Introduction

This article investigates notions of unknown truths and unknowable truths. A proposition is an *unknown truth*, if it is true but unknown; a proposition is an *unknowable truth*, if it is true but unknowable. The two notions are related to each other in e.g. Fitch's paradox of knowability, which states that if there is an unknown truth, then there is an unknowable truth.

The notion of unknown truths is important in formal epistemology. For example, it is related to verification thesis, which says that all truths can be known. By the thesis and two uncontroversial principles of knowledge, it follows from the notion that all truths are actually known. In this way, the notion gives rise to a well-known counterexample to the verification thesis. This is the so-called 'Fitch's paradox of knowability'. To take another example: the notion gives us an important type of Moore sentences, which is in turn crucial to Moore's paradox, saying that one cannot claim the paradoxical sentence "p but I do not know it". A well-known result is that such a Moore sentence is unsuccessful and self-refuting [11, 18].

The notion of unknowable truths is crucial in the dispute of realists and anti-realists. Anti-realists holds the aforementioned verificationist thesis, which states that all truths can be known, thereby denying the very possibility of unknowable truths. In contrast, realists believe that there are parts of reality, representable in some conceptually accessible language, that it is impossible for any agent ever to know. As guessed in [20, p. 119], "Perhaps, in truths that cannot be known lies 'the mystery at the heart of things'."

N. Alechina et al. (Eds.): LORI 2023, LNCS 14329, pp. 86–93, 2023.
https://doi.org/10.1007/978-3-031-45558-2_7

Note that the notion of unknowable truths also gives an important type of Moore sentences, since one cannot claim the paradoxical sentence "p is true but I cannot know it". We can also show that such a Moore sentence is unsuccessful and self-refuting.

Some researchers in the literature discuss about whether unknowable truths exist or not. For example, Edgington [10] admits that there are unknowable truths in the sense of Fitch's argument. Horsten [19] gives an interesting example of unknowable truths, by using a revision of the Gödel sentence. Cook [6] argues that even if it is impossible for it to be known of any particular sentence that it is both true and unknowable, we may still know that there are unknowable truths.[1] Besides, it is argued in [20, Sec. 2] that there are unknowable truths, that is, necessary limits to knowledge. As we will show below, under our interpretation of knowability, there are indeed unknowable truths. Thus our interpretation is in line with the aforementioned philosophical discussion.

In the literature, there have been several interpretations for the notion of knowability. For instance, in [3,5], knowability means 'known after an announcement'; in [7], knowability means the existence of proofs; in [13], knowability means capabilities to know; in [20], it is interpreted as a capacity-knowledge modality 'it can be known that'; in [21], knowability means 'known after an information update'; whereas in [17], knowability means 'dynamically possible knowledge of what was true before the update of the epistemic state of the agent'. In this paper, we follow [3,5] to interpret 'φ is knowable' as 'φ is known after some truthful public announcement', so that 'φ is an unknowable truth' is interpreted as 'φ is true and after any truthful public announcement, φ is unknown'.

Unknowable truths (and unknown truths) is a subjective concept, since it is possible that a proposition is an unknowable truth (an unknown truth, resp.) for an agent but not for another. For instance, consider the true proposition "it is raining but Ann does not know it". This proposition is an unknowable truth (an unknown truth, resp.) for Ann, but is *not* an unknowable truth (an unknown truth, resp.) for another agent Bob, who may be aware of Ann's ignorance. Therefore we move to the multiagent cases.

Although there have been some logical research on the notion of unknown truths (see e.g. [12,15]), and also some philosophical discussion on the two notions [6, 9,14,20], there seems to be no logical research on unknowable truths. For instance, there have been no axiomatization on the notion of unknowable truths. For another example, it is unnoticed that there are many similarities and relationship between the notions of unknown truths and unknowable truths. In this paper, we propose a logic of unknowable truths, investigate the logical properties of unknown truths and unknowable truths, which include the similarities and relationship between the two notions, and axiomatize the logic of unknowable truths.

The remainder of the article is organized as follows. Section 2 introduces the language and semantics of the logic of unknowable truths. Section 3 investigates logical properties of unknown truths and unknowable truths. Section 4 axiomatizes the logic of unknowable truths and shows its soundness and completeness. We conclude with some discussion in Sect. 5.

[1] However, it is shown in [9] to be unsuccessful since the argument depends on a paradoxical reasoning.

2 Syntax and Semantics

Throughout the paper, we let **P** be a denumerable set of propositional variables, and let **I** be a finite set of agents.

Definition 1. *Where* $p \in \boldsymbol{P}$, $i \in \boldsymbol{I}$, *the language of the logic of unknowable truths, denoted* **LUT**, *is defined recursively as follows:*

$$\boldsymbol{LUT} \quad \varphi ::= p \mid \neg\varphi \mid (\varphi \wedge \varphi) \mid K_i\varphi \mid [\varphi]\varphi \mid U_i\varphi$$

$K_i\varphi$ is read "agent i knows that φ", $[\psi]\varphi$ is read "after truthful public announcement of ψ, it is the case that φ", and $U_i\varphi$ is read "φ is an unknowable truth for i". Other connectives are defined as usual; in particular, $\bullet_i\varphi$, read "φ is an unknown truth for i", abbreviates $\varphi \wedge \neg K_i\varphi$, and $\langle\psi\rangle\varphi$ abbreviates $\neg[\psi]\neg\varphi$. Without the construct $U_i\varphi$, we obtain public announcement logic **PAL**; without the construct $[\varphi]\varphi$ further, we obtain epistemic logic **EL**. The language **LUT** is interpreted on models.

Definition 2. A model $\mathcal{M}$ *is a tuple* $\langle S, \{R_i \mid i \in \boldsymbol{I}\}, V\rangle$, *where* S *is a nonempty set of states, each* R_i *is the accessibility relation for* i, *and* V *is a valuation. We assume that* R *is reflexive.*[2] A frame *is a model without valuations. A* pointed model *is a pair of a model and a state in the model.*

Definition 3. *Given a model* $\mathcal{M} = \langle S, \{R_i \mid i \in \boldsymbol{I}\}, V\rangle$, *the semantics of* **LUT** *is defined recursively as follows.*

$$\begin{array}{|lcl|}
\hline
\mathcal{M}, s \vDash p & \Longleftrightarrow & s \in V(p) \\
\mathcal{M}, s \vDash \neg\varphi & \Longleftrightarrow & \mathcal{M}, s \nvDash \varphi \\
\mathcal{M}, s \vDash \varphi \wedge \psi & \Longleftrightarrow & \mathcal{M}, s \vDash \varphi \text{ and } \mathcal{M}, s \vDash \psi \\
\mathcal{M}, s \vDash K_i\varphi & \Longleftrightarrow & \text{for all } t, \text{ if } sR_it, \text{ then } \mathcal{M}, t \vDash \varphi \\
\mathcal{M}, s \vDash [\psi]\varphi & \Longleftrightarrow & \text{if } \mathcal{M}, s \vDash \psi \text{ then } \mathcal{M}|_\psi, s \vDash \varphi \\
\mathcal{M}, s \vDash U_i\varphi & \Longleftrightarrow & \mathcal{M}, s \vDash \varphi \text{ and for all } \psi \in \mathbf{EL}, \mathcal{M}, s \vDash [\psi]\neg K_i\varphi \\
\hline
\end{array}$$

where $\mathcal{M}|_\psi$ is the model restriction of $\mathcal{M}$ to ψ-states, and the restriction of announced formula in the semantics of U_i is to avoid circularity. We say that φ is *valid over a frame* F, if for all models $\mathcal{M}$ based on F and all states s, we have $\mathcal{M}, s \vDash \varphi$. We say that φ is *valid*, denoted $\vDash \varphi$, if it is valid over all (reflexive) frames. We say that φ is *satisfiable*, if there is a pointed model $(\mathcal{M}, s)$ such that $\mathcal{M}, s \vDash \varphi$.

Intuitively, $U_i\varphi$ means that φ is true and unknowable for i; in more details, φ is true and after each truthful public announcement, φ is unknown to i.

One may compute the semantics of $\bullet_i\varphi$ and $\langle\psi\rangle\varphi$ as follows:

$$\begin{array}{lcl}
\mathcal{M}, s \vDash \bullet_i\varphi & \Longleftrightarrow & \mathcal{M}, s \vDash \varphi \text{ and for some } t, sR_it \text{ and } \mathcal{M}, t \nvDash \varphi \\
\mathcal{M}, s \vDash \langle\psi\rangle\varphi & \Longleftrightarrow & \mathcal{M}, s \vDash \psi \text{ and } \mathcal{M}|_\psi, s \vDash \varphi.
\end{array}$$

[2] Note that here for simplicity, we only consider the minimal restriction of knowledge. We can also consider extra properties of knowledge, for instance, transitivity, Euclidicity, etc., in which cases we add the corresponding axioms for knowledge in the axiomatization.

3 Logical Properties

In this section, we investigate logical properties of unknown truths and unknowable truths. Due to space limitation, we omit almost all the proofs.

3.1 Similarities

First, we can summarize some similarities between the logical properties of unknown truths and the logical properties of unknowable truths in the following diagram.

	Unknown Truths	*Unknowable Truths*
validities	$\bullet_i\varphi \to \varphi$	$U_i\varphi \to \varphi$
	$\bullet_i(\varphi \to \psi) \to (\bullet_i\varphi \to \bullet_i\psi)$	$U_i(\varphi \to \psi) \to (U_i\varphi \to U_i\psi)$
	$\bullet_i\varphi \wedge \bullet_i\psi \to \bullet_i(\varphi \wedge \psi)$	$U_i\varphi \wedge U_i\psi \to U_i(\varphi \wedge \psi)$
	$\bullet_i\varphi \to \bullet_i\bullet_i\varphi$	$U_i\varphi \to U_iU_i\varphi$
	$\bullet_i\varphi \leftrightarrow \bullet_i\bullet_i\varphi$	$U_i\varphi \leftrightarrow U_iU_i\varphi$
	$\neg K_i\bullet_i\varphi$	$\neg K_iU_i\varphi$
	$\neg\bullet_iK_i\varphi$ (*transitive*)	$\neg U_iK_i\varphi$ (*transitive*)
	$\neg\bullet_i\neg K_i\varphi$ (*Euclidean*)	$\neg U_i\neg K_i\varphi$ (*Euclidean*)
	$[\bullet_ip]\neg\bullet_ip$	$[U_ip]\neg U_ip$
invalidities	$\neg\bullet_i\varphi \to \bullet_i\neg\bullet_i\varphi$	$\neg U_i\varphi \to U_i\neg U_i\varphi$

In what follows, we take some of the validities and invalidities as examples, and leave others to the reader.

The following result states that unknowable truths are themselves unknowable truths. One can show that it is equivalent to the statement that it is impossible for it to be known of any particular sentence that it is both true and unknowable, where the latter is argued by Cook [6].

Proposition 1. *$\vDash U_i\varphi \to U_iU_i\varphi$, and thus $\vDash U_i\varphi \leftrightarrow U_iU_i\varphi$.*

Intuitively, one cannot know the unknowable truths, since otherwise one would know the truths that cannot be known, which is impossible. In other words, unknowable truths are necessary limits to knowledge. This follows immediately from the result below.

Proposition 2. *$\vDash \neg K_iU_i\varphi$.*

Remark 1. If we assume the property of transitivity, then $\neg U_iK_i\varphi$ is valid, which says that all knowledge is not an unknowable truth. If we assume the property of Euclidicity, we can show that $\neg U_i\neg K_i\varphi$ is valid, which says that all non-knowledge is not an unknowable truth. Similar results apply to the case for the notion of unknown truths: $\neg\bullet_iK_i\varphi$ is valid over transitive frames, and $\neg\bullet_i\neg K_i\varphi$ is valid over Euclidean frames.

The following states that *not* all non-unknowable-truths are themselves unknowable truths.

Proposition 3. *$\nvDash \neg U_i\varphi \to U_i\neg U_i\varphi$.*

As mentioned in the introduction, Moore sentences such as $\bullet_i p$ and $U_i p$ are unsuccessful and self-refuting. It is shown in [11] that $\bullet_i p$ is unsuccessful and self-refuting. Here we show that $U_i p$ is also unsuccessful and self-refuting.

Proposition 4. *$\vDash [U_i p]\neg U_i p$, and thus $\nvDash [U_i p]U_i p$.*

Proof. Suppose that $\mathcal{M}, s \vDash U_i p$, to show that $\mathcal{M}|_{U_i p}, s \nvDash U_i p$. By supposition, $\mathcal{M}, s \vDash p$. Thus $\mathcal{M}|_{U_i p} \subseteq \mathcal{M}|_p$, and then $\mathcal{M}|_{U_i p}|_{U_i p} \subseteq \mathcal{M}|_p|_{U_i p}$. Since $\mathcal{M}|_p, s \vDash K_i p$, we obtain that $\mathcal{M}|_p, s \nvDash U_i p$, that is, $s \notin \mathcal{M}|_p|_{U_i p}$. It then follows that $s \notin \mathcal{M}|_{U_i p}|_{U_i p}$, and therefore $\mathcal{M}|_{U_i p}, s \nvDash U_i p$.

One may ask if there are any differences between the logical properties of unknown truths and the logical properties of unknowable truths. The answer is positive. We have seen from [11] that $\bullet_i(\varphi \wedge \psi) \to \bullet_i\varphi \vee \bullet_i\psi$ is valid. In contrast, $U_i(\varphi \wedge \psi) \to U_i\varphi \vee U_i\psi$ is invalid.

Proposition 5. *$\nvDash U_i(\varphi \wedge \psi) \to U_i\varphi \vee U_i\psi$.*

Proof. Consider the following model $\mathcal{M}$:

We show that $\mathcal{M}, s \vDash U_i(\neg K_i p \wedge \neg K_i q)$ but $\mathcal{M}, s \nvDash U_i\neg K_i p$ and $\mathcal{M}, s \nvDash U_i\neg K_i q$.

- $\mathcal{M}, s \vDash U_i(\neg K_i p \wedge \neg K_i q)$. Clearly, $\mathcal{M}, s \vDash \neg K_i p \wedge \neg K_i q$. Suppose, for reductio, that there exists $\psi \in \mathbf{EL}$ such that $\mathcal{M}, s \vDash \langle\psi\rangle K_i(\neg K_i p \wedge \neg K_i q)$, and then $\mathcal{M}, s \vDash \psi$ and $\mathcal{M}|_\psi, s \vDash K_i(\neg K_i p \wedge \neg K_i q)$. Because the announcement is interpreted via world-elimination, we consider the following three cases.
  - t is retained in the updated model $\mathcal{M}|_\psi$. On the one hand, since $\mathcal{M}|_\psi, s \vDash K_i(\neg K_i p \wedge \neg K_i q)$ and $sR_i t$, we obtain $\mathcal{M}|_\psi, t \vDash \neg K_i p$; on the other hand, as t can only see itself and $t \vDash p$, we have $\mathcal{M}|_\psi, t \vDash K_i p$. A contradiction.
  - u is retained in the updated model $\mathcal{M}|_\psi$. Similar to the first case, we can show that $\mathcal{M}|_\psi, u \vDash K_i q$ and $\mathcal{M}|_\psi, u \vDash \neg K_i q$, a contradiction.
  - Neither t nor u is retained in $\mathcal{M}|_\psi$. Then in $\mathcal{M}|_\psi$, there is only a state s which is reflexive. Similar to the first case, we can also show that $\mathcal{M}|_\psi, s \vDash K_i p$ and $\mathcal{M}|_\psi, s \vDash \neg K_i p$, a contradiction.
- $\mathcal{M}, s \nvDash U_i\neg K_i p$. We have seen that $\mathcal{M}, s \vDash \neg K_i p$. It suffices to show that for some $\psi \in \mathbf{EL}$ we have $\mathcal{M}, s \vDash \langle\psi\rangle K_i\neg K_i p$. This is indeed the case, because announcing q makes t be deleted, and in the remaining model, $\neg K_i p$ is true at s and u.
- $\mathcal{M}, s \nvDash U_i\neg K_i q$. The proof is analogous to that of $\mathcal{M}, s \nvDash U_i\neg K_i p$, by announcing p instead.

3.2 Interactions

In what follows, we study the interactions between unknown truths and unknowable truths. First, every unknowable truth is an unknown truth, but not vise versa. This indicates that the notion of unknowable truths is stronger than that of unknown truths.

Proposition 6. $\models U_i\varphi \to \bullet_i\varphi$, *and* $\not\models \bullet_i\varphi \to U_i\varphi$.

Also, intuitively, one cannot know the unknown truths. This is because unknown truths are themselves unknowable truths. In other words, if it is an unknown truth that p, it is an unknowable truth that it is an unknown truth that p, which is argued in [14, Theorem 2] (see also [22, p. 154]).

Proposition 7. $\models \bullet_i\varphi \to U_i\bullet_i\varphi$, *thus* $\models \bullet_i\varphi \leftrightarrow U_i\bullet_i\varphi$.

From Proposition 6 and Proposition 7, we can see that $\models U_i\varphi \to U_i\bullet_i\varphi$. Despite this, the converse formula is not valid. This follows directly from Proposition 6 and Proposition 7.

Proposition 8. $\not\models U_i\bullet_i\varphi \to U_i\varphi$.

The following states that unknowable truths are themselves unknown truths. In other words, if it is an unknowable truth that φ, then it is an unknown truth that it is an unknowable truth that φ.

Proposition 9. $\models U_i\varphi \to \bullet_i U_i\varphi$, *thus* $\models U_i\varphi \leftrightarrow \bullet_i U_i\varphi$.

Fitch's paradox of knowability states that if all truths are knowable, then all truths are actually known. This can be shown as follows.

Proposition 10. $\models \neg U_i\varphi$ *for all* φ, *then* $\models \neg\bullet_i\varphi$ *for all* φ.

Although not every truth (e.g. unknown truths) is knowable, every *logical truth* (namely, validity) is knowable.

Proposition 11. *If* $\models \varphi$, *then* $\models \neg U_i\varphi$.

4 Axiomatization and Completeness

In this section, we will give a proof system for **LUT** and show its soundness and completeness.

Definition 4. *The proof system* $\mathbb{LUT}$ *consists of the following axioms and inference rules.*

PL	*All instances of tautologies*
K	$K_i(\varphi \to \psi) \to (K_i\varphi \to K_i\psi)$
KA	$[\chi](\varphi \to \psi) \to ([\chi]\varphi \to [\chi]\psi)$
T	$K_i\varphi \to \varphi$
AP	$[\psi]p \leftrightarrow (\psi \to p)$
AN	$[\psi]\neg\varphi \leftrightarrow (\psi \to \neg[\psi]\varphi)$
AC	$[\psi](\varphi \wedge \chi) \leftrightarrow ([\psi]\varphi \wedge [\psi]\chi)$
AK	$[\psi]K_i\varphi \leftrightarrow (\psi \to K_i[\psi]\varphi)$
AA	$[\psi][\chi]\varphi \leftrightarrow [\psi \wedge [\psi]\chi]\varphi$
AU	$U_i\varphi \to \varphi \wedge [\psi]\neg K_i\varphi$, *where* $\psi \in$ ***EL***
MP	*From* φ *and* $\varphi \to \psi$ *infer* ψ
GEN	*From* φ *infer* $K_i\varphi$
GENA	*From* φ *infer* $[\chi]\varphi$
RU	*From* $\eta(\varphi \wedge [\psi]\neg K_i\varphi)$ *for all* $\psi \in$ ***EL*** *infer* $\eta(U_i\varphi)$,

where $\eta(\sharp)$ *is an* admissible form. *The notion of admissible forms is originated from [16, pp. 55–56], which is called 'necessity forms' in [1–4], and defined as follows.*

Let $\varphi \in$ ***LUT*** *and* $i \in$ ***I****. Admissible forms* $\eta(\sharp)$ *are defined recursively in the following.*

$$\eta(\sharp) ::= \sharp \mid \varphi \to \eta(\sharp) \mid K_i\eta(\sharp) \mid [\varphi]\eta(\sharp).$$

The novel thing is the axiom AU and inference rule RU, which together characterize the semantics of U_i in a certain way: AU says that if φ is an unknowable truth for i, then φ is true and after every truthful public announcement, φ is unknown to i; RU roughly says that if φ is true but unknown to i after every truthful public announcement, then φ is an unknowable truth for i. Other axioms and inference rules are familiar from public announcement logic, cf. e.g. [8].

We say a formula φ is *provable* in $\mathbb{LUT}$, notation: $\vdash \varphi$, if φ is either an instantiation of some axiom, or obtained from axioms with an application of some inference rule.

The soundness and completeness of $\mathbb{LUT}$ can be proved mutatis mutandis as in [4].

Theorem 1. $\mathbb{LUT}$ *is sound and complete with respect to the class of all frames. That is, if* $\models \varphi$*, then* $\vdash \varphi$*.*

With the completeness in hand, we can give a syntactic proof for Fitch's paradox of knowability, which is much simpler than those in the literature, e.g. in [20].

Proposition 12. *If* $\vdash \neg U_i\varphi$ *for all* φ*, then* $\vdash \neg\bullet_i\varphi$ *for all* φ*.*

Proof. Suppose that $\vdash \neg U_i\varphi$ for all φ, then $\vdash \neg U_i\bullet_i\varphi$. Then by Proposition 7 and Theorem 1, $\vdash \bullet_i\varphi \to U_i\bullet_i\varphi$, and therefore $\vdash \neg\bullet_i\varphi$ for all φ.

5 Conclusion and Discussion

In this paper, we proposed a logic of unknowable truths, investigated the logical properties of unknown truths and the logical properties of unknowable truths, which includes the similarities and relationship between the two notions, and finally axiomatized the logic of unknowable truths.

So far we have seen that the semantics of unknowable truths depends on that of propositional knowledge. However, we have mainly focused on the relationship between unknowable truths and unknown truths, instead of between unknowable truths and propositional knowledge, and the notion of unknown truths is weaker than that of propositional knowledge. So there seems to be some asymmetry between the semantics of unknowable truths and our concerns. Then a natural question is: is there any other semantics for unknowable truths which relates the notion to that of unknown truths more directly/properly? Our answer is positive. We will introduce this new semantics in the future work.

For another future work, a natural extension would be to add propositional quantifiers. This can increase the expressive power of the current logic. For instance, we can express Fitch's paradox of knowability in the new language as follows: $\forall p\neg U_ip \to \forall p\neg\bullet_ip$, or equivalently, $\exists p\bullet_ip \to \exists pU_ip$.

Acknowledgement. This article is supported by the Fundamental Research Funds for the Central Universities. The author thanks for the insightful comments from the audience of the Second International Workshop on Logics of Multiagent Systems (LMAS 2023), where an earlier version of this article is presented. The author also thanks four anonymous referees of LORI 2023 for their insightful comments.

References

1. Balbiani, P.: Putting right the wording and the proof of the truth lemma for APAL. J. Appl. Non-Classical Logics **25**(1), 2–19 (2015)
2. Balbiani, P., Baltag, A., van Ditmarsch, H., Herzig, A., Hoshi, T., Lima, T.D.: What can we achieve by arbitrary announcements? A dynamic take on Fitch's knowability. In: Samet, D. (ed.) Proceedings of TARK XI, pp. 42–51. Presses Universitaires de Louvain, Louvain-la-Neuve, Belgium (2007)
3. Balbiani, P., Baltag, A., van Ditmarsch, H., Herzig, A., Hoshi, T., Lima, T.D.: Knowable as known after an announcement. Rev. Symbolic Logic **1**(3), 305–334 (2008)
4. Balbiani, P., van Ditmarsch, H.: A simple proof of the completeness of APAL. Stud. Logic **8**(1), 65–78 (2015)
5. van Benthem, J.: What one may come to know. Analysis **64**(2), 95–105 (2004)
6. Cook, R.T.: Knights, knaves and unknowable truths. Analysis **66**(1), 10–16 (2006)
7. Dean, W., Kurokawa, H.: From the knowability paradox to the existence of proofs. Synthese **176**, 177–225 (2010)
8. van Ditmarsch, H., van der Hoek, W., Kooi, B.: Dynamic Epistemic Logic, Synthese Library, vol. 337. Springer, Dordrecht (2008). https://doi.org/10.1007/978-1-4020-5839-4
9. Duke-Yonge, J.: Unknowable truths: a reply to cook. Analysis **66**(4), 295–299 (2006)
10. Edgington, D.: The paradox of knowability. Mind **94**, 557–568 (1985)
11. Fan, J.: Bimodal logics with contingency and accident. J. Philos. Log. **48**, 425–445 (2019)
12. Fan, J.: Unknown truths and false beliefs: completeness and expressivity results for the neighborhood semantics. Stud. Logica. **110**, 1–45 (2022)
13. Fara, M.: Knowability and the capacity to know. Synthese **173**, 53–73 (2010). https://doi.org/10.1007/s11229-009-9676-8
14. Fitch, F.B.: A logical analysis of some value concepts. J. Symbolic Logic **28**(2), 135–142 (1963)
15. Gilbert, D., Venturi, G.: Neighborhood semantics for logics of unknown truths and false beliefs. Australas. J. Logic **14**(1), 246–267 (2017)
16. Goldblatt, R.: Axiomatising the Logic of Computer Programming, vol. 130. Springer, Heidelberg (1982)
17. Holliday, W.H.: Knowledge, time, and paradox: introducing sequential epistemic logic. In: van Ditmarsch, H., Sandu, G. (eds.) Jaakko Hintikka on Knowledge and Game-Theoretical Semantics. OCL, vol. 12, pp. 363–394. Springer, Cham (2018). https://doi.org/10.1007/978-3-319-62864-6_15
18. Holliday, W., Icard, T.: Moorean phenomena in epistemic logic. In: Advances in Modal Logic, vol. 8, pp. 178–199. College Publications (2010)
19. Horsten, L.: A Kripkean approach to unknowability and truth. Notre Dame J. Formal Logic **39**(3), 389–406 (1998)
20. Routley, R.: Necessary limits to knowledge: unknowable truths. Synthese **73**, 107–122 (2010)
21. Wen, X., Liu, H., Huang, F.: An alternative logic for knowability. In: van Ditmarsch, H., Lang, J., Ju, S. (eds.) LORI 2011. LNCS (LNAI), vol. 6953, pp. 342–355. Springer, Heidelberg (2011). https://doi.org/10.1007/978-3-642-24130-7_25
22. Williamson, T.: On knowledge of the unknowable. Analysis **47**(3), 154–158 (1987)

A Characterization of Lewisian Causal Models

Jingzhi Fang and Jiji Zhang(✉)

The Chinese University of Hong Kong, Shatin, NT, Hong Kong
{jingzhifang,jijizhang}@cuhk.edu.hk

Abstract. An important component in the interventionist account of causal explanation is an interpretation of counterfactual conditionals as statements about consequences of hypothetical interventions. The interpretation receives a formal treatment in the framework of functional causal models. In Judea Pearl's influential formulation, functional causal models are assumed to satisfy a "unique-solution" property; this class of Pearlian causal models includes the ones called recursive. Joseph Halpern showed that every recursive causal model is Lewisian, in the sense that from the causal model one can construct a possible worlds model in David Lewis's well-known semantics that satisfies the exact same formulas in a certain language. Moreover, he demonstrated that some Pearlian (non-recursive) models are not Lewisian in this sense. This raises the question regarding the exact contour of Lewisian causal models. In this paper, we provide a characterization of the class of Lewisian causal models and a complete axiomatization with respect to this class. Our results have philosophically interesting consequences, two of which are especially worth noting. First, the class of Stalnakerian causal models, a subclass of Lewisian causal models, is precisely the class of Pearlian models that do not contain any cycle of counterfactual dependence (in a sense of counterfactual dependence akin to Lewis's famous relation between distinct events). Second, a more natural class of causal models is actually a superclass of Lewisian causal models, the logic of which respects only weak centering rather than centering.

Keywords: Structural equation model · Counterfactual · Possible worlds semantics

1 Introduction

A major contender in the philosophical literature on causation and causal explanation is known as interventionism ([13,14]). In this broadly counterfactual approach, subjunctive or counterfactual conditionals are interpreted as statements about consequences of hypothetical interventions. This interpretation is formally developed in the framework of functional causal models or structural equation models (e.g., [1–4,6,7,15]). In Judea Pearl's [11] influential formulation, causal models are assumed to satisfy a "unique-solution" property. Joseph Halpern [5]

N. Alechina et al. (Eds.): LORI 2023, LNCS 14329, pp. 94–108, 2023.
https://doi.org/10.1007/978-3-031-45558-2_8

studied the relationship between Pearlian causal models and possible worlds models in David Lewis's [9] well known semantics of counterfactuals (which Halpern called counterfactual structures). Halpern showed that the class of recursive causal models, a proper subclass of Pearlian causal models, is Lewisian in the sense that for every recursive causal model, one can construct a possible worlds model in Lewis's theory that satisfies the exact same formulas in a certain language as the causal model does, but some non-recursive Pearlian models are not Lewisian in this sense. In view of these results, Halpern wrote: "My own feeling is that these arguments show that models [that are Pearlian but not recursive] are actually not good models for causality." [5, p. 318] Non-recursive Pearlian models may well be problematic models for causality, but if the consideration here is whether a causal model is compatible with Lewisian constraints, it is a little hasty to dismiss such models altogether, for many non-recursive models are also Lewisian. It will be useful, therefore, to better understand the class of Lewisian causal models.

In this paper we build on Halpern's illuminating analysis and provide a characterization of Lewisian causal models. In addition to revealing the exact contour of Lewisian causal models, our results have philosophically interesting consequences. Among other things, we show that the class of Stalnakerian causal models, a subclass of Lewisian causal models, is precisely the class of Pearlian causal models that are free of cycles of counterfactual dependence, in a sense of counterfactual dependence that is akin to the famous relation between distinct events defined by Lewis [8]. Another potentially significant implication is that a more natural class of causal models is actually a superclass of Lewisian causal models, the logic of which respects only weak centering rather than centering.

The rest of the paper proceeds as follows. After introducing the technical setup in Sect. 2, we present our characterization of Lewisian causal models in Sect. 3, and an axiomatization of the class in Sect. 4. We conclude in Sect. 5 with some discussions of the implications of our results.

2 Preliminaries

We now introduce the setup of the problem under attack, largely following Halpern's [5] formulations. Let a *signature* be a triple $\langle \mathbf{U}, \mathbf{V}, \mathcal{R} \rangle$, where $\mathbf{U}$ and $\mathbf{V}$ are disjoint finite sets of variables, and $\mathcal{R}$ associates with each variable $X \in \mathbf{U} \cup \mathbf{V}$ a finite set of values $\mathcal{R}(X)$. A causal model is defined over a signature:

Definition 1 (Causal Model). *A causal model over a signature* $S = \langle \mathbf{U}, \mathbf{V}, \mathcal{R} \rangle$ *is a tuple* $\langle S, \mathcal{F} \rangle$, *where* $\mathcal{F}$ *is a collection of functions such that for each* $X \in \mathbf{V}$, $\mathcal{F}$ *contains one and only one function* $f_X : \times_{Y \in \mathbf{U} \cup \mathbf{V} \setminus \{X\}} \mathcal{R}(Y) \to \mathcal{R}(X)$.

In words, f_X maps each value combination of $\mathbf{U} \cup \mathbf{V} \setminus \{X\}$ to a unique value of X. It is intended to represent a causal mechanism that determines the value of X according to the values of other variables. Notice that a causal model specifies functions only for variables in $\mathbf{V}$; the mechanisms for variables in $\mathbf{U}$ are not modelled. A value setting of $\mathbf{U}$ represents an external input to the system.

Once the value setting of $\mathbf{U}$ is given, the values of variables in $\mathbf{V}$ are then to be derived according to structural equations. For this reason, variables in $\mathbf{U}$ are called *exogenous* and those in $\mathbf{V}$ *endogenous*.

Such a causal model is also known as a *structural equation model*, where each function f_X can be written as a structural equation $X = f_X(\mathbf{U} \cup \mathbf{V} \backslash \{X\})$. Let $P(X) \subseteq \mathbf{U} \cup \mathbf{V} \backslash \{X\}$ be the set of non-redundant arguments in f_X. A causal model is naturally associated with a directed graph, where each variable in $\mathbf{U} \cup \mathbf{V}$ is represented as a vertex and an arrow is drawn from Y to X if and only if $Y \in P(X)$. A model is called *recursive* if the associated graph is acyclic.

For any $\mathbf{X} \subseteq \mathbf{U} \cup \mathbf{V}$, we will often abuse notation and treat $\mathbf{X}$ as a vector of distinct variables. A *value configuration* of $\mathbf{X}$ is a set or vector of values $\mathbf{x}$ that contains a unique value $x \in \mathcal{R}(X)$ for every $X \in \mathbf{X}$. We call a value configuration $\mathbf{u}$ of the exogenous variables $\mathbf{U}$ a *context* for the model.

Interventions can be easily represented in causal models. For the present purposes, we only consider interventions that force a set of endogenous variables to have a certain value configuration. Given a causal model $T = \langle \mathbf{U}, \mathbf{V}, \mathcal{R}, \mathcal{F} \rangle$, an intervention that forces $\mathbf{X} \subseteq \mathbf{V}$ to take a value configuration $\mathbf{x}$ is to be represented by a *submodel* in the following sense:

Definition 2 (Submodel). *Let $T = \langle \mathbf{U}, \mathbf{V}, \mathcal{R}, \mathcal{F} \rangle$, $\mathbf{X} \subseteq \mathbf{V}$, and $\mathbf{x}$ be a value configuration of $\mathbf{X}$. A submodel of T with respect to the intervention $\mathbf{X} = \mathbf{x}$, denoted by $T_{\mathbf{X}=\mathbf{x}}$, is $\langle \mathbf{U}, \mathbf{V}, \mathcal{R}, \mathcal{F}_{\mathbf{X}=\mathbf{x}} \rangle$, where $\mathcal{F}_{\mathbf{X}=\mathbf{x}}$ differs from $\mathcal{F}$ only regarding variables in $\mathbf{X}$: for every $X \in \mathbf{X}$, f_X in $\mathcal{F}$ is replaced by a constant function $X = x$, where x is the component value in $\mathbf{x}$ for X.*

In plain words, an intervention is supposed to be effective and local, forcing the target variables into the target values, without affecting the mechanisms for other variables. We stipulate that if $\mathbf{X} = \emptyset$, $T_{\mathbf{X}=\mathbf{x}} = T$.

Given a causal model $\langle \mathbf{U}, \mathbf{V}, \mathcal{R}, \mathcal{F} \rangle$, a *solution* to the model relative to a context $\mathbf{u}$ is a value configuration $\mathbf{v}$ of $\mathbf{V}$ such that $\mathbf{U} = \mathbf{u}$ and $\mathbf{V} = \mathbf{v}$ together are consistent with all functions in $\mathcal{F}$. In Pearl's [11] influential formulation, a "unique-solution" constraint is imposed on causal models:

Definition 3 (Pearlian Causal Model). *A causal model is called Pearlian if every submodel (including the model itself) has a unique solution relative to every context.*

It is easy to see that the class of recursive models is a proper subclass of the class of Pearlian causal models ([3–5]).

In this paper, we also allow models outside of the class of Pearlian models, but we require that a model have at least one solution relative to every context. In other words, we confine in the first place to what we call "Solutionful" causal models:

Definition 4 (Solutionful Causal Model). *A causal model is Solutionful if the model has at least one solution relative to every context.*

At this stage we do not require submodels to be also Solutionful, though as we will see, a necessary condition for a (Solutionful) causal model to be Lewisian is that all its submodels are also Solutionful, which agrees with a result from [15]. Our definition of "Lewisian causal model" below is more stringent than that of [15] and is more appropriate because the notion in [15] inherited a bug from [3], as pointed out by [5]. The definition is simplified if we consider only Solutionful causal models. Since being Solutionful is a necessary condition for even the more liberal notion of "Lewisian causal model" in [15], this restriction does not affect our purpose.

The object language we will consider is defined over a signature $S = \langle \mathbf{U}, \mathbf{V}, \mathcal{R} \rangle$, denoted by $\mathcal{L}(S)$. In addition to the truth-functional operators, an operator to form subjunctive conditionals is included, symbolized as '$\Box\!\!\rightarrow$'. The well-formed formulas (wffs) of $\mathcal{L}(S)$ are defined as follows:

- For every $X \in \mathbf{V}$ and every $x \in \mathcal{R}(X)$, $X = x$ is a wff.
- If α and β are wffs, so are $\neg\alpha$, $\alpha \wedge \beta$.
- If α is a wff of the form $(X_1 = x_1) \wedge ... \wedge (X_n = x_n)$, where all X_i's are distinct, $x_i \in \mathcal{R}(X_i)$, and β is a wff that does not contain '$\Box\!\!\rightarrow$', then $\alpha \Box\!\!\rightarrow \beta$ is a wff. A special case will be written as $true \Box\!\!\rightarrow \beta$ when α contains no X_i.

Notice that the form of subjunctive conditionals in $\mathcal{L}(S)$ is restricted: no nested conditionals or conditionals with disjunctive antecedents are allowed (see [2] for an attempt to relax the restrictions). For convenience, we will often write '$\mathbf{X} = \mathbf{x}$' as a shorthand for '$(X_1 = x_1) \wedge ... \wedge (X_n = x_n)$'. We define $\alpha \Diamond\!\!\rightarrow \beta$ as an abbreviation of $\neg(\alpha \Box\!\!\rightarrow \neg\beta)$. The truth-functional operators $\vee$, $\Rightarrow$ and $\Leftrightarrow$ are defined in the standard way.

Given a causal model T over S, let $Sol(T) = \{(\mathbf{u}, \mathbf{v}) | \mathbf{v}$ is a solution to T relative to context $\mathbf{u}\}$; that is, $Sol(T)$ is the set of value configurations for $\mathbf{U} \cup \mathbf{V}$ that are consistent with all functions in T. Each wff in $\mathcal{L}(S)$ is evaluated relative to T and $(\mathbf{u}, \mathbf{v}) \in Sol(T)$, according to the following rules:

- $T, (\mathbf{u}, \mathbf{v}) \models X = x$ iff $\mathbf{v}$ assigns value x to X.
- $T, (\mathbf{u}, \mathbf{v}) \models \neg\alpha$ iff $T, (\mathbf{u}, \mathbf{v}) \not\models \alpha$.
- $T, (\mathbf{u}, \mathbf{v}) \models \alpha \wedge \beta$ iff $T, (\mathbf{u}, \mathbf{v}) \models \alpha$ and $T, (\mathbf{u}, \mathbf{v}) \models \beta$.
- $T, (\mathbf{u}, \mathbf{v}) \models \mathbf{X} = \mathbf{x} \Box\!\!\rightarrow \beta$ iff $T_{\mathbf{X}=\mathbf{x}}, (\mathbf{u}, \mathbf{v}') \models \beta$ for every $(\mathbf{u}, \mathbf{v}') \in Sol(T_{\mathbf{X}=\mathbf{x}})$.

The last clause formulates an interventionist interpretation of subjunctive conditionals. In words, it says that a subjunctive conditional is true in a causal model relative to a context (and a solution to the model in that context) just in case the consequent holds in all solutions to the submodel where the antecedent is enforced relative to the context. Note that unlike [5], we take semantic evaluation to be relative to a solution rather than just a context. This does not make a difference if we are confined to the set of Pearlian causal models, i.e., models that have a unique solution relative to every context. However, since we go beyond Pearlian causal models, this change is important. We restrict to Solutionful models to facilitate such a formulation.

How this semantics is related to the popular Stalnaker-Lewis semantics for subjunctive conditionals is then an interesting question. Following [5], we adopt an ordering-semantic formulation of possible worlds models.

Definition 5 (Lewisian Possible Worlds Model). *A Lewisian possible worlds model over a signature* $S = \langle \mathbf{U}, \mathbf{V}, \mathcal{R} \rangle$ *is a triple* $\langle \Omega, R, \pi \rangle$, *where* Ω *is a set of worlds,* π *is an assignment that for each* $X \in \mathbf{U} \cup \mathbf{V}$ *and each* $w \in \Omega$, *assigns a unique value* $x \in \mathcal{R}(X)$ *to* X *(i.e.,* $\pi(X, w) = x$*), and* R *associates with each* $w \in \Omega$ *a total preorder* $\preceq_w$ *over* $\Omega_w \subseteq \Omega$, *such that* $w \in \Omega_w$ *and for every* $v \neq w \in \Omega_w$, $w \prec_w v$ *(i.e.,* $w \preceq_w v$ *and* $v \npreceq_w w$*).*

Intuitively, Ω_w represents the set of worlds that are accessible from w, and $\preceq_w$ represents an ordering of "overall comparative similarity": $u \preceq_w v$ represents that u is at least as similar or close to w as v is. The definition imposes a "centering" constraint on models in that w itself is required to be closer to w than any other world is.

Given a model M, a world w, and a formula of the form $\mathbf{X} = \mathbf{x}$, let $C_M(w, \mathbf{X} = \mathbf{x}) = \{v | v \in \Omega_w, \pi(\mathbf{X}, v) = \mathbf{x}$, and $v \preceq_w v'$ for every $v' \in \Omega_w$ such that $\pi(\mathbf{X}, v') = \mathbf{x}\}$; that is, $C_M(w, \mathbf{X} = \mathbf{x})$ is the set of $\mathbf{X} = \mathbf{x}$-worlds that are closest to w. Notice that $C_M(w, \mathbf{X} = \mathbf{x}) = \emptyset$ if and only if there is no $\mathbf{X} = \mathbf{x}$-world in Ω_w. Given a model $M = \langle \Omega, R, \pi \rangle$ and a world $w \in \Omega$, each wff in $\mathcal{L}(S)$ gets a truth value according to the following rules:

- $M, w \models X = x$ iff $\pi(X, w) = x$.
- $M, w \models \neg\alpha$ iff $M, w \not\models \alpha$.
- $M, w \models \alpha \wedge \beta$ iff $M, w \models \alpha$ and $M, w \models \beta$.
- $M, w \models \mathbf{X} = \mathbf{x} \mathrel{\Box\!\!\to} \beta$ iff $M, v \models \beta$ for every $v \in C_M(w, \mathbf{X} = \mathbf{x})$.

Now we can define the main target with which this paper is concerned.

Definition 6 (Lewisian Causal Model). *A (Solutionful) causal model* T *over a signature* $S = \langle \mathbf{U}, \mathbf{V}, \mathcal{R} \rangle$ *is called Lewisian if there is a Lewisian possible worlds model over* S, $M = \langle \Omega, R, \pi \rangle$, *and a function* $\mu : Sol(T) \to \Omega$, *such that for every* $\psi \in \mathcal{L}(S)$ *and* $(\mathbf{u}, \mathbf{v}) \in Sol(T)$, $T, (\mathbf{u}, \mathbf{v}) \models \psi$ *iff* $M, \mu(\mathbf{u}, \mathbf{v}) \models \psi$.

In plain words, a causal model is Lewisian if there is a corresponding Lewisian possible worlds model that satisfies the exact same formulas (in $\mathcal{L}$) as the causal model does. As already noted, this notion of a Lewisian causal model is not the same as the "Lewisian causal model" defined in [15]. The latter picks out a proper superclass of the class of causal models that concern us here and is arguably too wide. Let $\mathcal{T}_{\mathrm{L}}$ be the set of Lewisian causal models as defined here. We provide a characterization of $\mathcal{T}_{\mathrm{L}}$ in the next section.

3 Characterizing Lewisian Causal Models

We first establish some necessary conditions for a causal model to be Lewisian. As previously mentioned, a necessary condition is that all submodels are also Solutionful (which was first established in [15], but in a different setup).

Lemma 1. *A causal model is Lewisian only if all its submodels are Solutionful.*

Proof. Suppose that some submodel $T_{\mathbf{X}=\mathbf{x}}$ of a (Solutionful) causal model T is not Solutionful, i.e., $T_{\mathbf{X}=\mathbf{x}}$ relative to some context $\mathbf{u}$ has no solution. It follows that for any $(\mathbf{u}, \mathbf{v}) \in Sol(T)$, $T, (\mathbf{u}, \mathbf{v}) \models \mathbf{X} = \mathbf{x} \,\Box\!\!\rightarrow \neg\mathbf{X} = \mathbf{x}$. On the other hand, let $\mathbf{v}'$ be any value configuration of $\mathbf{V}$ that is consistent with $\mathbf{X} = \mathbf{x}$. It obviously holds that $T, (\mathbf{u}, \mathbf{v}) \models \mathbf{V} = \mathbf{v}' \Diamond\!\!\rightarrow \mathbf{X} = \mathbf{x}$.

Then it is easy to see that T is not Lewisian. Suppose for contradiction that it is. Then there is a Lewisian possible worlds model M and a solution-world mapping μ such that $T, (\mathbf{u}, \mathbf{v}) \models \psi$ iff $M, \mu(\mathbf{u}, \mathbf{v}) \models \psi$ for every $\psi \in \mathcal{L}(S)$. Let $w = \mu(\mathbf{u}, \mathbf{v})$, it follows that $M, w \models \mathbf{X} = \mathbf{x} \,\Box\!\!\rightarrow \neg\mathbf{X} = \mathbf{x}$ and $M, w \models \mathbf{V} = \mathbf{v}' \Diamond\!\!\rightarrow \mathbf{X} = \mathbf{x}$. There is a world $w' \in C_M(w, \mathbf{V} = \mathbf{v}')$ such that $M, w' \models \mathbf{X} = \mathbf{x}$. Therefore, in Ω_w there is a world satisfying $\mathbf{X} = \mathbf{x}$, which implies that $M, w \not\models \mathbf{X} = \mathbf{x} \,\Box\!\!\rightarrow \neg\mathbf{X} = \mathbf{x}$, contradiction.

The next necessary condition was also introduced in [15], in terms of the following property:

Definition 7 (Solution-Conservative). *A causal model T is said to be Solution-Conservative if for every $\mathbf{X} \subseteq \mathbf{V}$, every value configuration $\mathbf{x}$, and every context $\mathbf{u}$, if T has a solution relative to $\mathbf{u}$ that assigns value $\mathbf{x}$ to $\mathbf{X}$, then every solution to $T_{\mathbf{X}=\mathbf{x}}$ relative to $\mathbf{u}$ is also a solution to T relative to $\mathbf{u}$.*

In plain words, no new solution should arise if some variables were fixed by an intervention to values that could have obtained without the intervention.

Lemma 2. *A causal model is Lewisian only if all its submodels are Solution-Conservative.*

Proof. Suppose that some submodel $T_{\mathbf{X}=\mathbf{x}}$ of a causal model T is not Solution-Conservative. That is, one of the solutions to $T_{\mathbf{X}=\mathbf{x}}$ relative to some context $\mathbf{u}$ assigns $\mathbf{y}$ to $\mathbf{Y}$, but $T_{\mathbf{X}=\mathbf{x},\mathbf{Y}=\mathbf{y}}$ has a solution that is not a solution to $T_{\mathbf{X}=\mathbf{x}}$ relative to $\mathbf{u}$. Since $Sol(T_{\mathbf{X}=\mathbf{x}})$ is non-empty and finite, we can write it as $\{(\mathbf{u}, \mathbf{v^1}), ..., (\mathbf{u}, \mathbf{v^n})\}$; then $T, (\mathbf{u}, \mathbf{v}) \models \mathbf{X} = \mathbf{x} \,\Box\!\!\rightarrow (\mathbf{V} = \mathbf{v^1} \vee ... \vee \mathbf{V} = \mathbf{v^n})$, $T, (\mathbf{u}, \mathbf{v}) \models \mathbf{X} = \mathbf{x} \Diamond\!\!\rightarrow \mathbf{Y} = \mathbf{y}$, but $T, (\mathbf{u}, \mathbf{v}) \not\models (\mathbf{X} = \mathbf{x} \wedge \mathbf{Y} = \mathbf{y}) \,\Box\!\!\rightarrow (\mathbf{V} = \mathbf{v^1} \vee ... \vee \mathbf{V} = \mathbf{v^n})$.

Suppose for contradiction that T is Lewisian. Then there is a Lewisian possible worlds model M and a solution-world mapping μ such that $T, (\mathbf{u}, \mathbf{v}) \models \psi$ iff $M, \mu(\mathbf{u}, \mathbf{v}) \models \psi$ for any $\psi \in \mathcal{L}(S)$. Let $w = \mu(\mathbf{u}, \mathbf{v})$, then we have (1) $M, w \models \mathbf{X} = \mathbf{x} \Diamond\!\!\rightarrow \mathbf{Y} = \mathbf{y}$, (2) $M, w \models \mathbf{X} = \mathbf{x} \,\Box\!\!\rightarrow (\mathbf{V} = \mathbf{v^1} \vee ... \vee \mathbf{V} = \mathbf{v^n})$, and (3) $M, w \not\models (\mathbf{X} = \mathbf{x} \wedge \mathbf{Y} = \mathbf{y}) \,\Box\!\!\rightarrow (\mathbf{V} = \mathbf{v^1} \vee ... \vee \mathbf{V} = \mathbf{v^n})$. (1) and (2) imply that there is a world $w_1 \in C_M(w, \mathbf{X} = \mathbf{x})$ such that $M, w_1 \models \mathbf{Y} = \mathbf{y}$ and $M, w_1 \models \mathbf{V} = \mathbf{v^1} \vee ... \vee \mathbf{V} = \mathbf{v^n}$. However, (3) implies that there is a world $w_2 \in C_M(w, \mathbf{X} = \mathbf{x} \wedge \mathbf{Y} = \mathbf{y})$ such that $M, w_2 \not\models \mathbf{V} = \mathbf{v^1} \vee ... \vee \mathbf{V} = \mathbf{v^n}$. Since w_1 satisfies $\mathbf{X} = \mathbf{x} \wedge \mathbf{Y} = \mathbf{y}$, $w_2 \preceq_w w_1$, which entails that $w_2 \in C_M(w, \mathbf{X} = \mathbf{x})$ (because $w_1 \in C_M(w, \mathbf{X} = \mathbf{x})$). It then follows from (2) that $M, w_2 \models \mathbf{V} = \mathbf{v^1} \vee ... \vee \mathbf{V} = \mathbf{v^n}$, contradiction.

The next condition has to do with the Lewisian requirement of centering.

Definition 8 (Solution-Determinate). *A causal model is Solution-Determinate if the model has at most one solution relative to every context.*

Lemma 3. *A causal model is Lewisian only if it is Solution-Determinate.*

Proof. Suppose that a causal model T is not Solution-Determinate. Then relative to some context $\mathbf{u}$, T has at least two solutions. Let $\mathbf{v_1}$ and $\mathbf{v_2}$ be two different solutions to T relative to $\mathbf{u}$. We have $T, (\mathbf{u}, \mathbf{v_1}) \models (true \diamond\!\!\rightarrow \mathbf{V} = \mathbf{v_1}) \wedge (true \diamond\!\!\rightarrow \mathbf{V} = \mathbf{v_2})$.

Suppose for contradiction that T is Lewisian, that is, there is a Lewisian possible worlds model M and a function μ such that $T, (\mathbf{u}, \mathbf{v_1}) \models \psi$ iff $M, \mu(\mathbf{u}, \mathbf{v_1}) \models \psi$ for any $\psi \in \mathcal{L}(S)$. Let w be $\mu(\mathbf{u}, \mathbf{v_1})$. Then $M, w \models (true \diamond\!\!\rightarrow \mathbf{V} = \mathbf{v_1}) \wedge (true \diamond\!\!\rightarrow \mathbf{V} = \mathbf{v_2})$. However, since $w \in C_M(w, true)$, it cannot have any other element due to the requirement of centering in Definition 5. It follows that $M, w \models \mathbf{V} = \mathbf{v_1} \wedge \mathbf{V} = \mathbf{v_2}$, contradiction.

Finally, in view of Halpern's [5, pp. 317–8] example showing that some Pearlian models are not Lewisian (see also [16]), we need another condition.

Definition 9 (Solution-Transitive in Cycles). *A causal model T is Solution-Transitive in Cycles if for every $\mathbf{X_1},...,\mathbf{X_k} \subseteq \mathbf{V}$, every value configuration $\mathbf{x_i}$ of $\mathbf{X_i}$ ($i = 1,...,k$), and every context $\mathbf{u}$, if $T_{\mathbf{X_i}=\mathbf{x_i}}$ has a solution relative to $\mathbf{u}$ that is consistent with $\mathbf{X_{i+1}} = \mathbf{x_{i+1}}$ ($i = 1,...,k-1$) and $T_{\mathbf{X_k}=\mathbf{x_k}}$ has a solution relative to $\mathbf{u}$ that is consistent with $\mathbf{X_1} = \mathbf{x_1}$, then $T_{\mathbf{X_1}=\mathbf{x_1}}$ has a solution relative to $\mathbf{u}$ that is consistent with $\mathbf{X_k} = \mathbf{x_k}$.*

Lemma 4. *A causal model is Lewisian only if it is Solution-Transitive in Cycles.*

Proof. Suppose that a (Solutionful) causal model T is not Solution-Transitive in Cycles. Then there are $\mathbf{X_1},...,\mathbf{X_k} \subseteq \mathbf{V}$, some value configurations $\mathbf{x_1},...,\mathbf{x_k}$ and some context $\mathbf{u}$ such that $T_{\mathbf{X_i}=\mathbf{x_i}}$ has a solution relative to $\mathbf{u}$ that is consistent with $\mathbf{X_{i+1}} = \mathbf{x_{i+1}}$ and $T_{\mathbf{X_k}=\mathbf{x_k}}$ has a solution relative to $\mathbf{u}$ that is consistent with $\mathbf{X_1} = \mathbf{x_1}$, but no solution to $T_{\mathbf{X_1}=\mathbf{x_1}}$ relative to $\mathbf{u}$ is consistent with $\mathbf{X_k} = \mathbf{x_k}$. It follows that for any $(\mathbf{u}, \mathbf{v}) \in Sol(T)$,

$$T, (\mathbf{u}, \mathbf{v}) \models \bigwedge_{i=1}^{k-1} \mathbf{X_i} = \mathbf{x_i} \diamond\!\!\rightarrow \mathbf{X_{i+1}} = \mathbf{x_{i+1}} \wedge \mathbf{X_k} = \mathbf{x_k} \diamond\!\!\rightarrow \mathbf{X_1} = \mathbf{x_1} \tag{1}$$

but $T, (\mathbf{u}, \mathbf{v}) \not\models \mathbf{X_1} = \mathbf{x_1} \diamond\!\!\rightarrow \mathbf{X_k} = \mathbf{x_k}$.

Suppose for contradiction that T is Lewisian, that is, there is a Lewisian possible worlds model M and a function μ such that $T, (\mathbf{u}, \mathbf{v}) \models \psi$ iff $M, \mu(\mathbf{u}, \mathbf{v}) \models \psi$ for every $\psi \in \mathcal{L}(S)$. Let $w = \mu(\mathbf{u}, \mathbf{v})$, thus

$$M, w \models \bigwedge_{i=1}^{k-1} \mathbf{X_i} = \mathbf{x_i} \diamond\!\!\rightarrow \mathbf{X_{i+1}} = \mathbf{x_{i+1}} \wedge \mathbf{X_k} = \mathbf{x_k} \diamond\!\!\rightarrow \mathbf{X_1} = \mathbf{x_1}$$

but $M, w \not\models \mathbf{X_1} = \mathbf{x_1} \diamond\!\!\rightarrow \mathbf{X_k} = \mathbf{x_k}$. The former implies that there is some $w_i \in C_M(w, \mathbf{X_i} = \mathbf{x_i})$ such that $M, w_i \models \mathbf{X_{i+1}} = \mathbf{x_{i+1}}$ and there is some $w_k \in$

$C_M(w, \mathbf{X_k} = \mathbf{x_k})$ such that $M, w_k \models \mathbf{X_1} = \mathbf{x_1}$. Hence $w_1 \preceq_w w_k \preceq_w ... \preceq_w w_1$. Since $w_1 \in C_M(w, \mathbf{X_1} = \mathbf{x_1})$, it follows that $w_k \in C_M(w, \mathbf{X_1} = \mathbf{x_1})$. But then $M, w \models \mathbf{X_1} = \mathbf{x_1} \diamond\!\!\rightarrow \mathbf{X_k} = \mathbf{x_k}$, contradiction.

These four necessary conditions turn out to be jointly sufficient for a causal model to be Lewisian.

Theorem 1. *A causal model is Lewisian if and only if it is Solution-Determinate and Solution-Transitive in Cycles, and all its submodels are Solutionful and Solution-Conservative.*

Proof. The "only if" direction is established by Lemmas 1–4. The proof of the "if" direction is a bit lengthy and can be found in the Appendix.

4 Axiomatization

We now present an axiomatic system for the class of Lewisian causal models. The following results present characteristic axioms for the four conditions in the characterization of Lewisian causal models.

Lemma 5. *A causal model T is Solution-Determinate if and only if for every $(\mathbf{u}, \mathbf{v}) \in Sol(T)$, $T, (\mathbf{u}, \mathbf{v}) \models \bigvee_{x \in \mathcal{R}(X)} (true \;\Box\!\!\rightarrow X = x)$.*

Proof. From left to right, suppose that T is Solution-Determinate. Then for any context $\mathbf{u}$, the solutions of T are all the same (note that T is presupposed to be Solutionful). In other words, the solutions of T relative to $\mathbf{u}$ assign the same value to every endogenous variable. Thus $T, (\mathbf{u}, \mathbf{v}) \models \bigvee_{x \in \mathcal{R}(X)} (true \;\Box\!\!\rightarrow X = x)$. The other direction is similarly easy to verify.

Lemma 6. *A causal model T is Solution-Transitive in Cycles if and only if for every $(\mathbf{u}, \mathbf{v}) \in Sol(T)$, $T, (\mathbf{u}, \mathbf{v}) \models (\mathbf{X_1} = \mathbf{x_1} \diamond\!\!\rightarrow \mathbf{X_2} = \mathbf{x_2}) \wedge ... \wedge (\mathbf{X_{k-1}} = \mathbf{x_{k-1}} \diamond\!\!\rightarrow \mathbf{X_k} = \mathbf{x_k}) \wedge (\mathbf{X_k} = \mathbf{x_k} \diamond\!\!\rightarrow \mathbf{X_1} = \mathbf{x_1}) \Rightarrow (\mathbf{X_1} = \mathbf{x_1} \diamond\!\!\rightarrow \mathbf{X_k} = \mathbf{x_k})$.*

Proof. From left to right, suppose T is Solution-Transitive in Cycles. Assume that $T, (\mathbf{u}, \mathbf{v}) \models (\mathbf{X_1} = \mathbf{x_1} \diamond\!\!\rightarrow \mathbf{X_2} = \mathbf{x_2}) \wedge ... \wedge (\mathbf{X_{k-1}} = \mathbf{x_{k-1}} \diamond\!\!\rightarrow \mathbf{X_k} = \mathbf{x_k}) \wedge (\mathbf{X_k} = \mathbf{x_k} \diamond\!\!\rightarrow \mathbf{X_1} = \mathbf{x_1})$ for any context $\mathbf{u}$, then $T_{\mathbf{X_i} = \mathbf{x_i}}$ has a solution relative to $\mathbf{u}$ that is consistent with $T_{\mathbf{X_{i+1}} = \mathbf{x_{i+1}}}$ $(1 \leq i \leq k-1)$. It follows that $T_{\mathbf{X_1} = \mathbf{x_1}}$ has a solution relative to $\mathbf{u}$ that is consistent with $T_{\mathbf{X_k} = \mathbf{x_k}}$. That means $T, (\mathbf{u}, \mathbf{v}) \models \mathbf{X_1} = \mathbf{x_1} \diamond\!\!\rightarrow \mathbf{X_k} = \mathbf{x_k}$. The other direction is similarly straightforward.

Lemma 7. *Every submodel of a causal model T is Solutionful if and only if for every $(\mathbf{u}, \mathbf{v}) \in Sol(T)$, $T, (\mathbf{u}, \mathbf{v}) \models \bigvee_{y \in \mathcal{R}(Y)} \mathbf{X} = \mathbf{x} \diamond\!\!\rightarrow Y = y$.*

Lemma 8. *Every submodel of a causal model T is Solution-Conservative if and only if for every $(\mathbf{u}, \mathbf{v}) \in Sol(T)$, $T, (\mathbf{u}, \mathbf{v}) \models (\mathbf{X} = \mathbf{x} \diamond\!\!\rightarrow \mathbf{W} = \mathbf{w} \wedge (\mathbf{X} = \mathbf{x} \wedge \mathbf{W} = \mathbf{w}) \diamond\!\!\rightarrow \mathbf{Y} = \mathbf{y}) \Rightarrow \mathbf{X} = \mathbf{x} \diamond\!\!\rightarrow (\mathbf{W} = \mathbf{w} \wedge \mathbf{Y} = \mathbf{y})$.*

Lemmas 7 and 8 are essentially the same as Lemmas 4 and 5 in [15], respectively, and we omit the proofs.

Let $\mathcal{T}_{\mathrm{L}}(S)$ denote the class of Lewisian causal models relative to a signature S. An axiomatic system for $\mathcal{T}_{\mathrm{L}}(S)$ in the language $\mathcal{L}(S)$, $AX_{\mathrm{L}}(S)$, consists of the following axiom schemas:

L0 All instances of propositional tautologies

L1 $\bigvee_{y\in\mathcal{R}(Y)} \mathbf{X}=\mathbf{x} \diamond\!\!\to Y=y$

L2 $\mathbf{X}=\mathbf{x} \diamond\!\!\to Y=y \Rightarrow \neg\mathbf{X}=\mathbf{x} \diamond\!\!\to Y=y'$ where $\mathbf{X}=\mathbf{V}\backslash\{Y\}$ and $y\neq y'$

L3 $(\bigvee_{x\in\mathcal{R}(X)} X=x) \wedge (X=x \Rightarrow X\neq x')$ where $x\neq x'$

L4 $\mathbf{V}=\mathbf{v} \Rightarrow \mathbf{V}_{V_1}=\mathbf{v} \diamond\!\!\to V_1=v_1 \wedge \ldots \wedge \mathbf{V}_{V_n}=\mathbf{v} \diamond\!\!\to V_n=v_n$ where $\mathbf{V}_{V_i}=\mathbf{V}\backslash\{V_i\}$, $\mathbf{V}=\{V_1,\ldots,V_n\}$ and v_i is the value of V_i in $\mathbf{v}$

L5 $\mathbf{X}=\mathbf{x} \diamond\!\!\to (\beta_1 \vee \beta_2) \Leftrightarrow (\mathbf{X}=\mathbf{x} \diamond\!\!\to \beta_1) \vee (\mathbf{X}=\mathbf{x} \diamond\!\!\to \beta_2)$

L6 $\mathbf{X}=\mathbf{x} \diamond\!\!\to \mathbf{Y}=\mathbf{y} \Leftrightarrow \bigvee_{\mathbf{z}\in\mathcal{R}(\mathbf{Z})} \mathbf{X}=\mathbf{x} \diamond\!\!\to (\mathbf{Y}=\mathbf{y} \wedge \mathbf{Z}=\mathbf{z})$ where $\mathbf{Z}=\mathbf{V}\backslash(\mathbf{X}\cup\mathbf{Y})$

L7 $\mathbf{X}=\mathbf{x} \diamond\!\!\to (Y=y \wedge \mathbf{Z}=\mathbf{z}) \Rightarrow (\mathbf{X}=\mathbf{x} \wedge Y=y) \diamond\!\!\to \mathbf{Z}=\mathbf{z}$

L8 $((\mathbf{X}=\mathbf{x} \wedge Y=y) \diamond\!\!\to (W=w \wedge \mathbf{Z}=\mathbf{z}) \wedge (\mathbf{X}=\mathbf{x} \wedge W=w) \diamond\!\!\to (Y=y \wedge \mathbf{Z}=\mathbf{z})) \Rightarrow \mathbf{X}=\mathbf{x} \diamond\!\!\to (Y=y \wedge W=w \wedge \mathbf{Z}=\mathbf{z})$ where $\mathbf{Z}=\mathbf{V}\backslash(\mathbf{X}\cup\{Y,W\})$

L9 $(\mathbf{X}=\mathbf{x} \diamond\!\!\to \mathbf{W}=\mathbf{w} \wedge (\mathbf{X}=\mathbf{x} \wedge \mathbf{W}=\mathbf{w}) \diamond\!\!\to \mathbf{Y}=\mathbf{y}) \Rightarrow \mathbf{X}=\mathbf{x} \diamond\!\!\to (\mathbf{W}=\mathbf{w} \wedge \mathbf{Y}=\mathbf{y})$

L10 $(\mathbf{X_1}=\mathbf{x_1} \diamond\!\!\to \mathbf{X_2}=\mathbf{x_2}) \wedge \ldots \wedge (\mathbf{X_{k-1}}=\mathbf{x_{k-1}} \diamond\!\!\to \mathbf{X_k}=\mathbf{x_k}) \wedge (\mathbf{X_k}=\mathbf{x_k} \diamond\!\!\to \mathbf{X_1}=\mathbf{x_1}) \Rightarrow (\mathbf{X_1}=\mathbf{x_1} \diamond\!\!\to \mathbf{X_k}=\mathbf{x_k})$

L11 $\bigvee_{x\in\mathcal{R}(X)} (true \;\Box\!\!\to X=x)$

And two rules of inference:

MP From $\vdash \alpha \Rightarrow \beta$ and $\vdash \alpha$, infer $\vdash \beta$

RE From $\vdash \beta_1 \Leftrightarrow \beta_2$, infer $\vdash (\mathbf{X}=\mathbf{x} \diamond\!\!\to \beta_1) \Leftrightarrow (\mathbf{X}=\mathbf{x} \diamond\!\!\to \beta_2)$

We can think of $AX_{\mathrm{L}}(S)$ as an extension of Halpern's system for the class of all causal models over S [4, pp. 325–6]. Specifically, if we use the language $\mathcal{L}(S)$ and the satisfaction relation defined above, the original system for all causal models in [4] corresponds to the system consisting of L0, L2-L8, MP, RE and a special L1 where $\mathbf{X}=\mathbf{V}\backslash\{Y\}$. Naturally, the axiomatic system for Lewisian causal models can be obtained by adding the axioms L1, L9, L10 and L11, the characteristic formulas for the four conditions characterizing Lewisian causal models, into the system for all causal models.

Note that the form of L2 is different from the corresponding Functionality axiom (D1) in Halpern's system for the full class of causal models [4]. We choose this form to slightly simplify the completeness proof. It is also worth mentioning that L5, L6 and RE were not explicitly listed by Halpern, but he seemed to have implicitly used them in his proof of completeness.

Theorem 2. *$AX_L(S)$ is sound and complete with respect to $\mathcal{T}_L(S)$.*

Proof. See the appendix.

5 Conclusion

Given the prominent status of the Stalnaker-Lewis semantics of counterfactuals and the increasing influence of the causal modelling approach, it is instructive and useful to gain a better understanding of the connections between the two. We have added to this understanding by providing a characterization of Lewisian causal models, as well as an axiomatization of this class of causal models, following Halpern's [4,5] seminal work.

Let us end with two observations on the implications of our results, one about a proper subclass and the other about a proper superclass of the class of Lewisian causal models. First, Pearlian causal models obviously satisfy the conditions of Theorem 1, except possibly that of solution-transitivity in cycles. So an easy corollary is that Pearlian causal models are Lewisian if and only if they are solution-transitive in cycles. In fact, given the "unique solution" property of Pearlian causal models, models that are both Pearlian and Lewisian are Stalnakerian ([12]), for which the total preorder $\preceq_w$ required in Lewisian possible worlds models can be strengthened into a linear order. The characterization of Stalnakerian causal models is thus simply Pearlian models that are solution-transitive in cycles. Interestingly, since Pearlian models do not allow any mutual counterfactual dependence between distinct events due to their satisfaction of the so-called reversibility axiom ([16]), the further requirement of solution-transitivity in cycles in effect rules out all cycles of counterfactual dependence, for any such cycle would yield mutual counterfactual dependence. An implication is that although a Stalnakerian causal model can be non-recursive at the variable level, it in a sense must be recursive at the value-of-variable level.

Second, our characterization in Theorem 1 reveals a seemingly peculiar feature of Lewisian causal models. They must be solution-determinate even though their submodels need not be. If we take away the requirement of solution-determinateness, we get a superclass that seems more "natural" in its characterization. As the proof of Lemma 3 shows, solution-determinateness corresponds to the Lewisian requirement of centering. It is then easy to show that the more "natural" class of causal models corresponds to a relaxation of centering to weak centering. It remains controversial in the philosophical literature, but there have been several proposals to relax centering to weak centering in broadly counterfactual accounts of causation, especially in connection to mental or supervenient causation (e.g., [10,17]). It is therefore interesting to observe that our characterization suggests that weak centering is in a way more natural than centering in the context of causal models.

Appendix: Proofs of Theorems 1 and 2

Proof of Theorem 1

The "only if" direction is established by Lemmas 1–4. Here we focus on the "if" direction. Suppose that a causal model T over signature S is Solution-Determinate and Solution-Transitive in Cycles, and all its submodels are Solutionful and Solution-Conservative. We first construct a possible worlds model

$M = \langle \Omega, R, \pi \rangle$ over S and prove it is Lewisian. Let Ω be the set of all possible value assignments to $\mathbf{U} \cup \mathbf{V}$. The definition of π is obvious.

To define R, we use the following notations. Since T is Solution-Determinate and every submodel of T is Solutionful, we use $\mathbf{s(u)}$ to denote the unique solution to T relative to a context $\mathbf{u}$. Let $\mathbf{s(u, x \mid v)}$ denote one solution $\mathbf{v}$ to $T_{\mathbf{X}=\mathbf{x}}$ relative to context $\mathbf{u}$. Let $\Omega_{\mathbf{s(u)}}$ be the set of worlds of the form $\mathbf{s(u, x \mid v)}$ where $\mathbf{x}$ is the value configuration of $\mathbf{X}$ and $\mathbf{v}$ is one of the solutions of $T_{\mathbf{X}=\mathbf{x}}$ relative to context $\mathbf{u}$.[1] For a world $\mathbf{s(u)}$, we define $\preceq^0_{\mathbf{s(u)}}$ over $\Omega_{\mathbf{s(u)}}$ as follows[2]: $\mathbf{s(u, x \mid v)} \preceq^0_{\mathbf{s(u)}} \mathbf{s(u, y \mid v')}$ iff $\mathbf{s(u, y \mid v')}$ assigns $\mathbf{x}$ to $\mathbf{X}$. Let $\preceq^1_{\mathbf{s(u)}}$ be the transitive closure of $\preceq^0_{\mathbf{s(u)}}$. Then we inductively define $\preceq_{\mathbf{s(u)}}$ as below. If $\preceq^i_{\mathbf{s(u)}}$ is not yet strongly-connected, let w_{ia} and w_{ib} be two incomparable worlds, and let $A_i = \{w \in \Omega_{\mathbf{s(u)}} \mid w \preceq^i_{\mathbf{s(u)}} w_{ia}\}$ and $B_i = \{w \in \Omega_{\mathbf{s(u)}} \mid w_{ib} \preceq^i_{\mathbf{s(u)}} w\}$. Then define $\preceq^{i+1}_{\mathbf{s(u)}} := \preceq^i_{\mathbf{s(u)}} \cup (A_i \times B_i)$. Let $\preceq_{\mathbf{s(u)}}$ be the first in this process that is strongly-connected.

We assert that R associates with each $\mathbf{s(u)}$ a total preorder $\preceq_{\mathbf{s(u)}}$ over $\Omega_{\mathbf{s(u)}}$ such that $\mathbf{s(u)} \in \Omega_{\mathbf{s(u)}}$ and $\mathbf{s(u)} \prec_{\mathbf{s(u)}} v$ for every $v \neq \mathbf{s(u)} \in \Omega_{\mathbf{s(u)}}$. $\preceq_{\mathbf{s(u)}}$ is strongly-connected by its construction. We now show that $\preceq^i_{\mathbf{s(u)}}$ is transitive by induction. The base case $\preceq^1_{\mathbf{s(u)}}$ is obvious. Assume that $\preceq^k_{\mathbf{s(u)}}$ is transitive, we show that so is $\preceq^{k+1}_{\mathbf{s(u)}}$. Given any $\mathbf{s(u, x \mid v)}, \mathbf{s(u, y \mid v')}, \mathbf{s(u, z \mid v'')} \in \Omega_{\mathbf{s(u)}}$, suppose $\mathbf{s(u, x \mid v)} \preceq^{k+1}_{\mathbf{s(u)}} \mathbf{s(u, y \mid v')}$ and $\mathbf{s(u, y \mid v')} \preceq^{k+1}_{\mathbf{s(u)}} \mathbf{s(u, z \mid v'')}$. There are four cases to consider:

Case 1: $\mathbf{s(u, x \mid v)} \preceq^k_{\mathbf{s(u)}} \mathbf{s(u, y \mid v')}$ and $\mathbf{s(u, y \mid v')} \preceq^k_{\mathbf{s(u)}} \mathbf{s(u, z \mid v'')}$. By the inductive hypothesis, $\mathbf{s(u, x \mid v)} \preceq^k_{\mathbf{s(u)}} \mathbf{s(u, z \mid v'')}$. So $\mathbf{s(u, x \mid v)} \preceq^{k+1}_{\mathbf{s(u)}} \mathbf{s(u, z \mid v'')}$.

Case 2: $\mathbf{s(u, x \mid v)} \preceq^k_{\mathbf{s(u)}} \mathbf{s(u, y \mid v')}$ and $(\mathbf{s(u, y \mid v')}, \mathbf{s(u, z \mid v'')}) \in A_k \times B_k$. There are two $\preceq^k_{\mathbf{s(u)}}$-incomparable worlds w_{ka} and w_{kb} such that $\mathbf{s(u, y \mid v')} \preceq^k_{\mathbf{s(u)}} w_{ka}$ and $w_{kb} \preceq^k_{\mathbf{s(u)}} \mathbf{s(u, z \mid v'')}$, thus $\mathbf{s(u, x \mid v)} \preceq^k_{\mathbf{s(u)}} w_{ka}$. It follows that $(\mathbf{s(u, x \mid v)}, \mathbf{s(u, z \mid v'')}) \in A_k \times B_k$, and so $\mathbf{s(u, x \mid v)} \preceq^{k+1}_{\mathbf{s(u)}} \mathbf{s(u, z \mid v'')}$.

Case 3: $(\mathbf{s(u, x \mid v)}, \mathbf{s(u, y \mid v')}) \in A_k \times B_k$ and $\mathbf{s(u, y \mid v')} \preceq^k_{\mathbf{s(u)}} \mathbf{s(u, z \mid v'')}$. Similar to Case 2.

Case 4: $(\mathbf{s(u, x \mid v)}, \mathbf{s(u, y \mid v')}) \in A_k \times B_k$ and $(\mathbf{s(u, y \mid v')}, \mathbf{s(u, z \mid v'')}) \in A_k \times B_k$. Then $w_{kb} \preceq^k_{\mathbf{s(u)}} \mathbf{s(u, y \mid v')}$ and $\mathbf{s(u, y \mid v')} \preceq^k_{\mathbf{s(u)}} w_{ka}$, and so $w_{kb} \preceq^k_{\mathbf{s(u)}} w_{ka}$, which contradicts the supposition that w_{ka} and w_{kb} are $\preceq^k_{\mathbf{s(u)}}$-incomparable.

Therefore, $\preceq^{k+1}_{\mathbf{s(u)}}$ is also transitive. By construction, so is $\preceq_{\mathbf{s(u)}}$.

Next, we prove the other Lewisian constraints on R. Obviously, $\mathbf{s(u)} \in \Omega_{\mathbf{s(u)}}$, and for every $v \in \Omega_{\mathbf{s(u)}}$, $\mathbf{s(u)} \preceq^0_{\mathbf{s(u)}} v$, so $\mathbf{s(u)} \preceq_{\mathbf{s(u)}} v$. Suppose that there is some

[1] $\mathbf{X}$ could be empty so that $\mathbf{s(u)}$ is in $\Omega_{\mathbf{s(u)}}$.

[2] For the other worlds $\mathbf{s(u, x \mid v)} \in \Omega$, we can find a $\Omega_{\mathbf{s(u,x|v)}}$ such that $\mathbf{s(u, x \mid v)} \in \Omega_{\mathbf{s(u,x|v)}}$ and arbitrarily define a total preorder $\preceq_{\mathbf{s(u,x|v)}}$ such that $\mathbf{s(u, x \mid v)} \prec_{\mathbf{s(u,x|v)}} v$ for any $v \neq \mathbf{s(u, x \mid v)} \in \Omega_{\mathbf{s(u,x|v)}}$.

$\mathbf{s(u,x \mid v)}$ such that $\mathbf{s(u,x \mid v)} \neq \mathbf{s(u)}$ and $\mathbf{s(u,x \mid v)} \preceq_{\mathbf{s(u)}} \mathbf{s(u)}$. That means either there exist $\mathbf{s(u,x_1 \mid v_1)},\ldots, \mathbf{s(u,x_m \mid v_m)}$ such that $\mathbf{s(u,x \mid v)} \preceq^0_{\mathbf{s(u)}} \mathbf{s(u,x_1 \mid v_1)} \preceq^0_{\mathbf{s(u)}} \cdots \preceq^0_{\mathbf{s(u)}} \mathbf{s(u,x_m \mid v_m)} \preceq^0_{\mathbf{s(u)}} \mathbf{s(u)}$ or there are two $\preceq^0_{\mathbf{s(u)}}$-incomparable worlds in the chain from $\mathbf{s(u,x \mid v)}$ to $\mathbf{s(u)}$. For the former case, as $\mathbf{s(u)}$ assigns $\mathbf{x_m}$ to $\mathbf{X_m}$ and T is Solution-Conservative, $\mathbf{v_m}$ is the solution to T relative to $\mathbf{u}$ due to Solution-Determinateness. Thus $\mathbf{s(u,x_m \mid v_m)} = \mathbf{s(u)}$. Repeat the same argument, we would have $\mathbf{s(u,x \mid v)} = \mathbf{s(u)}$, a contradiction. For the latter case, for any world $\mathbf{s(u,x_k \mid v_k)}$ such that $\mathbf{s(u,x_k \mid v_k)} \preceq^0_{\mathbf{s(u)}} \mathbf{s(u)}$, $\mathbf{s(u,x_k \mid v_k)} = \mathbf{s(u)}$. Assume that the pair of $\preceq^0_{\mathbf{s(u)}}$-incomparable worlds that are closest to $\mathbf{s(u)}$ is $s \preceq_{\mathbf{s(u)}} t$ with $t = \mathbf{s(u)}$, then s and t would be $\preceq^0_{\mathbf{s(u)}}$-comparable, also a contradiction.

So the constructed possible worlds model is Lewisian as desired. What remains to be shown is that $T, (\mathbf{u},\mathbf{v}) \models \psi$ iff $M, \mu(\mathbf{u},\mathbf{v}) \models \psi$ for every $\psi \in \mathcal{L}(S)$, where μ assigns $\mathbf{s(u)}(\in \Omega)$ to $(\mathbf{u},\mathbf{v})(\in \mathit{Sol}(T))$. We prove this claim by induction on the structure of ψ.

At first we show that for any $\mathcal{L}(S)$ formula β that does not contain '$\Box\!\!\rightarrow$', $T_{\mathbf{X}=\mathbf{x}}, (\mathbf{u}',\mathbf{v}') \models \beta$ iff $M, \mathbf{s(u',x \mid v')} \models \beta$. For the case of $Y = y$, suppose $T_{\mathbf{X}=\mathbf{x}}, (\mathbf{u}',\mathbf{v}') \models Y = y$, then $\mathbf{v}'$ assigns y to Y. Therefore, we have $M, \mathbf{s(u',x \mid v')} \models Y = y$. Conversely, if $M, \mathbf{s(u',x \mid v')} \models Y = y$, then $Y = y$ is consistent with $\mathbf{v}'$. Thus, $T_{\mathbf{X}=\mathbf{x}}, (\mathbf{u}',\mathbf{v}') \models Y = y$. The Boolean cases are routine.

From the above result, we have $T, (\mathbf{u},\mathbf{v}) \models Y = y$ iff $M, \mu(\mathbf{u},\mathbf{v}) \models Y = y$. We then show that $T, (\mathbf{u},\mathbf{v}) \models \mathbf{X} = \mathbf{x} \Box\!\!\rightarrow \beta$ iff $M, \mathbf{s(u)} \models \mathbf{X} = \mathbf{x} \Box\!\!\rightarrow \beta$. The other cases are straightforward by inductive hypothesis.

From left to right, if $T, (\mathbf{u},\mathbf{v}) \models \mathbf{X} = \mathbf{x} \Box\!\!\rightarrow \beta$, then for all $(\mathbf{u},\mathbf{v^i}) \in \mathit{Sol}(T_{\mathbf{X}=\mathbf{x}})$, $T_{\mathbf{X}=\mathbf{x}}, (\mathbf{u},\mathbf{v^i}) \models \beta$. As β is a formula that does not contain '$\Box\!\!\rightarrow$', we have $M, \mathbf{s(u,x \mid v^i)} \models \beta$ for $(\mathbf{u},\mathbf{v^i})$. Since $\mathbf{s(u,x \mid v^i)} \in \Omega_{\mathbf{s(u)}}$ and $\mathbf{s(u,x \mid v^i)} \preceq^0_{\mathbf{s(u)}} v$ for any $v \in \Omega_{\mathbf{s(u)}}$ such that v assigns $\mathbf{x}$ to $\mathbf{X}$, then $\mathbf{s(u,x \mid v^i)} \preceq_{\mathbf{s(u)}} v$, so $\mathbf{s(u,x \mid v^i)} \in C_M(\mathbf{s(u)}, \mathbf{X} = \mathbf{x})$. Suppose that there is a world $\mathbf{s(u,y \mid v')} \in \Omega_{\mathbf{s(u)}}$ such that $\mathbf{s(u,y \mid v')} \neq \mathbf{s(u,x \mid v^i)}$ where $\mathbf{s(u,x \mid v^i)}$ is the corresponding world of $(\mathbf{u},\mathbf{v^i}) \in \mathit{Sol}(T_{\mathbf{X}=\mathbf{x}})$, and $\mathbf{s(u,y \mid v')} \in C_M(\mathbf{s(u)}, \mathbf{X} = \mathbf{x})$. Then $\mathbf{s(u,y \mid v')}$ assigns $\mathbf{x}$ to $\mathbf{X}$ and $\mathbf{s(u,y \mid v')} \preceq_{\mathbf{s(u)}} \mathbf{s(u,x \mid v^i)}$. As $\mathbf{s(u,x \mid v^i)} \preceq^0_{\mathbf{s(u)}} \mathbf{s(u,y \mid v')}$ according to the definition of $\preceq^0_{\mathbf{s(u)}}$, they cannot be $\preceq^1_{\mathbf{s(u)}}$-incomparable and then cannot be $\preceq^j_{\mathbf{s(u)}}$-incomparable. So, based on the definition of $\preceq_{\mathbf{s(u)}}$, it can only be the case that $\mathbf{s(u,y \mid v')} \preceq^1_{\mathbf{s(u)}} \mathbf{s(u,x \mid v^i)}$ from the fact that $\mathbf{s(u,y \mid v')} \preceq_{\mathbf{s(u)}} \mathbf{s(u,x \mid v^i)}$. Thus there exist $\mathbf{s(u,z_1 \mid v_1)},\ldots, \mathbf{s(u,z_k \mid v_k)}$ such that $\mathbf{s(u,y \mid v')} \preceq^0_{\mathbf{s(u)}} \mathbf{s(u,z_1 \mid v_1)} \preceq^0_{\mathbf{s(u)}} \cdots \preceq^0_{\mathbf{s(u)}} \mathbf{s(u,z_k \mid v_k)} \preceq^0_{\mathbf{s(u)}} \mathbf{s(u,x \mid v^i)}$. Together with the fact that $\mathbf{s(u,x \mid v^i)} \preceq^0_{\mathbf{s(u)}} \mathbf{s(u,y \mid v')}$, that means $T_{\mathbf{X}=\mathbf{x}}$ relative to $\mathbf{u}$ has a solution $\mathbf{v^i}$ that is consistent with $\mathbf{Z_k} = \mathbf{z_k}$,..., $T_{\mathbf{Y}=\mathbf{y}}$ relative to $\mathbf{u}$ has a solution that is consistent with $\mathbf{X} = \mathbf{x}$. Since T is Solution-Transitive in Cycles, $T_{\mathbf{X}=\mathbf{x}}$ has a solution relative to $\mathbf{u}$ that is consistent with $\mathbf{Y} = \mathbf{y}$. Suppose that the solution is $\mathbf{v^k}$, $\mathbf{s(u,x \mid v^k)}$ assigns $\mathbf{y}$ to $\mathbf{Y}$. As $\mathbf{s(u,y \mid v')}$ assigns $\mathbf{x}$ to $\mathbf{X}$, the intervention $\mathbf{X} = \mathbf{x}$ and $\mathbf{Y} = \mathbf{y}$ do not contradict with each other. We have the world $\mathbf{s}(\mathbf{u}, \mathbf{x}, \mathbf{y} \mid \mathbf{v}')$ $(= \mathbf{s}(\mathbf{u}, \mathbf{y}, \mathbf{x} \mid \mathbf{v}'))$. For every submodel of T is Solution-Conservative, $\mathbf{v}'$ is one of the solutions to $T_{\mathbf{X}=\mathbf{x}}$ relative to $\mathbf{u}$, but $\mathbf{s}(\mathbf{u}, \mathbf{y} \mid \mathbf{v}') \neq \mathbf{s}(\mathbf{u}, \mathbf{x} \mid \mathbf{v^i})$, contradiction. Therefore $C_M(\mathbf{s}(\mathbf{u}), \mathbf{X} = \mathbf{x})$ only has the worlds of the form $\mathbf{s}(\mathbf{u}, \mathbf{x} \mid \mathbf{v^i})$. Then we have that for all $v \in C_M(\mathbf{s}(\mathbf{u}), \mathbf{X} = \mathbf{x})$, $M, v \models \beta$, so $M, \mathbf{s}(\mathbf{u}) \models \mathbf{X} = \mathbf{x} \,\Box\!\!\rightarrow \beta$.

From right to left, suppose $M, \mathbf{s}(\mathbf{u}) \models \mathbf{X} = \mathbf{x} \,\Box\!\!\rightarrow \beta$. As we have proved that the elements in $C_M(\mathbf{s}(\mathbf{u}), \mathbf{X} = \mathbf{x})$ are the worlds of the form $\mathbf{s}(\mathbf{u}, \mathbf{x} \mid \mathbf{v^i})$, then for each $\mathbf{s}(\mathbf{u}, \mathbf{x} \mid \mathbf{v^i})$, $M, \mathbf{s}(\mathbf{u}, \mathbf{x} \mid \mathbf{v^i}) \models \beta$. It follows that $T_{\mathbf{X}=\mathbf{x}}, (\mathbf{u}, \mathbf{v^i}) \models \beta$. Then for all $(\mathbf{u}, \mathbf{v^i}) \in Sol(T_{\mathbf{X}=\mathbf{x}})$, $T_{\mathbf{X}=\mathbf{x}}, (\mathbf{u}, \mathbf{v^i}) \models \beta$. Hence, $T, (\mathbf{u}, \mathbf{v}) \models \mathbf{X} = \mathbf{x} \,\Box\!\!\rightarrow \beta$.

Proof of Theorem 2

The soundness of $AX_{\mathrm{L}}(S)$ is easy to verify and omitted to save space. To prove completeness, we follow the canonical model approach. That is, for an $AX_{\mathrm{L}}(S)$-consistent formula φ, we construct a Lewisian causal model from a maximally $AX_{\mathrm{L}}(S)$-consistent set containing φ and prove φ is satisfied in that model.

Given an $AX_{\mathrm{L}}(S)$-consistent formula φ, we can extend it into a maximally $AX_{\mathrm{L}}(S)$-consistent set C. According to the formulas in C, we define structural equations for the canonical model as follows: for any endogenous variable X, $f_X(\mathbf{u}, \mathbf{y}) = x$ iff $\mathbf{Y} = \mathbf{y} \Diamond\!\!\rightarrow X = x \in C$ (well-defined by L1 and L2). As L3 is in C, we can determine a value configuration $\mathbf{v}^c$ for $\mathbf{V}$ which is not relative to any context. Then the canonical model is denoted as $T^c, (\mathbf{u}, \mathbf{v}^c)$ for every context $\mathbf{u}$.

Before we prove that φ is true in $T^c, (\mathbf{u}, \mathbf{v}^c)$, we shall show that T^c is a solutionful causal model, that is, T^c has at least one solution given every context $\mathbf{u}$. Since L4 and $\mathbf{V} = \mathbf{v^c}$ are in C, $\mathbf{V}_{V_1} = \mathbf{v^c} \Diamond\!\!\rightarrow V_1 = v_1 \wedge \ldots \wedge \mathbf{V}_{V_n} = \mathbf{v^c} \Diamond\!\!\rightarrow V_n = v_n \in C$. That means $f_{V_i}(\mathbf{u}, \mathbf{v^c}_{V_i}) = v_i$, where $\mathbf{v^c}_{V_i}$ and v_i are the respective values of $\mathbf{V}_{V_i}$ and V_i in $\mathbf{v^c}$ for any context $\mathbf{u}$. Hence, given any context $\mathbf{u}$, $\mathbf{v^c}$ can solve all the functions. That is to say, T^c has one solution $\mathbf{v^c}$ relative to every context, and hence is a solutionful causal model.

Now we can prove that $\psi \in C$ iff $T^c, (\mathbf{u}, \mathbf{v}^c) \models \psi$ for every context $\mathbf{u}$ and every $\psi \in \mathcal{L}(S)$ by induction on the structure of ψ. If ψ is $X = x$, suppose that $X = x \in C$, then $\mathbf{V} = \mathbf{v^c}$ is consistent with $X = x$ and thus $T^c, (\mathbf{u}, \mathbf{v}^c) \models X = x$. The other direction is similar. To prove the case of $\mathbf{X} = \mathbf{x} \Diamond\!\!\rightarrow \beta$, we follow the strategy to reduce it into simpler formulas by applying some axioms and rules. Basically, due to L0 and RE, β can be written as a disjunctive normal form. Then thanks to L5, $\mathbf{X} = \mathbf{x} \Diamond\!\!\rightarrow \beta$ can be separated into several formulas $\mathbf{X} = \mathbf{x} \Diamond\!\!\rightarrow \beta_i$ where β_i is a conjunction of formulas of the form $Y = y$ or its negation. According to L3, we have $X \neq x \Leftrightarrow \bigvee_{x' \in \mathcal{R}(X) \setminus \{x\}} X = x' \in C$. After applying the rule RE with L5 repeatedly, we can delete the negations in β_i and reduce $\mathbf{X} = \mathbf{x} \Diamond\!\!\rightarrow \beta$ into formulas of the form $\mathbf{X} = \mathbf{x} \Diamond\!\!\rightarrow \mathbf{Y} = \mathbf{y}$. According to L6, to prove the case of $\mathbf{X} = \mathbf{x} \Diamond\!\!\rightarrow \mathbf{Y} = \mathbf{y}$, it suffices to show that $\mathbf{X} = \mathbf{x} \Diamond\!\!\rightarrow \mathbf{W} = \mathbf{w} \in C$ iff $T^c, (\mathbf{u}, \mathbf{v}^c) \models \mathbf{X} = \mathbf{x} \Diamond\!\!\rightarrow \mathbf{W} = \mathbf{w}$ in which $\mathbf{W} = \mathbf{V} \setminus \mathbf{X}$.

We establish the above clause by induction on $|\mathbf{V} \setminus \mathbf{X}|$. When $|\mathbf{V} \setminus \mathbf{X}| = 1$, suppose $\mathbf{X} = \mathbf{x} \Diamond\!\!\rightarrow W = w \in C$, we have $f_W(\mathbf{u}, \mathbf{x}) = w$ and thus there is a solution $(\mathbf{x}, w)$ to $T^c_{\mathbf{X}=\mathbf{x}}$ relative to $\mathbf{u}$. Hence $T^c, (\mathbf{u}, \mathbf{v}^c) \models \mathbf{X} = \mathbf{x} \Diamond\!\!\rightarrow W = w$. The other direction is similar. Assume that the above clause holds for $|\mathbf{V} \setminus \mathbf{X}| < k$, we now show the case of $|\mathbf{V} \setminus \mathbf{X}| = k$. As $k \geq 2$, we can write $\mathbf{X} = \mathbf{x} \Diamond\!\!\rightarrow$

$\mathbf{W} = \mathbf{w}$ as $\mathbf{X} = \mathbf{x} \diamond\!\!\rightarrow (W_1 = w_1 \wedge W_2 = w_2 \wedge \mathbf{W_3} = \mathbf{w_3})$. Suppose that $\mathbf{X} = \mathbf{x} \diamond\!\!\rightarrow (W_1 = w_1 \wedge W_2 = w_2 \wedge \mathbf{W_3} = \mathbf{w_3}) \in C$, then $(\mathbf{X} = \mathbf{x} \wedge W_1 = w_1) \diamond\!\!\rightarrow (W_2 = w_2 \wedge \mathbf{W_3} = \mathbf{w_3}) \in C$ due to L7. According to the inductive hypothesis, $T^c, (\mathbf{u}, \mathbf{v^c}) \models (\mathbf{X} = \mathbf{x} \wedge W_1 = w_1) \diamond\!\!\rightarrow (W_2 = w_2 \wedge \mathbf{W_3} = \mathbf{w_3})$. Similarly, $T^c, (\mathbf{u}, \mathbf{v^c}) \models (\mathbf{X} = \mathbf{x} \wedge W_2 = w_2) \diamond\!\!\rightarrow (W_1 = w_1 \wedge \mathbf{W_3} = \mathbf{w_3})$. Since L8 holds in any causal model, we have $T^c, (\mathbf{u}, \mathbf{v^c}) \models \mathbf{X} = \mathbf{x} \diamond\!\!\rightarrow (W_1 = w_1 \wedge W_2 = w_2 \wedge \mathbf{W_3} = \mathbf{w_3})$. Conversely, if $T^c, (\mathbf{u}, \mathbf{v^c}) \models \mathbf{X} = \mathbf{x} \diamond\!\!\rightarrow (W_1 = w_1 \wedge W_2 = w_2 \wedge \mathbf{W_3} = \mathbf{w_3})$, then $T^c, (\mathbf{u}, \mathbf{v^c}) \models (\mathbf{X} = \mathbf{x} \wedge W_1 = w_1) \diamond\!\!\rightarrow (W_2 = w_2 \wedge \mathbf{W_3} = \mathbf{w_3})$ as L7 holds in any causal model. According to the inductive hypothesis, $(\mathbf{X} = \mathbf{x} \wedge W_1 = w_1) \diamond\!\!\rightarrow (W_2 = w_2 \wedge \mathbf{W_3} = \mathbf{w_3}) \in C$. Similarly, $(\mathbf{X} = \mathbf{x} \wedge W_2 = w_2) \diamond\!\!\rightarrow (W_1 = w_1 \wedge \mathbf{W_3} = \mathbf{w_3}) \in C$. Since L8 is in C, $\mathbf{X} = \mathbf{x} \diamond\!\!\rightarrow (W_1 = w_1 \wedge W_2 = w_2 \wedge \mathbf{W_3} = \mathbf{w_3}) \in C$.

It is then easy to show that $\mathbf{X} = \mathbf{x} \diamond\!\!\rightarrow \beta \in C$ iff $T^c, (\mathbf{u}, \mathbf{v^c}) \models \mathbf{X} = \mathbf{x} \diamond\!\!\rightarrow \beta$, using the aforementioned strategy of reducing $\mathbf{X} = \mathbf{x} \diamond\!\!\rightarrow \beta$ via $\mathbf{X} = \mathbf{x} \diamond\!\!\rightarrow \beta^{dnf}$, where β^{dnf} is a disjunctive normal form of β.

The further cases of the inductive step concern Boolean combinations of formulas of the form $X = x$ and formulas of the form $\mathbf{X} = \mathbf{x} \diamond\!\!\rightarrow \beta$. These cases are very straightforward. Therefore, for every $\psi \in \mathcal{L}(S)$, $\psi \in C$ iff $T^c, (\mathbf{u}, \mathbf{v^c}) \models \psi$ for every context $\mathbf{u}$.

Finally, we need to show that $T^c \in \mathcal{T}_{\text{L}}(S)$. This is trivial given what has been shown. Since L1, L9, L10 and L11 are in C, they hold relative to $T^c, (\mathbf{u}, \mathbf{v^c})$. Hence, by Lemmas 5–8, T^c is solution-determinate and solution-transitive in cycles, and every submodel of T^c is solutionful and solution-conservative. That is, T^c is indeed a Lewisian causal model.

References

1. Bareinboim, E., Correa, J.D., Ibeling, D., Icard, T.: On Pearl's hierarchy and the foundations of causal inference. In: Probabilistic and Causal Inference: The Works of Judea Pearl, pp. 507–556 (2022)
2. Briggs, R.: Interventionist counterfactuals. Philos. Stud.: Int. J. Philos. Anal. Tradit. **160**(1), 139–166 (2012). http://www.jstor.org/stable/23262477
3. Galles, D., Pearl, J.: An axiomatic characterization of causal counterfactuals. Found. Sci. **3**(1), 151–182 (1998). https://doi.org/10.1023/a:1009602825894
4. Halpern, J.Y.: Axiomatizing causal reasoning. J. Artif. Intell. Res. **12**, 317–337 (2000)
5. Halpern, J.Y.: From causal models to counterfactual structures. Rev. Symb. Logic **6**(2), 305–322 (2013)
6. Huber, F.: Structural equations and beyond. Rev. Symb. Logic **6**(4), 709–732 (2013)
7. Ibeling, D., Icard, T.: On open-universe causal reasoning. In: Adams, R.P., Gogate, V. (eds.) Proceedings of the 35th Uncertainty in Artificial Intelligence Conference. Proceedings of Machine Learning Research, vol. 115, pp. 1233–1243. PMLR (2020). http://proceedings.mlr.press/v115/ibeling20a.html
8. Lewis, D.: Causation. J. Philos. **70**, 556–567 (1973)
9. Lewis, D.: Counterfactuals. Blackwell, Malden (1973)
10. List, C., Menzies, P.: Nonreductive physicalism and the limits of the exclusion principle. J. Philos. **106**(9), 475–502 (2009). https://doi.org/10.5840/jphil2009106936

11. Pearl, J.: Causality: Models, Reasoning, and Inference. Cambridge University Press, New York (2009)
12. Stalnaker, R.: A theory of conditionals. In: Rescher, N. (ed.) Studies in Logical Theory (American Philosophical Quarterly Monographs 2), pp. 98–112. Blackwell, Oxford (1968)
13. Woodward, J.: Making Things Happen: A Theory of Causal Explanation. Oxford University Press, Oxford, U.K. (2003)
14. Woodward, J., Hitchcock, C.: Explanatory generalizations, part i: A counterfactual account. Noûs **37**(1), 1–24 (2003). http://www.jstor.org/stable/3506202
15. Zhang, J.: A Lewisian logic of causal counterfactuals. Mind. Mach. **23**(1), 77–93 (2013). https://doi.org/10.1007/s11023-011-9261-z
16. Zhang, J., Lam, W.Y., De Clercq, R.: A peculiarity in Pearl's logic of interventionist counterfactuals. J. Philos. Logic **42**(5), 783–794 (2013). http://www.jstor.org/stable/42001257
17. Zhong, L.: Sophisticated exclusion and sophisticated causation. J. Philos. **111**(7), 341–360 (2014). http://www.jstor.org/stable/43820849

The Expressive Power of Revised Datalog on Problems with Closure Properties

Shiguang Feng(✉)

School of Computer Science and Engineering, Sun Yat-sen University, Guangzhou 510006, China
fengshg3@mail.sysu.edu.cn

Abstract. In this paper, we study the expressive power of revised Datalog on the problems that are closed under substructures. We show that revised Datalog cannot define all the problems that are in PTIME and closed under substructures. As a corollary, LFP cannot define all the extension-closed problems that are in PTIME.

Keywords: Datalog · preservation theorem · closure property · expressive power

1 Introduction

Datalog and its variants are widely used in artificial intelligence and other fields, such as deductive database, knowledge representation, data integration, cloud computing, etc [8,11,13,18,19,23]. As a declarative programming language, it is often used to perform data analysis and create complex queries. The complexity and expressive power is an important issue of the study [1,5,19,21,22]. With the recursive computing ability, Datalog is more powerful than first-order logic. It defines exactly the polynomial time computable queries on ordered finite structures [9]. Hence, Datalog captures the complexity class PTIME on ordered finite structures. While on all finite structures, the expressive power of Datalog is very limited. It even cannot define the parity of a set [9]. A Datalog program is constituted of a set of Horn clauses. The characteristics of syntax determine the monotonicity properties of its semantics. That is, every Datalog (resp., positive Datalog, the fragment of Datalog where no negated atomic formula occurs in the body of any clauses) definable query is preserved under extensions [3] (resp., homomorphisms [4]). It is natural to ask from the point of view of descriptive complexity that whether Datalog (resp., positive Datalog) captures the polynomial time computable problems that are closed under extensions (resp., homomorphisms). The answer is negative by the work of Afrati et al. who showed that positive Datalog cannot express all monotone queries computable in polynomial time, and the perfect squares problem that is in polynomial time and closed under extensions is not expressible in Datalog [3].

In model theory, many preservation theorems are proved to show the relationship between the closure properties and the syntactic properties of formulas.

N. Alechina et al. (Eds.): LORI 2023, LNCS 14329, pp. 109–125, 2023.
https://doi.org/10.1007/978-3-031-45558-2_9

Most of these preservation theorems fail when restricted to finite structures. A lot of research about the preservation theorems on Datalog, first-order logic (FO) and least fixpoint logic (LFP) have been conducted on finite structures. Ajtai and Gurevich showed that a positive Datalog formula is bounded iff it is definable in positive existential first-order logic, and every first-order logic expressible positive Datalog formula is bounded [4], where a Datalog formula is bounded if there exists a number n such that the fixpoint of the formula can be reached for any finite structure within n steps. Dawar and Kreutzer showed that the homomorphism preservation theorem fails for LFP, both in general and in restriction to finite structures [6]. That is, there is an LFP formula that is preserved under homomorphisms (in the finite) but is not equivalent (in the finite) to a Datalog formula. The paper [16] studied Datalog with negation and monotonicity, and the expressive power with respect to monotone and homomorphism properties.

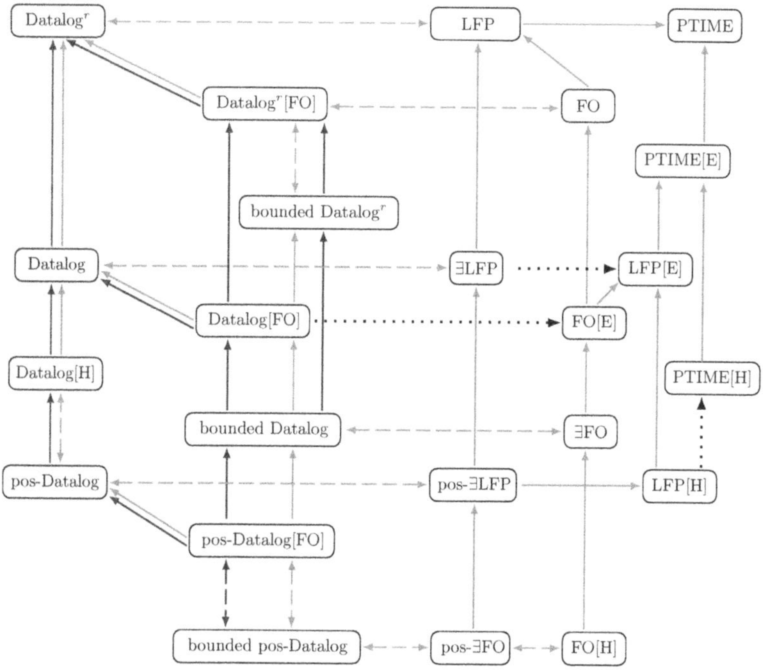

Fig. 1. The relationship of FO, LFP, PTIME, Datalog and its variants. Datalog^r denotes revised Datalog. pos-$\mathcal{L}$ denotes the positive fragment of $\mathcal{L}$. $\exists\mathcal{L}$ denotes the existential fragment of $\mathcal{L}$. $\mathcal{L}$[FO] denotes the set of $\mathcal{L}$ formulas that are first-order definable. $\mathcal{L}$[H] (resp., $\mathcal{L}$[E]) denotes the set of $\mathcal{L}$ formulas (or problems computable in $\mathcal{L}$) that are preserved under homomorphisms (resp., extensions). The blue arrow shows the containment relationship on Datalog and its variants. The red arrow shows the relationship about the expressive power. The solid arrow implies that the relationship is strict, and the dashed bidirectional arrow implies the equality relationship. The black dotted arrow means whether the relationship is strict is still open.

The papers [7,20] studied the preservation results under extensions for FO and Datalog. All the results are summarized in Fig. 1.

Revised Datalog (Datalog^r) is an extension of Datalog, where universal quantification over intensional relations is allowed in the body of rules. Abiteboul and Vianu first introduced the idea that the body of a rule in Datalog can be universally quantified [2]. The author of the paper showed that Datalog^r equals LFP on all finite structures [10]. In the paper, we study the expressive power of Datalog^r on problems with closure properties, i.e., closed under substructures (or extensions). We conclude that a Datalog^r formula is equivalent to a first-order formula iff it is equivalent to a bounded Datalog^r formula. As the main result of the paper, we show that Datalog^r cannot define all the problems that are in PTIME and closed under substructures. Since Datalog^r equals LFP, and the complement of a substructure-closed problem is extension-closed, as a corollary, LFP cannot define all the extension-closed problems that are in PTIME. This result contributes the strict containment LFP[E] $\subsetneq$ PTIME[E] in Fig. 1. A technique of tree encodings for arbitrary structures is used in the proof. For an arbitrary set of structures $\mathcal{K} \in$ EXPTIME, we can encode them into a set of substructure-closed structures $\mathcal{K}'$, where the tree used to encode the structure in $\mathcal{K}$ is exponentially larger. Therefore, $\mathcal{K}'$ is in PTIME. For every structure in $\mathcal{K}'$, there is a characteristic structure $S_{\mathbf{T}}$ of it such that they are equivalent with respect to Datalog^r-transformations. Since $S_{\mathbf{T}}$ can be computed from the structure in $\mathcal{K}$ in logspace, this implies that $\mathcal{K}$ is also in PTIME, contrary to the time hierarchy theorem. Figure 2 shows the sketch of the proof.

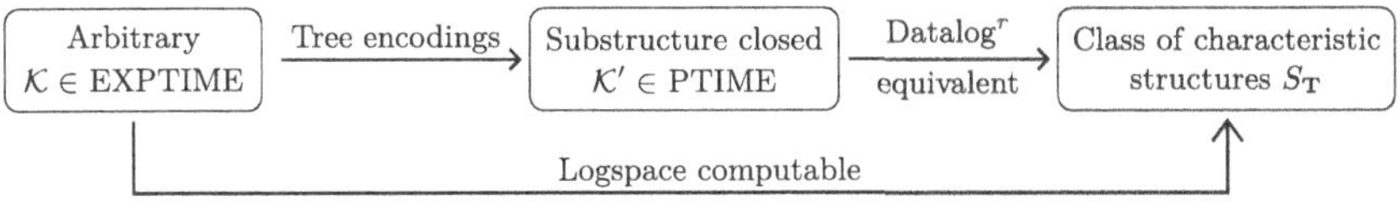

Fig. 2. The idea of the proof for the nondefinability of Datalog^r.

The paper is organized as follows: In Sect. 2, we give the basic definitions and notations. In Sect. 3, we recall invariant relations on perfect binary trees, and introduce the technique of tree encodings for arbitrary structures. And we prove the nondefinability results for Datalog^r on substructure-closed problems. Finally, we conclude the paper in Sect. 4.

2 Preliminaries

Let $\tau = \{\mathbf{c}_1, \dots, \mathbf{c}_m, P_1, \dots, P_n\}$ be a vocabulary, where $\mathbf{c}_1, \dots, \mathbf{c}_m$ are constant symbols and $P_1, \dots, P_n$ are relation symbols. A τ-structure is a tuple $\mathbf{A} = \langle A, \mathbf{c}_1^A, \dots, \mathbf{c}_m^A, P_1^A, \dots, P_n^A \rangle$ where A is the domain, and $\mathbf{c}_1^A, \dots, \mathbf{c}_m^A$, $P_1^A, \dots, P_n^A$ are interpretations of the constant and relation symbols over A, respectively. We assume the equality relation "=" is contained in every vocabulary, and omit the superscript "A" when it is clear from context. We call $\mathbf{A}$ *finite* if its domain A is a finite set. Unless otherwise stated, all structures considered in this paper are finite. We use $arity(R)$ to denote the arity of a relation R, and use "$|\ |$" to indicate the cardinality of a set or the arity of a tuple, e.g., $|A|$ denotes the cardinality of A and $|(x_1, x_2, x_3)| = 3$. A finite structure is *ordered* if it is equipped with a linear order relation "$\leq$", and the successor relation "SUCC", the constants "**min**" and "**max**" for the minimal and maximal elements, respectively, with respect to "$\leq$". Let $\mathbf{A} = \langle A, \mathbf{c}_1^A, \ldots, \mathbf{c}_m^A, P_1^A, \ldots, P_n^A \rangle$ and $\mathbf{B} = \langle B, \mathbf{c}_1^B, \ldots, \mathbf{c}_m^B, P_1^B, \ldots, P_n^B \rangle$ be two structures. If $B \subseteq A$, $\mathbf{c}_i^A = \mathbf{c}_i^B$ $(1 \leq i \leq m)$, and $P_j^B = P_j^A \cap B^{arity(P_j)}$ $(1 \leq j \leq n)$, then we say that $\mathbf{B}$ is a *substructure* of $\mathbf{A}$, and $\mathbf{A}$ is an *extension* of $\mathbf{B}$.

An r-ary *global relation* R of a vocabulary τ is a mapping that assigns to every τ-structure $\mathbf{A}$ an r-ary relation R^A over A such that for every isomorphism $\pi : \mathbf{A} \simeq \mathbf{B}$ and every $a_1, \ldots, a_r \in A$, $\mathbf{A} \models R^A a_1 \ldots a_r$ iff $\mathbf{B} \models R^B \pi(a_1) \ldots \pi(a_r)$. A *query* is a global relation. We say that a query $\mathcal{Q}$ is expressible in a logic $\mathcal{L}$ if there is an $\mathcal{L}$-formula that defines $\mathcal{Q}$. Two formulas are *equivalent* if they define the same query. Given two logics $\mathcal{L}_1$ and $\mathcal{L}_2$, we use $\mathcal{L}_1 \leq \mathcal{L}_2$ to denote that every $\mathcal{L}_1$-formula is equivalent to an $\mathcal{L}_2$-formula. If $\mathcal{L}_1 \leq \mathcal{L}_2$ and $\mathcal{L}_2 \leq \mathcal{L}_1$, then we denote it by $\mathcal{L}_1 \equiv \mathcal{L}_2$.

Suppose that a relation symbol X occurs positively in $\varphi(\bar{x})$ and $|\bar{x}| = arity(X)$. Given a structure $\mathbf{A}$, we can define a monotonic sequence $X_0, X_1, X_2, \ldots$, where $X_0 = \emptyset$ and $X_{i+1} = \{\bar{a} \mid (\mathbf{A}, X_i) \models \varphi[\bar{a}]\}$ for $i \geq 0$, such that $X_i \subseteq X_{i+1}$. Since $\mathbf{A}$ is finite, the sequence will eventually reach a fixpoint.

Definition 1. *The least fixpoint logic* LFP *is an extension of first-order logic by adding the following rule [9]:*

- *If φ is an* LFP *formula, X occurs positively in φ, and $|\bar{x}| = |\bar{u}| = arity(X)$, then* $[\mathrm{LFP}_{\bar{x},X}\varphi]\bar{u}$ *is an* LFP *formula.*

Given an LFP formula $[\mathrm{LFP}_{\bar{x},X}\varphi]\bar{u}$, for any structure $\mathbf{A}$ and $\bar{a} \in A^{arity(\bar{a})}$, we have $\mathbf{A} \models [\mathrm{LFP}_{\bar{x},X}\varphi]\bar{a}$ iff $\bar{a}$ is in the fixpoint of the sequence induced by X and φ on $\mathbf{A}$.

Proposition 1. *[15, 24]* LFP *captures* PTIME *on ordered finite structures.*

Definition 2. *Let τ be a vocabulary. A* Datalog *program* Π *over τ is a finite set of rules of the form*

$$\beta \leftarrow \alpha_1, \ldots, \alpha_l$$

where $l \geq 0$ and

(1) each α_i is either an atomic formula or a negated atomic formula,
(2) β is an atomic formula $R\bar{x}$, where R doesn't occur negatively in any rule of Π.

β is the head of the rule and the sequence $\alpha_1, \ldots, \alpha_l$ constitute the body. Every relation symbol occurring in the head of some rule of Π is intensional, and the other symbols in τ are extensional. We use $(\tau, \Pi)_{\text{int}}$ and $(\tau, \Pi)_{\text{ext}}$ to denote the set of intensional and extensional symbols, respectively. We also allow 0-ary relation symbols. If Q is a 0-ary relation, its value is from $\{\emptyset, \{\emptyset\}\}$. $Q = \emptyset$ means that Q is FALSE and $Q = \{\emptyset\}$ means that Q is TRUE. We use the least fixpoint semantics for Datalog programs. A Datalog formula has the form $(\Pi, P)\bar{x}$, where P is an r-ary intensional relation symbol and $\bar{x} = (x_1, \ldots, x_r)$ are variables that do not occur in Π. For a $(\tau, \Pi)_{\text{ext}}$-structure $\mathbf{A}$ and $\bar{a} = (a_1, \ldots, a_r) \in A^r$,

$$\mathbf{A} \models (\Pi, P)\bar{x}[\bar{a}] \text{ iff } (a_1, \ldots, a_r) \in P_{(\infty)},$$

where $P_{(\infty)}$ is the least fixpoint for relation P when Π is evaluated on $\mathbf{A}$. If P is 0-ary, then $\mathbf{A} \models (\Pi, P)$ iff $P_{(\infty)} = \{\emptyset\}$.

3 Datalog^r on Problems with Closure Properties

3.1 Revised Datalog Programs

Definition 3. *In Definition 2, if we replace Condition (1) by*

(1′) each α_i is either an atomic formula, or a negated atomic formula, or a formula $\forall\bar{y}R\bar{y}\bar{z}$, where R occurs in the head of some rule,

then we call this logic program revised Datalog *program, denoted by* Datalog^r.

Example 1. Let $G = \langle V, E\rangle$ be a directed acyclic graph, and the set of nodes V partitioned into two disjointed sets V_{uni} and V_{exi}. The nodes in V_{uni} (resp., V_{exi}) are universal (resp., existential). The notion of alternating path is defined recursively. There is an alternating path from $\mathbf{s}$ to $\mathbf{t}$ in G if

- $\mathbf{s} = \mathbf{t}$; or
- $\mathbf{s} \in V_{\text{exi}}$, $\exists x \in V$ such that $(\mathbf{s}, x) \in E$ and there is an alternating path from x to $\mathbf{t}$; or
- $\mathbf{s} \in V_{\text{uni}}$, $\exists x \in V$ such that $(\mathbf{s}, x) \in E$, and $\forall y \in V$, if $(\mathbf{s}, y) \in E$ then there is an alternating path from y to $\mathbf{t}$.

The alternating graph accessibility problem is defined as follows:

Input: A directed acyclic graph $G = \langle V_{\text{uni}} \cup V_{\text{exi}}, E\rangle$ and two nodes $\mathbf{s}, \mathbf{t}$.
Output: Yes if there is an alternating path from $\mathbf{s}$ to $\mathbf{t}$ in G, otherwise no.

This problem is P-complete [14]. The following Datalog^r program Π defines the alternating graph accessibility problem

$$P_{\text{alt}}xy \leftarrow x = y;$$
$$P_{\text{alt}}xy \leftarrow \neg V_{\text{uni}}x, Exz, P_{\text{alt}}zy;$$
$$P_{\text{uni}}x \leftarrow V_{\text{uni}}x, Exy;$$
$$Qxzy \leftarrow P_{\text{uni}}x, \neg Exz;$$
$$Qxzy \leftarrow P_{\text{uni}}x, Exz, P_{\text{alt}}zy;$$
$$P_{\text{alt}}xy \leftarrow P_{\text{uni}}x, \forall zQxzy;$$
$$P \leftarrow P_{\text{alt}}\mathbf{st}.$$

We have $(\tau, \Pi)_{\text{int}} = \{P_{\text{alt}}, Q, P_{\text{uni}}, P\}$ and $(\tau, \Pi)_{\text{ext}} = \{E, V_{\text{uni}}, \mathbf{s}, \mathbf{t}\}$. The relation P_{uni} saves the nodes in V_{uni} that have a successor. The relation P_{alt} saves the pairs (x, y) such that there is an alternating path from x to y. We use $Qxzy$ to denote that for any $x \in P_{\text{uni}}$, either there is no edge from x to z, or there is an alternating path from z to y. For any directed acyclic $(\tau, \Pi)_{\text{ext}}$-structure $\mathbf{A}$, we have $\mathbf{A} \models (\Pi, P)$ iff there is an alternating path from $\mathbf{s}$ to $\mathbf{t}$.

The Datalog formulas are preserved under extensions [7], i.e., if a structure $\mathbf{B}$ satisfies a Datalog formula φ and $\mathbf{A}$ is an extension of $\mathbf{B}$, then $\mathbf{A}$ also satisfies φ. A directed acyclic graph with an alternating path from $\mathbf{s}$ to $\mathbf{t}$ can be extended to a directed acyclic graph without any alternating path from $\mathbf{s}$ to $\mathbf{t}$ by adding new nodes. So Datalog cannot define the alternating graph accessibility problem, which implies that Datalog^r is strictly more expressive than Datalog. Allowing universal quantification over intensional relations is essential for Datalog^r to increase its expressive power. With the help of it, every FO(LFP) formula can be transformed into an equivalent Datalog^r formula.

Proposition 2. *[10]* Datalog^r $\equiv$ LFP *on all finite structures.*

A Datalog program is *positive* if no negated atomic formula occurs in the body of any rule. A Datalog formula $(\Pi, P)\bar{t}$ is *bounded* if there is an $n \geq 0$ such that $P_{(n)} = P_{(\infty)}$ for all structures. A bounded (positive) Datalog formula is equivalent to an existential (positive) first-order formula, and vice versa [9]. Furthermore, a positive Datalog formula is bounded iff it is equivalent to a first-order formula. The statement is false for all Datalog formulas. There is an unbounded Datalog formula that is equivalent to an FO formula, but no bounded Datalog formula is equivalent to it [4]. Unlike Datalog, if an unbounded Datalog^r formula is equivalent to an FO formula, then it must be equivalent to a bounded Datalog^r formula.

Proposition 3. *A* Datalog^r *formula is equivalent to a first-order formula iff it is equivalent to a bounded* Datalog^r *formula.*

Proof. Suppose that a Datalog^r formula is equivalent to a first-order formula φ. Using the method in [10] we can construct a bounded Datalog^r formula that is equivalent to φ. For the other direction, the proof in [9] which shows that every bounded Datalog formula is equivalent to an FO formula remains valid for bounded Datalog^r formulas.

3.2 Invariant Relations on Perfect Binary Trees

In [17], Lindell introduced invariant relations that are defined on perfect binary trees, and showed that there are queries computable in PTIME but not definable in LFP. A perfect binary tree is a binary tree in which all internal nodes have two children and all leaf nodes are in the same level. Let $T = \langle V, E, \mathbf{root} \rangle$ be a perfect binary tree, where V is the set of nodes, E is the set of edges and **root** is the root node. Suppose that R is an r-ary relation on V and f is an automorphism of T. Given a tuple $\bar{a} = (a_1, \ldots, a_r) \in R$, we write $f(\bar{a}) = (f(a_1), \ldots, f(a_r))$ and $f[R] = \{(f(a_1), \ldots, f(a_r)) \mid (a_1, \ldots, a_r) \in R\}$. We say that R is an invariant relation if for every automorphism f, $R = f[R]$. It is easily seen that the equality $=$ and E are invariant relations.

First we give several technical lemmas. The proofs of Lemmas 1, 2, 3, and 5 can be found in the full arXiv version of the paper.

Lemma 1. *If R_1 and R_2 are r-ary invariant relations, then $\neg R_1$, $R_1 \cap R_2$ and $R_1 \cup R_2$ are also invariant relations.*

Lemma 2. *Suppose that R is an r-ary invariant relation, R' is a k-ary invariant relation and g is a permutation of $\{1, \ldots, r\}$. Define*

$$R_1 = \{(a_{g(1)}, \ldots, a_{g(r)}) \mid (a_1, \ldots, a_r) \in R\},$$
$$R_2 = \{(a_1, \ldots, a_r, b_1, \ldots, b_k) \mid (a_1, \ldots, a_r) \in R \text{ and } (b_1, \ldots, b_k) \in R'\}.$$

Then R_1 and R_2 are also invariant relations.

Lemma 3. *Suppose that R is a $(k + r)$-ary invariant relation. Define*

$$R_1 = \{(a_1, \ldots, a_r) \mid (b_1, \ldots, b_k, a_1, \ldots, a_r) \in R \text{ for all nodes } b_1, \ldots, b_k\}$$
$$R_2 = \{(a_1, \ldots, a_r) \mid \exists b_1, \ldots, b_k \text{ such that } (b_1, \ldots, b_k, a_1, \ldots, a_r) \in R\}.$$

Then R_1 and R_2 are also invariant relations.

Let a, b be two nodes of a perfect binary tree T, we use $a \barwedge b$ and $d(a)$ to denote the least common ancestor of a, b and the depth of a, respectively. For example, in Fig. 3 there is a perfect binary tree in which $d(\mathbf{root}) = 0$, $d(a) = 1$, $d(c) = d(e) = 2$, and $c \barwedge e = \mathbf{root}$.

Let $(a_1, \ldots, a_r)$ be an r-ary tuple of nodes, its characteristic tuple is defined as

$$\begin{aligned}(a_1, \ldots, a_r)^* = (&d(a_1), d(a_1 \barwedge a_2), \ldots, d(a_1 \barwedge a_r),\\ &d(a_2), d(a_2 \barwedge a_3), \ldots, d(a_2 \barwedge a_r),\\ &\ldots, d(a_r))\end{aligned}$$

which is a $\frac{r(r+1)}{2}$-ary tuple of numbers. Let R be an invariant relation, the characteristic relation of R is defined to be

$$R^* = \{(a_1, \ldots, a_r)^* \mid (a_1, \ldots, a_r) \in R\}.$$

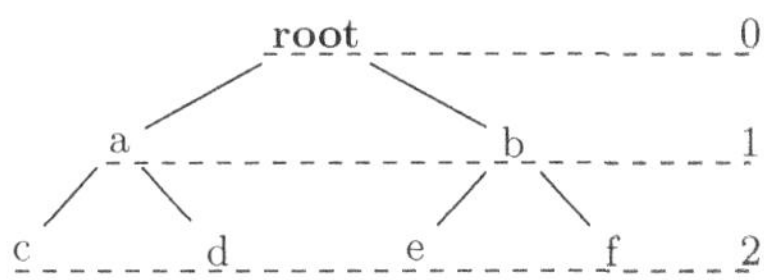

Fig. 3. A perfect binary tree of depth 3.

Proposition 4. *[17] Let $\bar{a} = (a_1, \dots, a_r)$ and $\bar{b} = (b_1, \dots, b_r)$ be two tuples, and R an r-ary invariant relation of a perfect binary tree T.*

- *$(a_1, \dots, a_r)^* = (b_1, \dots, b_r)^*$ iff there is an automorphism f of T such that $f(\bar{a}) = \bar{b}$.*
- *If $(a_1, \dots, a_r)^* = (b_1, \dots, b_r)^*$, then $\bar{a} \in R$ iff $\bar{b} \in R$.*

For any two invariant relations R_1 and R_2, $R_1 = R_2$ iff $R_1^* = R_2^*$.

3.3 Tree Encodings and Characteristic Structures

This section is devoted to the definitions of tree encodings and characteristic structures, and the propositions about the equivalent relationship between them on Datalog^r programs, which will be used in the next section.

Definition 4. *Let T be a perfect binary tree, R an r-ary relation on T. R is a saturated relation if for any nodes $a_1, \dots, a_r$, $b_1, \dots, b_r$, whenever $d(a_i) = d(b_i)\,(1 \leq i \leq r)$, then $(a_1, \dots, a_r) \in R$ iff $(b_1, \dots, b_r) \in R$.*

The following proposition can be proved easily from the definitions of invariant relations and saturated relations.

Proposition 5. *A saturated relation is also an invariant relation.*

From now on we make the assumption: τ is the vocabulary $\{R_1, \dots, R_k\}$, and $\tau' = \tau \cup \{\mathbf{root}, E\}$, where **root** is a constant symbol and E is a binary relation symbol that is not in τ. We define a class of τ'-structures

$$\mathcal{T} = \{\langle V, \mathbf{root}, E, R_1, \dots, R_k\rangle \mid \langle V, E, \mathbf{root}\rangle \text{ is a perfect binary tree, } R_1, \dots, R_k \text{ are saturated relations on it}\}.$$

Definition 5. *Let $\mathbf{A} = \langle \{0, 1, \dots, h-1\}, R_1^A, \dots, R_k^A\rangle$ be a τ-structure. The tree encoding of $\mathbf{A}$ is a τ'-structure $C(\mathbf{A}) = \langle V, \mathbf{root}, E, R_1^T, \dots, R_k^T\rangle \in \mathcal{T}$, such that $\langle V, E, \mathbf{root}\rangle$ is a perfect binary tree of depth h, and for any relation symbol $R_i\,(1 \leq i \leq k)$ and any nodes $a_1, \dots, a_{r_i} \in V$,*

$$C(\mathbf{A}) \models R_i^T a_1 \cdots a_{r_i} \text{ iff } \mathbf{A} \models R_i^A d(a_1) \cdots d(a_{r_i})$$

where r_i is the arity of R_i, and $d(a_j)\,(1 \leq j \leq r_i)$ is the depth of a_j.

Roughly speaking, $C(\mathbf{A})$ encodes $\mathbf{A}$ in a tree, but its size is exponentially larger. Conversely, given a τ'-structure $\mathbf{T} = \langle V, \mathbf{root}, E, R_1^T, \dots, R_k^T\rangle \in \mathcal{T}$, we can compute the τ-structure $\mathbf{A}$ encoded by $\mathbf{T}$ as follows:

(1) The domain is $\{0, \dots, h-1\}$, where h is the depth of $\mathbf{T}$;
(2) For each $i = 1, \dots, k$,

$$R_i^A = \{(d(a_1), \dots, d(a_{r_i})) \mid \exists a_1, \dots, a_{r_i} \in V \text{ such that } \mathbf{T} \models R_i^T a_1 \cdots a_{r_i}\}.$$

We use $C^{-1}(\mathbf{T})$ to denote the corresponding τ-structure $\mathbf{A}$ encoded by $\mathbf{T}$. Let $\mathrm{FUL}_m = V^m$ be a relation of arity m, where $m \geq 1$ and V is the domain of $\mathbf{T}$. Define the vocabulary

$$\sigma = \{\mathbf{0}, \mathrm{SUCC}, R_{\neq}, R_{\neg e}, \mathrm{FUL}_m^*\} \cup \{R_1^*, \ldots, R_k^*, (\neg R_1)^*, \ldots, (\neg R_k)^*\}$$

where FUL_m^* has arity $\frac{m(m+1)}{2}$, $R_{\neq}$ and $R_{\neg e}$ have arity 3, R_i^* or $(\neg R_1)^*$ has arity $\frac{r_i(r_i+1)}{2}$ $(1 \leq i \leq k$ and r_i is the arity of $R_i)$.

Definition 6. *Given a τ'-structure $\mathbf{T} = \langle V, \mathbf{root}, E, R_1, \ldots, R_k\rangle \in \mathcal{T}$, the characteristic structure $S_{\mathbf{T}}$ of $\mathbf{T}$ is a σ-structure*

$$\langle \{0, 1, \ldots, h-1\}, \mathbf{0}, \mathrm{SUCC}, R_{\neq}, R_{\neg e}, \mathrm{FUL}_m^*, R_1^*, \ldots, R_k^*, (\neg R_1)^*, \ldots, (\neg R_k)^*\rangle$$

where h is the depth of $\mathbf{T}$, $\mathbf{0}$ is a constant interpreted by 0, SUCC is the successor relation on the domain, and $R_{\neq}, R_{\neg e}, \mathrm{FUL}_m^*$, $R_1^*, \ldots, R_k^*$, $(\neg R_1)^*, \ldots, (\neg R_k)^*$ are the characteristic relations of $\neq, (\neg E)$, $\mathrm{FUL}_m, R_1, \ldots, R_k$, $\neg R_1, \ldots, \neg R_k$, respectively.

In the following we show that for every Datalog^r program Π on the tree encodings, there is a Datalog^r program Π^* on the corresponding characteristic structures such that Π^* simulates Π. More precisely, Π^* handles the characteristic relations of the relations in Π. Let $\Pi = \{\gamma_1, \ldots, \gamma_s\}$ be a Datalog^r program on $\mathcal{T}$. Suppose $X_1, \ldots, X_w$ are all intensional relation symbols in Π and for each rule γ_i, let n_{γ_i} be the number of free variables occurring in γ_i. Set

$$m = \max\{n_{\gamma_1}, \ldots, n_{\gamma_s}, \mathrm{arity}(R_1), \ldots, \mathrm{arity}(R_k), \mathrm{arity}(X_1), \ldots, \mathrm{arity}(X_w)\}.$$

We shall construct, based on Π, a Datalog^r program Π^* such that for any Datalog^r formula (Π, P), there exists a Datalog^r formula (Π^*, P^*), and $\mathbf{T} \models (\Pi, P)$ iff $S_{\mathbf{T}} \models (\Pi^*, P^*)$ for any $\mathbf{T} \in \mathcal{T}$, where P and P^* are 0-ary.

Every element of $\mathbf{T}$ is a node of a perfect binary tree, while every element of $S_{\mathbf{T}}$ is a number which can be treated as the depth of some node. Hence, for each variable x in Π, we introduce a new variable i_x, and for any two variables x_1, x_2 in Π, we introduce a new variable $i_{x_1 \barwedge x_2}$. For a tuple of variables $\bar{x} = x_1 \cdots x_r$, we use the following abbreviations:

$$(\bar{x})^* = i_{x_1} i_{x_1 \barwedge x_2} \cdots i_{x_1 \barwedge x_r} i_{x_2} i_{x_2 \barwedge x_3} \cdots i_{x_{r-1} \barwedge x_r} i_{x_r},$$
$$\forall(\bar{x})^* = \forall i_{x_1} \forall i_{x_1 \barwedge x_2} \cdots \forall i_{x_1 \barwedge x_r} \forall i_{x_2} \forall i_{x_2 \barwedge x_3} \cdots \forall i_{x_{r-1} \barwedge x_r} \forall i_{x_r}.$$

Without loss of generality, we treat $i_{u \barwedge v}$ and $i_{v \barwedge u}$ as the same variable. Additionally, we assume that the 0-ary relation is also an invariant relation, and the characteristic relation of a 0-ary relation is itself.

First we construct a quasi-Datalog^r program Π' as follows. For each rule $\beta \leftarrow \alpha_1, \ldots, \alpha_l$ in Π, suppose that $v_1, \ldots, v_n$ are the free variables in it, we add the formula

$$\mathrm{FUL}_m v_1 v_2 \cdots v_{n-1} v_n v_n \cdots v_n$$

to the body and obtain a new rule

$$\beta \leftarrow \alpha_1, \ldots, \alpha_l, \mathrm{FUL}_m v_1 v_2 \cdots v_{n-1} v_n \cdots v_n.$$

For each new rule, we

- replace $x = y$ by $i_x = i_{x \bar{\wedge} y}, i_{x \bar{\wedge} y} = i_y$ (reason: $d(x) = d(x \bar{\wedge} y) = d(y)$), for constant **root**, we replace $i_{\mathbf{root}}$ by constant $\mathbf{0}$, and replace $i_{\mathbf{root} \bar{\wedge} x}$ also by $\mathbf{0}$, since $\mathbf{root} \bar{\wedge} a = \mathbf{root}$ for any node a;
- replace Exy by $i_x = i_{x \bar{\wedge} y}, \mathrm{SUCC} i_{x \bar{\wedge} y} i_y$ (reason: $d(y) = d(x \bar{\wedge} y) + 1 = d(x) + 1$);
- replace $x \neq y$ by $R_{\neq} i_x i_{x \bar{\wedge} y} i_y$ (reason: $R_{\neq}$ is the characteristic relation of $\neq$);
- replace $\neg Exy$ by $R_{\neg e} i_x i_{x \bar{\wedge} y} i_y$ (reason: $R_{\neg e}$ is the characteristic relation of $\neg E$);
- replace $P\bar{x}$ by $P^*(\bar{x})^*$, where P is in $\{R_1, \ldots, R_k, \mathrm{FUL}_m\}$, or an intensional relation symbol;
- replace $\neg R\bar{x}$ by $(\neg R)^*(\bar{x})^*$, where R is a symbol in $\{R_1, \ldots, R_k\}$;
- replace $\forall y_1 \cdots \forall y_t P y_1 \cdots y_t z_1 \cdots z_s$ by

$$\Psi_P = \begin{pmatrix} \mathrm{FUL}_m^*(z_1 z_2 \cdots z_{s-1} z_s \cdots z_s)^* \wedge \\ \forall (y_1 \cdots y_t)^* \forall i_{y_1 \bar{\wedge} z_1} \cdots \forall i_{y_1 \bar{\wedge} z_s} \forall i_{y_2 \bar{\wedge} z_1} \cdots \forall i_{y_t \bar{\wedge} z_{s-1}} \forall i_{y_t \bar{\wedge} z_s} \\ (\mathrm{FUL}_m^*(y_1 \cdots y_t z_1 \cdots z_s \cdots z_s)^* \rightarrow P^*(y_1 \cdots y_t z_1 \cdots z_s)^*) \end{pmatrix}$$

  where P is an intensional relation symbol.

By adding FUL_m to each rule of Π and replacing it with FUL_m^* in Π', we can restrict to characteristic tuples. Π' is not a Datalog^r program because of Ψ_P. Note that Ψ_P is equivalent to the Datalog^r formula $(\Pi_1, Q_2)\bar{t}$, where

$$\begin{aligned} \Pi_1 : Q(y_1 \cdots y_t z_1 \cdots z_s)^* &\leftarrow \neg \mathrm{FUL}_m^*(y_1 \cdots y_t z_1 \cdots z_s z_s \cdots z_s)^*; \\ Q(y_1 \cdots y_t z_1 \cdots z_s)^* &\leftarrow P^*(y_1 \cdots y_t z_1 \cdots z_s)^*; \\ Q_1(z_1 \cdots z_s)^* &\leftarrow \forall (y_1 \cdots y_t)^* \forall i_{y_1 \bar{\wedge} z_1} \cdots \forall i_{y_1 \bar{\wedge} z_s} \\ &\qquad \forall i_{y_2 \bar{\wedge} z_1} \cdots \forall i_{y_t \bar{\wedge} z_{s-1}} \forall i_{y_t \bar{\wedge} z_s} Q(y_1 \cdots y_t z_1 \cdots z_s)^*; \\ Q_2(z_1 \cdots z_s)^* &\leftarrow Q_1(z_1 \cdots z_s)^*, \mathrm{FUL}_m^*(z_1 z_2 \cdots z_{s-1} z_s \cdots z_s)^*. \end{aligned}$$

The Datalog^r program Π^* can be obtained by adding Π_1 to Π' and changing Ψ_P to $Q_2(z_1 \cdots z_s)^*$.

Remark 1. We cannot replace $\forall y_1 \cdots \forall y_t P y_1 \cdots y_t z_1 \cdots z_s$ directly by

$$\forall (y_1 \cdots y_t)^* \forall i_{y_1 \bar{\wedge} z_1} \cdots \forall i_{y_1 \bar{\wedge} z_s} \forall i_{y_2 \bar{\wedge} z_1} \cdots \forall i_{y_t \bar{\wedge} z_{s-1}} \forall i_{y_t \bar{\wedge} z_s} P^*(y_1 \cdots y_t z_1 \cdots z_s)^*$$

since there may be $\mathbf{T} \in \mathcal{T}$, $\bar{a} \in \mathbf{T}$, and an invariant relation P such that

$$\begin{aligned} &\mathbf{T} \models \forall y_1 \cdots \forall y_t P y_1 \cdots y_t z_1 \cdots z_s[\bar{a}], \text{ and} \\ &S_{\mathbf{T}} \not\models \forall (y_1 \cdots y_t)^* \forall i_{y_1 \bar{\wedge} z_1} \cdots \forall i_{y_1 \bar{\wedge} z_s} \forall i_{y_2 \bar{\wedge} z_1} \cdots \forall i_{y_t \bar{\wedge} z_{s-1}} \forall i_{y_t \bar{\wedge} z_s} \\ &\qquad P^*(y_1 \cdots y_t z_1 \cdots z_s)^*[(\bar{a})^*]. \end{aligned}$$

For example, let $\mathbf{T}$ be the structure with the perfect binary tree of Fig. 3 and relation

$$P = \{(\mathbf{root}, \mathbf{root}), (\mathbf{root}, a), (\mathbf{root}, b), (\mathbf{root}, c), (\mathbf{root}, d), (\mathbf{root}, e), (\mathbf{root}, f)\}.$$

Obviously, we have $\mathbf{T} \models \forall y P(\mathbf{root}, y)$. But $S_{\mathbf{T}} \not\models P^*(\mathbf{0}, 1, 1)$ since $(\mathbf{0}, 1, 1)$ is not the characteristic tuple of any tuple in $\mathbf{T}$. This problem can be solved by changing $\forall y_1 \cdots \forall y_t P y_1 \cdots y_t z_1 \cdots z_s$ to the equivalent formula

$$\mathrm{FUL}_m z_1 \cdots z_{s-1} z_s \cdots z_s \wedge \forall y_1 \cdots \forall y_t (\mathrm{FUL}_m y_1 \cdots y_t z_1 \cdots z_s \cdots z_s \rightarrow P y_1 \cdots y_t z_1 \cdots z_s).$$

We replace FUL_m and P by their characteristic relations FUL_m^* and P^*, respectively, to obtain Ψ_P. This guarantees that only characteristic tuples are considered.

Example 2. The following Datalog^r program Π computes the transitive closure R of edges E

$$\begin{aligned} \Pi : R x_1 x_2 &\leftarrow E x_1 x_2; \\ R x_1 x_3 &\leftarrow R x_1 x_2, E x_2 x_3. \end{aligned}$$

The corresponding Datalog^r program Π^* below computes the characteristic relation R^* of R.

$$\begin{aligned} \Pi^* : R^* i_{x_1} i_{x_1 \bar{\wedge} x_2} i_{x_2} &\leftarrow i_{x_1} = i_{x_1 \bar{\wedge} x_2}, \mathrm{SUCC} i_{x_1 \bar{\wedge} x_2} i_{x_2}, \\ &\quad \mathrm{FUL}_3^* i_{x_1} i_{x_1 \bar{\wedge} x_2} i_{x_1 \bar{\wedge} x_2} i_{x_2} i_{x_2} i_{x_2}; \\ R^* i_{x_1} i_{x_1 \bar{\wedge} x_3} i_{x_3} &\leftarrow R^* i_{x_1} i_{x_1 \bar{\wedge} x_2} i_{x_2}, i_{x_2} = i_{x_2 \bar{\wedge} x_3}, \mathrm{SUCC} i_{x_2 \bar{\wedge} x_3} i_{x_3}, \\ &\quad \mathrm{FUL}_3^* i_{x_1} i_{x_1 \bar{\wedge} x_2} i_{x_1 \bar{\wedge} x_3} i_{x_2} i_{x_2 \bar{\wedge} x_3} i_{x_3}. \end{aligned}$$

Lemma 4. *Given $\psi_P = \forall y_1 \cdots \forall y_t P y_1 \cdots y_t z_1 \cdots z_s$, a structure $\mathbf{T} \in \mathcal{T}$, let Ψ_P be defined as above, and $Q_1 = \{\bar{a} \mid \mathbf{T} \models \psi_P[\bar{a}]\}$, $Q_2 = \{\bar{e} \mid S_{\mathbf{T}} \models \Psi_P[\bar{e}]\}$. If P is an invariant relation on $\mathbf{T}$, then $(Q_1)^* = Q_2$.*

Proof. Because P is an invariant relation, by Lemma 3 and the definition of Q_1, we know that Q_1 is also an invariant relation. We first show that $(Q_1)^* \subseteq Q_2$. Suppose that $\bar{e} \in (Q_1)^*$ for some $\bar{e} \in S_{\mathbf{T}}$, there must exist a tuple $\bar{a}$ from $\mathbf{T}$ such that $\bar{a} \in Q_1$, $(\bar{a})^* = \bar{e}$, and $\bar{b}\bar{a} \in P$ for all tuples $\bar{b}$ of $\mathbf{T}$, i.e.,

$$\begin{aligned} \mathbf{T} \models \Big(&\mathrm{FUL}_m z_1 \cdots z_{s-1} z_s z_s \cdots z_s \wedge \\ &\forall y_1 \cdots \forall y_t (\mathrm{FUL}_m y_1 \cdots y_t z_1 \cdots z_s z_s \cdots z_s \rightarrow P y_1 \cdots y_t z_1 \cdots z_s)\Big)[\bar{a}]. \end{aligned}$$

By the definition of Ψ_P we see that $S_{\mathbf{T}} \models \Psi_P[(\bar{a})^*]$, which implies $\bar{e} \in Q_2$.

To prove $Q_2 \subseteq (Q_1)^*$, consider an arbitrary tuple $\bar{e} \in S_{\mathbf{T}}$ such that $\bar{e} \in Q_2$. By the definition of Q_2 and Ψ_P, we have $S_{\mathbf{T}} \models \mathrm{FUL}_m(z_1 \cdots z_s z_s \cdots z_s)^*[\bar{e}]$, so there exists a tuple $\bar{a}$ of $\mathbf{T}$ such that $\bar{e} = (\bar{a})^*$. On the contrary, assume $\bar{a} \notin Q_1$, then there is a tuple $\bar{b}$ such that $\bar{b}\bar{a} \notin P$. Because P is an invariant relation, for any tuples $\bar{b}'$ and $\bar{a}'$, if $(\bar{b}\bar{a})^* = (\bar{b}'\bar{a}')^*$ then $\bar{b}'\bar{a}' \notin P$. Combing that P^* is the characteristic relation of P we conclude that

$$S_{\mathbf{T}} \not\models P^*(y_1 \cdots y_t z_1 \cdots z_s)^*[(\bar{b}\bar{a})^*], \text{and} \tag{1}$$
$$S_{\mathbf{T}} \models \mathrm{FUL}_m^*(y_1 \cdots y_t z_1 \cdots z_s z_s \cdots z_s)^*[(\bar{b}\bar{a})^*]. \tag{2}$$

(1) and (2) give $S_{\mathbf{T}} \not\models \Psi_P[(\bar{a})^*]$. Hence, $\bar{e} \notin Q_2$, contrary to the assumption that $\bar{e} \in Q_2$. Therefore, $\bar{a}$ must be in Q_1, which implies $\bar{e} \in (Q_1)^*$. □

Let P be an intensional relation symbol in Π, and $\mathbf{T}$ a structure in $\mathcal{T}$. We use $P_{(n)}$ $(n > 0)$ to denote the relation obtained in the n-th evaluation of Π on $\mathbf{T}$ for P, and $P^{\mathbf{T}[\Pi]}$ to denote the relation obtained by applying Π on $\mathbf{T}$ for P, i.e., the fixpoint of the sequence $P_{(0)}, P_{(1)}, P_{(2)}, \ldots$

Proposition 6. *For any intensional relation symbol P in Π and any $\mathbf{T} \in \mathcal{T}$, $P^{\mathbf{T}[\Pi]}$ is an invariant relation on $\mathbf{T}$ and $(P^{\mathbf{T}[\Pi]})^* = (P^*)^{S_{\mathbf{T}}[\Pi^*]}$. Moreover, if P is a 0-ary intensional relation symbol, then $\mathbf{T} \models (\Pi, P)$ iff $S_{\mathbf{T}} \models (\Pi^*, P^*)$.*

Proof. We first show that if P is an intensional relation symbol in Π and $\mathbf{T}$ is a structure in $\mathcal{T}$, then $P^{\mathbf{T}[\Pi]}$ is an invariant relation on $\mathbf{T}$. Let $P^1, \ldots, P^{m'}$ be all intensional relation symbols in Π. Consider the following formula constructed for each P^i

$$\phi_{P^i}(\bar{x}_{P^i}) = \bigvee\{\exists\bar{v}(\alpha_1 \wedge \cdots \wedge \alpha_l) \mid P^i\bar{x}_{P^i} \leftarrow \alpha_1, \ldots, \alpha_l \in \Pi \text{ and } \bar{v} \text{ are the free variables in } \alpha_1 \wedge \cdots \wedge \alpha_l \text{ that are different from } \bar{x}_{P^i}\}.$$

If the relation defined by each α_s is an invariant relation, then by Lemmas 1, 2 and 3, we know that the relation defined by ϕ_{P^i} is also an invariant relation. Each α_s is either an atomic (or negated atomic) formula with the relation symbol from $\{=, E, R_1, \ldots, R_k\}$ where the relations defined by them are all invariant relations, or an atomic formula $P^j\bar{x}$, or a formula $\forall\bar{y}P^j\bar{y}\bar{z}$ $(1 \leq j \leq m')$.

When computing the fixpoint of $P^1, \ldots, P^{m'}$, we set $P^i_{(0)} = \emptyset$ $(1 \leq i \leq m')$, where $\emptyset$ is an invariant relation. By Lemma 3 we know that if P^j is an invariant relation then the relation defined by $\forall\bar{y}P^j\bar{y}\bar{z}$ is also an invariant relation. We proceed by induction on n. Suppose that $P^1_{(n)}, \ldots, P^{m'}_{(n)}$ are invariant relations, then each

$$P^i_{(n+1)} = \{\bar{a} \mid (\mathbf{T}, P^1_{(n)}, \ldots, P^{m'}_{(n)}) \models \phi_{P^i}(\bar{x}_{P^i})[\bar{a}]\}, \text{or}$$
$$P^i_{(n+1)} = \{\emptyset \mid (\mathbf{T}, P^1_{(n)}, \ldots, P^{m'}_{(n)}) \models \phi_{P^i}\}$$

is also an invariant relation. Therefore, the fixpoints $P^1_{(\infty)}, \ldots, P^{m'}_{(\infty)}$ are invariant relations, i.e., $P^{\mathbf{T}[\Pi]}$ is an invariant relation on $\mathbf{T}$.

Next we shall show that $(P^{\mathbf{T}[\Pi]})^* = (P^*)^{S_{\mathbf{T}}[\Pi^*]}$. It suffices to prove that $(P^i_{(n)})^* = (P^i)^*_{(n)}$ $(1 \leq i \leq m')$ for each $n \geq 0$. The proof is by induction on n.

Basis: If $n = 0$, then $P^i_{(0)} = \emptyset$, $(P^i)^*_{(0)} = \emptyset$ $(1 \leq i \leq m')$. We have $(P^i_{(0)})^* = (P^i)^*_{(0)}$ $(1 \leq i \leq m')$.

Inductive Step: Assuming $(P^i_{(k)})^* = (P^i)^*_{(k)}$ $(1 \leq i \leq m')$, we show that $(P^i_{(k+1)})^* = (P^i)^*_{(k+1)}$ $(1 \leq i \leq m')$. The case where P^i is 0-ary is trivial, in the following we only consider the relation P^i of no 0-ary.

To prove $(P^i_{(k+1)})^* \subseteq (P^i)^*_{(k+1)}$, suppose $\bar{e} \in (P^i_{(k+1)})^*$ for some $\bar{e} \in S_{\mathbf{T}}$. There must be a tuple $\bar{a}$ of $\mathbf{T}$ such that $\bar{a} \in P^i_{(k+1)}$ and $\bar{e} = (\bar{a})^*$. By the semantics of Datalog^r we know that

$$P^i_{(k+1)} = \{\bar{a} \mid (\mathbf{T}, P^1_{(k)}, \ldots, P^{m'}_{(k)}) \models \phi_{P^i}(\bar{x}_{P^i})[\bar{a}]\}.$$

By the definition of ϕ_{P^i}, there is a rule $P^i\bar{x}_{P^i} \leftarrow \alpha_1, \ldots, \alpha_l$ in Π such that

$$\langle \mathbf{T}, P^1_{(k)}, \ldots, P^{m'}_{(k)} \rangle \models \exists \bar{v}(\alpha_1 \wedge \cdots \wedge \alpha_l)[\bar{a}].$$

Thus, there exists some $\bar{b}$ such that

$$\langle \mathbf{T}, P^1_{(k)}, \ldots, P^{m'}_{(k)} \rangle \models (\alpha_1 \wedge \cdots \wedge \alpha_l)[\bar{a}\bar{b}].$$

Because $P^i\bar{x}_{P^i} \leftarrow \alpha_1, \ldots, \alpha_l$ is a rule of Π, we can infer that

$$(P^i)^*(\bar{x}_{P^i})^* \leftarrow \alpha'_1, \ldots, \alpha'_l, \mathrm{FUL}^*_m(\bar{x}_{P^i}\bar{v}\tilde{v}')^*$$

is a rule of Π^*, where $\alpha'_1, \ldots, \alpha'_l$ and $\mathrm{FUL}^*_m(\bar{x}_{P^i}\bar{v}\tilde{v}')^*$ are obtained by replacing $\alpha_1, \ldots, \alpha_l, \mathrm{FUL}_m$ with the corresponding formulas respectively in the construction of Π^*. Note that we replace $\forall \bar{y} P \bar{y}\bar{z}$ by Ψ_P, and by Lemma 4 the relation defined by Ψ_P is the characteristic relation of that defined by $\forall \bar{y} P \bar{y}\bar{z}$. By the definition of $S_{\mathbf{T}}$ and the induction hypothesis $(P^i_{(k)})^* = (P^i)^*_{(k)}$ $(1 \le i \le m')$ we deduce that

$$\langle S_{\mathbf{T}}, (P^1)^*_{(k)}, \ldots, (P^{m'})^*_{(k)} \rangle \models (\alpha'_1 \wedge \cdots \wedge \alpha'_l \wedge \mathrm{FUL}^*_m(\bar{x}_{P^i}\bar{v}\tilde{v}')^*)[(\bar{a}\bar{b})^*], \text{i.e.,}$$
$$\langle S_{\mathbf{T}}, (P^1)^*_{(k)}, \ldots, (P^{m'})^*_{(k)} \rangle \models \exists \bar{u}(\alpha'_1 \wedge \cdots \wedge \alpha'_l \wedge \mathrm{FUL}^*_m(\bar{x}_{P^i}\bar{v}\tilde{v}')^*)[(\bar{a})^*]$$

where $\bar{u}$ are the free variables in $\alpha'_1 \wedge \cdots \wedge \alpha'_l \wedge \mathrm{FUL}^*_m(\bar{x}_{P^i}\bar{v}\tilde{v}')^*$ that are different from $(\bar{x}_{P^i})^*$. Combining $\bar{e} = (\bar{a})^*$ we obtain $\bar{e} \in (P^i)^*_{(k+1)}$.

To prove $(P^i)^*_{(k+1)} \subseteq (P^i_{(k+1)})^*$, suppose $\bar{e} \in (P^i)^*_{(k+1)}$ for some $\bar{e} \in S_{\mathbf{T}}$. There must exist a rule

$$(P^i)^*(\bar{x}_{P^i})^* \leftarrow \alpha'_1, \ldots, \alpha'_l, \mathrm{FUL}^*_m(\bar{x}_{P^i}\bar{v}\tilde{v}')^* \tag{3}$$

in Π^* such that

$$\langle S_{\mathbf{T}}, (P^1)^*_{(k)}, \ldots, (P^{m'})^*_{(k)} \rangle \models \exists \bar{u}(\alpha'_1 \wedge \cdots \wedge \alpha'_l \wedge \mathrm{FUL}^*_m(\bar{x}_{P^i}\bar{v}\tilde{v}')^*)[\bar{e}].$$

Hence there exists a tuple $\bar{f} \in S_{\mathbf{T}}$ such that

$$(S_{\mathbf{T}}, (P^1)^*_{(k)}, \ldots, (P^{m'})^*_{(k)}) \models (\alpha'_1 \wedge \cdots \wedge \alpha'_l \wedge \mathrm{FUL}^*_m(\bar{x}_{P^i}\bar{v}\tilde{v}')^*)[\bar{e}\bar{f}].$$

The formula $\mathrm{FUL}^*_m(\bar{x}_{P^i}\bar{v}\tilde{v}')^*$ guarantees that $(\bar{a}\bar{b})^* = \bar{e}\bar{f}$ and $(\bar{a})^* = \bar{e}$ for some tuple $\bar{a}\bar{b}$ of $\mathbf{T}$. Because $P^1_{(k)}, \ldots, P^{m'}_{(k)}$ are invariant relations, by the induction hypothesis $(P^i_{(k)})^* = (P^i)^*_{(k)}$ $(1 \le i \le m')$ we know that

$$\langle \mathbf{T}, P^1_{(k)}, \ldots, P^{m'}_{(k)} \rangle \models (\alpha_1 \wedge \cdots \wedge \alpha_l)[\bar{a}\bar{b}]$$

where $\alpha_1, \ldots, \alpha_l$ occur in the rule $P^i\bar{x}_{P^i} \leftarrow \alpha_1, \ldots, \alpha_l$ that is the original of (3) in Π. Hence $\bar{a} \in P^i_{(k+1)}$, which implies $\bar{e} \in (P^i_{(k+1)})^*$. This completes the proof. □

3.4 Nondefinability Results for Datalog^r

The complexity class EXPTIME contains the decision problems decidable by a deterministic Turing machine in $O(2^{n^c})$ time. By the time hierarchy theorem, we know that PTIME is a proper subset of EXPTIME. In this section, for every class of structures $\mathcal{K} \in$ EXPTIME, we construct a class $\mathcal{K}'$ of structures that is in PTIME and closed under substructures, and show that if $\mathcal{K}'$ is definable by a Datalog^r formula, then $\mathcal{K}$ is in P, which is impossible.

Let c be a constant, and $\mathbf{A} = \langle\{0,\ldots,h-1\}, R_1^A,\ldots,R_k^A\rangle$ a τ-structure. The *trivial extension* $\mathbf{A}^+ = \langle\{0,\ldots,h-1,h,\ldots,h+h^c-1\}, R_1^A,\ldots,R_k^A\rangle$ of $\mathbf{A}$ is a τ-structure obtained by adding h^c dummy elements to the domain of $\mathbf{A}$ and keeping all other relations unchanged.

For technical reasons we introduce a new unary relation symbol U and let $\tau_U = \tau \cup \{U\}$, $\tau_U' = \tau \cup \{\mathbf{root}, E, U\}$. From now on when we speak of a τ_U'-structure

$$\mathbf{G} = \langle V, \mathbf{root}, E, U, R_1, \cdots, R_k\rangle$$

we assume that

(1) $\langle V, E, \mathbf{root}\rangle$ is a directed acyclic graph and the nodes reachable from **root** form a binary tree, and

(2) all relations $U, R_1, \cdots, R_k$ are saturated relations restricted on $T(\mathbf{G})$, which is the largest perfect binary subtree of $\mathbf{G}$ with **root** as the root.

It is easy to check that if a τ_U'-structure $\mathbf{G}$ satisfies the aforementioned two conditions, then all its substructures also satisfy the two conditions.

Definition 7. *Let $\mathcal{K}$ be a class of τ-structures. Define a class $\mathcal{K}'$ of τ_U'-structures such that, for any $\mathbf{G} = \langle V, \mathbf{root}, E, U, R_1,\ldots,R_k\rangle$, let h be the largest number where all nodes in the first h levels of $T(\mathbf{G})$ are marked by U, $\mathbf{G} \in \mathcal{K}'$ iff the following Condition (1) or Condition (2) holds.*

Condition (1)

- **(a)** *The depth of $T(\mathbf{G})$ is $h + h^c$.*
- **(b)** *The relations $R_1,\ldots,R_k$ do not hold on any tuple that contains a node in the last h^c consecutive levels of $T(\mathbf{G})$.*
- **(c)** *$C^{-1}(T(\mathbf{G}))$ is the trivial extension of $C^{-1}(T_h(\mathbf{G}))$, where $T_h(\mathbf{G})$ is the subtree of $T(\mathbf{G})$ by restricting to the first h levels.*
- **(d)** *$C^{-1}(T_h(\mathbf{G})) \in \mathcal{K}$ when ignoring the relation U.*

Condition (2)

- **(a)** *The depth of $T(\mathbf{G})$ is strictly less than $h + h^c$.*

Proposition 7. *Let $\mathcal{K}$ be an arbitrary class of τ-structures decidable in 2^{n^c} time, where n is the cardinality of the structure's domain, and $\mathcal{K}'$ defined as above. Then*

(i) *$\mathcal{K}'$ is closed under substructures;*
(ii) *$\mathcal{K}'$ is decidable in* PTIME.

Proof. To prove (i), suppose that $\mathbf{G}$ is a τ'_U-structure in $\mathcal{K}'$, then it satisfies either Condition (1) or Condition (2) in Definition 7. Let $\mathbf{H}$ be an arbitrary substructure of $\mathbf{G}$. If $\mathbf{G}$ satisfies Condition (2), then $\mathbf{H}$ also satisfies Condition (2), and is in $\mathcal{K}'$. Suppose that $\mathbf{G}$ satisfies Condition (1), then the perfect binary tree $T(\mathbf{H})$ either equals $T(\mathbf{G})$, which implies $\mathbf{H}$ satisfies Condition (1), or the depth of $T(\mathbf{H})$ is less than that of $T(\mathbf{G})$, which implies $\mathbf{H}$ satisfies Condition (2). Altogether, $\mathbf{H} \in \mathcal{K}'$.

To prove (ii), let $\mathbf{G}$ be an arbitrary τ'_U-structure, we just need to do the following steps to check whether $\mathbf{G} \in \mathcal{K}'$:

(1) Check that $\langle V, E \rangle$ is a directed acyclic graph.
(2) Check that all nodes reachable from **root** form a binary tree.
(3) Compute $T(\mathbf{G})$, the largest perfect binary subtree with **root** as root.
(4) Check that $U, R_1, \ldots, R_k$ are saturated relations on $T(\mathbf{G})$.
(5) Compute the largest number h such that all nodes in the first h levels of $T(\mathbf{G})$ have property U.
(6) Check whether the depth of $T(\mathbf{G})$ is less than $h + h^c$.
(7) If the depth of $T(\mathbf{G})$ is $h + h^c$, then check whether (b), (c) and (d) in Condition (1) of Definition 7 hold.

Note that $C^{-1}(T_h(\mathbf{G}))$ has h elements and $\mathcal{K}$ is decidable in 2^{n^c} time, the statement (d) in Condition (1) of Definition 7 can be verified in polynomial time since if the depth of $T(\mathbf{G})$ is $h + h^c$ then the input size is at least 2^{h+h^c}. □

Let $\mathcal{K}$ and $\mathcal{K}'$ be defined as in Proposition 7. For an arbitrary τ-structure $\mathbf{A}$, let $\mathbf{A}_U$ be the τ_U-structure obtained by marking every element in $\mathbf{A}$ by U, $\mathbf{A}_U^+$ the trivial extension of $\mathbf{A}_U$ by adding $|A|^c$ elements, and $\mathbf{T}$ the τ'_U-structure such that $C^{-1}(\mathbf{T}) = \mathbf{A}_U^+$. If $\mathcal{K}'$ is axiomatizable by a Datalog^r formula (Π, Q), then

$$\mathbf{A} \in \mathcal{K} \quad \text{iff} \quad \mathbf{T} \in \mathcal{K}' \quad \text{iff} \quad \mathbf{T} \models (\Pi, Q). \tag{4}$$

Define the vocabulary

$$\sigma_U = \{\mathbf{0}, \mathrm{SUCC}, R_{\neq}, R_{\neg e}, \mathrm{FUL}_m^*, U^*, R_1^*, \ldots, R_k^*, (\neg U)^*, (\neg R_1)^*, \ldots, (\neg R_k)^*\}.$$

By Definition 6, we can compute $\mathbf{T}$'s characteristic structure that is a σ_U-structure

$$S_{\mathbf{T}} = \big\langle \{0, 1, \ldots, |A| + |A|^c - 1\}, \mathbf{0}, \mathrm{SUCC}, R_{\neq}, R_{\neg e}, \mathrm{FUL}_m^*, U^*, \\ R_1^*, \ldots, R_k^*, (\neg U)^*, (\neg R_1)^*, \ldots, (\neg R_k)^* \big\rangle$$

where $\mathbf{0}$ is interpreted by 0, SUCC is the successor relation on the domain and $R_{\neq}, R_{\neg e}$, FUL_m^*, U^*, $R_1^*, \ldots, R_k^*, (\neg U)^*, (\neg R_1)^*, \ldots, (\neg R_k)^*$ are the characteristic relations of $\neq, (\neg E)$, FUL_m, U, $R_1, \ldots$, R_k, $\neg U, \neg R_1, \ldots, \neg R_k$, respectively. By Proposition 6, we know there is a Datalog^r formula (Π^*, Q^*) such that $\mathbf{T} \models (\Pi, Q)$ iff $S_{\mathbf{T}} \models (\Pi^*, Q^*)$. Combining (4), we have $\mathbf{A} \in \mathcal{K}$ iff $S_{\mathbf{T}} \models (\Pi^*, Q^*)$.

Lemma 5. *$S_{\mathbf{T}}$ is logspace computable from* **A**.

By Lemma 5, we know that $S_{\mathbf{T}}$ is computable from **A** in polynomial time. Hence, $\mathcal{K}$ is in PTIME. Since $\mathcal{K}$ is an arbitrary class in EXPTIME, this would imply EXPTIME=PTIME, which contradicts the time hierarchy theorem. So we must have:

Proposition 8. *There is a problem in* PTIME *and closed under substructures but not definable in* Datalog^r.

If a problem is closed under substructures, then its complement is closed under extensions. By Proposition 2, we can obtain the following corollary.

Corollary 1. DATALOG^r[E] = LFP[E] $\subsetneq$ PTIME[E].

4 Conclusion

Revised Datalog is an extension of Datalog by allowing universal quantification over intensional relations in the body of rules. On all finite structures, Datalog^r is strictly more expressive than Datalog, and has the same expressive power as that of LFP. In classical model theory, the closure properties of a formula are usually related to some syntactic properties. Many preservation theorems have been proven to reflect this relationship. When restricted to finite structures, most of these preservation theorems fail. Due to the syntax and semantics of Datalog, we can treat it as the dual of SO-HORN logic, which is closed under substructures [12]. It follows that Datalog is closed under extensions. A lot of work has been conducted between Datalog and FO (or LFP) to study the closure property. In this paper, we study the expressive power of revised Datalog on the problems that are closed under substructures. We show that Datalog^r cannot define all the problems that are in PTIME and closed under substructures. As a corollary, LFP cannot define all the extension-closed problems that are in PTIME. A method of tree encodings for arbitrary structures is used in the proof. If we replace the extension closure property by the homomorphism closure property, it is still open whether the statement also holds. This is desirable for future work.

References

1. Abiteboul, S., Hull, R., Vianu, V.: Foundations of Databases. Addison-Wesley, Boston (1995)
2. Abiteboul, S., Vianu, V.: Datalog extensions for database queries and updates. J. Comput. Syst. Sci. **43**(1), 62–124 (1991)
3. Afrati, F.N., Cosmadakis, S.S., Yannakakis, M.: On datalog vs polynomial time. J. Comput. Syst. Sci. **51**(2), 177–196 (1995)
4. Ajtai, M., Gurevich, Y.: Datalog vs first-order logic. J. Comput. Syst. Sci. **49**(3), 562–588 (1994)
5. Dantsin, E., Eiter, T., Gottlob, G., Voronkov, A.: Complexity and expressive power of logic programming. ACM Comput. Surv. **33**(3), 374–425 (2001)

6. Dawar, A., Kreutzer, S.: On datalog vs. LFP. In: Aceto, L., Damgård, I., Goldberg, L.A., Halldórsson, M.M., Ingólfsdóttir, A., Walukiewicz, I. (eds.) ICALP 2008. LNCS, vol. 5126, pp. 160–171. Springer, Heidelberg (2008). https://doi.org/10.1007/978-3-540-70583-3_14
7. Dawar, A., Sankaran, A.: Extension preservation in the finite and prefix classes of first order logic. In: Baier, C., Goubault-Larrecq, J. (eds.) 29th EACSL Annual Conference on Computer Science Logic, CSL 2021, 25–28 January 2021, Ljubljana, Slovenia. LIPIcs, vol. 183, pp. 18:1–18:13. Schloss Dagstuhl - Leibniz-Zentrum für Informatik (2021)
8. Doan, A., Halevy, A., Ives, Z.: Principles of Data Integration. Elsevier, Amsterdam (2012)
9. Ebbinghaus, H.D., Flum, J.: Finite Model Theory. Springer-Verlag, Berlin (1995)
10. Feng, S., Zhao, X.: The complexity and expressive power of second-order extended logic. Stud. Logic **5**(1), 11–34 (2012)
11. Fernandes, A.A., Paton, N.W.: Databases. In: Meyers, R.A. (ed.) Encyclopedia of Physical Science and Technology (Third Edition), pp. 213–228. Academic Press, New York (2003)
12. Grädel, E.: The expressive power of second order horn logic. In: Choffrut, C., Jantzen, M. (eds.) STACS 1991. LNCS, vol. 480, pp. 466–477. Springer, Heidelberg (1991). https://doi.org/10.1007/BFb0020821
13. Gurevich, Y.: Datalog: a perspective and the potential. In: Barceló, P., Pichler, R. (eds.) Datalog 2.0 2012. LNCS, vol. 7494, pp. 9–20. Springer, Heidelberg (2012). https://doi.org/10.1007/978-3-642-32925-8_2
14. Immerman, N.: Number of quantifiers is better than number of tape cells. J. Comput. Syst. Sci. **22**(3), 384–406 (1981)
15. Immerman, N.: Relational queries computable in polynomial time. In: Fourteenth Annual ACM Symposium on Theory of Computing. pp. 147–152. ACM (1982)
16. Ketsman, B., Koch, C.: Datalog with negation and monotonicity. In: Lutz, C., Jung, J. (eds.) 23rd International Conference on Database Theory, ICDT 2020. Leibniz International Proceedings in Informatics, LIPIcs, Schloss Dagstuhl-Leibniz-Zentrum fur Informatik GmbH, Dagstuhl Publishing (2020)
17. Lindell, S.: An analysis of fixed-point queries on binary trees. Theoret. Comput. Sci. **85**(1), 75–95 (1991)
18. Marinescu, D.C.: Chapter 6 - cloud data storage. In: Marinescu, D.C. (ed.) Cloud Computing (Second Edition), pp. 195–233. Morgan Kaufmann (2018)
19. Revesz, P.Z.: Constraint databases: a survey. In: Thalheim, B., Libkin, L. (eds.) SiD 1995. LNCS, vol. 1358, pp. 209–246. Springer, Heidelberg (1997). https://doi.org/10.1007/BFb0035010
20. Rosen, E.: Finite Model Theory and Finite Variable Logics. Ph.D. thesis, University of Pennsylvania (1995)
21. Schlipf, J.S.: Complexity and undecidability results for logic programming. Ann. Math. Artif. Intell. **15**, 257–288 (1995)
22. Shmueli, O.: Decidability and expressiveness aspects of logic queries. In: Proceedings of the Sixth ACM SIGACT-SIGMOD-SIGART Symposium on Principles of Database Systems, PODS 1987, pp. 237–249. Association for Computing Machinery, New York (1987)
23. Siekmann, J.H.: Computational Logic, Handbook of the History of Logic, vol. 9. North Holland (2014)
24. Vardi, M.Y.: The complexity of relational query languages. In: Fourteenth Annual ACM Symposium on Theory of Computing, pp. 137–146. ACM (1982)

Of Temporary Coalitions in Terms of Concurrent Game Models, Announcements, and Temporal Projection

Dimitar P. Guelev(✉)

Institute of Mathematics and Informatics, Bulgarian Academy of Sciences, Sofia, Bulgaria
gelevdp@math.bas.bg

Abstract. We use Concurrent Game Models (CGM) in which simple conditional promises are assigned the role of negotiation steps aiming to represent the formation of temporary coalitions and their agendas. By transforming these extended CGMs into equivalent CGMs with incomplete information, established methods for rational synthesis become enabled. The interpretation of promises is compatible with that of announcements as in dynamic epistemic logics. To accommodate requirements on plays that are written wrt the runs of the original model, we use temporal projection that hides negotiation steps.

Keywords: concurrent game models · rational synthesis · temporary coalitions · announcements · temporal projection

Introduction

This paper is about reducing temporary coalitions in the context of *rational synthesis*. A concrete setting for rational synthesis for temporal objectives, without *temporary* coalitions, was originally proposed [14]. This work is a revision of the approach to the topic in [16], where it was assumed that negotiation could be abstracted away entirely and coalitions were supposed to be pairwise disjoint. Here we commit to a simple epistemic logic-based protocol for agreements on single moves in CGMs, a variant of that from [1], where, unlike *requests*, *grants*, *refusals*, etc., [23,25,29], conditional *promises* which eventually lead to *agreements*, are analyzed. Like [16], agreements that last longer than a single move can *emerge* due to voluntary renewals, a sign of *stability*. Long term agreements that are sealed ahead of all implementation are left outside this paper, to facilitate reasonable self-containedness. Unlike [16], starting from a CGM for the given game, here we do the obvious thing by observing that negotiation moves are interleaved with the given game's *implementation* ones. Negotiation actions are unilateral conditional promises, and (overlapping) temporary coalitions (agreements) are formed by the exchange of promises. E.g., the timetable and price list of a transportation service, which are conditional promises on behalf of that

N. Alechina et al. (Eds.): LORI 2023, LNCS 14329, pp. 126–134, 2023.
https://doi.org/10.1007/978-3-031-45558-2_10

service, together with the purchase of a ticket, which makes the promise's condition true, lead to a temporary agreement with the service earning a fare and the passenger making a leg of her journey to a point where she will possibly rely on the viability of another agreement for spending a night and further agreements for the following legs of the journey, each of these agreements being expected to be desirable to the parties involved *at their respective starting times*. Implementation actions are suppressed in moves with negotiation actions in them and *filibustering* is avoided by requiring promises to be increasingly restrictive. Several agreements can be entered simultaneously as long as all of them can be honoured, which is straightforward to check mechanically. At this point this model is markedly simpler than reality because (a) checking promises requires awareness such as what only the promisor is certain to have, and (b) it is usual of, e.g. a bank, to be ultimately unable to deliver in full. Since negotiation is about players' mental states, the resulting CGM is always a partial information one. Voluntary renewals restrict the need for co-ordination to accounting of other players' promises as a distinct type of activity. This setting brings a reduction of temporary coalitions to rational synthesis for independent players, e.g. with ordered objectives as in [8].

We subscribe to this approach here but stayed away from it in [16]. The balance between [16] and this paper was tipped for us by [12] where the possibility to simultaneously participate in several coalitions was considered, de-facto generalizing coalitions to agreement signatory sets with players allowed to enter multiple agreements each. Along with announcements [1,4,5,15,22], where the idea of epistemic actions that restrict mental states maps to coalition agreements restricting future behaviour, the approach also concurs with *epistemic planning* [3,7,26], where epistemic and implementation actions are unified. In our setting, within a period of negotiation, promises (offers as in auctions) 'stand'. Once an implementation action occurs, 'things change' and promises become obsolete. Our work can be related to the system proposed and investigated in [10] too, where axiomatic systems are centerstage and the semantic compatibility arises from the use of epistemic models.

We complement the reduction of problems about temporary coalitions to the respective thoroughly studied rational synthesis problems by an off-the-shelf translation based on temporal projection of temporal conditions on the given game's model, such as players' objectives, into conditions on the model with negotiation moves interleaved.

1 Preliminaries

Concurrent Game Models (CGM). An (incomplete information) CGM for some given sets of *players* $\Sigma = \{1, \ldots, N\}$ and *atomic propositions* AP is a tuple

$$M \hat{=} \langle W, W_I, \langle Act_i : i \in \Sigma \rangle, \langle \sim_i : i \in \Sigma \rangle, \langle P_i : i \in \Sigma \rangle, o, V \rangle \tag{1}$$

where W is the set of *states*, $W_I \subseteq W$ is the set of the *initial* states, Act_i is the set of *actions* of player $i \in \Sigma$, $Act_\Gamma \hat{=} \prod_{i \in \Gamma} Act_i$, $\sim_i$, $i \in \Sigma$, are equivalence relations on W, $P_i : W \to \mathcal{P}(Act_i)\backslash\{\emptyset\}$ is the *protocol* of player i, $i \in \Sigma$, $o : W \times Act_\Sigma \to W$ is the *outcome* function, and $V \subseteq W \times AP$ is the *valuation* relation. Given a $\mathbf{w} \in W^+$, we put

$$R_M^{\mathrm{inf}}(\mathbf{w}) \hat{=} \{\mathbf{v} \in W^\omega : \mathbf{v}^0 \ldots \mathbf{v}^{|\mathbf{w}|-1} = \mathbf{w}, (\forall k)(\exists a \in \prod_{i \in \Sigma} P_i(\mathbf{v}^k))(\mathbf{v}^{k+1} = o(\mathbf{v}^k, a))\}.$$

for the set of all the infinite plays in M which are continuations of $\mathbf{w}$. The set $R_M^{\mathrm{fin}}(\mathbf{w}) \subseteq W^+$ of the *finite* continuations of $\mathbf{w}$ is defined similarly. $P_i(w') = P_i(w'')$ is required for $w', w'' \in W$ such that $w' \sim_i w''$. We denote $\mathbf{w}^{k'} \ldots \mathbf{w}^{k''}$ by $\mathbf{w}[k'..k'']$. Only immediate observation ability is captured by $\sim_i$. Under the *perfect recall* assumption, $\mathbf{w}_1 \sim_i \mathbf{w}_2 \hat{=} |\mathbf{w}_1| = |\mathbf{w}_2| \wedge \bigwedge_{k<|\mathbf{w}_1|} \mathbf{w}_1^k \sim_i \mathbf{w}_2^k$. We assume AP and Act_i, $i \in \Sigma$, to be pairwise disjoint for notational convenience.

Temporal Objectives are (TL-definable) subsets of W^ω. Comprehensive studies on such objectives with (pre-)orders to represent players' preferences such as [8] focus on *regular* conditions and include algorithmic solutions.

Epistemic Temporal Logic on CGMs. The relations $\sim_i$ enable the interpretation of epistemic modalities and reasoning about the mental states of players in epistemic logics for strategic ability as in [17,30,31] and their underlying temporal epistemic logics [13,20,21,32,33]. The system of epistemic TL that we use in this paper to specify promises and objectives is a CTL*. It has linear time past formulas ψ in the role usually played by state formulas in systems of CTL* and future formulas φ as the operands of the path quantifier:

$$\varphi ::= \bot \mid p \mid \varphi \Rightarrow \varphi \mid \mathsf{K}_i\varphi \mid \mathsf{C}_\Gamma\varphi \mid \exists\psi \mid \ominus\varphi \mid (\varphi \mathsf{S} \varphi) \quad \psi ::= \varphi \mid \psi \Rightarrow \psi \mid \bigcirc\psi \mid (\psi \mathsf{U} \psi)$$

where $p \in AP$, $i \in \Sigma$ and $\Gamma \subseteq \Sigma$.

Satisfaction has the form $M, \mathbf{w}, k \models \ldots$ for all formulas, with $\mathbf{w}$ being an infinite play and $k < \omega$:

$M, \mathbf{w}, k \models p$ iff $V(p, \mathbf{w}^k)$;
$M, \mathbf{w}, k \models \mathsf{C}_\Gamma\varphi$ iff $M, \mathbf{v}, k \models \psi$ for all $\mathbf{v}$ such that $\mathbf{v}[0..k](\bigcup_{i \in \Gamma} \sim_i)^* \mathbf{w}[0..k]$;
$M, \mathbf{w}, k \models \exists\psi$ iff $M, \mathbf{v}, k \models \psi$ for some $\mathbf{v} \in R_M^{\mathrm{inf}}(\mathbf{w}[0..k])$.

The clauses for $\bot$, $\Rightarrow$, $\bigcirc$, $\ominus$ ('yesterday'), (.S.), and (.U.) are as usual. $\top$, $\neg$, $\vee$, $\wedge$, $\equiv$ and $\forall$, are defined as usual. We write K_i for $\mathsf{C}_{\{i\}}$ and $\mathsf{P}_i\varphi$ for $\neg\mathsf{K}_i\neg\varphi$.

Storing Latest Moves. Any CGM M can be transformed into an equivalent CGM M^1 where *moves* $a \in Act_\Sigma$ are stored in their destination states. We let

$$M^1 \hat{=} \langle W^1, W^1_I, \langle Act_i : i \in \Sigma\rangle, \langle \sim^1_i : i \in \Sigma\rangle, \langle P^1_i : i \in \Sigma\rangle, o^1, V^1\rangle$$

where

$$\begin{array}{l} W^1 \hat{=} W \times (Act_\Sigma \cup \{*\}), \quad W^1_I \hat{=} W_I \times \{*\}; \\ \langle w', a'\rangle \sim^1_i \langle w'', a''\rangle \hat{=} w' \sim_i w'' \wedge (a' = a'' = * \vee a'_i = a''_i); \\ P^1_i(\langle w, a\rangle) \hat{=} P_i(w), \quad o^1(\langle w, b\rangle, a) \hat{=} \langle o(w, a), a\rangle. \end{array}$$

Here $* \notin \bigcup_{i \in \Sigma} Act_i$. M^1's vocabulary is $AP^{Act} \hat{=} AP \cup \bigcup_{i \in \Sigma} Act_i$. We let

$$V^1(x, \langle w, b\rangle) \hat{=} \begin{cases} b \neq * \wedge b_i = x, & \text{if } x \in Act_i; \\ V(x, w), & \text{if } x \in AP. \end{cases}$$

M^1 is similar to the *bound-1 unwinding* of a Kripke model, cf. e.g. [6].

2 From CGMs to CGMs with Negotiation

The building blocks of negotiation are unilateral promises where *a player* $i \in \Sigma \backslash \Gamma$ *promises to some set of prospective partners* $\Gamma \subseteq \Sigma \backslash \{i\}$ *that it will choose an action from* $B \subset Act_i$, *if the players from* Γ *promise to choose an action profile from* $A \subset Act_\Gamma$. Let $\hat{X} \hat{=} \bigvee_{x \in X} \bigwedge_{j \in \Delta} x_j$ for $X \subseteq Act_\Delta$ and $\Delta \subseteq \Sigma$. Let

$$\mathsf{promise}_{i,\Gamma}(A, B) \hat{=} \mathsf{C}_{\{i\} \cup \Gamma}(\mathsf{K}_i \hat{A} \Rightarrow \mathsf{C}_\Gamma \hat{B}) \; . \tag{2}$$

Then the above promise is fulfilled in M^1 wherever $\forall \bigcirc \mathsf{promise}_{i,\Gamma}(A, B)$ holds. Mutually binding agreements are the logical corollaries of the promises exchanged. Observe that the *knowledge* that a move $a \in A$ is to be played, and not actually playing the move makes promise (2) work. Given any CGM M, using M's corresponding M^1, a corresponding negotiation CGM $\check{M}$ can be defined with the additional facility of logging promises and rejecting promises which are not honorable. The following technical notions provide for this.

Definition 1. *Let* $\mathcal{S}_i$ *be the set of the formulas of the form (2),* $\mathcal{S}^\wedge_i \hat{=} \{ \bigwedge_{k<n} s_k : s_k \in \mathcal{S}_i, k < n, n < \omega\}$, $i \in \Sigma$. *Let* $\mathcal{S}^\wedge \hat{=} \{ \bigwedge_{k<n} s_k : s_k \in \mathcal{S}_i, k < n, n < \omega, i \in \Sigma\}$.

Definition 2. *Compound promise* $s \in \mathcal{S}^\wedge$ *is* honorable *if,*

(i) for every $a \in Act_\Sigma$, *either* $s \cup \{C_\Sigma a_i : i \in \Sigma\}$ *is consistent, or* $s \cup \{C_\Sigma a_i\}$ *is inconsistent for some* i.
(ii) there exists an $a \in Act_\Sigma$ *such that* $s \cup \{C_\Sigma a_i : i \in \Sigma\}$ *is consistent.*

Condition (i) states that an $a \in Act_\Sigma$ either fulfills all the promises in place, or a_i alone is a breach of i's promises by some $i \in \Sigma$. Condition (ii) ensures that what is promised is doable. Unlike just consistency of s, honorability implies that every agent can verify its ability to honour its promises using its own knowledge. Building on the example of AGM-style update operations [2], the *promise update* operation $\star$ prevents accepting promises that are not honorable:

Definition 3. *Given* $s_0 \in \mathcal{S}^\wedge$ *and* $s \in \mathcal{S}_i^\wedge$, $s_0 \star s \hat{=} s_0 \wedge s$, *if* $s_0 \wedge s$ *is honorable. Otherwise* $s_0 \star s \hat{=} s_0$.

In the sequel we call moves a such that $a_i \in \mathcal{S}_i^\wedge$ for some $i \in \Sigma$ *negotiation* moves. We call moves $a \in \prod_{i \in \Sigma} Act_i$ *implementation* moves. The protocols of a state $w \in W$ can be expressed as de-facto *unconditional* promises by putting $\mathsf{p}(w) \hat{=} \bigwedge_{i \in \Sigma} \mathsf{C}_\Sigma \bigvee P_i(w)$ and i's knowledge on the promises that are logical corollaries of some $s \in \mathcal{S}^\wedge$ can be expressed by $\mathsf{k}_i(s) \hat{=} \{s' \in \mathcal{S}^\wedge : \models s \Rightarrow \mathsf{K}_i s'\}$.

Definition 4. *The* Negotiation CGM (NCGM) *of CGM (1) is*

$$\check{M} \hat{=} \langle \check{W}, \check{W}_I, \langle \check{Act}_i : i \in \Sigma \rangle, \langle \check{\sim}_i : i \in \Sigma \rangle, \langle \check{P}_i : i \in \Sigma \rangle, \check{o}, \check{V} \rangle \tag{3}$$

where

$$\begin{array}{l}
\check{W} \hat{=} W^{\mathbf{1}} \times \mathcal{S}^\wedge, \quad \check{W}_I \hat{=} \{\langle w, \mathsf{p}(w) \rangle : w \in W_I^{\mathbf{1}}\}; \quad \check{Act}_i \hat{=} Act_i \cup \mathcal{S}_i^\wedge; \\
\langle w', a', s' \rangle \check{\sim}_i \langle w'', a'', s'' \rangle \hat{=} \langle w', a' \rangle \sim_i^{\mathbf{1}} \langle w'', a'' \rangle \wedge \mathsf{k}_i(s') = \mathsf{k}_i(s''); \\
\check{P}_i(\langle w, a, s \rangle) \hat{=} P_i^{\mathbf{1}}(\langle w, a \rangle) \cup \mathcal{S}_i^\wedge = P_i(w) \cup \mathcal{S}_i^\wedge; \\
\check{o}(\langle w, b, s \rangle, a) \hat{=} \left\{ \begin{array}{l} \langle o(w,a), a, \mathsf{p}(o(w,a)) \rangle, \text{ if } a \text{ is implementation;} \\ \langle w, b, (\ldots(s \star p_1) \star \ldots \star p_{K-1}) \star p_K \rangle, \\ \quad \text{if } a \text{ is negotiation and } p_1, \ldots, p_K \text{ are } a\text{'s promise actions;} \end{array} \right. \\
\check{V}(x, \langle w, a, s \rangle) \hat{=} V^{\mathbf{1}}(x, \langle w, a \rangle), \; x \in AP^{Act}.
\end{array}$$

The definition of $\check{o}$ encodes the accumulation of promises along negotiation, and the effect of implementation moves as in the given game. Observe that, if $i \neq j$, then $s \star p \star q = s \star q \star p$ for $p \in \mathcal{S}_i^\wedge$, $q \in \mathcal{S}_j^\wedge$ as a promise can only be rejected if inconsistent with previous ones *by the same player*. Hence the ordering of the promises $p_1, \ldots, p_K$ in Definition 4 is irrelevant.

3 From NCGMs to Honest Play CGMs

Given the CGM (1), its corresponding *honest play* CGM $\bar{M}$ is obtained from the corresponding $\check{M}$ by redefining $\check{P}_i$ to enable precisely those runs (plays) in which promises are honoured.

Definition 5. *Run* $\langle w^0, *, s^0 \rangle \langle w^1, a^1, s^1 \rangle \ldots \in R_{\check{M}}^{\inf}(\langle w^0, *, s^0 \rangle)$ *is* honest, *if, for all* k,

(i) $s^k \wedge \mathsf{P}_i a_i^{k+1}$ *is consistent for all* i *such that* $a_i^{k+1} \in \mathcal{S}_i^\wedge$;

(ii) $s^k \wedge \mathsf{P}_i \bigwedge\limits_{j\in\Sigma} a_j^{k+1}$ *is consistent, if* $a^k \in Act_\Sigma$, *i.e.,* a^k *is implementation;*
(iii) $s^k \wedge \neg(s^k \star a_i^{k+1})$ *is consistent for all* i *such that* $a_i^{k+1} \in \mathcal{S}_i^\wedge$.

In finite M, Condition (iii) rules out *filibustering*, that is, infinite negotiations. All the other components of $\bar{M}$ being as in $\check{M}$, we put

$$\bar{P}_i(\langle w, b, s\rangle) \hat{=} \{a \in P_i(w) : \not\models s \Rightarrow \mathsf{K}_i\neg a\} \cup \{S \in \mathcal{S}_i^\wedge : \not\models s \Rightarrow \neg S, \not\models s \Rightarrow S\}.$$

If $\langle w', a', s'\rangle \tilde{\sim}_i \langle w'', a'', s''\rangle$, then $\mathsf{k}_i(s') = \mathsf{k}_i(s'')$. The honorability and the net novelty of a promise by player i can be determined from $\mathsf{k}_i(s)$. Hence $\bar{P}_i$ respects $\tilde{\sim}_i$-equivalence as required by the definition of CGM.

In $\bar{M}$, promises work like announcements in the sense of dynamic epistemic logic [1,5,22]: $\bar{M}, \mathbf{w} \models \forall\bigcirc s^{|\mathbf{w}-1|}$ where $\langle s^{|\mathbf{w}-1|}, a^{|\mathbf{w}-1|}, s^{|\mathbf{w}-1|}\rangle \hat{=} \mathbf{w}^{|\mathbf{w}-1|}$, concurs with the satisfaction of $[\varphi]\mathsf{K}\varphi$ about truthful announcements. Hence computing $\bar{P}_i$ can be viewed as a special case of epistemic update, with every announcement (promise) restricting the continuations of the reference finite play from its latest state on.

Let $\eta \hat{=} \bigvee\limits_{i\in\Sigma} \bigvee \mathcal{S}_i^\wedge$, where $s \in \mathcal{S}_i^\wedge$ appear as action names. Then η indicates that the latest move was a negotiation one. Then, since protocols allow only implementation actions that are consistent with promises in $\bar{M}$, we have:

Proposition 1. *Let* $\langle w^0, *, \mathsf{p}(w^0)\rangle \in \bar{W}_I$, $\mathbf{w} \in R^{\inf}_{\bar{M}}(\langle w^0, *, \mathsf{p}(w^0)\rangle)$ *and* $\langle w^k, a^k, s^k\rangle \hat{=} \mathbf{w}^k$, $k \geq 1$. *Then* $\bar{M}, \mathbf{w}, k \models \forall\bigcirc(\eta\mathsf{U}(s^k \wedge \neg\eta))$ *for all* $k < \omega$. *Every honest run in* $R^{\inf}_{\check{M}}(w_I)$ *appears in* $R^{\inf}_{\bar{M}}(w_I)$ *too.*

4 Writing Player Objectives for NCGMs

Reasoning about players' objectives which are written for the given game's CGM M can be facilitated by translating them into equivalent ones where the presence of negotiation moves is accounted of and abstracted away. The translation is based on Keller's work [24]; variants for LTL, regular expressions, and discrete time ITL [9,27] can be found in [11,18], and [28], respectively. It is based on the *state projection* temporal operator Π of ITL from [19]. Here follows a definition of Π on LTL (CTL* path) formulas in CGMs:

Definition 6. *Given a run* $\mathbf{w}$ *in CGM* M *and an* LTL *formula* ψ, *let* $\mathbf{w}|_\psi$ *be obtained from* $\mathbf{w}$ *by deleting the states* $\mathbf{w}^n$ *such that* $M, \mathbf{w}, n \not\models \psi$, $n < |\mathbf{w}|$. *Let* $d(\mathbf{w}, k)$ *be the number of deletions in* $\mathbf{w}[0..k]$, $k < |\mathbf{w}|$. *Then*

$$M, \mathbf{w}, k \models \psi_1 \Pi \psi_2 \text{ iff } |\mathbf{w}|_{\psi_1}| = \omega,\ d(\mathbf{w}, k) \leq k,\ \text{and } M, \mathbf{w}|_{\psi_1}, k - d(\mathbf{w}, k) \models \psi_2.$$

In general, $\mathbf{w}|_{\psi_1}$ is not guaranteed to be a play, nor to be infinite, even if $\mathbf{w}$ is. However, both requirements are met in $\bar{M}$, in case ψ_1 is $\neg\eta$ from Proposition 1: Deleting negotiation moves correctly links the remaining states by the respective implementation moves because s^k, which restrict protocols in $\bar{M}$, become reset by implementation moves. Infinite (filibustering) sequences of negotiation moves, which could render $\mathbf{w}|_{\neg\eta}$ finite, are ruled out in finite CGM M by Condition (iii) from Definition 5. Hence an objective L written for M would be achieved along run $\mathbf{w} \in R_M^{\mathrm{inf}}(w^0)$, $w^0 \in W_I$, iff there exists a $\mathbf{v} \in R_{\bar{M}}^{\mathrm{inf}}(\langle w^0, *, \mathsf{p}(w^0)\rangle)$ such that $\bar{M}, \mathbf{v}, 0 \models (\neg\eta)\Pi L$ and $\mathbf{w}$ consists of the w-components of $\mathbf{v}|_{\neg\eta}$.

The use of (.U.) in Proposition 1 can be viewed as the translation of $(\neg\eta)\Pi s^k$.

Concluding Remarks

We have highlighted the possibility to combine some fundamental techniques into a setting that allows the reduction of temporary coalitions with the players bound by self-interest. We find the merit of this study to be that it emphasizes that conceptual complexity can be contained. Features that appear to be necessary to add to CGMs for reasoning about temporary coalitions can be enjoyed at the specification stage and then processed so that known techniques can be used in their original forms. The key observation is that, with coalitions assumed to be depending for their longevity on self-interest all the way, negotiation can be largely restricted to single move agendas. There is no really bad obstacle for the forging of long term agreements to be put in the same framework too. The approach of [10] confirms this. However, since infinitely many long term promises are possible even for finite Act_i, filibustering may require special attention.

Acknowledgement. The author wishes to thank Ben Moszkowski for his comments on this work.

References

1. Ågotnes, T., van Ditmarsch, H.P.: Coalitions and announcements. In: Proceedings of the AAMAS 2008, IFAAMAS, vol. 2, pp. 673–680 (2008)
2. Alchourrón, C.E., Gärdenfors, P., Makinson, D.: On the logic of theory change: partial meet contraction and revision functions. J. Symb. Log. **50**(2), 510–530 (1985)
3. Andersen, M.B., Bolander, T., Jensen, M.H.: Conditional epistemic planning. In: del Cerro, L.F., Herzig, A., Mengin, J. (eds.) JELIA 2012. LNCS (LNAI), vol. 7519, pp. 94–106. Springer, Heidelberg (2012). https://doi.org/10.1007/978-3-642-33353-8_8
4. Balbiani, P., Baltag, A., van Ditmarsch, H., Herzig, A., Hoshi, T., de Lima, T.: Knowable as known after an announcement. Rev. Symb. Log. **1**(3), 305–334 (2008)
5. Baltag, A., Moss, L.S., Solecki, S.: The logic of public announcements and common knowledge and private suspicions. In: Proceedings of TARK-98, Morgan Kaufmann, pp. 43–56 (1998)
6. Biere, A., Kröning, D.: SAT-based model checking. In: Handbook of Model Checking, pp. 277–303. Springer, Cham (2018). https://doi.org/10.1007/978-3-319-10575-8_10
7. Bolander, T., Jensen, M.H., Schwarzentruber, F.: Complexity results in epistemic planning. In: Proceedings of IJCAI 2015, pp. 2791–2797. AAAI Press (2015)

8. Bouyer, P., Brenguier, R., Markey, N., Ummels, M.: Concurrent games with ordered objectives. In: Birkedal, L. (ed.) FoSSaCS 2012. LNCS, vol. 7213, pp. 301–315. Springer, Heidelberg (2012). https://doi.org/10.1007/978-3-642-28729-9_20
9. Cau, A., Moszkowski, B., Zedan, H.: ITL web pages. http://www.antonio-cau.co.uk/ITL/
10. de Lima, T.: Alternating-time temporal dynamic epistemic logic. J. Log. Comput. **24**(6), 1145–1178 (2014)
11. Eisner, C., Fisman, D., Havlicek, J., McIsaac, A., Van Campenhout, D.: The definition of a temporal clock operator. In: Baeten, J.C.M., Lenstra, J.K., Parrow, J., Woeginger, G.J. (eds.) ICALP 2003. LNCS, vol. 2719, pp. 857–870. Springer, Heidelberg (2003). https://doi.org/10.1007/3-540-45061-0_67
12. Enqvist, S., Goranko, V.: The temporal logic of coalitional goal assignments in concurrent multi-player games. CoRR abs/2012.14195 (2020)
13. Fagin, R., Halpern, J., Moses, Y., Vardi, M.: Reasoning about Knowledge. MIT Press, Cambridge (1995)
14. Fisman, D., Kupferman, O., Lustig, Y.: Rational synthesis. In: Esparza, J., Majumdar, R. (eds.) TACAS 2010. LNCS, vol. 6015, pp. 190–204. Springer, Heidelberg (2010). https://doi.org/10.1007/978-3-642-12002-2_16
15. Galimullin, R., Alechina, N., van Ditmarsch, H.: Model checking for coalition announcement logic. In: Trollmann, F., Turhan, A.-Y. (eds.) KI 2018. LNCS (LNAI), vol. 11117, pp. 11–23. Springer, Cham (2018). https://doi.org/10.1007/978-3-030-00111-7_2
16. Guelev, D.P.: Reasoning about temporary coalitions and ltl-definable ordered objectives in infinite concurrent multiplayer games. CoRR abs/2011.03724 (2020)
17. Guelev, D.P., Dima, C., Enea, C.: An alternating-time temporal logic with knowledge, perfect recall and past: axiomatisation and model-checking. J. Appl. Non-Class. Logics **21**(1), 93–131 (2011)
18. Guelev, D.P., Ryan, M.D., Schobbens, P.Y.: Synthesising features by games. In Automated Verification of Critical Systems (AVoCS 2005), vol. 145 of ENTCS, pp. 79–93. Elsevier (2006)
19. Halpern, J., Manna, Z., Moszkowski, B.: A hardware semantics based on temporal intervals. In: Diaz, J. (ed.) ICALP 1983. LNCS, vol. 154, pp. 278–291. Springer, Heidelberg (1983). https://doi.org/10.1007/BFb0036915
20. Halpern, J., van der Meyden, R., Vardi, M.: Complete axiomatizations for reasoning about knowledge and time. SIAM J. Comput. **33**(3), 674–703 (2004)
21. Van Ditmarsch, H., van der Hoek, W., Halpern, J.Y., Kooi, B.: Handbook of Epistemic Logic. College Publications (2015)
22. Van Ditmarsch, H., van Der Hoek, W., Kooi, B.: Dynamic Epistemic Logic. Springer, Cham (2007)
23. Jennings, N.R., Faratin, P., Lomuscio, A.R., Parsons, S., Sierra, C., Wooldridge, M.: Automated negotiation: prospects, methods and challenges. Int. J. Group Decis. Negot. **10**(2), 199–215 (2001)
24. Keller, R.M.: Formal verification of parallel programs. Commun. ACM **19**(7), 371–384 (1976)
25. Kraus, S.: Strategic Negotiation in Multiagent Environments. MIT Press, Cambridge (2001)
26. Li, Y.: Multi-agent conformant planning with distributed knowledge. In: Ghosh, S., Icard, T. (eds.) LORI 2021. LNCS, vol. 13039, pp. 128–140. Springer, Cham (2021). https://doi.org/10.1007/978-3-030-88708-7_10
27. Moszkowski, B.: Temporal logic for multilevel reasoning about hardware. IEEE Comput. **18**(2), 10–19 (1985)

28. Moszkowski, B., Guelev, D.P.: An application of temporal projection to interleaving concurrency. Formal Aspects Comput. **29**(4), 705–750 (2017). https://doi.org/10.1007/s00165-017-0417-3
29. Rosenschein, J.S., Zlotkin, G.: Rules of Encounter - Designing Conventions for Automated Negotiation among Computers. MIT Press, Cambridge (1994)
30. Schobbens, P.Y.: Alternating-time logic with imperfect recall. ENTCS **85**(2), 82–93 (2004)
31. van der Hoek, W., Wooldridge, M.: Cooperation, knowledge and time: alternating-time temporal epistemic logic and its applications. Stud. Logica. **75**, 125–157 (2003)
32. van der Meyden, R., Wong, K.S.: Complete axiomatizations for reasoning about knowledge and branching time. Stud. Logica. **75**(1), 93–123 (2003)
33. van der Meyden, R., Shilov, N.V.: Model checking knowledge and time in systems with perfect recall. In: Rangan, C.P., Raman, V., Ramanujam, R. (eds.) FSTTCS 1999. LNCS, vol. 1738, pp. 432–445. Springer, Heidelberg (1999). https://doi.org/10.1007/3-540-46691-6_35

Epistemic Monadic Boolean Algebras

Juntong Guo(✉) and Minghui Ma

Institute of Logic and Cognition, Department of Philosophy, Sun Yat-sen University, Guangzhou, China
guojt5@mail2.sysu.edu.cn, mamh6@mail.sysu.edu.cn

Abstract. Epistemic monadic Boolean algebras are obtained by enriching monadic Boolean algebras with a knowledge operator. Epistemic monadic logic as the monadic fragment of first-order epistemic logic is introduced for talking about knowing things. A Halmos-style representation of epistemic monadic Boolean algebras is established. Relativizations of epistemic monadic algebras are given for modelling updates. These logics are semantically complete.

Keywords: Monadic Boolean algebra · epistemic logic · relativization

1 Introduction

Epistemic logic introduced by Hintikka [10] studies modal principles of knowledge and belief, which are widely applied in many areas. Dynamic epistemic logic enriches epistemic logic with logical dynamics, which model the update of knowledge and belief (cf. e.g. [1,2,5]). In the model-theoretic semantics, updates are viewed in general as a kind of relativization, and they extend static epistemic logic in such a way that explicit reduction axioms are admitted in the mechanism of dynamics (cf. e.g. [2]). From the algebraic perspective, updates are interpreted as quotient maps which are obtained by congruence relations with respect to information encoding epistemic actions (cf. e.g. [11–13,16]). Philosophical reflections on these works show that the existing symbolisms of dynamic epistemic logic mainly concern knowledge and belief on *propositions* which represent the information content conveyed by sentences. They study little about knowing or believing *things*. One feature of them is that they are not quantificational.

When quantification is involved in knowledge or belief, first-order modal logic seems to be a natural formalism. For example, let $T(x)$ be the predicate that stands for the property of being a tiger. Then, using a simple first-order epistemic language (cf. e.g. [6]), we can formalize the sentence that Bob knows that there are tigers simply as $K_{\mathrm{Bob}}\exists xT(x)$. An immediate philosophical question arises from the the famous distinction between *de re* and *de dicto* modality. However, the formula $\exists xK_{\mathrm{Bob}}T(x)$ seems not allowing a *de re* reading because Bob's knowing that a thing is tiger cannot be a property of that thing. Thus,

Supported by Chinese National Funding of Social Science (Grant no. 18ZDA033).

N. Alechina et al. (Eds.): LORI 2023, LNCS 14329, pp. 135–148, 2023.
https://doi.org/10.1007/978-3-031-45558-2_11

the *de re*/*de dicto* distinction cannot occur in an appropriate first-order epistemic logic. A second philosophical question concerns further the semantics of first-order epistemic logic. One possible choice is that each epistemic state in a model is assigned a domain of objects. But the constant domain assumption seems plausible due to the intuition that being an object at an epistemic state is certainly independent of the epistemic attitudes of agents. Thus, we should take the constant domain semantics and corresponding first-order epistemic logic into consideration.

There are some related works on first-order epistemic logics in the literature (cf. e.g. [15,17]). In the present paper, we work with the monadic fragment of first-order epistemic logic and use Halmos' monadic (Boolean) algebra as a tool (cf. [7]). We investigate first-order epistemic logic within the algebraic setting. A monadic Boolean algebra is a pair $(A, \exists)$ where A is a Boolean algebra and $\exists$ is an existential quantifier. The universal quantifier $\forall$ is dually defined. When a knowledge operator $\Diamond$ and its dual $\Box$ are introduced, we can express the sentence that Bob knows tigers as $\Box_{\text{Bob}}T$ where T is a unary predicate standing for tigers. We restrict to the single-agent case, and what has been done can be easily extended to multi-agent scenarios. With the knowledge operator $\Box$, the sentence that an agent knows all tigers are cats can be formalized as $\Box\forall(T \rightarrow C)$ where $\rightarrow$ is the Boolean implication and C stands for cats. We can prove $\Box T \rightarrow \Box\exists T$ which means knowing tigers implies knowing the existence of tigers.

Like Halmos' representation theorem of monadic algebras (cf. [7,8]), we prove that every epistemic monadic algebra $(A, \exists, \Diamond)$ is isomorphic to a functional one. After that, we introduce epistemic monadic logics, and a syntactically defined logic is given as a particular epistemic monadic logic. Furthermore, we present updates on epistemic monadic algebras by relativizations. Given a certain element s in an epistemic monadic algebra $(A, \exists, \Diamond)$, the relativization $(A_s, \exists_s, \Diamond_s)$ and corresponding updated epistemic monadic logic are developed. Note that these relativizations differ from the standard public announcements in the sense that iterated relativization does not produce new knowledge. Finally we give syntactically defined relativized epistemic monadic logics.

2 Monadic Boolean Algebras

An algebra $(A, \wedge, \neg, 0, \exists)$ is called a *monadic (Boolean) algebra* ('MA' for short) if $(A, \wedge, \neg, 0)$ is a Boolean algebra and $\exists$ is a unary operator on A such that the following conditions hold for all $p, q \in A$:

(M1) $\exists 0 = 0$.
(M2) $p \leq \exists p$.
(M3) $\exists(p \wedge \exists q) = \exists p \wedge \exists q$.

The lattice order $\leq$ on A is defined by: $p \leq q$ if and only if $p = p \wedge q$, or equivalently $q = p \vee q$. For all $p, q \in A$, we define $1 := \neg 0$ and the operators $p \vee q := \neg(\neg p \wedge \neg q)$, $p \rightarrow q := \neg(p \wedge \neg q)$ and $p \leftrightarrow q := (\neg(p \wedge \neg q)) \wedge (\neg(q \wedge \neg p))$. Let $\forall p := \neg\exists\neg p$. The operator $\exists$ on A is called an *existential quantifier*, and its dual $\forall$ is called a *universal quantifier*. We write $(A, \exists)$ for a monadic algebra, A is assumed to be a Boolean algebra. Let MA be the variety of all monadic algebras.

A polynomial $\exists p$ means that p exists. It is well-known that monadic algebras are algebras for the monadic fragment of first-order logic. In such a fragment, we can express the existence or universality of a set of objects. For example, "Tigers exist" can be expressed by $\exists T$ where T stands for the set of all tigers.

Fact 1 (cf. [7]). *Let $(A, \exists)$ be a monadic algebra. The following hold for all $p, q \in A$: (1) $\exists 1 = 1$; (2) $\exists(p \vee q) = \exists p \vee \exists q$; (3) if $p \leq \exists q$, then $\exists p \leq \exists q$ and $\forall p \leq \forall q$; (4) $\forall p \leq p$, $\exists\exists p = \exists p$ and $\forall\forall p = \forall p$; (5) $\exists\forall p = \forall p$ and $\forall\exists p = \exists p$.*

Lemma 1. *Let A be a Boolean algebra. Then $(A, \exists)$ is a monadic algebra if and only if for every $p, q \in A$, the following conditions hold:* (M2) $p \leq \exists p$, (M4) $\exists\exists p \leq \exists p$ *and* $(\mathrm{Adj}_{\exists,\forall})$ $\exists p \leq q$ *if and only if* $p \leq \forall q$.

Proof. Assume $(A, \exists)$ is a monadic algebra. We show that $(\mathrm{Adj}_{\exists,\forall})$ holds. Suppose $\exists p \leq q$. By Fact 1, $p \leq \exists p = \forall\exists p \leq \forall q$. Assume $p \leq \forall q$. Then $\exists p \leq \exists\forall q = \forall q \leq q$. Conversely, it suffices to derive (M1) and (M3) from (M2), (M4) and $(\mathrm{Adj}_{\exists,\forall})$. Clearly $0 \leq \forall 0$. By $(\mathrm{Adj}_{\exists,\forall})$, $\exists 0 = 0$. Let $s, r \in A$. Clearly $s \leq \forall\exists s$. Hence, if $s \leq r$, then $\exists s \leq \exists r$ and $\forall s \leq \forall r$. By (M2) and $(\mathrm{Adj}_{\exists,\forall})$, $\forall(s \wedge r) = \forall s \wedge \forall r$ and $\exists(s \vee r) = \exists s \vee \exists r$. Clearly $\exists(p \wedge \exists q) \leq \exists p \wedge \exists\exists q \leq \exists p \wedge \exists q$. Suppose $\exists(p \wedge \exists q) \leq r$. Then $p \wedge \exists q \leq \forall r$. Then $p \leq \forall\neg q \vee \forall r$. Then $\exists p \leq \exists(\forall\neg q \vee \forall r) = \forall\exists q \rightarrow \exists\forall r$. By (M4) and $(\mathrm{Adj}_{\exists,\forall})$, $\exists q \leq \forall\exists q$. Then $\exists p \leq \exists q \rightarrow \exists\forall r$. Then $\exists p \wedge \exists q \leq \exists\forall r$. Clearly $\exists\forall r \leq \forall r$. By (M2), $\forall r \leq r$. Then $\exists p \wedge \exists q \leq r$. Hence $\exists p \wedge \exists q \leq \exists(p \wedge \exists q)$. □

By Lemma 1, the pair $\langle \exists, \forall \rangle$ of quantifiers on a monadic algebra satisfies the adjointness condition. By (M2) and (M4), $\exists$ is a closure operator and $\forall$ is an interior operator on A. The variety MA is exactly the variety of modal $\mathsf{S5}$-algebras (cf. e.g. [4]). For the representation of monadic algebras, the modal approach uses relational structures while Halmos [7] introduced functional monadic algebras.

Definition 2. *Let A be a Boolean algebra and $X \neq \varnothing$ (domain). Let A^X be the set of all functions from X to A such that (i) $(A^X, \wedge, \neg, 0)$ is a Boolean algebra with respect to pointwise Boolean operations, namely $0(x) = 0$, $(\neg p)(x) = \neg p(x)$ and $(p \wedge q)(x) = p(x) \wedge q(x)$ for all $p, q \in A^X$ and $x \in A$; (ii) if $p \in A^X$, the supremum $\bigvee R(p)$ and the infimum $\bigwedge R(p)$ exist in A where $R(p) = \{p(x) : x \in X\}$. For all $p \in A^X$ and $x \in X$, let $(\exists p)(x) = \bigvee R(p)$ and $(\forall p)(x) = \bigwedge R(p)$. Every subalgebra of $(A^X, \wedge, \neg, 0, \exists)$ is called a* functional monadic algebra.

Note that the operator $\exists$ on A^X is an existential quantifier on A. We can also check that $\forall p = \neg\exists\neg p$ for all $p \in A^X$. Halmos [8] introduced a representation theorem for monadic algebras which says that every monadic algebra is isomorphic to a functional one. Moreover, for the definition of a logic in the setting of monadic algebras, Halmos [7] proposed that a monadic Boolean logic and its model are defined as a monadic algebra with an ideal.

Definition 3. *I is a monadic ideal if I is an ideal of B, and if $b \in I$, then $\exists b \in I$. A* monadic logic *is a pair (A, I) where A is a monadic algebra and I is a proper monadic ideal in A. An element $p \in A$ is* refutable *if $p \in I$. A* model *is a pair $(A, \{0\})$ where A is a monadic algebra. An* interpretation *of a monadic logic (A, I) in a model $(B, \{0\})$ is a MA-homomorphism $f : A \to B$ such that $f(p) = 0$ for every $p \in I$. An element $p \in A$ is* universally invalid *if $f(p) = 0$ for every interpretation f. A monadic logic (A, I) is* semantically complete *if every universally invalid element in A is refutable.*

We develop further a syntactically defined monadic logic. Let $\mathbb{V} = \{P_i : i \in \omega\}$ be a denumerable set of variables. The *term algebra* $\mathcal{T}$ is defined as follows:

$$\mathcal{T} \ni \varphi ::= P \mid \bot \mid \neg\varphi \mid (\varphi_1 \wedge \varphi_2) \mid (\varphi_1 \vee \varphi_2) \mid \exists\varphi$$

The connectives $\top, \to, \leftrightarrow$ and $\forall$ are defined as usual. Let $var(\varphi)$ be the set of all variables in φ. We write $\varphi(P_1, \ldots, P_n)$ for the term φ such that $var(\varphi) \subseteq \{P_1, \ldots, P_n\}$. For a monadic algebra $(A, \exists)$, a n-ary term $\varphi(P_1, \ldots, P_n)$ induces a n-ary polynomial $\varphi^A(p_1, \ldots, p_n)$ on A. A term is *valid* (notation: $\mathsf{MA} \models \varphi$) if $\varphi^A(p_1, \ldots, p_n) = 1$ for all $p_1, \ldots, p_n \in A$ and monadic algebra $(A, \exists)$.

Definition 4. *The Hilbert-style axiomatic system H consists of the following axiom schemata and inference rules:*

(Tau) *All instances of Boolean valid terms.*
$(T_\exists)$ $\varphi \to \exists\varphi$
$(4_\exists)$ $\exists\exists\varphi \to \exists\varphi$

$$\frac{\varphi \to \psi \quad \varphi}{\psi}\,(\mathrm{MP}) \qquad \frac{\exists\varphi \to \psi}{\varphi \to \forall\psi}\,(\mathrm{Adj}_{\exists,\forall})$$

The double line in $(\mathrm{Adj}_{\exists,\forall})$ *means that both terms are derivable from each other. A term φ is* derivable *in H (notation: $\vdash_{\mathsf{H}} \varphi$) if there is a derivation of φ in H.*

A term φ is H-*equivalent to* ψ (notation: $\varphi \sim_{\mathsf{H}} \psi$) if $\vdash_{\mathsf{H}} \varphi \leftrightarrow \psi$. The relation $\sim_{\mathsf{H}}$ is a congruence on $\mathcal{T}$. For every term φ, let φ^ϵ be the equivalence class of φ module $\sim_{\mathsf{H}}$. Let A^{H} be the Tarski-Lindenbaum quotient algebra of $\mathcal{T}$ which is a monadic algebra. For every $\varphi, \psi \in \mathcal{T}$, $\varphi^\epsilon \leq \psi^\epsilon$ if and only if $\vdash_{\mathsf{H}} \varphi \to \psi$. Then the least element in A^{H} is $\bot^\epsilon$. With these observation we get the fact that MA is an adequate algebraic semantics for H.

Fact 5. *For every term $\varphi \in \mathcal{T}$, $\vdash_{\mathsf{H}} \varphi$ if and only if $\mathsf{MA} \models \varphi$.*

Let $I^{\mathsf{H}} = \{\varphi^\epsilon : \vdash_{\mathsf{H}} \neg\varphi\}$. Clearly I^{H} is a monadic ideal in A^{H}. Then $(A^{\mathsf{H}}, I^{\mathsf{H}})$ is a monadic logic (cf. Definition 3) in the sense of Halmos.

Theorem 1. *$(A^{\mathsf{H}}, I^{\mathsf{H}})$ is semantically complete.*

Proof. Assume φ^ϵ is universally invalid. Let $(B, \{0\})$ be a model and f an interpretation of $(A^{\mathsf{H}}, I^{\mathsf{H}})$ in $(B, \{0\})$. Then $f(\varphi^\epsilon) = 0$. Hence $\varphi^\epsilon = \bot^\epsilon$. Then $\vdash_{\mathsf{H}} \psi \leftrightarrow \bot$. Then $\vdash_{\mathsf{H}} \neg\varphi$ and so $\varphi^\epsilon \in I^{\mathsf{H}}$. □

Since monadic algebras are exactly modal $\mathsf{S5}$-algebras, we have the relational semantics for MA by introducing frames with equivalence relation (cf. e.g. [4]).

3 Epistemic Monadic Algebras

As in the study of epistemic logic, we introduce an epistemic operator on a monadic algebra $(A, \exists)$. For each $p \in A$, the agent knows p is expressed as $\Box p$. If an agent knows p, then the agent knows the existence of p. Bob's knowing tigers implies Bob knows there are tigers. This principle is expressed as $\Box p \to \Box\exists p$. Dually $\Diamond p$ means that the agent reckons the possibility of p. As what is done in the standard epistemic logic, we define epistemic monadic algebras by introducing principles of knowledge (cf. e.g. [10]). We shall extend Halmos' representation of monadic algebras to epistemic monadic algebras.

Definition 6. *An* epistemic monadic algebra *('EMA' for short) is an algebra $(A, \exists, \Diamond)$ where $(A, \exists)$ is a monadic algebra and $\Diamond$ is a unary operator on A such that the following principles hold for all $p, q \in A$:*

(E1) $\Diamond 0 = 0$.
(E2) $p \leq \Diamond p$.
(E3) $\Diamond(p \wedge \Diamond q) = \Diamond p \wedge \Diamond q$.
(E4) $\exists\Diamond p = \Diamond\exists p$.

An operator $\Diamond$ satisfying these principles is called a possibility knowledge operator. *The dual operator $\Box$ defined by $\Box p := \neg\Diamond\neg p$ is called a* necessity knowledge operator. *For every epistemic monadic algebra $(A, \exists, \Diamond)$, we say an element $p \in A$ is* $\exists$-closed *if $\exists p = p$; and p is* $\Diamond$-closed *if $\Diamond p = p$. Let $\exists(A) = \{p \in A : \exists p = p\}$, $\Diamond(A) = \{p \in A : \Diamond p = p\}$ and $\mathrm{Cl}(A) = \Diamond(A) \cup \exists(A)$.*

By (E1)–(E3), the operator $\Diamond$ is an existential quantifier and hence $\Box$ is a universal quantifier. (E4) means that, there exists something which the agent reckons possible to be p if and only if the agent reckons that there exists p. This principle is plausible in the constant semantics of epistemic monadic logic. The properties of quantifiers hold for knowledge operators.

Fact 7. *Let $(A, \exists, \Diamond)$ be an EMA. For every $p \in A$, the following hold:* (1) *$\Box p \leq \Box\exists p$ and $\Diamond p \leq \Diamond\exists p$;* (2) *$\Box p \leq p$ and $\Box p \leq \Box\Box p$;* (3) *$\neg\Box p \leq \Box\neg\Box p$;* (4) *$\forall\Box p = \Box\forall p$; and* (5) ($\mathrm{Adj}_{\Diamond,\Box}$) *$\Diamond p \leq q$ if and only if $p \leq \Box q$.*

Lemma 2. *Let $(A, \exists)$ be a monadic algebra. Then $(A, \exists, \Diamond)$ is an EMA if and only if for all $p, q \in A$,* (E2), (E4), ($4_\Diamond$) *$\Diamond\Diamond p \leq \Diamond p$ and* ($\mathrm{Adj}_{\Diamond,\Box}$) *hold.*

Proof. The proof is quite similar to Lemma 1. □

The principle $\Box p \leq p$ expresses the truth of knowledge, and we have the principles of positive introspection $\Box p \leq \Box\Box p$ and negative introspection $\neg\Box p \leq \Box\neg\Box p$. The principle $\forall\Box p = \Box\forall p$ means the universality of knowing is the knowing of universality. This is the monadic version of the Barcan formula, which is valid in constant domain semantics (cf. e.g. [6, p.108]). Epistemic monadic algebras are algebras for the monadic fragment of a first-order modal logic. Since we have two kinds of S5-modalities $\Diamond$ and $\exists$, an EMA is indeed a bi-modal S5-algebra.

Lemma 3. *Let $(A, \exists, \Diamond)$ be an epistemic monadic algebra. Then $\exists(A)$ and $\Diamond(A)$ are subalgebras of A.*

Proof. Clearly $0 \in \exists(A)$. Assume $p \in \exists(A)$. Then $\exists p = p$. By $p \leq \forall\exists p = \forall p$, we have $\exists\neg p \leq \neg p$. Then $\neg p \in \exists(A)$. Assume $p, q \in \exists(A)$. Then $\exists(p \wedge q) \leq \exists p \wedge \exists q = p \wedge q$. Then $p \wedge q \in \exists(A)$. Assume $p \in \exists(A)$. By $\exists\exists p = \exists p$, we have $\exists p \in \exists(A)$. Moreover, $\exists\Diamond p = \Diamond\exists p = \Diamond p$ and so $\Diamond p \in \exists(A)$. Hence $\exists(A)$ is a subalgebra of A. Similarly $\Diamond(A)$ is a subalgebra of A. □

Let A be a Boolean algebra. A quantifier Q on A is *simple* if $\mathsf{Q}0 = 0$ and $\mathsf{Q}p = 1$ for all $p \in A \setminus \{0\}$. Let $(A, \exists, \Diamond)$ be an epistemic monadic algebra. A (Boolean) ideal I in A is *proper* if $1 \notin I$. An ideal I in A is *epistemic monadic* if $\exists(I) \cup \Diamond(I) \subseteq I$. Let $\mathcal{I}_{em}(A)$ and $\mathcal{I}^p_{em}(A)$ be the set of all epistemic monadic ideals and proper epistemic monadic ideals respectively. An epistemic monadic algebra A is *simple* if $\{0\}$ is the only proper epistemic monadic ideal in A.

Lemma 4. *An epistemic monadic algebra $(A, \exists, \Diamond)$ is simple if and only if $\exists$ and $\Diamond$ are simple quantifiers on A.*

Proof. Assume A is simple. Suppose $p \in A$ and $p \neq 0$. Let $I = \{q \in A : q \leq \exists p\}$ and $J = \{s \in A : s \leq \Diamond p\}$. Clearly I, J are ideals and $p \in I \cap J$. Then $I = J = A$. Then $1 \in I \cap J$ and so $\exists p = 1 = \Diamond p$. Conversely, assume $\exists$ and $\Diamond$ are simple. Suppose I is a proper monadic ideal in A. Clearly $0 \in I$. Let $0 \neq p \in A$. Then $\exists p = 1 = \Diamond p$. If $p \in I$, then $\exists p = \Diamond p = 1 \in I$. Hence $I = \{0\}$. □

Now we introduce functional epistemic monadic algebras. Note that we have two quantifiers in an epistemic monadic algebra. According to the fact $A^{W \times D} \cong (A^D)^W$, the domain that we choose should be the form $W \times D$ where W and D are disjoint nonempty sets. Intuitively, W is the domain for $\Diamond$ and D is the domain for $\exists$.

Definition 8. *Let $(A, \wedge, \neg, 0)$ be a Boolean algebra, $D \cap W = \varnothing$ and $Z = W \times D$. For all $p \in A^Z$, $w \in W$ and $x \in D$, we write $p_w(x)$ for $p(w)(x)$. Let*

$$R^w_D(p) = \{p_w(x) : x \in D\} \quad \text{and} \quad R^x_W(p) = \{p_w(x) : w \in W\}.$$

Let $(A^Z, \wedge, \vee, \neg, 0)$ be the Boolean algebra with respect to pointwise Boolean operations such that the supremums $\bigvee R^w_D(p)$, $\bigvee R^x_W(p)$ and the infimums $\bigwedge R^w_D(p)$, $\bigwedge R^x_W(p)$ exist in A for all $p \in A^Z$, $x \in D$ and $w \in W$. For every $p \in A^Z$, $w \in W$ and $x \in D$, we define

$$(\exists^* p)_w(x) = \bigvee R^w_D(p) \qquad (\forall^* p)_w(x) = \bigwedge R^w_D(p)$$

$$(\Box^* p)_w(x) = \bigwedge R^x_W(p) \qquad (\Diamond^* p)_w(x) = \bigvee R^x_W(p).$$

Lemma 5. *Let $(A, \wedge, \neg, 0)$ be a Boolean algebra, $D \cap W = \varnothing$, $Z = W \times D$. Then $\exists^*$ is an existential quantifier and $\Diamond^*$ is a possibility knowledge operator on A^Z.*

Proof. Let $p, q \in A^Z$, $w \in W$ and $x \in D$. By Halmos [7], it is easy to show $\exists^*$ is an existential quantifier. For (E2), we have $(\Diamond^* p)_w(x) = \bigvee R^x_W(p) = \bigvee_{w \in W} p_w(x)$. Hence $p_w(x) \leq (\Diamond^* p)_w(x)$. For ($4_\Diamond$), we have

$$(\Diamond^*\Diamond^* p)_w(x) = \bigvee R^x_W(\Diamond^* p) = \bigvee_{w \in W} (\Diamond^* p)_w(x) = \bigvee_{w \in W} p_w(x) = (\Diamond^* p)_w(x).$$

For (E4), we have

$$(\exists^*\Diamond^* p)_w(x) = \bigvee_{x \in D} (\Diamond^* p)_w(x) = \bigvee_{w \in W}^{x \in D} p_w(x) = \bigvee_{w \in W} (\exists^* p)_w(x) = (\Diamond^*\exists^* p)_w(x).$$

For ($\mathrm{Adj}_{\Diamond,\Box}$), assume $\Diamond^* p \leq q$. For all $w \in W$ and $x \in D$, $(\Diamond^* p)_w(x) = \bigvee R^x_W(p) = \bigvee\{p_w(x) : w \in W\} \leq q_w(x)$. Note that $(\Box^* q)_w(x) = \bigwedge R^x_W(q) = \bigwedge\{q_w(x) : w \in W\}$. Suppose $u \in W$. Then $p_w(x) \leq (\Diamond^* p)_w(x) = (\Diamond^* p)_u(x) \leq q_u(x)$. Hence $p_w(x) \leq (\Box^* q)_w(x)$. Assume $p \leq \Box^* q$. For all $w \in W$ and $x \in D$, $p_w(x) \leq (\Box^* q)_w(x)$. Then $(\Diamond^* p)_w(x) \leq (\Box^* q)_w(x)$. Clearly $(\Box^* q)_w(x) \leq q_w(x)$. Hence $(\Diamond^* p)_w(x) \leq q_w(x)$. By Lemma 2, $\Diamond^*$ is a possibility knowledge operator. □

Every subalgebra of the algebra $(A^Z, \wedge, \neg, 0, \exists^*, \Diamond^*)$ is called a *functional epistemic monadic algebra* on A.

Theorem 2. *An epistemic monadic algebra is simple if and only if it is isomorphic to a functional EMA on the two-element Boolean algebra* $\mathbf{2}$.

Proof. Assume $(A, \exists, \Diamond)$ is a subalgebra of $\mathbf{2}^Z$ where $Z = W \times D$ and $D \cap W = \varnothing$. Let $0 \neq p \in A$. Then there exist $w \in W$ and $x \in D$ such that $p_w(x) = 1$. Then $1 \in R^w_D(p) \cap R^x_W(p)$. Then $\exists^* p = 1 = \Diamond^* p$. This means that $\exists^*$ and $\Diamond^*$ are simple. By Lemma 4, A is simple. Conversely, assume $(A, \exists, \Diamond)$ is simple. Then A is a Boolean algebra. By Stone's representation, there exists a set $D \neq \varnothing$ such that A is embedded into $\mathbf{2}^D$. Let $W = \{w\}$ be a singleton set such that $D \cap W = \varnothing$ and $Z = W \times D$. Then A is Boolean isomorphic to a subalgebra B of $\mathbf{2}^Z$. By Lemma 4, $\exists$ and $\Diamond$ are simple. Hence A is also EMA-isomorphic to B. □

Next we show a Halmos-style representation theorem for epistemic monadic algebras. For a set X, let id_X be the identity function on X. We start from an appropriate definition of constants and richness with respect to Boolean algebras.

Definition 9. *Let* $(A, \exists, \Diamond)$ *be an epistemic monadic algebra and* B, C *be Boolean algebras such that* $\mathrm{Cl}(A) \subseteq B \cap C$.

- *An* epistemic C-constant *of* B *is a Boolean homomorphism* $g : B \to C$ *with* $g{\restriction}\mathrm{Cl}(A) = id_{\mathrm{Cl}(A)}$. *Let* $C_\Diamond(B)$ *be the set of all epistemic* C*-constants of* B.
- *An* existential B-constant *of* A *is a Boolean homomorphism* $f : A \to B$ *with* $f{\restriction}\mathrm{Cl}(A) = id_{\mathrm{Cl}(A)}$. *Let* $B_\exists(A)$ *be the set of all existential* B*-constants of* A.

We say A *is* $\langle B, C\rangle$-rich *if for every* $p \in A$, *(i) if* $f \in B_\exists(A)$, *then there exists* $g \in C_\Diamond(B)$ *with* $\Diamond p = g(f(p))$; *and (ii) there exists* $f \in B_\exists(A)$ *with* $\exists p = f(p)$.

Theorem 3 (Halmos representation). *Let $(A, \exists, \Diamond)$ be a $\langle B, C\rangle$-rich epistemic monadic algebra where B and C are Boolean algebras with $\mathrm{Cl}(A) \subseteq B \cap C$. Let $W = C_{\Diamond}(B)$, $D = B_{\exists}(A)$ and $Z = W \times D$. Then A is isomorphic to a functional epistemic monadic algebra on C^Z.*

Proof. The function $\pi : A \to C^Z$ is defined as $p \mapsto p^\pi$ where $p^\pi \in C^Z$ is given as follows: for every $g \in W$ and $f \in D$, $p^\pi_g(f) = g(f(p))$. First, we observe that π is a Boolean homomorphism by the following equations:

$$0^\pi_g(f) = f(0) \wedge g(0) = 0$$
$$(\neg p)^\pi_g(f) = g(f(\neg p)) = \neg g(f(p)) = \neg p^\pi_g(f)$$
$$(p \wedge q)^\pi_g(f) = g(f(p \wedge q)) = g(f(p)) \wedge g(f(q)) = p^\pi_g(f) \wedge q^\pi_g(f).$$

Now we show π is an EMA-homomorphism. By the definition, $(\exists p)^\pi_g(f) = g(f(\exists p)) = g(\exists p)$ and we have

$$(\exists^* p^\pi)_g(f) = \bigvee R^g_D(p^\pi) = \bigvee_{k \in D} p^\pi_g(k) = \bigvee_{k \in D} g(k(p)).$$

By the richness of A, there exists $h \in D$ such that $\exists p = h(p)$. Then $(\exists p)^\pi_g(f) = g(h(p))$. Then $(\exists p)^\pi_g(f) \leq (\exists^* p^\pi)_g(f)$. Let $k \in D$. By $p \leq \exists p$, $g(k(p)) \leq g(k(\exists p)) = g(\exists p)$. Then $(\exists^* p^\pi)_g(f) \leq (\exists p)^\pi_g(f)$. Hence $(\exists p)^\pi_g(f) = (\exists^* p^\pi)_g(f)$. By the definition, $(\Diamond p)^\pi_g(f) = g(f(\Diamond p)) = g(\Diamond p) = \Diamond p$ and we have

$$(\Diamond^* p^\pi)_g(f) = \bigvee R^f_W(p^\pi) = \bigvee_{s \in W} p^\pi_s(f) = \bigvee_{s \in W} s(f(p)).$$

By the richness of A, there exists $s \in W$ with $\Diamond p = s(f(p))$. Then $(\Diamond p)^\pi_g(f) \leq (\Diamond^* p^\pi)_g(f)$. Let $t \in W$. By $p \leq \Diamond p$, $t(f(p)) \leq t(f(\Diamond p)) = t(\Diamond p) = \Diamond p$. Then $(\Diamond^* p^\pi)_g(f) \leq (\Diamond p)^\pi_g(f)$. Hence $(\Diamond p)^\pi_g(f) = (\Diamond^* p^\pi)_g(f)$. Finally, we show π is injective. Assume $0 \neq r \in A$. By the richness of A, there is $f \in D$ with $\exists r = f(r)$. Let $g \in W$. Then $r^\pi_g(f) = g(f(r)) = g(\exists r) = \exists r \neq 0$. Hence $r^\pi \neq 0$. Suppose $p \not\leq q \in A$. Then $p \wedge \neg q \neq 0$. Then $(p \wedge \neg q)^\pi = p^\pi \wedge \neg q^\pi \neq 0$. Then $p^\pi \not\leq q^\pi$. □

Like the Definition 3, an *epistemic monadic logic* is defined as a pair (A, I) where A is an epistemic monadic algebra and $I \in \mathcal{I}_{em}(A)$. Now we give a Hilbert-style axiomatic system which is the syntactically defined epistemic monadic logic. The set of terms $\mathcal{T}_{\Diamond}$ is obtained from $\mathcal{T}$ by adding $\Diamond$.

Definition 10. *The Hilbert-style axiomatic system* E *is obtained from* H *by adding the axiom schemata and rules:*

$$(T_{\Diamond})\ \varphi \to \Diamond\varphi \quad (4_{\Diamond})\ \Diamond\Diamond\varphi \to \Diamond\varphi \quad (\mathrm{BC})\ \Diamond\exists\varphi \leftrightarrow \exists\Diamond\varphi \quad \frac{\Diamond\varphi \to \psi}{\varphi \to \Box\psi}\,(\mathrm{Adj}_{\Diamond,\Box})$$

Let $\vdash_{\mathsf{E}} \varphi$ denote that φ is derivable *in* E.

For a term $\varphi \in \mathcal{T}_{\Diamond}$, let $\mathsf{EMA} \models \varphi$ denote that φ is valid in EMA. We obtain again the *Tarski-Lindenbaum* algebra $(A^{\mathsf{E}}, \exists^{\mathsf{E}}, \Diamond^{\mathsf{E}})$ which belongs to EMA. For every $\varphi \in \mathcal{T}_{\Diamond}$, $\vdash_{\mathsf{E}} \varphi$ if and only if $\mathsf{EMA} \models \varphi$. Let $I^{\mathsf{E}} = \{\varphi^{\epsilon} : \vdash_{\mathsf{E}} \neg\varphi\}$. Then I^{E} is an epistemic monadic ideal in A^{E} and so $(A^{\mathsf{E}}, I^{\mathsf{E}})$ is an epistemic monadic logic.

Theorem 4. *$(A^{\mathsf{E}}, I^{\mathsf{E}})$ is semantically complete.*

Proof. The proof is similar to the proof of Theorem 1. □

4 Relativizations of Epistemic Monadic Algebras

In this section, we introduce relativizations of epistemic monadic algebras, which algebraically model dynamics on epistemic monadic logics. Given an epistemic monadic logic (A, I), we define the relativized logic of (A, I) when a piece of new information comes to the agent. Similarly we consider relativized models for an epistemic monadic logic. Here, we only treat the single-agent case.

Definition 11. *Let $(A, \wedge, \neg, 0, \exists, \Diamond)$ be an epistemic monadic algebra and $s \in \Diamond(A) \cap \exists(A)$. Let $A_s = [0, s] = \{p \in A : p \leq s\}$. For all $p, q \in A_s$, we define*

$$\neg_s p = s \wedge \neg p; \quad 0_s = 0; \quad p \wedge_s q = p \wedge q; \quad \exists_s p = s \wedge \exists p; \quad \Diamond_s p = s \wedge \Diamond p.$$

The operations $1_s, \vee_s, \rightarrow_s$ and $\leftrightarrow_s$ are defined as usual. Moreover, let $\forall_s p := \neg_s \exists_s \neg_s p$ and $\Box_s p := \neg_s \Diamond_s \neg_s p$.

Clearly $1_s = \neg_s 0_s = s \wedge \neg 0 = s \wedge 1 = s$. Moreover, $p \vee_s q = \neg_s(\neg_s p \wedge_s \neg_s q) = \neg_s((s \wedge \neg p) \wedge (s \wedge \neg q)) = s \wedge \neg(s \wedge \neg(p \vee q)) = s \wedge (p \vee q) = (s \wedge p) \vee (s \wedge q) = p \vee q$. We have $p \rightarrow_s q = \neg_s p \vee_s q = (s \wedge \neg p) \vee q = (s \wedge \neg p) \vee (s \wedge q) = s \wedge (p \rightarrow q)$. Furthermore, $\forall_s p = s \wedge \neg(\exists_s(s \wedge \neg p)) = s \wedge \neg(s \wedge \exists(s \wedge \neg p)) = s \wedge \neg\exists(s \wedge \neg p) = s \wedge \forall(s \rightarrow p)$. Similarly $\Box_s p = s \wedge \Box(s \rightarrow p)$.

Lemma 6. *If $(A, \exists, \Diamond)$ is an epistemic monadic algebra and $s \in \exists(A) \cap \Diamond(A)$, then $(A_s, \exists_s, \Diamond_s)$ is an epistemic monadic algebra.*

Proof. Note that $(A_s, \wedge_s, \neg_s, 0_s)$ is a Boolean algebra and the proof is omitted here. Now we show $\exists_s$ is an existential quantifier. Clearly $\exists_s 0_s = s \wedge \exists 0 = s \wedge 0 = 0$. Let $p, q \in A_s$. Then $p \leq s$ and $q \leq s$. Then $p \leq s \wedge \exists p = \exists_s p$. Moreover, $\exists_s(p \wedge_s \exists_s q) = s \wedge \exists(p \wedge (s \wedge \exists q) = s \wedge \exists(p \wedge \exists q) = s \wedge (\exists p \wedge \exists q) = (s \wedge \exists p) \wedge (s \wedge \exists q) = \exists_s p \wedge_s \exists_s q$. Similarly $\Diamond_s$ is also an existential quantifier. Clearly $\Diamond_s \exists_s p = s \wedge \Diamond(s \wedge \exists p)$ and $\exists_s \Diamond_s p = s \wedge \exists(s \wedge \Diamond p)$. By $\Diamond(s \wedge \exists p) \leq \Diamond\exists p = \exists\Diamond p$ and $\exists(s \wedge \Diamond p) \leq \exists\Diamond p = \Diamond\exists p$, we have $s \wedge \Diamond(s \wedge \exists p) \leq s \wedge \exists s \wedge \Diamond\exists p = s \wedge \exists s \wedge \exists\Diamond p = s \wedge \exists(\exists s \wedge \Diamond p) = s \wedge \exists(s \wedge \Diamond p)$ and $s \wedge \exists(s \wedge \Diamond p) \leq s \wedge \Diamond s \wedge \exists\Diamond p = s \wedge \Diamond s \wedge \Diamond\exists p = s \wedge \Diamond(\Diamond s \wedge \exists p) = s \wedge \Diamond(s \wedge \exists p)$. Hence $\Diamond_s \exists_s p = \exists_s \Diamond_s p$. Then $\Diamond_s$ is a possibility knowledge operator. □

Lemma 7. *Let $(A, \exists, \Diamond)$ be an epistemic monadic algebra and $s \in \exists(A) \cap \Diamond(A)$. Then $(A_s, \exists_s, \Diamond_s)$ is an EMA-homomorphic image of $(A, \exists, \Diamond)$.*

Proof. The function $\eta : A \to A_s$ is defined by $\eta(p) = p \wedge s$ for every $p \in A$. Clearly η is surjective. Now we show η is an EMA-homomorphism. Obviously $\eta(0) = 0 \wedge s = s$ and $\eta(1) = 1 \wedge s = s$. Moreover, $\eta(p \wedge q) = p \wedge q \wedge s = (p \wedge s) \wedge_s (q \wedge s) = \eta(p) \wedge_s \eta(q)$. We have $\eta(\neg p) = \neg p \wedge s$ and then $\neg_s \eta(p) = \neg_s(p \wedge s) = s \wedge \neg(p \wedge s) = (s \wedge \neg p) \vee (s \wedge \neg s) = s \wedge \neg p = \eta(\neg p)$. Clearly $\eta(\exists p) = s \wedge \exists p = s \wedge (\exists s \wedge \exists p) = s \wedge \exists(p \wedge \exists s) = s \wedge \exists(s \wedge p)$ and then $\exists_s \eta(p) = \exists_s(s \wedge p) = s \wedge \exists(s \wedge p) = \eta(\exists p)$. Similarly $\eta(\Diamond p) = \Diamond_s \eta(p)$. □

For every epistemic monadic algebra $(A, \exists, \Diamond)$, let $\mathscr{C}(A) = \exists(A) \cap \Diamond(A)$. For every $s \in \mathscr{C}(A)$, the algebra $(A_s, \exists_s, \Diamond_s)$ is called the *s-relativization* of A. By Lemma 6, $(A_s, \exists_s, \Diamond_s)$ is an epistemic monadic algebra.

Lemma 8. *Let $(A, \exists, \Diamond)$ be an epistemic monadic algebra and $I \in \mathcal{I}_{em}(A)$. For every $s \in \mathscr{C}(A)$, the set $I_s = \{p \in I : p \leq s\}$ belongs to $\mathcal{I}_{em}(A_s)$.*

Proof. Assume $s \in \mathscr{C}(A)$. Clearly I_s is a Boolean ideal. Assume $p \in I_s$. Then $p \in I$ and $p \leq s$. Then $\exists p \leq \exists s = s$ and $\Diamond p \leq \Diamond s = s$. Since $I \in \mathcal{I}_{em}(A)$, we have $\exists p, \Diamond p \in I$. Then $\exists p, \Diamond p \in I_s$. □

Let (A, I) be an epistemic monadic logic where $(A, \exists, \Diamond)$ is an epistemic monadic algebra. For every $s \in \mathscr{C}(A)$, by Lemma 8, I_s is an epistemic monadic ideal in $(A_s, \exists_s, \Diamond_s)$. Then (A_s, I_s) is called the *s-relativization* of the logic (A, I).

Theorem 5. *Let (A, I) be an epistemic monadic logic and $s \in \mathscr{C}(A)$. If (A, I) is semantically complete, then (A_s, I_s) is semantically complete.*

Proof. Assume that (A, I) is semantically complete. Let $p \in A_s$ be universally invalid. Let $(B, \{0\})$ be a model and $f : A \to B$ be an interpretation of (A, I). We define a function $f_s : A_s \to B$ by setting $f_s(q) = f(q)$ for all $q \in A_s$. Now we show that f_s is an interpretation of (A_s, I_s). Clearly f_s is a Boolean homomorphism. We have $f_s(\exists_s q) = f_s(s \wedge \exists q) = f_s(s) \wedge f_s(\exists q) = f(s) \wedge f(\exists q)$. Since $q \in A_s$, $q \leq s$ and so $\exists q \leq \exists s = s$. Then $f(\exists q) \leq f(s)$ and so $f_s(\exists_s q) = f(\exists q) = \exists f(q)$. Similarly $f_s(\Diamond_s q) = f(\Diamond q) = \Diamond f(q)$. Hence f_s is an EMA-homomorphism. Let $q \in I_s$. Then $q \in I$ and $q \leq s$. Then $f(q) = 0 = f_s(q)$. Hence f_s is an interpretation of (A_s, I_s). By the assumption that $p \in A_I$ is universally invalid, we have $f_s(p) = 0$. Then $p = 0 \in I_s$. Hence p is refutable. □

By Theorem 5, every s-relativization of a semantically complete epistemic monadic logic is also semantically complete. Now we specialize the relativization to syntactically defined relativizing epistemic monadic logics. We first give the relativizing operators. Let $\mathsf{O} = \{\varphi \in \mathcal{T}_\Diamond : \vdash_{\mathsf{E}} \varphi \leftrightarrow \Diamond\varphi \text{ and } \vdash_{\mathsf{E}} \varphi \leftrightarrow \exists\varphi\}$. For every $\varphi \in \mathsf{O}$, we call $\langle\varphi\rangle$ a *relativizing operator*. The set of relativizing epistemic monadic terms $\mathcal{T}_{\mathrm{RE}}^{\varphi}$ is obtained from $\mathcal{T}_\Diamond$ by adding $\langle\varphi\rangle$. Let $[\varphi]\alpha := \neg\langle\varphi\rangle\neg\alpha$. The *complexity* $c(\alpha)$ of a term $\alpha \in \mathcal{T}_{\mathrm{RE}}^{\varphi}$ is defined inductively as follows:

$$c(P) = 0 = c(\bot).$$
$$c(\odot\alpha) = c(\alpha) + 1, \text{ where } \odot \in \{\neg, \exists, \Diamond, \langle\varphi\rangle\}.$$
$$c(\alpha \wedge \beta) = \max\{c(\alpha), c(\beta)\} + 1.$$

Note that φ is a fixed term in the relativizing operator $\langle\varphi\rangle$.

Definition 12. *The Hilbert-style axiomatic system* RE_φ *is obtained from* E *by adding the following axiom schemata and inference rules:*

(1) (Ax_φ) $\varphi \to \langle\varphi\rangle\varphi$
(2) *Reduction axioms:*

$$(\mathrm{RA_{at}})\ \langle\varphi\rangle P \leftrightarrow \varphi \wedge P \qquad (\mathrm{RA}_\bot)\ \langle\varphi\rangle\bot \leftrightarrow \bot$$
$$(\mathrm{RA}_\neg)\ \langle\varphi\rangle\neg\alpha \leftrightarrow \varphi \wedge \neg\langle\varphi\rangle\alpha \qquad (\mathrm{RA}_\wedge)\ \langle\varphi\rangle(\alpha\wedge\beta) \leftrightarrow \langle\varphi\rangle\alpha \wedge \langle\varphi\rangle\beta$$
$$(\mathrm{RA}_\exists)\ \langle\varphi\rangle\exists\alpha \leftrightarrow \varphi \wedge \exists\langle\varphi\rangle\alpha \qquad (\mathrm{RA}_\Diamond)\ \langle\varphi\rangle\Diamond\alpha \leftrightarrow \varphi \wedge \Diamond\langle\varphi\rangle\alpha$$
$$(\mathrm{RA}_{\langle\varphi\rangle})\ \langle\varphi\rangle\langle\varphi\rangle\alpha \leftrightarrow \langle\varphi\rangle\alpha$$

(3) *Inference rules:*

$$\frac{\alpha \to \beta}{\langle\varphi\rangle\alpha \to \langle\varphi\rangle\beta}\ (\mathrm{Mon}) \qquad \frac{\alpha}{[\varphi]\alpha}\ (\mathrm{Gen}_{[\varphi]})$$

Let $\vdash_{\mathsf{DE}_\varphi} \alpha$ *denote that* α *is provable in* DE_φ.

Lemma 9. *If* $\vdash_{\mathsf{RE}_\varphi} \alpha \leftrightarrow \beta$, *then* $\vdash_{\mathsf{RE}_\varphi} \gamma \leftrightarrow \gamma(\beta/\alpha)$ *where* $\gamma(\beta/\alpha)$ *is the term obtained from* γ *by replacing one or more occurrences of* β *in* γ *by* α.

Proof. Assume $\vdash_{\mathsf{RE}_\varphi} \alpha \leftrightarrow \beta$. The proof proceeds by induction on the complexity of $\alpha \in \mathcal{T}^\varphi_{\mathrm{RE}}$. If $\gamma = P$ or $\gamma = \beta$, then $\gamma(\beta/\alpha) = \alpha$ and so $\vdash_{\mathsf{RE}_\varphi} \gamma \leftrightarrow \gamma(\beta/\alpha)$. The Boolean and modal cases are shown easily by induction hypothesis. Suppose $\gamma = \langle\varphi\rangle\chi$. By induction hypothesis and (Mon), $\vdash_{\mathsf{RE}_\varphi} \gamma \leftrightarrow \gamma(\beta/\alpha)$. □

Lemma 10. *For every term* $\alpha \in \mathcal{T}^\varphi_{\mathrm{RE}}$, *the following hold in* RE_φ: (1) $\vdash_{\mathsf{RE}_\varphi} \langle\varphi\rangle\alpha \to \varphi$; (2) $\vdash_{\mathsf{RE}_\varphi} [\varphi]\varphi$; (3) $\vdash_{\mathsf{RE}_\varphi} [\varphi]\Box\varphi$; (4) $\vdash_{\mathsf{RE}_\varphi} [\varphi]\forall\varphi$; (5) $\vdash_{\mathsf{RE}_\varphi} \langle\varphi\rangle\varphi \leftrightarrow \varphi$.

Proof. For (1), the proof proceeds by induction on the complexity of α. The only interesting case is $\alpha = \langle\varphi\rangle\beta$. By induction hypothesis, $\vdash_{\mathsf{RE}_\varphi} \langle\varphi\rangle\beta \to \varphi$. By $(\mathrm{RA}_{\langle\varphi\rangle})$, $\vdash_{\mathsf{RE}_\varphi} \langle\varphi\rangle\langle\varphi\rangle\beta \to \varphi$. For (2), by (Ax_φ), $\vdash_{\mathsf{RE}_\varphi} \varphi \to \langle\varphi\rangle\varphi$. Then $\vdash_{\mathsf{RE}_\varphi} \neg(\varphi \wedge \neg\langle\varphi\rangle\varphi)$. By $(\mathrm{RA}_\neg)$, $\vdash_{\mathsf{RE}_\varphi} \neg\langle\varphi\rangle\neg\varphi$. For (3), we have $\vdash_{\mathsf{RE}_\varphi} \Box[\varphi]\varphi$ by (2). Then $\vdash_{\mathsf{RE}_\varphi} \neg(\varphi \wedge \Diamond\langle\varphi\rangle\neg\varphi)$. By $(\mathrm{RA}_\Diamond)$, $\vdash_{\mathsf{RE}_\varphi} \neg\langle\varphi\rangle\Diamond\neg\varphi$. Then $\vdash_{\mathsf{RE}_\varphi} [\varphi]\Box\varphi$. Note that Lemma 9 is used in the whole proof. Similarly $\vdash_{\mathsf{RE}_\varphi} [\varphi]\forall\varphi$. For (4), it is obtained by (1) and (Ax_φ). □

The term $[\varphi]\Box\varphi$ means that the agent knows φ under the relativization φ. The term $[\varphi]\forall\varphi$ says that φ is universal under the relativization φ.

Lemma 11. *For every term* $\alpha \in \mathcal{T}^\varphi_{\mathrm{RE}}$, *there exists* $\beta \in \mathcal{T}_\Diamond$ *such that* $\vdash_{\mathsf{RE}_\varphi} \alpha \leftrightarrow \beta$.

Proof. The proof proceeds by induction on the complexity of α. The atomic, Boolean and epistemic cases are easily done by induction hypothesis. Assume $\alpha = \langle\varphi\rangle\chi$. The proof is given easily by subinduction on the complexity of χ. In each case we use the reduction axioms. Details are omitted. □

The axiomatic system RE_φ is not talking about public announcements but relativization. When a piece of information φ is revieved, the agent accepts φ and relativizes its knowledge with respect to φ. In the algebraic form, instead of using updates on algebra by public announcements in [12], we use relativizations of epistemic monadic algebras. For a term $\alpha(P_1, \dots, P_n) \in \mathcal{T}^\varphi_{\mathrm{RE}}$, a EMA $(A, \exists, \Diamond)$ and $p_1, \dots, p_n \in A$, we write $A \models \alpha[p_1, \dots, p_n]$ for $\alpha^A(p_1, \dots, p_n) = 1$.

Definition 13. *Let $\varphi(P_1, \dots, P_n) \in \mathcal{T}_\Diamond$, $\vdash_{\mathsf{E}} \varphi \leftrightarrow \Diamond\varphi$ and $\vdash_{\mathsf{E}} \varphi \leftrightarrow \exists\varphi$. Let $(A, \exists, \Diamond)$ be an EMA and $\alpha(P_1, \dots, P_n) \in \mathcal{T}^\varphi_{\mathrm{RE}}$. For all $p_1, \dots, p_n \in A$, $A \models \langle\varphi\rangle\alpha[p_1, \dots, p_n]$ if and only if $A \models \varphi[p_1, \dots, p_n]$ and $A_s \models \alpha[p_1 \wedge s, \dots, p_n \wedge s]$ where $s = \varphi^A(p_1, \dots, p_n)$.*

By the Definition 13, we have $A \models [\varphi]\alpha[p_1, \dots, p_n]$ if and only if $A \models \varphi[p_1, \dots, p_n]$ implies $A_s \models \alpha[p_1 \wedge s, \dots, p_n \wedge s]$.

Lemma 12. *For every $\alpha \in \mathcal{T}^\varphi_{\mathrm{RE}}$, if $\vdash_{\mathsf{RE}_\varphi} \alpha$, then $\mathsf{EMA} \models \alpha$.*

Proof. The proof proceeds by induction on a derivation of $\alpha(P_1, \dots, P_n)$ in RE_φ. If α is an axiom of E, then $\mathsf{EMA} \models \alpha$. Let $(A, \exists, \Diamond)$ be an EMA. By Lemma 6, $(A_s, \exists_s, \Diamond_s)$ is an EMA. We show $\mathsf{EMA} \models \varphi \to \langle\varphi\rangle\varphi$. Assume $A \models \varphi[p_1, \dots, p_n]$. Then $\varphi^A(p_1, \dots, p_n) = 1$. By Lemma 7, $\eta(\varphi^A(p_1, \dots, p_n)) = \varphi^{A_s}(p_1 \wedge s, \dots, p_n \wedge s) = s$. Then $A_s \models \varphi[p_1 \wedge s, \dots, p_n \wedge s]$. Similarly we obtain that all reduction axioms are valid in EMA. Here we check only $\mathsf{EMA} \models \mathrm{RA}_{\langle\varphi\rangle}$ and the proof of others is omitted. Note that $(A_s)_s$ is isomorphic to A_s. Suppose $(\langle\varphi\rangle\langle\varphi\rangle\alpha)^A(p_1, \dots, p_n) = 1$. Then $\varphi^A(p_1, \dots, p_n) = 1$ and $(\langle\varphi\rangle\alpha)^{A_s}(p_1 \wedge s, \dots, p_n \wedge s) = s$. Note that $\varphi^{A_s}(p_1 \wedge s, \dots, p_n \wedge s) = s$. Then $\alpha^{A_s}(p_1 \wedge s, \dots, p_n \wedge s) = s$. Hence $(\langle\varphi\rangle\alpha)^A(p_1, \dots, p_n) = 1$. The other direction is shown similarly. It is easy to see (Mon) and $(\mathrm{Gen}_{[\varphi]})$ preserve validity in EMA. □

Theorem 6. *For every $\alpha \in \mathcal{T}^\varphi_{\mathrm{RE}}$, if $\mathsf{EMA} \models \alpha$, then $\vdash_{\mathsf{RE}_\varphi} \alpha$.*

Proof. Assume $\mathsf{EMA} \models \alpha$. By Lemma 12 and Lemma 11, there exists $\beta \in \mathcal{T}_\Diamond$ such that $\mathsf{EMA} \models \alpha \leftrightarrow \beta$. Then $\mathsf{EMA} \models \beta$. Then $\vdash_{\mathsf{E}} \beta$. Then $\vdash_{\mathsf{RE}_\varphi} \beta$. By Lemma 11 again, $\vdash_{\mathsf{RE}_\varphi} \alpha \leftrightarrow \beta$ and hence $\vdash_{\mathsf{RE}_\varphi} \alpha$. □

Return to the Halmos' style epistemic monadic logics. Take the Lindenbaum-Tarski algebra $(A^{\mathsf{E}}, \Diamond, \exists)$ for the system E. Let $\varphi \in \mathcal{T}_\Diamond$. Assume $\vdash_{\mathsf{E}} \varphi \leftrightarrow \Diamond\varphi$ and $\vdash_{\mathsf{E}} \varphi \leftrightarrow \exists\varphi$. Then $\varphi^\epsilon \in \mathscr{C}(A^{\mathsf{E}})$. Now consider the φ^ϵ-relativization $(A^{\mathsf{E}}_{\varphi^\epsilon}, I^{\mathsf{E}}_{\varphi^\epsilon})$ which can be viewed as the algebraization of the system RE_φ.

Corollary 1. *$(A^{\mathsf{E}}_{\varphi^\epsilon}, I^{\mathsf{E}}_{\varphi^\epsilon})$ is semantically complete.*

Proof. By Theorem 4 and Theorem 5. □

The reduction axiom $\mathrm{RA}_{\langle\varphi\rangle}$ says that double relativization is equal to one. This point differs from the standard public announcement logic dramatically. But if we take all relativizing operators $\langle\varphi\rangle$ with $\varphi \in \mathsf{O}$, we get a bunch of relativizing monadic epistemic logics. The more fundamental feature of these relativizations is that they provide algebraic models for one-step dynamics of agent's knowing sets of things.

5 Concluding Remarks

There are two main contributions of the present paper. Firstly, we extend Halmos' monadic Boolean algebras to epistemic monadic algebras. Epistemic monadic logics are also provided. We demonstrate the Halmos-style representation theorem for these algebras in particular. Secondly, we show how relativizations of epistemic monadic algebras work for talking about updates. Undoubtedly, there are several ways to expand on the current work. The results for the single-agent case can be clearly extended to the multi-agent cases. Then they can be extended to polyadic algebras (cf. [9]), and thus we have algebras for the full first-order epistemic logic, and relativizations are considered again. Moreover, as is shown in [12], we can deal with intuitionistic epistemic logic and extend our work to monadic Heyting algebras (cf. [3]). These directions need more philosophical considerations.

References

1. Baltag, A., Moss, L., Solecki, S.: The logic of public announcements: common knowledge and private suspicions. In: Gilboa, I., (ed.) Proceedings of the 7th Conference on Theoretical Aspects of Rationality and Knowledge, pp. 43–56. Morgan Kaufmann Publishers, San Francisco (1998)
2. van Benthem, J.: Logical Dynamics of Information and Interaction. Cambridge University Press, New York (2011)
3. Bezhanishvili, G.: Varieties of monadic Heyting algebras. Part I. Stud. Logica **61**(3), 367–402 (1998)
4. Blackburn, P., de Rijke, M., Venema, Y.: Modal Logic. Cambridge University Press, Cambridge (2001)
5. van Ditmarsch, H.P., van der Hoek, W., Kooi, B.P.: Dynamic epistemic logic and knowledge puzzles. In: Priss, U., Polovina, S., Hill, R. (eds.) ICCS-ConceptStruct 2007. LNCS (LNAI), vol. 4604, pp. 45–58. Springer, Heidelberg (2007). https://doi.org/10.1007/978-3-540-73681-3_4
6. Fitting, M., Mendelsohn, R.: First-order Modal Logic. Kluwer Academic Publishers, Dordrecht (1999)
7. Halmos P.: Algebraic logic I. monadic boolean algebras. Compo. Math. **12**, 217–249 (1954–1956)
8. Halmos, P.: The representation of monadic Boolean algebras. Duke Math. J. **26**(3), 447–454 (1959)
9. Halmos, P.: Algebraic Logic. Dover Publications, New York (2016)
10. Hintikka, J.: Knowledge and Belief: An Introduction to the Logic of the Two Notions. Cornell University Press, Ithaca (1962)
11. Ma, M.: Mathematics of public announcements. In: van Ditmarsch, H., Lang, J., Ju, S. (eds.) LORI 2011. LNCS (LNAI), vol. 6953, pp. 193–205. Springer, Heidelberg (2011). https://doi.org/10.1007/978-3-642-24130-7_14
12. Ma, M., Palmigiano, A., Sadrzadeh, M.: Algebraic semantics and model completeness for intuitionistic public announcement logic. Ann. Pure Appl. Logic **165**, 963–995 (2014)
13. Palmigiano, A., Kurz, A.: Epistemic updates on algebras. Log. Methods Comput. Sci. **9**(4:17), 1–28 (2013)

14. Plaza, J.: Logics of public announcements. In: Proceedings 4th International Symposium on Methodologies for Intelligent Systems, pp. 201–216 (1989)
15. Su, Y., Sano, K.: A first-order expansion of Artemov and Protopopescu's intuitionistic epistemic logic. Studia Logica (2023). https://doi.org/10.1007/s11225-023-10037-6
16. Conradie, W., Frittella, S., Palmigiano, A., Tzimoulis, A.: Probabilistic epistemic updates on algebras. In: van der Hoek, W., Holliday, W.H., Wang, W. (eds.) LORI 2015. LNCS, vol. 9394, pp. 64–76. Springer, Heidelberg (2015). https://doi.org/10.1007/978-3-662-48561-3_6
17. Wolter, F.: First order common knowledge logics. Stud. Logica. **65**, 249–271 (2000)

Knowing the Value of a Predicate

Bo Hong(✉)

Department of Philosophy and Religious Studies, Peking University, Beijing, China
hongbo@stu.pku.edu.cn

Abstract. In 1989, Plaza introduced the "knowing value" operator $\mathsf{Kv}_i d$ to characterize the "Mr. Sum and Mr. Product" puzzle in propositional modal logic. Previous research had primarily focused on the $\mathsf{Kv}d$ operator, which captures the idea of knowing the value of a designator d. This paper expands the scope of application for the Kv operator beyond designators to include predicates, interpreting the $\mathsf{Kv}P$ operator as denoting knowledge of the value of a predicate P. Additionally, we present two distinct semantics - MS (Mention-Some) semantics and MA (Mention-All) semantics - for the $\mathsf{Kv}P$ operator, and prove the strong completeness theorem for two axiom systems containing only the $\mathsf{Kv}P$ operator, as well as two axiom systems containing both the $\mathsf{Kv}P$ and $\mathsf{Kv}d$ operators.

Keywords: Knowing-value Logic · Epistemic Logic · Knowing-value of Predicates · Completeness

1 Introduction

Epistemic Logic is a branch of modal logic that studies possibilities and necessities from a knowledge perspective. In Epistemic Logic, the modal operator K can be used to express the knowledge of a person or agent regarding a proposition, such as "someone knows that φ is true" (referred to as "knowing that" in [15]). This language or notation can help people better understand and analyze the structure, characteristics, and influence of knowledge. Epistemic Logic provides tools and techniques for describing and handling semantic concepts such as possibility, necessity, belief and knowledge, which can be applied in fields such as philosophy, computer science, cryptography and linguistics [5].

Although research on "knowing that" constitutes a significant portion of Epistemic Logic, it can be observed from natural language expressions that people also use other forms of knowledge expressions in daily life, such as "knowing whether" ("I know whether this proposition is correct"), "knowing what" ("I know what your password is"), and "knowing how" ("I know how to swim"). These constructions were called *knowing-wh*: *know* followed by a *wh*-question word [15]. Within the scope of logical research on "knowing what", there is a more precise research area, which is "knowing value", or knowing what the value is. The statement mentioned above "I know what your password is" can be rephrased as "I know what the value of your password is" or "I know the value of your password".

N. Alechina et al. (Eds.): LORI 2023, LNCS 14329, pp. 149–166, 2023.
https://doi.org/10.1007/978-3-031-45558-2_12

In 1983, Ma and Guo [11] introduced the formula $S * c$ to denote "S knows what the term c is", and provided an equivalent definition using existential quantification, namely $\exists x \mathsf{K}_S(c = x)$. Plaza [13], on the other hand, proposed the "knowing value" operator $\mathsf{Kv}_i d$ to describe the well-known puzzle of Mr. Sum and Mr. Product. Based on the axiomatization $\mathbb{EL}$ of epistemic logic, Plaza presented two valid axioms for the Kv operator (we name them as Kv4 and Kv5), and form the axiomatization $\mathbb{S5ELKV}$ for the logic **S5ELKv** [13]:

The axiom system $\mathbb{S5ELKV}$			
Axioms		Rules	
Taut	All instances of tautologies	*MP*	$\dfrac{\varphi \quad \varphi \to \psi}{\psi}$
DistK	$\mathsf{K}_i(\varphi \to \psi) \to (\mathsf{K}_i\varphi \to \mathsf{K}_i\psi)$	*NecK*	$\dfrac{\vdash \varphi}{\vdash \mathsf{K}_i\varphi}$
T	$\mathsf{K}_i\varphi \to \varphi$		
4	$\mathsf{K}_i\varphi \to \mathsf{K}_i\mathsf{K}_i\varphi$		
5	$\neg\mathsf{K}_i\varphi \to \mathsf{K}_i\neg\mathsf{K}_i\varphi$		
Kv4	$\mathsf{Kv}_i d \to \mathsf{K}_i\mathsf{Kv}_i d$		
Kv5	$\neg\mathsf{Kv}_i d \to \mathsf{K}_i\neg\mathsf{Kv}_i d$		

Although Plaza [13] stated in Remark 1.10, item 5 that the axiom system is sound and complete, the proposition was not proven. However, the K-version rather than $S5$-version of this proposition is proved by Gu and Wang. By proving the completeness of the axiom system $\mathbb{SMLKV}$, Gu and Wang indirectly proved that the axiom system $\mathbb{ELKV}$ ($\mathbb{S5ELKV}$-*T*-*4*-*5*-*Kv4*-*Kv5*) is strongly complete with respect to the logic **ELKv** [8].

In recent years, the concept of "knowing value" has continuously driven the development of various branches related to *knowing-wh* in Epistemic Logic. Other logical areas related to "knowing value", such as public announcement [1,16], dependency relations [2,4,14], and public inspection [1,3,6], have also received a great deal of attention from scholars. Previous studies about Kv operators itself have primarily focused on the $\mathsf{Kv}d$ operator (knowing the value of a designator), or its relativized version $\mathsf{Kv}_i^r(\varphi, d)$ [16,17]. What should we do if we want to express an agent i knows the value of a **concept**?

For example, from the extension of the concept, when we view the concept "cafes within 2 km" as a set, it is obvious that we know the value of the concept "cafes within 2 km" as long as we know which elements are in the set "cafes within 2 km". But how should we express concepts in logical languages? A convenient way is to **use predicates to represent concepts**. In first-order predicate logic, when we give an interpretation I based on a first-order language $\mathcal{L}$ and a $\mathcal{L}$-structure $\mathcal{M} = (M, I)$ (M is the domain of $\mathcal{M}$), we need to interpret the predicates in this language. Usually, we interpret an n-ary predicate P as a subset of M^n. In particular, if we use a unary predicate P_c^1 to represent "cafes within 2 km" in a structure with the domain of all cafes, then the interpretation of P_c^1 is a subset of the large set of all cafes. When we say "I know the value of the concept 'cafes within 2 km'," it means "I know the value of the predicate 'cafes within 2 km'." To formally describe these contents, we will introduce a new operator called the "knowing predicate value" operator $\mathsf{Kv}P$ in this paper.

To indicate that we know the value of a predicate P, it is evident that we need to be able to answer the question "what is the value of P?" In 1976, Hamblin suggested a question sets up a choice-situation between a set of propositions, namely, those propositions that count as answers to it [9]. Furthermore, Groenendijk and Stokhof proposed two different interpretations of interrogatives: **mention-some interpretation** and **mention-all interpretation** [7]. To answer a question, the respondent only needs to provide at least one possible option of the question under mention-some interpretation, while the mention-all interpretation requires the respondent to exhaustively list all the options satisfying the question's properties. They use two formulas to differentiate two interpretations: $\exists x \mathsf{K}_I Px$ for mention-some, and $\forall x(Px \to \mathsf{K}_I Px)$ for mention-all.[1] For instance, if the predicate P represents "cafes within 2 km", we may ask under what circumstances we can say that someone (denoted as a) "knows the value of the predicate 'cafes within 2 km'." Clearly, a must be capable of correctly answering the question "what cafes are within 2 km". Suppose there are only two cafes in 2 km, Starbucks and Costa, then both "Starbucks" and "Starbucks and Costa" could be counted as answers to the question under mention-some interpretation, but only "Starbucks and Costa" can be counted as an answer under mention-all interpretation. Therefore, a "knowing predicate value" operator, $\mathsf{Kv}P$, can have both Mention-Some (MS) and Mention-All (MA) semantics.

In this paper, we axiomatize the single-agent logics $\mathbf{MSELKvP}_1$ and $\mathbf{MAELKvP}_1$, which only contain the $\mathsf{Kv}P$ operator, as well as $\mathbf{MSELKv}_1$ and $\mathbf{MAELKv}_1$, which contain both the $\mathsf{Kv}P$ and $\mathsf{Kv}d$ operators. By proving the strong completeness of axiomatizations of these four logics, we hope to gain a better understanding of the essence of the "knowing value" operator.

This paper is structured into four sections. Section 2 provides an introduction to the single-agent version of epistemic logic with Kv operator. Section 3 focuses on the construction of four axiom systems about $\mathsf{Kv}P$ operator and their corresponding completeness proofs. Finally, in Sect. 4, we provide our conclusion and outlook, summarizing the research results of this paper and highlighting some potential future research directions.

[1] However, the mention-all interpretation proposed by Groenendijk and Stokhof does not imply exhaustiveness, which means the respondent of a question may contain false belief. For example, suppose there are only two cafes in 2 km, Starbucks and Costa. The answer "Starbucks, Costa and Peet's" may also count as an answer of the question "What is the value of the predicate 'cafes within 2 km'?" under Groenendijk and Stokhof's mention-all interpretation. To avoid this, in this article, when we refer to "mention-all interpretation", we actually mean the "strongly exhaustive interpretation" proposed in [10]. In this situation, the formulation of mention-all interpretation should be $\forall x(\hat{K}Px \to \mathsf{K}_I Px)$.

2 Preliminaries

Given a countable set $\mathbf{P}$ of atomic propositions and a countable set of nonrigid designators $\mathbf{D}$, the language of epistemic logic with Kv operator (single-agent version) $\mathcal{L}_{ELKv_1}(\mathbf{P}, \mathbf{D})$ is defined as follows:

$$\varphi ::= \top \mid p \mid \mathsf{Kv}d \mid \neg\varphi \mid (\varphi \wedge \varphi) \mid \mathsf{K}\varphi$$

where $p \in \mathbf{P}$, $d \in \mathbf{D}$. The formula $\mathsf{Kv}d$ means "knows the value of d". Define $\bot := \neg\top$, $\varphi \vee \psi := \neg(\neg\varphi \wedge \neg\psi)$, $\varphi \to \psi := \neg\varphi \vee \psi$, $\hat{\mathsf{K}}\varphi := \neg\mathsf{K}\neg\varphi$. In the following, we fix $\mathbf{P}$ and $\mathbf{D}$ so that we simply write $\mathcal{L}_{ELKv_1}$ for $\mathcal{L}_{ELKv_1}(\mathbf{P}, \mathbf{D})$.

It should be noted that the designators considered in this paper are **nonrigid** designators, which means that in different possible worlds, the same designator may refer to different objects. For example, the designator t_{beijing} represents "the temperature (°C) of Beijing". If t_{beijing} is a nonrigid designator, then it is possible that in w, the value of t_{beijing} is 23, while in u, the value of t_{beijing} is 5.

An epistemic model for $\mathcal{L}_{ELKv_1}$ is a tuple $\mathcal{M} = \langle S, \sim, O, V, V_{\mathbf{D}} \rangle$ where S is a non-empty set of possible worlds, $\sim$ is an equivalence relation over S, O is a non-empty set of objects, V is a valuation function assigning a set of worlds $V(p) \subseteq S$ to each $p \in \mathbf{P}$, and $V_{\mathbf{D}} : \mathbf{D} \times S \to O$ is an assignment function.

Given a pointed epistemic model $\mathcal{M}, s$, the semantic of $\mathsf{Kv}d$ is defined as follows:

$\mathcal{M}, s \vDash \mathsf{Kv}d \iff$ for any $t_1, t_2 \in S$: if $s \sim t_1$ and $s \sim t_2$, then $V_{\mathbf{D}}(d, t_1) = V_{\mathbf{D}}(d, t_2)$.

3 Axiom Systems About $\mathbf{Kv}P$ Operator

3.1 Axiom Systems $\mathbb{MSELKVP}_1$ and $\mathbb{MAELKVP}_1$ for $\mathcal{L}_{ELKvP_1}$

As we may quote what Plaza had said [13]: *"An agent is said to know the value of a designator d if d has the same value in all worlds indistinguishable from the actual one."* Previous investigations of Kv operators have predominantly concentrated on the $\mathsf{Kv}d$ operator, which pertains to knowledge of the value of a designator, or its relativized variant, $\mathsf{Kv}_i^r(\varphi, d)$. In this section, we extend the Kv operator to apply to both designators and predicates. We confine our attention exclusively to the **single-agent** versions of logics and models based on $S5$ **frames** (in other words, epistemic models).

To better understand the semantics of $\mathsf{Kv}P$ operator, we consider the following example:

Suppose Pd means "d attended the lecture that was held in the library yesterday afternoon", then $\mathsf{Kv}P$ (i.e., "I know the value of 'the person who attended the lecture that was held in the library yesterday afternoon"') can be understood in two ways:

- *Mention-Some (MS): I know someone who attended the lecture that was given in the library yesterday afternoon.* In MS semantics, for the expression to be true, the speaker only needs to know that a particular person (e.g. the speaker himself) attended the lecture.

- *Mention-All (MA): I know everyone who attended the lecture in the library yesterday afternoon.* In MA semantics, for the expression to be true, the speaker must know everyone who attended the lecture yesterday afternoon.

Based on the idea of the above example, we can give the following informal definition of semantics: Suppose for any unary predicate P and any possible world w, $V(P, w)$ denotes the set of objects that are assigned to P in w. Then we say $\mathsf{Kv}P$ is true in MS semantics if there exists an object o such that for every possible world u accessible from w, $o \in V(P, u)$, while in MA semantics all $V(P, u)$ must be equal.

Now we start to formalize the language, model and semantics.

Given a countable set $\mathbf{P}$ of **predicates** and a countable set of nonrigid designators $\mathbf{D}$, the language $\mathcal{L}_{ELKvP_1}(\mathbf{P}, \mathbf{D})$ is defined as follows:

$$\varphi ::= \top \mid Pd_1...d_n \mid \neg\varphi \mid (\varphi \wedge \varphi) \mid \mathsf{K}\varphi \mid \mathsf{Kv}P$$

where $P \in \mathbf{P}$ is an n-ary predicate, and $d \in \mathbf{D}$ is a nonrigid designator. In the following, we simply write $\mathcal{L}_{ELKvP_1}(\mathbf{P}, \mathbf{D})$ as $\mathcal{L}_{ELKvP_1}$. In particular, we allow the existence of 0-ary predicates $P_1^0, P_2^0, \ldots$ in the language. For any 0-ary predicate P^0, we can simply understand it as a propositional variable p.

An $\mathbf{S5ELKvP_1}$ model is a tuple $\mathcal{M} = \langle W, \sim, O, V\rangle$ that satisfies:

- W is a non-empty set of possible worlds.
- $\sim$ is an equivalence relation on W.
- O is a non-empty set of objects, and we call O the domain of $\mathcal{M}$.
- V is an assignment function on $S5$ frame $\mathcal{F} = \langle W, \sim, O\rangle$, where:
  - for any n-ary predicate $P \in \mathbf{P}$, $V(P, w) \subseteq O^n$[2], and;
  - for any possible world w and any nonrigid designator $d \in \mathbf{D}$, $V(d, w) \in O$.

For any $\mathbf{S5ELKvP_1}$ model $\mathcal{M} = \langle W, \sim, O, V\rangle$, MS semantics are defined as follows:

$\mathcal{M}, w \models_{\mathrm{MS}} \top$	$\iff$	always holds
$\mathcal{M}, w \models_{\mathrm{MS}} Pd_1.....d_n$	$\iff$	$\langle V(d_1, w), ..., V(d_n, w)\rangle \in V(P, w)$
$\mathcal{M}, w \models_{\mathrm{MS}} \neg\varphi$	$\iff$	$\mathcal{M}, w \not\models_{\mathrm{MS}} \varphi$
$\mathcal{M}, w \models_{\mathrm{MS}} \varphi \wedge \psi$	$\iff$	$\mathcal{M}, w \models_{\mathrm{MS}} \varphi$ and $\mathcal{M}, w \models_{\mathrm{MS}} \psi$
$\mathcal{M}, w \models_{\mathrm{MS}} \mathsf{K}\varphi$	$\iff$	for any possible world $u \in W$: $w \sim u$ implies $\mathcal{M}, u \models_{\mathrm{MS}} \varphi$
$\mathcal{M}, w \models_{\mathrm{MS}} \mathsf{Kv}P$ (P is an n-ary predicate)	$\iff$	there exist objects $o_1, ..., o_n \in O$ such that for every possible world $t \sim w$ we have $\langle o_1, ..., o_n\rangle \in V(P, t)$

In MA semantics, only the semantic of $\mathsf{Kv}P$ differs from that in MS semantics:

$$\mathcal{M}, w \models_{\mathrm{MA}} \mathsf{Kv}P \iff \text{for any possible world } t \sim w,\ V(P, t) = V(P, w)$$

[2] It should be particularly noted that for the 0-ary predicate P^0, $V(P^0, w) \subseteq O^0 = \{\varnothing\}$.

For MS semantics, we say that a formula φ is valid in a model $\mathcal{M}$ (written as $\mathcal{M} \vDash_{\mathrm{MS}} \varphi$) if for any possible world $w \in W$ in the model $\mathcal{M} = \langle W, \sim, O, V \rangle$, we have $\mathcal{M}, w \vDash_{\mathrm{MS}} \varphi$. We say that a formula φ is valid in a frame $\mathcal{F}$ (written as $\mathcal{F} \vDash_{\mathrm{MS}} \varphi$) if for any model $\mathcal{M} = \langle \mathcal{F}, V \rangle$ based on the frame $\mathcal{F}$, we have $\mathcal{M} \vDash_{\mathrm{MS}} \varphi$. We make similar definitions for the MA semantics.[3] We have two different logics, $\mathbf{MSELKvP_1}$ and $\mathbf{MAELKvP_1}$, corresponding to the two semantics.

Now consider a 0-ary predicate P^0, what does it mean when we say $\mathsf{Kv}P^0$ in both MS and MA semantics? In MS semantics $\mathsf{Kv}P^0$ is equal to $\mathsf{K}P^0$. In MA semantics, however, $\mathsf{Kv}P^0$ is equivalent to $\mathsf{K}P^0 \vee \mathsf{K}\neg P^0$, which is essentially a "knowing whether" operator $\mathsf{Kw}P^0$.[4]

Compare to the axioms for $\mathsf{Kv}_i d$ in $\mathbb{S5ELKV}$ mentioned in Sect. 1 (the axioms *Kv4* and *Kv5*), it is not hard to see that $\mathsf{Kv}P$ operator also obeys the positive and negative introspections, which means *KvP4* axiom ($\mathsf{Kv}P \to \mathsf{KKv}P$) and *KvP5* axiom ($\neg\mathsf{Kv}P \to \mathsf{K}\neg\mathsf{Kv}P$) are valid.

According to [11–13,15], we can equivalently express $\mathsf{Kv}_i d$ as a first-order modal formula $\exists x \mathsf{K}_i(d = x)$. Likewise, we can also express $\mathsf{Kv}P$ in MS and MA semantics as $\exists x_1 ... \exists x_n \mathsf{K}Px_1...x_n$ and $\forall x_1 ... \forall x_n (\hat{\mathsf{K}}Px_1...x_n \to \mathsf{K}Px_1...x_n)$, respectively. By analogy with the introduction rule for universal quantifiers in first-order logic, we can formulate the following two rules regarding $\mathsf{Kv}P$:

- *MS'*: $\dfrac{\psi \to \hat{\mathsf{K}}\neg Pd_1...d_n}{\psi \to \neg\mathsf{Kv}P}$, where d_i does not occur in ψ
- *MA'*: $\dfrac{\psi \to (\hat{\mathsf{K}}Pd_1....d_n \to \mathsf{K}Pd_1....d_n)}{\psi \to \mathsf{Kv}P}$, where d_i does not occur in ψ

Based on the single-agent version of $\mathbb{S5ELKV}$, we propose two axiom systems $\mathbb{MSELKVP}_1$ and $\mathbb{MAELKVP}_1$ (See next page).

[3] As for 0-ary predicate P^0, the semantics for formulas with P^0 are:

- $\mathcal{M}, w \vDash_{\mathrm{MS}} P^0$ if and only if $\mathcal{M}, w \vDash_{\mathrm{MA}} P^0$, if and only if $\varnothing \in V(P, w)$,
- $\mathcal{M}, w \vDash_{\mathrm{MS}} \mathsf{Kv}P$ if and only if for every possible world $t \sim w$, we have $\varnothing \in V(P, t)$,
- $\mathcal{M}, w \vDash_{\mathrm{MA}} \mathsf{Kv}P$ if and only if for any possible world $t \sim w$, we have $V(P, t) = V(P, w)$.

[4] However, this does not imply that one operator has stronger expressive power than the other in MS semantics. Although the Kv operator can be combined with 0-ary predicates, it cannot express combinations of predicates within its scope (such as conjunction $\wedge$, disjunction $\vee$, etc.). Meanwhile, although logical connectives can be used within the scope of the K operator, we cannot express sentences such as "knowing the value of an n-ary predicate P" only using K.

The axiom system $\mathbb{MSELKVP}_1$			
Axioms		**Rules**	
Taut	All instances of tautologies	*MP*	$\dfrac{\varphi \quad \varphi \to \psi}{\psi}$
DistK	$\mathsf{K}(\varphi \to \psi) \to (\mathsf{K}\varphi \to \mathsf{K}\psi)$	*NecK*	$\dfrac{\vdash \varphi}{\vdash \mathsf{K}\varphi}$
T	$\mathsf{K}\varphi \to \varphi$	*MS'*	$\dfrac{\vdash \psi \to \hat{\mathsf{K}}\neg Pd_1...d_n}{\vdash \psi \to \neg\mathsf{Kv}P}$
4	$\mathsf{K}\varphi \to \mathsf{K}\mathsf{K}\varphi$	*RE*	$\dfrac{\vdash \psi \leftrightarrow \chi}{\vdash \varphi \leftrightarrow \varphi[\psi/\chi]}$
5	$\neg\mathsf{K}\varphi \to \mathsf{K}\neg\mathsf{K}\varphi$		
KvP4	$\mathsf{Kv}P \to \mathsf{K}\mathsf{Kv}P$		

Note that in *MS'* rule, d_i does not occur in ψ.

The axiom system $\mathbb{MAELKVP}_1$			
Axioms		**Rules**	
Taut	All instances of tautologies	*MP*	$\dfrac{\varphi \quad \varphi \to \psi}{\psi}$
DistK	$\mathsf{K}_i(\varphi \to \psi) \to (\mathsf{K}_i\varphi \to \mathsf{K}_i\psi)$	*NecK*	$\dfrac{\vdash \varphi}{\vdash \mathsf{K}_i\varphi}$
T	$\mathsf{K}_i\varphi \to \varphi$	*MA'*	$\dfrac{\psi \to (\hat{\mathsf{K}}Pd_1....d_n \to \mathsf{K}Pd_1....d_n)}{\psi \to \mathsf{Kv}P}$
4	$\mathsf{K}_i\varphi \to \mathsf{K}_i\mathsf{K}_i\varphi$	*RE*	$\dfrac{\vdash \psi \leftrightarrow \chi}{\vdash \varphi \leftrightarrow \varphi[\psi/\chi]}$
5	$\neg\mathsf{K}_i\varphi \to \mathsf{K}_i\neg\mathsf{K}_i\varphi$		
KvP4	$\mathsf{Kv}P \to \mathsf{K}\mathsf{Kv}P$		

Note that in *MA'* rule, d_i does not occur in ψ.

As we can see, *KvP5* axiom is omitted in both axiom systems. The reason for this omission is that *KvP5* axiom can be derived from *KvP4*, *T*, *5* axioms and *RE* rules.

Before proving the soundness result, some definitions should be made:

Definition 1. *Given a frame $\mathcal{F} = \langle W, \sim, O\rangle$ and its assignment function V, an object $o \in O$ and a designator $d \in \mathbf{D}$, $V[(d,w)/o]$ is an assignment function on $\mathcal{F}$ where:*

- *For any n-ary predicate $P \in \mathbf{P}$ and possible world w', $V[(d,w)/o](P,w') = V(P,w')$;*
- *For any designator $d' \in \mathbf{D}$ and possible world w', if $d' \neq d$ or $w' \neq w$, then $V[(d,w)/o](d',w') = V(d',w')$;*
- *$V[(d,w)/o](d,w) = o$.*

We shorten $V[(d_1,w)/o_1][(d_2,w)/o_2]...[(d_n,w)/o_n]$ to $V[(\bar{d},w)/\bar{o}]$, and shorten $V[(\bar{d},u_1)/\bar{o}]$ $[(\bar{d},u_2)/\bar{o}]...$ to $V[\bar{d}/\bar{o},w]$ for brevity (where $u_1, u_2, ...$ is the enumeration of all successors of w, that is, $u_i \sim w$ for all i).

By induction on formulas, we can prove the following proposition:

Proposition 1. *For any φ and designators $d_1, ..., d_n \in \mathbf{D}$ that do not occur in φ, if $\mathcal{F}, V, w \vDash \varphi$, then for any objects $o_1, ..., o_n \in O$ and possible world* *u (u needs not to be different from w), we have* $\mathcal{F}, V[(\bar{d}, u)/\bar{o}], w \models \varphi$, *hence* $\mathcal{F}, V[\bar{d}/\bar{o}, w], w \models \varphi$.

Now we prove the soundness of $\mathbb{MSELKVP}_1$ and $\mathbb{MAELKVP}_1$:

Theorem 1 (Soundness Theorem for $\mathbb{MSELKVP}_1$ and $\mathbb{MAELKVP}_1$). *Under MS (MA) semantics,* $\mathbb{MSELKVP}_1$ *(*$\mathbb{MAELKVP}_1$*) is sound with respect to the class of* $S5$ *frames.*

Proof. All we need to check is that *KvP4* is valid and *MS'(MA')* is validity-preserving under MS (MA) semantics. We show that cases for MSP and MAP'.

Now, we prove *KvP4* is valid under MS semantics and *MA'* is validity-preserving under MA semantics. The other cases are similar.

- Suppose for any arbitrary model $\mathcal{M} = \langle W, \sim, O, V\rangle$ and an arbitrary world w on W, if $\mathcal{M}, w \models_{\text{MS}} \mathsf{Kv}P$, then by the definition of MS semantics, there exist objects $o'_1, ..., o'_n \in O$ such that for any $v \sim w$, $\langle o'_1, ..., o'_n\rangle \in V(P, v)$. Since $\mathcal{M}$ underlies an $S5$ frame, for any world u satisfying $u \sim w$, we have $v \sim u$. Therefore, for any world u satisfying $u \sim w$, there exist objects $o'_1, ..., o'_n \in O$ such that for any $v \sim u$, $\langle o'_1, ..., o'_n\rangle \in V(P, v)$, that is, $\mathcal{M}, u \models_{\text{MS}} \mathsf{Kv}P$. From the arbitrariness of u, we have $\mathcal{M}, w \models_{\text{MS}} \mathsf{KKv}P$. Therefore, *KvP4* is valid on the class of $S5$ frames under MS semantics.
- Suppose *MA'* is not validity-preserving, then there must be formulas ψ and $\hat{\mathsf{K}}Pd_1....d_n \to \mathsf{K}Pd_1....d_n$ (d_i does not appear in ψ), $S5$ frame $\mathcal{F} = \langle W, \sim, O\rangle$, assignment function V on $\mathcal{F}$ and some possible world w on W such that $\psi \to (\hat{\mathsf{K}}Pd_1....d_n \to \mathsf{K}Pd_1....d_n)$ is valid in every $S5$ frame, but $\mathcal{F}, V, w \not\models_{\text{MA}} \psi \to \mathsf{Kv}P$. Since $\mathcal{F}, V, w \not\models_{\text{MA}} \psi \to \mathsf{Kv}P$, $\mathcal{F}, V, w \models_{\text{MA}} \psi$ and $\mathcal{F}, V, w \not\models_{\text{MA}} \mathsf{Kv}P$. According to MA semantics, there is a possible world $u \sim w$ with $V(P, u) \neq V(P, w)$. It can be assumed that there exist $o_1, ..., o_n \in O$ such that $\langle o_1, ..., o_n\rangle \in V(P, u)$ but $\langle o_1, ..., o_n\rangle \notin V(P, w)$ (the other case is the same). And since $\mathcal{F}, V, w \models_{\text{MA}} \psi$ and none of d_i occurs in ψ, it follows from Proposition 1 that $\mathcal{F}, V[\bar{d}/\bar{o}, w], w \models_{\text{MA}} \psi$. Because $\psi \to (\hat{\mathsf{K}}Pd_1....d_n \to \mathsf{K}Pd_1....d_n)$ is valid in every $S5$ frame, we have $\mathcal{F}, V[\bar{d}/\bar{o}, w], w \models_{\text{MA}} \hat{\mathsf{K}}Pd_1....d_n \to \mathsf{K}Pd_1....d_n$. Because $\langle o_1, ..., o_n\rangle \in V[\bar{d}/\bar{o}, w](P, u)$ (since $\langle o_1, ..., o_n\rangle \in V(P, u)$), $\mathcal{F}, V[\bar{d}/\bar{o}, w], u \models_{\text{MA}} Pd_1....d_n$, thus $\mathcal{F}, V[\bar{d}/\bar{o}, w], w \models_{\text{MA}} \hat{\mathsf{K}}Pd_1....d_n$. Therefore, by $\mathcal{F}, V[\bar{d}/\bar{o}, w], w \models_{\text{MA}} \hat{\mathsf{K}}Pd_1....d_n \to \mathsf{K}Pd_1....d_n$, we have $\mathcal{F}, V[\bar{d}/\bar{o}, w], w \models_{\text{MA}} \mathsf{K}Pd_1....d_n$. Since $w \sim w$, $\mathcal{F}, V[\bar{d}/\bar{o}, w], w \models_{\text{MA}} Pd_1....d_n$, i.e. $\langle o_1, ..., o_n\rangle \in V[\bar{d}/\bar{o}, w](P, w)$, which contradicts the assumption that $\langle o_1, ..., o_n\rangle \notin V(P, w)$. Therefore, *MA*' is validity-preserving on the class of $S5$ frames under MA semantics. □

Now we start our completeness proof. We adopt a similar approach as the completeness proof in first-order logic (**building Henkin sets** and **constructing witnesses**) to prove the strong completeness of $\mathbb{MSELKVP}_1$. We make some special treatment when constructing Henkin sets:

Definition 2. *For any n-ary predicate P and set Γ of $\mathcal{L}_{ELKvP_1}$ formulas, we say that* $\mathsf{Kv}P$ ***has MSP witness in*** Γ *if* $\mathsf{Kv}P \in \Gamma$ ***implies*** *the existence of n designators $d_1, ..., d_n$ such that $\mathsf{K}Pd_1....d_n \in \Gamma$. We say that Γ is a* ***MSP-Henkin set*** *if Γ is maximal $\mathbb{MSELKVP}_1$-consistent, and $\mathsf{Kv}P$ has MSP witness in Γ for any $\mathsf{Kv}P \in \Gamma$.*

Here, we have a slightly different definition of witness compared to first-order logic. In first-order logic, if $\exists x Px \in \Gamma$, we say that the formula $\exists x Px$ has a witness in Γ if and only if there exists a term t such that $Pt \in \Gamma$. However, in the above definition, we do not use the "if and only if" definition but only keep one direction.

The reason why we only keep one direction is partly because we only need one direction to prove the strong completeness of $\mathbb{MSELKVP}_1$, and partly because we cannot prove the other direction, i.e., the implication from $\mathsf{K}Pd_1...d_n \in \Gamma$ to $\mathsf{Kv}P \in \Gamma$ is not necessarily valid if there exist n designators $d_1, ..., d_n$.[5]

The following proposition would be of great help in the completeness proof:

Proposition 2. *Let Γ be a set of $\mathcal{L}_{ELKvP_1}$ formulas, P be any n-ary predicate, and $d_1, ..., d_n$ be any designators. If Γ is $\mathbb{MSELKVP}_1$-consistent with $\mathsf{Kv}P$, then for any designators $d_1, ..., d_n$ that do not occur in Γ, Γ is also $\mathbb{MSELKVP}_1$-consistent with $\mathsf{K}Pd_1...d_n$.*

Proof. Suppose Γ is $\mathbb{MSELKVP}_1$-consistent with $\mathsf{Kv}P$. Let $d_1, ..., d_n$ be any designators that do not occur in $\Gamma \cup \{\mathsf{Kv}P\}$. If Γ is not $\mathbb{MSELKVP}_1$-consistent with $\mathsf{K}Pd_1...d_n$, then $\Gamma \vdash_{\mathbb{MSELKVP}_1} \neg\mathsf{K}Pd_1...d_n$. Therefore, there exist $\gamma_1, ..., \gamma_m \in \Gamma$ such that $\vdash_{\mathbb{MSELKVP}_1} (\gamma_1 \wedge ... \wedge \gamma_m) \to \neg\mathsf{K}Pd_1...d_n$. According to the *MS'* rule, we have $\vdash_{\mathbb{MSELKVP}_1} (\gamma_1 \wedge ... \wedge \gamma_m) \to \neg\mathsf{Kv}P$. Hence, $\Gamma \vdash_{\mathbb{MSELKVP}_1} \neg\mathsf{Kv}P$, which contradicts our assumption that Γ is $\mathbb{MSELKVP}_1$-consistent with $\mathsf{Kv}P$. □

Apparently, not all maximal $\mathbb{MSELKVP}_1$-consistent sets in $\mathcal{L}_{ELKvP_1}$ are MSP-Henkin sets. To address this issue, we need to enrich the existing language $\mathcal{L}_{ELKvP_1}$ to ensure that we can construct an MSP-Henkin set Γ^+ based on a consistent set Γ in $\mathcal{L}_{ELKvP_1}$.

We add a countably infinite number of new designators $c_1, c_2, ... \in \mathbf{C}$ into $\mathcal{L}_{ELKvP_1}$ to form an expanding language $\mathcal{L}^+_{ELKvP_1} = \mathcal{L}_{ELKvP_1}(\mathbf{P}, \mathbf{D} \cup \mathbf{C})$. Then we can prove the following proposition:

Proposition 3. *Every maximal $\mathbb{MSELKVP}_1$-consistent set Γ in $\mathcal{L}_{ELKvP_1}$ can be extended to an MSP-Henkin set Γ^+ in $\mathcal{L}^+_{ELKvP_1}$.*

Proof. Let $\mathsf{Kv}P_1, \mathsf{Kv}P_2, ...$ be the sequence of all $\mathcal{L}_{ELKvP_1}$ formulas of the form $\mathsf{Kv}P$. We define a sequence of new designator tuples $\bar{e}_1, \bar{e}_2, ...$ as follows: for each $k \geq 1$, let $\bar{e}_k = \langle c'_1, ..., c'_{\mathrm{ary}(P_k)} \rangle$, where $\mathrm{ary}(P_k)$ denotes the arity of predicate P_k, and $c'_1, ..., c'_{\mathrm{ary}(P_k)} \in \mathbf{C}$ are the first $\mathrm{ary}(P_k)$ new designators that do not appear in $\bar{e}_1, ..., \bar{e}_{k-1}$. We abbreviate $P_k\bar{e}_k$ as $P_k c'_1...c'_{\mathrm{ary}(P_k)}$. Define the sequence of

[5] From a semantic perspective, if there exist $d_1, ..., d_n$ such that $\mathcal{M}, w \vDash_{\mathrm{MS}} \mathsf{K}Pd_1...d_n$, then since d_i is a nonrigid designator, we may not have $\mathcal{M}, w \vDash_{\mathrm{MS}} \mathsf{Kv}P$.

formula sets in $\mathcal{L}^{+}_{ELKvP_1}$ as $\Gamma_0 \subseteq \Gamma_1 \subseteq \cdots$, where: (1) $\Gamma_0 = \Gamma$; (2) for each $k \geq 0$, $\Gamma_{k+1} = \Gamma_k \cup \{\mathsf{Kv}P_k \to \mathsf{K}P_k\bar{e}_k\}$.

According to the assumption, Γ_0 is consistent. Suppose that $\Gamma_0, ..., \Gamma_k$ are all consistent, but Γ_{k+1} is not consistent. Then, Γ_k and $\mathsf{Kv}P_k \to \mathsf{K}P_k\bar{e}_k$ are not consistent, so Γ_k is not consistent with both $\neg\mathsf{Kv}P_k$ and $\mathsf{K}P_k\bar{e}_k$. Since Γ_k is consistent and not consistent with $\neg\mathsf{Kv}P_k$, we have that Γ_k is consistent with $\mathsf{Kv}P_k$. Now, we have that Γ_k is consistent with $\mathsf{Kv}P_k$ but not consistent with $\mathsf{K}P_k\bar{e}_k$, which contradicts the assumption and Proposition 2. Therefore, each Γ_{k+1} is consistent.

Let $\Gamma_\omega = \bigcup_{n<\omega} \Gamma_n$. Then, Γ_ω is consistent. Hence by Lindenbaum's Lemma, it could be extended to an MSP-Henkin set Γ^+ in $\mathcal{L}^{+}_{ELKvP_1}$. □

Definition 3. *The canonical model $\mathcal{M}^c$ for $\mathbb{MSELKVP}_1$ is a tuple $\langle W^c, \sim^c, O^c, V^c\rangle$, where:*

- *$W^c = MCS \times \{0, 1\}$, where MCS is the set of all MSP-Henkin sets in $\mathcal{L}^{+}_{ELKvP_1}$. If the formula φ is in the maximal consistent set contained in s, then we write $\varphi \in s$. If $\varphi \in s$ and $\varphi \in t$, then we write $\varphi \in s \cap t$.*
- *$s \sim^c t \iff \{\varphi \mid \mathsf{K}\varphi \in s\} \subseteq t$.*
- *$O^c = (\mathbf{D} \cup \mathbf{C}) \times W^c$.*
- *V^c is the assignment function that satisfies the following conditions:*
  - *for any nonrigid designator $d \in \mathbf{D} \cup \mathbf{C}$, $V^c(d, s) = (d, s)$;*
  - *for any n-ary predicate P and possible world $s \in W^c$,*

$$V^c(P, s) = \begin{cases} \bigcup_{t \sim^c s} \{\langle (d_1, t), ..., (d_n, t)\rangle \mid Pd_1....d_n \in t\}, & \mathsf{Kv}P \in s. \\ \{\langle (d_1, s), ..., (d_n, s)\rangle \mid Pd_1....d_n \in s\}, & \mathsf{Kv}P \notin s. \end{cases}$$

Clearly, the canonical model $\mathcal{M}^c$ is an **S5ELKvP₁** model. It should be noted that in the construction of possible worlds, we do not simply use MSP-Henkin sets in $\mathcal{L}^{+}_{ELKvP_1}$ as our possible worlds. For each MSP-Henkin set, we construct a copy (i.e., for each MSP-Henkin set, we have two different possible worlds corresponding to it). The reason for this modification is to ensure that when $\mathsf{Kv}P$ is not included in the MSP-Henkin set contained in some possible world s ($\mathsf{Kv}P \notin s$), the formula $\mathsf{Kv}P$ is not true at $\mathcal{M}^c, s$ (if $\mathsf{Kv}P \notin s$, then $\mathcal{M}^c, s \nvDash_{\mathrm{MS}} \mathsf{Kv}P$).

Based on *KvP4* axiom and the construction of $V^c(P, s)$, we can easily prove the followings:

Proposition 4. *$s \sim^c t$ if and only if $\{\varphi \mid \mathsf{K}\varphi \in s\} = \{\varphi \mid \mathsf{K}\varphi \in t\}$, hence $\sim^c$ is an equivalence relation.*

Proposition 5. *For any $s \sim^c t$, $\mathsf{Kv}P \in s$ if and only if $\mathsf{Kv}P \in t$.*

Proposition 6. *If $\mathsf{Kv}P \in s$, then $V^c(P, s) = V^c(P, t)$ for all $t \sim^c s$. If $\mathsf{Kv}P \notin s$, then $V^c(P, s) \cap V^c(P, t) = \varnothing$ for all $t \sim^c s$ such that $t \neq s$.*

However, the above propositions alone are not enough to prove the Truth Lemma. Given an MSP-Henkin set $s \in W^c$ in $\mathcal{L}^+_{ELKvP_1}$ and assuming that $\hat{\mathsf{K}}\varphi \in s$, we can certainly construct a possible world s' such that $\varphi \in s'$ and s' is a maximal consistent set, but we cannot directly conclude that s' is necessarily an MSP-Henkin set in $\mathcal{L}^+_{ELKvP_1}$ (at least this conclusion is not obvious). Therefore, we still need to prove the following Existence Lemma:

Lemma 1 (Existence Lemma). *Given an MSP-Henkin set $s \in W^c$ in $\mathcal{L}^+_{ELKvP_1}$, if $\hat{\mathsf{K}}\varphi \in s$, then there exists an MSP-Henkin set $t \in W^c$ in $\mathcal{L}^+_{ELKvP_1}$ such that $s \sim^c t$ and $\varphi \in t$.*

Proof. Let $t' = \{\psi \mid \mathsf{K}\psi \in s\} \cup \{\varphi\}$. It is easy to show that t' is consistent. By Lindenbaum's Lemma, we can extend t' to a maximal consistent set t in $\mathcal{L}^+_{ELKvP_1}$.

We now show that t is an MSP-Henkin set: for any $\mathsf{Kv}P \in t$, assume that $\neg\mathsf{Kv}P \in s$. Then, by T and $KvP4$ axiom, we have $\mathsf{Kv}P \leftrightarrow \mathsf{KKv}P \in s$, so $\neg\mathsf{KKv}P \in s$. By 5 axiom and $\neg\mathsf{KKv}P \in s$, we have $\mathsf{K}\neg\mathsf{KKv}P \in s$. Moreover, according to $\mathsf{Kv}P \leftrightarrow \mathsf{KKv}P \in s$, we have $\mathsf{K}\neg\mathsf{Kv}P \in s$. By the construction of t', we have $\neg\mathsf{Kv}P \in t$, which contradicts $\mathsf{Kv}P \in t$. Therefore, for any $\mathsf{Kv}P \in t$, we have $\mathsf{Kv}P \in s$. Since s is an MSP-Henkin set in $\mathcal{L}^+_{ELKvP_1}$, there exist n designators $d_1, ..., d_n$ such that $\mathsf{K}Pd_1...d_n \in s$. According to 4 axiom and the construction of t', we also have $\mathsf{K}Pd_1...d_n \in t$.

Therefore, t is an MSP-Henkin set in $\mathcal{L}^+_{ELKvP_1}$. By Proposition 4, $s \sim^c t$ and $\varphi \in t$. □

Now we prove the Truth Lemma and the completeness result for $\mathbb{MSELKVP}_1$:

Lemma 2 (Truth Lemma for $\mathbb{MSELKVP}_1$). *For any $\mathcal{L}^+_{ELKvP_1}$ formula φ, we have $\varphi \in s$ if and only if $\mathcal{M}^c, s \vDash_{MS} \varphi$.*

Proof. We only need to prove the cases where $\varphi = Pd_1...d_n$ or $\varphi = \mathsf{Kv}P$.

- If $\varphi = Pd_1...d_n$, consider whether $\mathsf{Kv}P \in s$ holds or not.
  - If $\mathsf{Kv}P \in s$, then $V^c(P, s) = \bigcup_{t \sim^c s} \langle(d_1, t), ..., (d_n, t)\rangle \mid Pd_1...d_n \in t\}$. We have $\mathcal{M}^c, s \vDash_{\mathrm{MS}} Pd_1...d_n$ if and only if $\langle(d_1, s), ..., (d_n, s)\rangle \in V^c(P, s)$, if and only if $Pd_1...d_n \in s$ ($s \sim^c s$).
  - If $\mathsf{Kv}P \notin s$, then $V^c(P, s) = \{\langle(d_1, s), ..., (d_n, s)\rangle \mid Pd_1...d_n \in s\}$. We have $\mathcal{M}^c, s \vDash_{\mathrm{MS}} Pd_1...d_n$ if and only if $\langle(d_1, s), ..., (d_n, s)\rangle \in V^c(P, s)$, if and only if $Pd_1...d_n \in s$.

  Therefore, $Pd_1...d_n \in s$ if and only if $\mathcal{M}^c, s \vDash_{\mathrm{MS}} Pd_1...d_n$.
- If $\varphi = \mathsf{Kv}P$, we also consider whether $\mathsf{Kv}P \in s$ holds or not.
  - If $\mathsf{Kv}P \in s$, then $V^c(P, s) = \bigcup_{t \sim^c s} \langle(d_1, t), ..., (d_n, t)\rangle \mid Pd_1...d_n \in t\}$. Since s is an MSP-Henkin set, there exist designators $c_1, ..., c_n \in \mathbf{D} \cup \mathbf{C}$ such that $\mathsf{K}Pc_1...c_n \in s$. By T axiom, we have $Pc_1...c_n \in s$, so $\langle(c_1, s), ..., (c_n, s)\rangle \in V^c(P, s)$, which means $V^c(P, s)$ is non-empty. By Proposition 6 and the MS semantics, we have $\mathcal{M}^c, s \vDash_{\mathrm{MS}} \mathsf{Kv}P$.

- If $\mathsf{Kv}P \notin s$, then $V^c(P,s) = \{\langle(d_1,s),...,(d_n,s)\rangle \mid Pd_1...d_n \in s\}$. By Proposition 6, we only need to prove that there exists a non-self successor of s. Assume $s = (\Gamma, x)$, where Γ is an MSP-Henkin set and $x \in \{0,1\}$. Let $s' = (\Gamma, 1-x)$. Obviously, by the definition of $\sim^c$ and T axiom, we have $s \sim^c s'$ and $s \neq s'$. Therefore, we have $\mathcal{M}^c, s \nvDash_{\mathrm{MS}} \mathsf{Kv}P$.

Therefore, for any $\mathcal{L}^+_{ELKvP_1}$ formula φ, we have $\varphi \in s$ if and only if $\mathcal{M}^c, s \vDash_{\mathrm{MS}} \varphi$. □

Suppose V_1 is an assignment function in $\mathcal{L}^+_{ELKvP_1}$ and V_2 is in $\mathcal{L}_{ELKvP_1}$, and both functions are on frame $\mathcal{F}$. We say that V_1 and V_2 agree on φ, if φ is an $\mathcal{L}_{ELKvP_1}$ formula, and for any designator d, predicate P in φ and any possible world w, we have $V_1(d,w) = V_2(d,w)$ and $V_1(P,w) = V_2(P,w)$.

By induction on formulas, it would not be hard to prove the Coincidence Lemma:

Lemma 3 (Coincidence Lemma). *If V_1 and V_2 agree on φ, then $\mathcal{F}, V_1, w \vDash_{MS} \varphi$ if and only if $\mathcal{F}, V_2, w \vDash_{MS} \varphi$.*

We call $\langle \mathcal{F}, V_2\rangle$ a **$\mathcal{L}_{ELKvP_1}$-reduct** of $\langle \mathcal{F}, V_1\rangle$ if and only if V_1 and V_2 agree on all $\mathcal{L}_{ELKvP_1}$ formulas.

With Coincidence Lemma, we can easily prove the completeness theorem for $\mathbb{MSELKVP}_1$:

Theorem 2 (Completeness Theorem for $\mathbb{MSELKVP}_1$). *Under MS semantics, $\mathbb{MSELKVP}_1$ is strongly complete with respect to the class of S5 frames.*

Proof. Suppose Γ is an arbitrary $\mathbb{MSELKVP}_1$- consistent set of $\mathcal{L}_{ELKvP_1}$ formulas, then by Proposition 3, can be extended to an MSP-Henkin set Γ^+ in $\mathcal{L}^+_{ELKvP_1}$. By Lemma 2, $\mathcal{M}^c, \Gamma^+ \vDash_{\mathrm{MS}} \Gamma$. Let $\mathcal{M}^- = \langle W^c, \sim^c, O^c, V^-\rangle$, where V^- is an assignment function in $\mathcal{L}_{ELKvP_1}$ and for any designator d, predicate P and possible world w, $V^-(d,w) = V^c(d,w)$ and $V^-(P,w) = V^c(P,w)$. Then $\mathcal{M}^-$ is a $\mathcal{L}_{ELKvP_1}$-reduct of M^c, and by Lemma 3, $\mathcal{M}^-, \Gamma^+ \vDash_{\mathrm{MS}} \Gamma$, which means Γ is **MSELKvP$_1$**-satisfiable. Hence we have the strong completeness result for $\mathbb{MSELKVP}_1$. □

When proving the completeness result for $\mathbb{MAELKVP}_1$, we need to slightly change the definition of MAP-Henkin set and canonical model, and have the followings:

Definition 4. *For any n-ary predicate P and set Γ of $\mathcal{L}_{ELKvP_1}$ formulas, we say that $\neg\mathsf{Kv}P$ **has MAP witness in** Γ if $\neg\mathsf{Kv}P \in \Gamma$ **implies** the existence of n designators $d_1,...,d_n$ such that $\hat{\mathsf{K}}\neg Pd_1...d_n \wedge \hat{\mathsf{K}}Pd_1...d_n \in \Gamma$. We say that Γ is a **MAP-Henkin set** if Γ is maximal $\mathbb{MAELKVP}_1$-consistent, and $\neg\mathsf{Kv}P$ has MAP witness in Γ for $\neg\mathsf{Kv}P \in \Gamma$.*

Proposition 7. *Every maximal $\mathbb{MAELKVP}_1$-consistent set Γ in $\mathcal{L}_{ELKvP_1}$ can be extended to an MAP-Henkin set Γ^+ in $\mathcal{L}^+_{ELKvP_1}$.*

Proof. We show the main difference between this proof and the proof of Proposition 3. Define the sequence of formula sets in $\mathcal{L}^+_{ELKvP_1}$ as $\Gamma_0 \subseteq \Gamma_1 \subseteq \cdots$, where: (1) $\Gamma_0 = \Gamma$; (2) for each $k \geq 0$, $\Gamma_{k+1} = \Gamma_k \cup \{\neg \mathsf{Kv}P_k \rightarrow (\hat{\mathsf{K}}\neg P_k\bar{e}_k \wedge \hat{\mathsf{K}}P_k\bar{e}_k)\}$. □

Definition 5. *The canonical model* $\mathcal{M}^c$ *for* $\mathbb{MAELKVP}_1$ *is a tuple* $\langle W^c, \sim^c, O^c, V^c\rangle$*, where:*

- W^c *is the set of all MAP-Henkin sets in* $\mathcal{L}^+_{ELKvP_1}$.[6]
- $s \sim^c t \iff \{\varphi \mid \mathsf{K}\varphi \in s\} \subseteq t$.
- $O^c = (\mathbf{D} \cup \mathbf{C}) \times W^c$.
- V^c *is the assignment function that satisfies the following conditions:*
  - *for any nonrigid designator* $d \in \mathbf{D} \cup \mathbf{C}$, $V^c(d, s) = (d, s)$;
  - *for any n-ary predicate P and possible world* $s \in W^c$,

$$V^c(P, s) = \begin{cases} \bigcup\limits_{t \sim^c s} \{\langle (d_1, t), ..., (d_n, t)\rangle \mid Pd_1....d_n \in t\}, & \mathsf{Kv}P \in s. \\ \{\langle (d_1, s), ..., (d_n, s)\rangle \mid Pd_1....d_n \in s\}, & \mathsf{Kv}P \notin s. \end{cases}$$

It would be easy to prove the following completeness theorem according to the proof of Theorem 2:

Theorem 3 (Completeness Theorem for $\mathbb{MAELKVP}_1$**).** *Under MA semantics,* $\mathbb{MAELKVP}_1$ *is strongly complete with respect to the class of S5 frames.*

3.2 Axiom Systems $\mathbb{MSELKV}_1$ and $\mathbb{MAELKV}_1$ for $\mathcal{L}_{ELKvdP_1}$

In this part, we will add the "knowing value" operator $\mathsf{Kv}d$ for single agent on the basis of language $\mathcal{L}_{ELKvP_1}$, and obtain the following language $\mathcal{L}_{ELKvdP_1}$:

$$\varphi ::= \top \mid \mathsf{Kv}d \mid Pd_1...d_n \mid \neg\varphi \mid (\varphi \wedge \varphi) \mid \mathsf{K}\varphi \mid \mathsf{Kv}P$$

The definitions of MS and MA semantics are the same as those in $\mathcal{L}_{ELKvP_1}$, where the semantics of the $\mathsf{Kv}d$ operator is defined in the same way as in the language $\mathcal{L}_{ELKv_1}$. We use shorthand notation $\bigwedge \mathsf{Kv}d_i$ to represent the formula $\bigwedge\limits_{1 \leq i \leq n} \mathsf{Kv}d_i$. We present two axiom systems $\mathbb{MSELKV}_1$ and $\mathbb{MAELKV}_1$.

[6] It should be noted that we do not need to construct possible world copies for MAP-Henkin sets here. This is because the definition of MAP-Henkin set ensures that every $\neg\mathsf{Kv}P \in s$ in an MAP-Henkin set s has a corresponding MAP witness $\hat{\mathsf{K}}\neg Pd_1...d_n \wedge \hat{\mathsf{K}}Pd_1...d_n \in s$, which naturally leads to at least two distinct successor possible worlds of s.

The axiom system $\mathbb{MSELKV}_1$			
Axioms		**Rules**	
Taut	All instances of tautologies	*MP*	$\dfrac{\varphi \quad \varphi \to \psi}{\psi}$
DistK	$\mathsf{K}_i(\varphi \to \psi) \to (\mathsf{K}_i\varphi \to \mathsf{K}_i\psi)$	*NecK*	$\dfrac{\vdash \varphi}{\vdash \mathsf{K}_i\varphi}$
T	$\mathsf{K}_i\varphi \to \varphi$	*MS'*	$\dfrac{\vdash \psi \to (\bigwedge \mathsf{Kv}d_i \to \hat{\mathsf{K}}\neg Pd_1...d_n)}{\vdash \psi \to \neg\mathsf{Kv}P}$
4	$\mathsf{K}_i\varphi \to \mathsf{K}_i\mathsf{K}_i\varphi$	*RE*	$\dfrac{\vdash \psi \leftrightarrow \chi}{\vdash \varphi \leftrightarrow \varphi[\psi/\chi]}$
5	$\neg\mathsf{K}_i\varphi \to \mathsf{K}_i\neg\mathsf{K}_i\varphi$		
Kv4	$\mathsf{Kv}d \to \mathsf{KKv}d$		
KvP4	$\mathsf{Kv}P \to \mathsf{KKv}P$		
MS	$\bigwedge \mathsf{Kv}d_i \wedge \mathsf{K}Pd_1...d_n \to \mathsf{Kv}P$		

Note that in *MS'* rule, d_i does not occur in ψ.

The axiom system $\mathbb{MAELKV}_1$			
Axioms		**Rules**	
Taut	All instances of tautologies	*MP*	$\dfrac{\varphi \quad \varphi \to \psi}{\psi}$
DistK	$\mathsf{K}_i(\varphi \to \psi) \to (\mathsf{K}_i\varphi \to \mathsf{K}_i\psi)$	*NecK*	$\dfrac{\vdash \varphi}{\vdash \mathsf{K}_i\varphi}$
T	$\mathsf{K}_i\varphi \to \varphi$	*MA'*	$\dfrac{\vdash \psi \to (\bigwedge \mathsf{Kv}d_i \wedge \hat{\mathsf{K}}Pd_1...d_n \to \mathsf{K}Pd_1...d_n)}{\vdash \psi \to \mathsf{Kv}P}$
4	$\mathsf{K}_i\varphi \to \mathsf{K}_i\mathsf{K}_i\varphi$	*RE*	$\dfrac{\vdash \psi \leftrightarrow \chi}{\vdash \varphi \leftrightarrow \varphi[\psi/\chi]}$
5	$\neg\mathsf{K}_i\varphi \to \mathsf{K}_i\neg\mathsf{K}_i\varphi$		
Kv4	$\mathsf{Kv}d \to \mathsf{KKv}d$		
KvP4	$\mathsf{Kv}P \to \mathsf{KKv}P$		
MA	$\bigwedge \mathsf{Kv}d_i \wedge \hat{\mathsf{K}}Pd_1...d_n \wedge \mathsf{Kv}P \to \mathsf{K}Pd_1...d_n$		

Note that in *MA'* rule, d_i does not occur in ψ.

It could be easily found out that we have added MS and MA axioms in $\mathbb{MSELKV}_1$ and $\mathbb{MAELKV}_1$, respectively. This is because by adding the formula $\bigwedge \mathsf{Kv}d_i$, we can "convert" $d_1, ..., d_n$ from nonrigid designators to rigid ones, which enables us to better characterize the relationship between the $\mathsf{Kv}P$ operator and formulas such as $Pd_1...d_n$. With the $\mathsf{Kv}d$ operator, we can further obtain the following conclusions:

Proposition 8. *For any frame $\mathcal{F}$, possible world w, designator tuple $\bar{d} = \langle d_1, ..., d_n\rangle$ and object tuple $\bar{o} = \langle o_1, ..., o_n\rangle$, we have $\mathcal{F}, V[\bar{d}/\bar{o}, w], w \vDash_{MS} \bigwedge \mathsf{Kv}d_i$ and $\mathcal{F}, V[\bar{d}/\bar{o}, w], w \vDash_{MA} \bigwedge \mathsf{Kv}d_i$.*

It would not be difficult to prove the soundness theorem for the two axiom systems.

We give out some important definitions for both axiom systems, and prove the Truth Lemma for $\mathsf{Kv}d$ formulas in MS (MA) semantics. Based on the earlier parts of this paper, it is not difficult to infer the remaining content of the proof.

Definition 6. *For any n-ary predicate P and set Γ of $\mathcal{L}_{ELKvdP_1}$ formulas, we say that $\mathsf{Kv}P$* **has MS witness in** *Γ if $\mathsf{Kv}P \in \Gamma$* **if and only if** *there exist n designators $d_1, ..., d_n$ such that $\bigwedge \mathsf{Kv}d_i \wedge \mathsf{K}Pd_1...d_n \in \Gamma$. We say that Γ is a* **MS-Henkin set** *if Γ is maximal $\mathbb{MSELKV}_1$-consistent, and $\mathsf{Kv}P$ has MS witness in Γ for $\mathsf{Kv}P \in \Gamma$.*

Definition 7. *The canonical model $\mathcal{M}^c$ for $\mathbb{MSELKV}_1$ is a tuple $\langle W^c, \sim^c, O^c, V^c \rangle$, where:*

- $W^c = MCS \times \{0, 1\}$, *where MCS is the set of all MS-Henkin sets in $\mathcal{L}^+_{ELKvdP_1}$.*
- $s \sim^c t \iff \{\varphi \mid \mathsf{K}\varphi \in s\} \subseteq t$.
- $O^c = (\mathbf{D} \cup \mathbf{C}) \times W^c$.
- V^c *is the assignment function that satisfies the following conditions:*
  - *For any nonrigid designator $d \in \mathbf{D} \cup \mathbf{C}$, we have $V^c(d, s) = |(d, s)|_R$, where* $R = \{((d, s), (e, t)) \mid d = e,\ s \sim^c t \text{ and } \mathsf{Kv}d \in s\} \cup \{((d, s), (d, s)) \mid d \in \mathbf{D} \cup \mathbf{C}, s \in W^c\}$;
  - *For any n-ary predicate P and possible world $s \in W^c$, we have* $V^c(P, s) = \{\langle |(d_1, s)|_R, ..., |(d_n, s)|_R \rangle \mid Pd_1...d_n \in s\}$.

Definition 8. *For any n-ary predicate P and set Γ of $\mathcal{L}_{ELKvdP_1}$ formulas, we say that $\neg\mathsf{Kv}P$* **has MA witness in** *Γ if $\neg\mathsf{Kv}P \in \Gamma$* **if and only if** *there exist n designators $d_1, ..., d_n$ such that $\bigwedge \mathsf{Kv}d_i \wedge \hat{\mathsf{K}}\neg Pd_1...d_n \wedge \hat{\mathsf{K}}Pd_1...d_n \in \Gamma$. We say that Γ is a* **MA-Henkin set** *if Γ is maximal $\mathbb{MAELKV}_1$-consistent, and $\neg\mathsf{Kv}P$ has MA witness in Γ for $\neg\mathsf{Kv}P \in \Gamma$.*

Definition 9. *The canonical model $\mathcal{M}^c$ for $\mathbb{MAELKV}_1$ is a tuple $\langle W^c, \sim^c, O^c, V^c \rangle$, where:*

- $W^c = MCS \times \{0, 1\}$, *where MCS is the set of all MA-Henkin sets in $\mathcal{L}^+_{ELKvdP_1}$.*
- $s \sim^c t \iff \{\varphi \mid \mathsf{K}\varphi \in s\} \subseteq t$.
- $O^c = (\mathbf{D} \cup \mathbf{C}) \times W^c$.
- V^c *is the assignment function that satisfies the following conditions:*
  - *For any nonrigid designator $d \in \mathbf{D} \cup \mathbf{C}$, we have $V^c(d, s) = |(d, s)|_R$, where* $R = \{((d, s), (e, t)) \mid d = e,\ s \sim^c t \text{ and } \mathsf{Kv}d \in s\} \cup \{((d, s), (d, s)) \mid d \in \mathbf{D} \cup \mathbf{C}, s \in W^c\}$;
  - *For any n-ary predicate P and possible world $s \in W^c$, we have*
  $$V^c(P, s) = \begin{cases} \bigcup\limits_{t \sim^c s} \{\langle (d_1, t), ..., (d_n, t) \rangle \mid Pd_1....d_n \in t\}, & \mathsf{Kv}P \in s. \\ \{\langle (d_1, s), ..., (d_n, s) \rangle \mid Pd_1....d_n \in s\}, & \mathsf{Kv}P \notin s. \end{cases}$$

Proposition 9. *R is well-defined in both canonical models, i.e., R is an equivalence relation.*

As we can see, in the construction of W^c of M^c for both axiom systems, we build a copy for each MS-Henkin (MA-Henkin) set. The construction of copies here would be used when we prove the Truth Lemma for $\mathsf{Kv}d$ formulas ($\mathsf{Kv}d \in s \Leftrightarrow \mathcal{M}^c, s \vDash_{\text{MS}} \mathsf{Kv}d$). Take $\mathbb{MSELKV}_1$ for example.

($\Leftarrow$) If $\mathsf{Kv}d \notin s$, then assume $s = (\Gamma, x)$, where Γ is a $\mathbb{MSELKV}_1$-Henkin set and $x \in \{0, 1\}$. Let $s' = (\Gamma, 1 - x)$. Obviously, by the definition of $\sim^c$ and T axiom, we have $s \sim^c s'$. Suppose $|(d, s)|_R = |(d, s')|_R$. Then we have $\mathsf{Kv}d \in s$, which contradicts the assumption. Hence, $|(d, s)|_R \neq |(d, s')|_R$, i.e., $V_{\mathbf{D}}(d, s) \neq V_{\mathbf{D}}(d, s')$, so $\mathcal{M}^c, s \nvDash \mathsf{Kv}d$.

($\Rightarrow$) If $\mathsf{Kv}d \in s$, then $(d, s)R(d, t)$ for any t such that $s \sim^c t$, i.e., $|(d, s)|_R = |(d, t)|_R$ and $V_{\mathbf{D}}(d, s) = V_{\mathbf{D}}(d, t)$. Therefore, we have $\mathcal{M}^c, s \vDash \mathsf{Kv}d$.

With the approach similar to the proof of $\mathbb{MSELKVP}_1$ and $\mathbb{MAELKVP}_1$, we can easily prove the completeness theorem for $\mathbb{MSELKV}_1$ and $\mathbb{MAELKV}_1$:

Theorem 4 (Completeness Theorem for $\mathbb{MSELKV}_1$ and $\mathbb{MAELKV}_1$). *Under MS (MA) semantics, $\mathbb{MSELKV}_1$ ($\mathbb{MAELKV}_1$) is strongly complete with respect to the class of $S5$ frames.*

4 Conclusion and Future Work

This paper focuses on the axiomatization problem of logics with the "knowing predicate value" operator $\mathsf{Kv}P$. We introduce $\mathsf{Kv}P$ operator to characterize the notion of "knowing the value of a predicate P", and provides two different semantics - MS semantics and MA semantics - for the $\mathsf{Kv}P$ operator based on two linguistic interpretations of problems (Mention-Some and Mention-All interpretations) at the beginning of the paper. In the remaining parts of the paper, we prove the strong completeness theorem for two sets of axiom systems containing only the $\mathsf{Kv}P$ operator, as well as two sets of axiom systems containing both the $\mathsf{Kv}P$ and $\mathsf{Kv}d$ operators.

Based on the work done by this paper, we believe that there are still many directions that deserve further research and exploration:

- The multi-agent version for $\mathbb{MSELKVP}_1$, $\mathbb{MAELKVP}_1$, $\mathbb{MSELKV}_1$ and $\mathbb{MAELKV}_1$. As all four axiom systems presented in this paper are single-agent versions, proving their corresponding multi-agent versions would be a good research direction. To solve this problem, we can consider two approaches: one is to abstract the semantics of the $\mathsf{Kv}P$ operator and prove the completeness of the multi-agent version of the abstract language, following the method in [8]; the other is to extend the existing canonical model proof method and prove the multi-agent version following the proof method in [17].
- The integration of MS and MA semantics. The only difference between MS and MA semantics are the interpretation of $\mathsf{Kv}P$ formulas. Therefore, by integrating MS and MA semantics, we may find some interesting interactions between the two $\mathsf{Kv}P$ operators.

- The modification of MS and MA semantics. In MS and MA semantics, we would have $\{\langle V(d_1, w), ..., V(d_n, w)\rangle \mid \mathcal{M}, w \models_{\text{MS}} Pd_1...d_n\} \subseteq V(P, w)$, but not necessarily that $\{\langle V(d_1, w), ..., V(d_n, w)\rangle \mid \mathcal{M}, w \models_{\text{MS}} Pd_1...d_n\} = V(P, w)$. This seems a little bit controversial. If we try to modify the semantics so that $\{\langle V(d_1, w), ..., V(d_n, w)\rangle \mid \mathcal{M}, w \models Pd_1...d_n\} = V(P, w)$ holds all the time, the axiom systems might have some interesting transformations.

In addition to the above directions, in recent years, topics such as public inspection [6], de re updates [3], and "knowing-function" operator [4,14], which are related to the concept of "knowing value", have also received much attention from scholars. The work of this paper can also be combined with the above topics, such as considering a function as a predicate, so we can express "knowing-function" in terms of the "knowing predicate value" operator, and thus create some new interactions with Dependence Logic.

References

1. Baltag, A.: To know is to know the value of a variable. In: Advances in Modal Logic (2016)
2. Baltag, A., van Benthem, J.: A simple logic of functional dependence. J. Philos. Log. **50**(5), 939–1005 (2021)
3. Cohen, M., Tang, W., Wang, Y.: De re updates. In: Theoretical Aspects of Rationality and Knowledge (2021)
4. Ding, Y.: Epistemic logic with functional dependency operator. CoRR abs/1706.02048 (2017). http://arxiv.org/abs/1706.02048
5. van Ditmarsch, H., Halpern, J., van der Hoek, W., Kooi, B. (eds.): Handbook of Epistemic Logic. College Publications (2015)
6. van Eijck, J., Gattinger, M., Wang, Y.: Knowing values and public inspection. In: Ghosh, S., Prasad, S. (eds.) ICLA 2017. LNCS, vol. 10119, pp. 77–90. Springer, Heidelberg (2017). https://doi.org/10.1007/978-3-662-54069-5_7
7. Groenendijk, J., Stokhof, M.: Studies on the semantics of questions and the pragmatics of answers. Ph.D. thesis, University of Amsterdam (1984)
8. Gu, T., Wang, Y.: "Knowing value" logic as a normal modal logic. In: Proceedings of the 11th Conference on Advances in Modal Logic, pp. 362–381. College Publications (2016)
9. Hamblin, C.L.: Questions in Montague English. In: Montague Grammar, pp. 247–259. Academic Press (1976)
10. Heim, I.: Interrogative semantics and Karttunen's semantics for 'know'. In: Proceedings of the Israeli Association for Theoretical Linguistics, pp. 128–144 (1994)
11. Ma, X., Guo, W.: W-js: a modal logic of knowledge. In: Proceedings of the Eighth International Joint Conference on Artificial Intelligence (I), pp. 398–401 (1983)
12. McCarthy, J.: First order theories of individual concepts and propositions. In: Machine Intelligence (1979)
13. Plaza, J.: Logics of public communications. In: Proceedings of the Fourth International Symposium on Methodologies for Intelligent Systems: Poster Session Program, pp. 201–216. Oak Ridge National Laboratory (1989)
14. Wang, X.: Epistemic logic with partial dependency operator. In: Blackburn, P., Lorini, E., Guo, M. (eds.) LORI 2019. LNCS, vol. 11813, pp. 385–398. Springer, Heidelberg (2019). https://doi.org/10.1007/978-3-662-60292-8_28

15. Wang, Y.: Beyond knowing that: a new generation of epistemic logics. In: van Ditmarsch, H., Sandu, G. (eds.) Jaakko Hintikka on Knowledge and Game-Theoretical Semantics. OCL, vol. 12, pp. 499–533. Springer, Cham (2018). https://doi.org/10.1007/978-3-319-62864-6_21
16. Wang, Y., Fan, J.: Knowing that, knowing what, and public communication: public announcement logic with KV operators. In: Proceedings of the 23rd International Joint Conference on Artificial Intelligence, pp. 1147–1154. IJCAI/AAAI (2013)
17. Wang, Y., Fan, J.: Conditionally knowing what. Adv. Modal Logic **10** (2014)

Metaphor Comprehension in Situations

Zhengyi Hong(✉)

Zhejiang University, Hangzhou, China
flourite@zju.edu.cn

Abstract. Since conceptual metaphor theory was put forward, the view that metaphors affect human cognition by constructing cognitive frames has been very influential. From the view of cognitive metaphor studies, the focus of metaphor formalization is no longer to describe the substitution of metaphorical meaning for literal meaning, but to consider the cognitive frames of concepts in conventional situations and the selection of information in specific contexts. Following the perspective of cognitive metaphor theory, this article constructs a logic based on situation semantics to provide a possible way to interpret the information transfer in the process of metaphor comprehension. It describes different concept domains by assigning each situation a unique theme and reflects an agent's belief by supporting relations between situations and propositions. Then we can distinguish between different kinds of information flow in metaphor comprehension.

Keywords: Cognitive Metaphor · Situation theory · Logical semantics

1 Introduction

In a broad sense, metaphors can be defined as using different types of concepts to explain an object. Sometimes they are used to make the expression more novel and vivid, and sometimes used to explain concepts that are difficult to convey directly. Cognitive metaphor theorists generally defend that "metaphor is pervasive in everyday life, not just in language but in thought and action. Our ordinary conceptual system, in terms of which we both think and act, is fundamentally metaphorical in nature" [12, p. 3]. With a familiar source domain, people can construct a cognitive frame and abstract and unfamiliar target concepts can be put into it for understanding. It is a basic way for humans to gain new knowledge. For example, when one says, "he is at a crossroads in his life". It does not means he is standing at some intersection, but he is facing choices that will shape his future and may be confused. Most of us have the experience of standing at a crossroads. At a crossroads, you have multiple choices, and different decisions will lead you in different directions. But how could information about travel can carry the information about life? Such a perspective no longer regards metaphor simply as a language skill but as a basic way of human cognition.

N. Alechina et al. (Eds.): LORI 2023, LNCS 14329, pp. 167–175, 2023.
https://doi.org/10.1007/978-3-031-45558-2_13

Metaphoric cognition is regarded as one of the three major topics in cognitive science in the twentieth century. It has been widely discussed in cognitive linguistics (e.g. Conceptual Metaphor Theory [11,13], Conceptual Integration Theory [7], The Career of Metaphor Theory [1,6], Deliberate Metaphor Theory [16]) and cognitive psychology [3,7]. Also, there is a booming interest in an interdisciplinary field interfacing with cognition and computation [5,8,9,15]. But relatively little work has been done on logical semantics, and most of them are piecemeal [14,17,18].

2 Background

After the introduction of information theory, linguistics have gradually recognized the similarity between the process of verbal communication and information exchange, and see information transfer as a basic functions of language. According to situation theory, ontologically, a situation is an entity as a part of the structured actual world that an agent recognizes and chooses [2, p. 31]. Technically, a situation is a set of infons. An infon is the basic unit of information, it can be denoted as a tuple $\sigma = \langle R, a_1, \ldots, a_n, \ell, i\rangle$, where R is a n-ary relation, $a_1, \ldots, a_n$ are individuals, ℓ is a space-time location, i called polarity is a element of $\{0, 1\}$. What an infon expresses is that at a certain space and certain time, some individuals has or does not have the property. The conjunction or disjunction of infons forms compound infons. Given a situation s and infon σ, s supports σ, which means that infon σ is satisfied in the situation s, denoted as $s \models \sigma$, $s \models \sigma$ if and only if $\sigma \in s$ [2, p. 31]. The interpretation of a sentence is considered to be a collection of several infons[1].

The correlation between events is to some extent universal. To describe such stable connection, elements in an infon can be abstracted as free variables. A situation containing variables is called an abstract infon, and concrete or abstract infons with some common elements can consist a situation type. Conversely, abstract infons can also be reversed to concrete infons by anchors. An anchor is a function f assigns to each parameter in the parameter set a concrete object in the object set [2, p. 52].

3 Logic of Metaphor Comprehension

Referring to [4] and considering the specificity of metaphorical comprehension, we give a language and semantics to formalize the information transfer and use it to describe metaphor understanding process. This logic of metaphor comprehension will be denoted as $\mathcal{L}_M$.

---

[1] The processing method with *infons* as the basic unit follows Devlin. In *Situations and Attitudes*, Barwise and Perry use *events* and *event types*.

3.1 Syntax

We assume a propositional language with finite constants $a_1, \ldots, a_n$, variables $x_1, \ldots, x_n$; n-ary relations $R_1, \ldots, R_n$; themes $t_1, \ldots, t_n$; classical logic operators $\neg, \sim, \rightarrow, \wedge$; a binary info operator $(.,.)$ applying two basic formulas or a basic formula and a theme results in a formula; and a unary preference operator P.

This is a language containing terms, basic formulas and formulas. the terms are individual variables and constants, and a basic formula can be defined as:

- If R is a n-ary relation, $\alpha_1, \ldots, \alpha_n$ are terms, then $p = R(\alpha_1, \ldots, \alpha_n)$ is a basic formula.
- If φ is a basic formula, then so are $\neg\varphi$, $\sim \varphi$.

After defining basic formulas, we can further give the formal definition of formulas.

Definition 1 (Formulas). *The formulas of $\mathcal{L}_M$ can be defined as following:*

- *A basic formula is a formula.*
- *A theme is a formula.*
- *If p is a basic formula and t is a theme, then (t, p) is a formula.*
- *If p and p' are basic formulas, then (p, p'), $M(p, p')$ are formulas.*
- *If φ is a formula, then so are $\neg\varphi$, $\sim \varphi$, $P\varphi$.*
- *If φ and ψ are formulas, then so are $\varphi \rightarrow \psi$.*

Here, we use a special element "theme". It is the topic of a expression, so it is a concept, and all infons with the same t constitute a concept domain about t. P is a preference operator. (p, p') says that information p conveys meaning p', $M(p, p')$ indicates that p is the literal information of a metaphor expression and p' is the actual meaning.

Since the information in a situation is partial, it is necessary to distinguish between "not contain this information" and "contain negative information". In the language, there are two different kinds of negation, $\neg\varphi$ is called a strong negation and $\sim \varphi$ is called a weak negation. Other connectives such as $\wedge$ and $\vee$ are introduced as abbreviations in the usual way, except that the negation used in these definitions are weak (i.e. $\varphi \vee \psi =_{df} \sim \varphi \rightarrow \psi$ and $\varphi \wedge \psi =_{df} \sim (\varphi \rightarrow \sim \psi)$).

3.2 Semantics

Let $\mathcal{D}$ be the domain of individuals, $\mathcal{R}$ be a set of relations, and $\{0, 1\}$ be a polarity set, $\mathcal{T}$ be a set of themes. An infon is an ordered n-tuple $\langle t, R, \alpha_1, \ldots, \alpha_n, i\rangle$, where $R \in \mathcal{R}, \alpha_1, \ldots, \alpha_n \in \mathcal{D}, i \in \{0, 1\}$. And denote in as a set of infons. A situation structure S is a pair $(in, \preceq)$, indicating some prioritized information. Especially, a situation is consisted of several infons denoted as $\{infon_1, \ldots, infon_n\}$. $f : INF \rightarrow T$ is a function on the set of infons, when the input in a infon a, the output $f(a)$ is the theme in this infon. In an exact situation, the agent may focus on a specific topic, the all infons may have a same theme (in the following we abbreviate the theme t in a situation s as t_s). $\preceq$ is a reflexive, transitive, strictly noetherian[2] relation on situations expressing the preference.

Definition 2 (Model). *A model $\mathfrak{M}$ is a tuple $\langle S, I\rangle$, where S is a situation structure. I is an interpretation, which maps n-ary relation symbols to n-ary relations, and individual constants to individuals. And let g be the variable assignment.*

Since the polarity can be 1 or 0, there are two satisfaction relations between point models and formulas in $\mathcal{L}_M$. $\mathfrak{M}, s, g \models_+ \varphi$ can be read as "φ is supported in the situation s(with respect to $\mathfrak{M}$ and g)", and $\mathfrak{M}, s, g \models_- \varphi$ can be read as "φ is rejected in the situation s (with respect to $\mathfrak{M}$ and g)". Compared with [4], this model differs in three ways:

(1) For metaphor comprehension, it is true that the specific spacial-temporal context is important, but it only affects our choice of which concepts to extract from the conceptual domain, rather than the common sense-based knowledge base of the concept. Here, we need information elements to reflect the information of general significance, and the selection of relevant information is reflected in preference reasoning, so information of time and space is not so necessary.
(2) It add themes to express one of the most important concept in CMT called "concept domains", since identifying metaphorical concepts is the basic premise for determining whether a sentence is a metaphor and understanding a metaphor.
(3) It add an operator P to express the cognitive preference. With the preference operator, we can describe the typical perceptions of this concept in our common knowledge base, which is called image schemes in cognitive metaphor research.

Definition 3 (Truth value). *The valuation of a term t_i can be defined as following:*

$$\|t\|^{\mathfrak{M},g} = \begin{cases} I(t), & \text{if } t_i \text{ is a constant;} \\ g(t), & \text{if } t_i \text{ is a variable.} \end{cases}$$

Then the truth value of $\mathcal{L}_M$ formulas in a situation s in M is defined as follows:

[2] A relation R is strictly noetherian if and only if there is no infinite sequence $w_0, w_1, \ldots$ such that $\langle w_0, w_1\rangle \in R, \langle w_1, w_2\rangle \in R, \ldots$ and such that $w_0 \neq w_1, w_1 \neq w_2$.

$$
\begin{array}{ll}
\mathfrak{M}, s, g \vDash_{+} R(t_1, \ldots, t_n) & \textit{iff } \langle I(R), \|t_1\|, \ldots, \|t_n\|, 1\rangle \in s \\
\mathfrak{M}, s, g \vDash_{-} R(t_1, \ldots, t_n) & \textit{iff } \langle I(R), \|t_1\|, \ldots, \|t_n\|, 0\rangle \in s \\
\mathfrak{M}, s, g \vDash_{+} t & \textit{iff } \forall inf \in s : f(inf) = t \\
\mathfrak{M}, s, g \vDash_{-} t & \textit{iff } \forall inf \in s : f(inf) \neq t \\
\mathfrak{M}, s, g \vDash_{+} \neg\varphi & \textit{iff } \mathfrak{M}, s, g \vDash_{-} \varphi \\
\mathfrak{M}, s, g \vDash_{-} \neg\varphi & \textit{iff } \mathfrak{M}, s, g \vDash_{+} \varphi \\
\mathfrak{M}, s, g \vDash_{+} \sim \varphi & \textit{iff } \mathfrak{M}, s, g \nvDash_{+} \varphi \\
\mathfrak{M}, s, g \vDash_{-} \sim \varphi & \textit{iff } \mathfrak{M}, s, g \vDash_{+} \varphi \\
\mathfrak{M}, s, g \vDash_{+} \varphi \to \psi & \textit{iff } \mathfrak{M}, s, g \vDash_{+} \varphi \;\Rightarrow\; \mathfrak{M}, s, g \vDash_{+} \psi \\
\mathfrak{M}, s, g \vDash_{-} \varphi \to \psi & \textit{iff } \mathfrak{M}, s, g \vDash_{+} \varphi \;\&\; \mathfrak{M}, s, g \vDash_{-} \psi \\
\mathfrak{M}, s, g \vDash_{+} P\varphi & \textit{iff } \forall s' : s' \preceq s \Rightarrow \mathfrak{M}, s', g \vDash_{+} \varphi \\
\mathfrak{M}, s, g \vDash_{-} P\varphi & \textit{iff } \exists s' : s' \preceq s \;\&\; \mathfrak{M}, s', g \vDash_{-} \varphi
\end{array}
$$

$$
\begin{array}{ll}
\mathfrak{M}, s, g \vDash_{+} (t, p) & \textit{iff } \forall s' : \mathfrak{M}, s', g \vDash_{+} t \;\Rightarrow\; \mathfrak{M}, s', g \vDash_{+} Pp \\
\mathfrak{M}, s, g \vDash_{-} (t, p) & \textit{iff } \forall s' : \mathfrak{M}, s', g \vDash_{+} t \;\Rightarrow\; \mathfrak{M}, s', g \nvDash_{+} Pp \\
\mathfrak{M}, s, g \vDash_{+} (p, p') & \textit{iff } \forall s' : ((\mathfrak{M}, s', g \vDash_{+} p \;\Rightarrow\; \mathfrak{M}, s' \vDash_{+} Pp') \\
& \qquad \& \; (\mathfrak{M}, s', g \vDash_{+} \neg p \;\Rightarrow\; \mathfrak{M}, s', g \vDash_{+} P\neg p')) \\
\mathfrak{M}, s, g \vDash_{-} (p, p') & \textit{iff } \forall s' : ((\mathfrak{M}, s', g \vDash_{+} p \;\Rightarrow\; \mathfrak{M}, s' \vDash_{-} Pp') \\
& \qquad \& \; (\mathfrak{M}, s', g \vDash_{+} \neg p \;\Rightarrow\; \mathfrak{M}, s', g \vDash_{-} P\neg p'))
\end{array}
$$

Then we can use information transfer language to define metaphor expressions.

Definition 4 (Metaphorical expression).
A metaphor expression can be divided into two categories: direct metaphorical expression $M_D(p, p')$ and indirect metaphorical expression $M_I(p, p')$.

- *$M_D(p, p')$ is a direct metaphorical information transfer in model $\mathfrak{M}$, respected to situation s and assignment g (denoted as $\mathfrak{M}, s, g \vDash_{+} M_D(p, p')$), if and only if for all situation s', if $s' \vDash p$, then $\mathfrak{M}, s', g \vDash (t_{s'}, p')$ and $t_s \neq t_{s'}$.*
- *$M_I(p, p')$ is an indirect metaphorical information transfer in model $\mathfrak{M}$, respected to situation s and assignment g (denoted as $\mathfrak{M}, s, g \vDash_{+} M_I(p, p')$), if and only if for all situation s', if $s' \vDash p$,then $\mathfrak{M}, s', g \vDash (t_{s'}, p'')$ and $\mathfrak{M}, s, g \vDash p'' \to p'$ and $t_s \neq t_{s'}$.*

Steen has distinguished between direct metaphor and implied metaphor. He called a metaphor that gives the source domain and target domain directly as the direct metaphor; otherwise, implied metaphor is a kind of metaphor whose framework associated with the source domain and target domain only exists in cognition. He mainly discussed the situation that the mapping from properties or objects in a source domain to a target domain. In fact, in some complex metaphor understanding, semantics are not obtained by direct cross-domain mapping, but need to integrate the contents of the two conceptual domains and continue to reason in the new conceptual domain. To describe this kind of metaphor, the core difference between direct and indirect metaphor given here is that the meaning of direct metaphor is directly mapped across domains, while indirect metaphor depends on further reasoning after mapping. The example "we are at a cross-roads" we mentioned before is a direct metaphor, since the target information "facing choices" can be inferred from "at a crossroads". While the following example is an indirect metaphor.

Example 1. (When talking about a surgeon, one says:) He is a butcher.

Let p denote the proposition "He is a butcher".; p' denote "He cuts limbs rudely".; p'' denote "His skills are poor".; t denote the theme "Surgeon"; and t' denote the theme "Butcher", then

1. $\mathfrak{M}, s, g \models_- p$.
2. $\mathfrak{M}, s, g \models_+ p'$, but p' not relevant information.
3. $\mathfrak{M}, s, g \models_+ p''$, it's what the information the expression wants to transfer.

That is, there is some s', it satisfies $\mathfrak{M}, s', g \models_+ p$, but $\mathfrak{M}, s' \models_- Pp'$, so $\mathfrak{M}, s, g \models_- (p, p')$, and $\mathfrak{M}, s, g \models_+ M(p, p'')$.

More precisely, now $s \not\models_+ M(p, p'')$ but $s \models_+ M(p, p')$, and in situations with theme t, $s_t \models_+ p' \rightarrow p''$. From $s \models_+ M(p, p')$ we can get $s \models_+ p'$, and therefore, $s \models_+ p''$. While in situations with theme t', $s_{t'} \not\models_+ p' \rightarrow p''$, actually, here $s_{t'} \models_- p' \rightarrow p''$, so $s \not\models_+ p''$.

From the truth value of $M_D(p, p')$, we can get requirements of being a direct metaphor. p conveys p' is metaphorical, if

1. p conveys p', i.e. $\models_+ p \rightarrow Pp'$ and $\models_+ \neg p \rightarrow P\neg p'$.
2. $\models_+ p \rightarrow t_s \neq t_{s'}$.

If condition (1) is not satisfied, there are two cases: a) When $\neg(p \rightarrow Pp')$ is not satisfied. It shows that the preference relationship between the two propositions does not hold. For example, in expressions like "I am feeling down /depressed" or "I am really low these days", "down", "depressed", "low" can express upset or sad, but cannot express happy, since under the frame of orientation metaphor "SAD IS DOWN". b) When $p \rightarrow Pp'$ is satisfied, but $\neg(\neg p \rightarrow P\neg p')$ is not satisfied. It shows that negation does not maintain a relationship between two propositions. For example, "He is a professor" usually infers "He is knowledgeable", but "He is not a professor" cannot infer "He is not knowledgeable".

And condition (2) says that for every situation supports p, the theme of this situation (a part of the source domain) is not the same as the one of the actual situation (a part of the target domain). This condition is to ensure the mapping is cross-domain (or say is constructed on different domains). Therefore, the literal and non-literal meanings are from different categories. For example, "It is not necessary to care about the destination of life", the category of "destination" is "travel", not "life". If condition (2) is not met, i.e. $\exists s' : \mathfrak{M}, s', g \models_+ p \;\&\; t_s = t'_s$, there is a situation supporting p with the same theme as the current situation. For example, the expression "Shanghai, the Paris of the Orient" is an analogy but not a metaphor, because both Shanghai and Paris belong to the category "city".

Proposition 1. *Given a model* $\mathfrak{M} = \langle S, I \rangle$, *for any situation* $s \in S$ *and assignment* g, $M(p, p') \wedge t \rightarrow \neg(t, p)$ *is true in the model.*

It indicates that for a metaphor, the literal information cannot be understood in the target domain.

Proposition 2. *If two propositions have an information transfer relation, their negations maintain this relation, or say, the existence of one property leading to the existence of another property is equivalent to the absence of this property leading to the absence of another one. i.e.* $(p, p') \leftrightarrow (\neg p, \neg p')$

The most common cross-domain mappings are built on similarity, or more precisely, refers to the similarity in the midst of difference [10]. See Example 2.

Example 2. (When talking about the current status of life, one says:) John is at a crossroads now.

Let Cro stands for "at a crossroads"; Con stands for "be confused"; Dif stands for "meet with difficulties"; Dri stands for "...drive..."; and j stands for "John".

The metaphorical concept of this expression is "LIFE IS JOURNEY". From this we can get two different themes: $t_1 = life, t_2 = journey$. It embraces the similarities between concept *life* and concept *journey*. And the literal information of the sentence is $Cro(j)$.

In a *journey* frame, "at a crossroads" usually means that there are multiple directions and one cannot decide. When this kind of property is replaced in the frame of *life*, it means that the choice of life cannot be determined and someone is confused about his or her life. So in this metaphor, the non-literal meaning "be confused" can be inferred from "at a crossroads".

Suppose $S = \{s_0, s_1, s_2, s_3, s_4\}$, where $s_0 = \{life, \langle \text{Con}, j, 1\rangle\}$ is the actual situation,

$s_1 = \{life, \langle \text{Con}, j, 0\rangle, \langle \text{Dif}, j, 1\rangle\}$, $s_2 = \{journey, \langle \text{Cro}, j, 1\rangle\}$,

$s_3 = \{journey, \langle \text{Cro}, j, 1\rangle, \langle \text{Con}, j, 1\rangle\}$, $s_4 = \{journey, \langle \text{Cro}, j, 0\rangle, \langle \text{Con}, j, 0\rangle\}$,

And considering in people's usual cognition, "at a crossroads" can be associated with "be confused", so the agent will prefer situations containing both these properties, preference $\preceq = \{(s_3, s_2), (s_3, s_3), (s_4, s_4)\}$. Then we can get $\mathfrak{M}, s_0, g \vDash_+ \sim Cro(j)$ and $\mathfrak{M}, s_0, g \vDash_+ \sim \neg Cro(j)$

Here, the actual situation is talking about *life*, "at a crossroads" is not related to this theme. So the listener could not associate it and cannot make a conclusion of whether it is true or false, that is, both $Cro(j)$ and $\neg Cro(j)$ are not supported in this situation.

A situation supports (p, p') here expresses that the situation satisfies a constraint in situation theory, that is, a situation that satisfies p' can be obtained from a situation that satisfies p. Here, information "at a crossroads" is to convey information "be confused".

$$\mathfrak{M}, s_0, g \vDash_+ (Cro(x), Con(x))$$

Then thinking about negative forms: "We are not at a crossroads". Since $\mathfrak{M}, s_0, g \models_+ (\neg Cro(x), \neg Con(x))$, and $\mathfrak{M}, s_0, g \models_+ M(\neg Cro(x), \neg Con(x))$

4 Conclusion

In this article, we have proposed a formalization from a cognitive perspective, especially based on Conceptual Metaphor Theory, to describe information transfer in a metaphor interpretation. It describes that a metaphor is a special kind of non-literal expression that depends on some accepted connections between concepts. And the transfer of negative information is equivalent to positive information, and the information association required for negative expression can be reliably obtained from the conventional association of positive information. And it reflect cross-domain mapping and the role of concept domains (or say, cognitive frames) in metaphor interpretation. Also, metaphorical understandings at different levels a can be distinguished.

In the future work, we hope to explain different roles of reasoning in the same concept domain and cross domains in metaphor understanding. In addition, it is necessary to consider metaphor cognition in a dynamic information environment. With the emergence of new information, the agent's knowledge of the situations, as well as the preference for related information may change, which will affect the results of metaphor cognition.

References

1. Bowdle, B., Gentner, D.: The career of metaphor. Psychol. Rev. **112**(1), 193–216 (2005)
2. Devlin, K.: Logic and Information. Cambridge University Press, Cambridge (1991)
3. Fauconnier, G., Turner, M.: Conceptual integration network. Cogn. Sci. **22**(2), 133–187 (1998)
4. Fenstad, J., Halvorsen, P., Langholm, T., van Benthem, J.: Situation, Language and Logic. D. Reidel Publishing Company, Dordrecht (1987)
5. Gentner, D., Bowdle, B.: Metaphor as structure-mapping. In: The Cambridge Handbook of Metaphor and Thought, pp. 109–128. Cambridge University Press (2008)
6. Gibbs, W.R.: Metaphor as dynamical-ecological performance. Metaphor. Symb. **34**(1), 33–44 (2019)
7. Glucksberg, S., McGlone, M.S.: When love is not a journey: what metaphors mean. J. Pragmat. **31**(12), 1541–1558 (1999)
8. Glucksberg, S., Keysar, B.: Understanding metaphorical comparisons: beyond similarity. Psychol. Rev. **97**(1), 3–18 (1990)
9. Goguen, J.: An introduction to algebraic semiotics, with application to user interface design. In: Nehaniv, C.L. (ed.) CMAA 1998. LNCS (LNAI), vol. 1562, pp. 242–291. Springer, Heidelberg (1999). https://doi.org/10.1007/3-540-48834-0_15
10. Isenberg, A.: On defining metaphor. J. Philos. **60**(21), 609–622 (1963). https://doi.org/10.2307/2023555
11. Lakoff, G., Johnson, M.: Metaphors We Live By. University of Chicago Press, Chicago (1980)

12. Lakoff, G., Johnson, M.: Philosophy in the Flesh: The Embodied Mind and Its Challenge to Western Thought. Basic Books, New York (1999)
13. Lakoff, G., Turner, M.: More than Cool Reason: A Field Guide to Poetic Metaphor. The University of Chicago Press, Chicago (1989)
14. Mori, T., Nakagawa, H.: A formalization of metaphor understanding in situation semantics. In: Barwise, J., Gawron, J.M., Plotkin, G., Tutiya, S. (eds.) Situation Theory and Its Applications, vol. 2, pp. 449–467. CSLI, Stanford (1991)
15. Shutova, E.: Design and evaluation of metaphor processing systems. Comput. Linguist. **41**(4), 579–623 (2015)
16. Steen, G.: The paradox of metaphor: why we need a three dimensional model of metaphor. Metaphor. Symb. **23**(4), 213–241 (2008)
17. Steinhart, E.C.: The Logic of Metaphor: Analogous Parts of Possible Worlds. Kluwer Academic Publisher, Dordrecht (2001)
18. Vogel, C.: Dynamic semantics for metaphor. Metaphor. Symb. **16**(1–2), 59–74 (2001). https://doi.org/10.1080/10926488.2001.9678886

A Temporal Logic for Successive Events

Yanjun Li(✉) and Jiajie Zhao

College of Philosophy, Nankai University, Tianjin, China
lyjlogic@gmail.com

Abstract. A succession of events is a sequence of events such that after one event is finished, the next one occurs successively. In this paper, we extended linear temporal logic with a new modality to capture the case that a sequence of events successively occurs. We compared the expressivity between this extended linear temporal logic and the standard linear temporal logic.

Keywords: Linear temporal logic · Successive events · Expressivity

1 Introduction

Temporal logic broadly refers to a family of modal logics for reasoning about temporal information. It originated from the logical framework of *tense logic* introduced in [17–19] by Prior. Prior's basic tense logic has been further discussed and extended, as a result, the whole of these developed logical frameworks is known as Temporal logic. Temporal logic has become a commonly used tool in various fields such as philosophy, computer science, artificial intelligence, and linguistics. For a detailed survey, see [7].

Linear time temporal logic is the most widely used type of temporal logic in computer science. It was first proposed in [16] and well studied in [4]. In linear temporal logic, time is conceived as a linear, discrete succession of time instants. In [16], Pnueli proposed to use linear temporal logic to specify and verify the properties of computer programs. For example, the property that "when p is true, next q will be true" can be formalized in linear temporal logic as

$$p \rightarrow \bigcirc q.$$

where the modality $\bigcirc$ means "on the next time instant it will be the case that". This property holds on a linear temporal logic model, if and only if for each time instant n, if p is true on n then q is true on the next time instant $n+1$.

A succession of events is a sequence of events occurring in succession. For example, first close the door, and next lock it. Firstly send a message, and next mark the message as "sent". A sequence of events occurring in succession means after one event is successfully executed and finished, the next event starts to be executed subsequently. Please note that the modality $\bigcirc$ in linear temporal

N. Alechina et al. (Eds.): LORI 2023, LNCS 14329, pp. 176–189, 2023.
https://doi.org/10.1007/978-3-031-45558-2_14

logic refers to the next time instant in time flow, but not to the next event in an event sequence. Thus, successive events cannot be naturally formalized in linear temporal logic.

There is a difference between the event that happens *next another event* and the event that happens on the *next time instant*. Let the current time instant be n. The event happens next the event ψ might not happen on the next time instant, i.e. $n+1$. The reason is that ψ might not be an instant event but a duration event. By instant events, we mean events that can be finished in one time instant, such as the event $p \wedge q$. By duration events, we mean events that need to be done in more than one time instants, such as the event $p \wedge \bigcirc p$. If ψ is the instant event $p \wedge q$, the event that happens next ψ will happen on the next time instant $n+1$. However, if ψ is the duration event $p \wedge \bigcirc p$, the event that happens next ψ will happen on the time instant $n+2$, because the time period of executing the event ψ will last from n to $n+1$. The modality *next* involved in successive events is the next event but not the next time instant. Thus, successive events cannot be naturally formalized in linear temporal logic. The following example will explain it in more details.

Consider the sentence S that "when the event ψ occurs, q will occur next after ψ". It will be inaccurate to formalize it in linear temporal logic as the formula

$$\phi = \psi \to \bigcirc q.$$

The reason will be discussed in the followings.

If ψ is an instant event, i.e. a propositional letter or a boolean formula of propositional letters, S indeed can be formalized as ϕ, because the next time instant after ψ happens to be the next time instant from now. However, if ψ is the formula $\bigcirc p$, which is a duration event, the next time instant after ψ is executed as the next-next time instant from now. In this case, S should be formalized as

$$\bigcirc p \to \bigcirc \bigcirc q.$$

Moreover, if ψ is the formula $p \vee \bigcirc p$, there will be two cases that ψ is successfully executed: either p or $\bigcirc p$. Thus, the next time instant after ψ is either the next time instant next p or the next time instant next $\bigcirc p$. Hence, if ψ is $p \vee \bigcirc p$, S should be formalized as

$$(p \to \bigcirc q) \vee (\bigcirc p \to \bigcirc \bigcirc q).$$

From the example above, we can see that successive events cannot be naturally expressed in linear temporal logic, due to the fact that the modality $\bigcirc$ in linear temporal logic refers to the next time instant, but not to the next event. In this paper, we proposed a new binary modality to express that "after one event finished, another event successively occurs". The logical language is interpreted on standard models of linear temporal logic, but we give a dynamic semantics for the new successive modality. Conceptually, the next time instants after ϕ is successfully executed are similar to the remaining states after ϕ is successfully announced in public announcement logic (see [5,15,21]). We also show that a formula with the new successive modality can be equivalently reduced to a formula in linear temporal logic, even if the reduced formula is unintuitive and cumbersome.

The paper is organized as follows. Section 2 introduces the syntax and the semantics of the temporal logic for successive events. Section 3 shows that the expressivity of the temporal logic for successive events is the same as the standard linear temporal logic. Section 4 discusses the succinctness of this temporal logic for successive events and briefly compares this logic with interval-based temporal logic. Section 5 concludes with some remarks.

2 Syntax and Semantics

In this section, we extend the basic linear temporal logic with a new binary modality and give a dynamic semantics for this extended language.

Let **P** be a set of propositional letters.

Definition 1 (Language). *The language $\mathcal{L}^{\bigcirc\langle\cdot\rangle}$ is defined by the following BNF rules (where $p \in \mathbf{P}$):*

$$\phi ::= \top \mid p \mid \neg\phi \mid (\phi \wedge \phi) \mid \bigcirc\phi \mid \langle\phi\rangle\phi$$

The auxiliary connectives $\bot, \rightarrow, \vee$ are defined as abbreviations as usual. Moreover, the formula $\neg\langle\phi\rangle\neg\psi$ is abbreviated as $[\phi]\psi$.

The formula $\bigcirc\phi$ intuitively means that event ϕ will occur at the next time instant. The formula $\langle\phi\rangle\psi$ means that after ϕ is finished, ψ will occur successively.

The language $\mathcal{L}^{\bigcirc\langle\cdot\rangle}$ is interpreted on linear temporal models defined as follows.

Definition 2 (Models). *A (linear temporal) model is a triple $\sigma = \langle\mathbb{N}, <, V\rangle$ where:*

- *$\mathbb{N}$ is the set of natural numbers;*
- *$<$ is the less-than relation on $\mathbb{N}$;*
- *$V : \mathbb{N} \to 2^{\mathbf{P}}$ is a valuation that assigns a set of propositional letters to each natural number.*

For each $n \in \mathbb{N}$, (σ, n) is a pointed model.

It is known that a linear model σ can also be seen as an infinite word in the alphabet $2^{\mathbf{P}}$, that is, $\sigma : \mathbb{N} \to 2^{\mathbf{P}}$. Thus, we will write $V(n)$ as $\sigma(n)$, where V is the valuation in σ.

The intuition of a successive formula $\langle\phi\rangle\psi$ to be true at the time instant n is that ψ is true at some time instant m where m is the next time instant when ϕ is successfully finished. Hence, in order to define the semantics, we need to define the notion that "next time instants" after ϕ is successfully finished.

Before the formal definitions, we first give some intuitive examples. Let the current time instant be n.

- After the event p is done, the immediate next time instant is $n+1$.
- After the event $\bigcirc p$ is done, the immediate next time instant is $n+2$.
- After the event $p \wedge \bigcirc p$ is done, the immediate next time instant also is $n+2$. This means that if the next time instant after ϕ is m and the next time instant after ψ is k, then the next time instant after $\phi \wedge \psi$ should be the max number of m and k.
- After the event $p \vee \bigcirc p$ is done, the immediate next time instant is either $n+1$ or $n+2$. This means that the next time instant after ϕ might not be a single time instant but a set of time instants, and that each time instant in this set is a well qualified next time instant after ϕ.
- After the event $(p \vee \bigcirc \bigcirc p) \wedge \bigcirc p$ is done, the immediate next time instant is either $n+2$ or $n+3$. The reason is that $n+2$ is the next time instant after the event $p \wedge \bigcirc p$ and $n+3$ is the next time instant after the event $\bigcirc \bigcirc p \wedge \bigcirc p$.

Before formally defining the time instants after doing ϕ, we need the auxiliary operation $\Cap$. The intuition of the operation $\Cap$ is to merge two sets of time instants.

Definition 3 ($\Cap$). *Let A and B be two sets of natural numbers. The set $A \Cap B$ is defined as $\{max(n, m) \mid n \in A, m \in B\}$.*

Please note that if either A or B is emptyset then $A \Cap B$ is emptyset.

For example, if the next time instant after p is $\{n+1\}$ and the next time instant after $\bigcirc p$ is $\{n+2\}$, then the next time instant after $p \wedge \bigcirc p$ should be the time instant $\{n+2\}$, i.e. $\{n+1\} \Cap \{n+2\}$.

If the next time instants after doing the event ϕ is the set $A = \{n+1, n+3\}$, and the next time instants after doing the event ψ is the set $B = \{n+2, n+4\}$, then the next time instants after doing the event $\phi \wedge \psi$ is mergence of the set A and B, that is, $A \Cap B = \{n+2, n+3, n+4\}$.

Table 1. $n|_\phi$

$$
\begin{aligned}
n|_\top &= \{n+1\} \\
n|_p &= \begin{cases} n+1 & p \in \sigma(n) \\ \emptyset & p \notin \sigma(n) \end{cases} \\
n|_{\phi \vee \psi} &= n|_\phi \cup n|_\psi \\
n|_{\bigcirc \phi} &= (n+1)|_\phi \\
n|_{\langle \phi \rangle \psi} &= \textstyle\bigcup_{n' \in n|_\phi} n'|_\psi
\end{aligned}
\qquad
\begin{aligned}
n|_{\neg \top} &= \emptyset \\
n|_{\neg p} &= \begin{cases} n+1 & p \notin \sigma(n) \\ \emptyset & p \in \sigma(n) \end{cases} \\
n|_{\neg\neg\phi} &= n|_\phi \\
n|_{\neg(\phi \vee \psi)} &= n|_{\neg\phi} \Cap n|_{\neg\psi} \\
n|_{\neg \bigcirc \phi} &= (n+1)|_{\neg\phi} \\
n|_{\neg \langle \phi \rangle \psi} &= n|_{\neg\phi} \cup (\textstyle\Cap_{n' \in n|_\phi} n'|_{\neg\psi})
\end{aligned}
$$

Next we are going to define the notion of the time instant set after successfully doing ϕ on the time instant n.

Definition 4 ($n|_\phi$). *Given a pointed model (σ, n) and a formula $\phi \in \mathcal{L}^{\bigcirc\langle\cdot\rangle}$, the set of time instants, $n|_\phi$, is defined in Table 1, which intuitively means the set of the next time instants after the event ϕ is successfully finished.*

The intuition of this definition is that $n|_\phi$ is set of time instants after ϕ is successfully done in n.

The set $n|_p$ is $\{n+1\}$ if p is true on n. Otherwise, $n|_p$ is the empty set. Generally, if ϕ is not true on ϕ, then $n|_\phi$ is the empty set.

The set $n|_{\phi\vee\psi}$ is the union of the set $n|_\phi$ and the set $n|_\psi$, which means that a next time instant after doing $\phi \vee \psi$ is either a next time instant after doing ϕ or a next time instant after doing ψ.

The set $n|_{\neg\phi\wedge\neg\psi}$ is the mergence of the set $n|_{\neg\phi}$ and $n|_{\neg\psi}$, which means that a next time instant after doing the event $\neg\phi\wedge\neg\psi$ is a next time instant on which both $\neg\phi$ and $\neg\psi$ are successfully done.

A next time instant n'' after ϕ and ψ are successively done on n, i.e. $n'' \in n|_{\langle\phi\rangle\psi}$, is a next time instant after ψ is done on the time instant which is a next time instant after ϕ is done on n, i.e. $n'' \in \bigcup_{n' \in n|_\phi} n'|_\psi$.

If the events ϕ and ψ do not successively happen on n, it means two cases: the first case is that ϕ does not happens on n; The second case is that even though ϕ successively happens on n, but ψ does not happen on each next time instant after ϕ (in other words, $\neg\psi$ happens on each next time instant after ϕ).

Example 1. Let the model σ_0 be defined as follows:

$$\begin{array}{ccccccccc} 0 & \longrightarrow & 1 & \longrightarrow & 2 & \longrightarrow & 3 & \longrightarrow & \cdots \\ \{p\} & & \{p,q\} & & \{p,q\} & & \{r\} & & \end{array}$$

We then have the followings:

- $0|_{p\vee\bigcirc p} = \{1,2\}$
- $0|_{(p\vee\bigcirc\bigcirc p)\wedge\bigcirc p} = \{2,3\}$
- $0|_{\langle p\rangle q} = \{2\}$
- $0|_{\langle p\vee\bigcirc p\rangle q} = \{2,3\}$
- $0|_{\neg\langle p\vee\bigcirc p\rangle r} = \{3\}$

Now we are ready to define the semantics of $\mathcal{L}^{\bigcirc\langle\cdot\rangle}$.

Definition 5 (Semantics). *The satisfaction relation between pointed models and formulas is defined in Table 2.*

By the semantics, it is obvious that $\models \bigcirc\phi \leftrightarrow \langle\top\rangle\phi$, which means that the modality $\bigcirc$ can be defined by the modality $\langle\cdot\rangle$. We will write $\sigma, n \models \phi$ as $n \models \phi$, when the model σ is obvious from the context.

Example 2. Let the model σ be defined as follows:

$$\begin{array}{ccccccccc} 0 & \longrightarrow & 1 & \longrightarrow & 2 & \longrightarrow & 3 & \longrightarrow & \cdots \\ \{p\} & & \{p\} & & \{q\} & & \{p\} & & \end{array}$$

We then have that $0|_{p\vee\bigcirc q} = \{1\}$ and $1|_{p\vee\bigcirc q} = \{2,3\}$. And it is easy to check that

$$0 \models \langle p\vee\bigcirc q\rangle\langle p\vee\bigcirc q\rangle p.$$

Table 2. The semantics of $\mathcal{L}^{\bigcirc\langle\cdot\rangle}$

$$\begin{array}{ll}
\sigma, n \vDash \top & \\
\sigma, n \vDash p & \iff p \in \sigma(n) \\
\sigma, n \vDash \neg\phi & \iff \sigma, n \nvDash \phi \\
\sigma, n \vDash \phi \vee \psi & \iff \sigma, n \vDash \phi \text{ or } \sigma, n \vDash \psi \\
\sigma, n \vDash \bigcirc\phi & \iff \sigma, n+1 \vDash \phi \\
\sigma, n \vDash \langle\phi\rangle\psi & \iff \sigma, n \vDash \phi \text{ and } \sigma, n' \vDash \psi \text{ for some } n' \in n|_\phi
\end{array}$$

Proposition 1. *Given a pointed model (σ, n), we have that for each $\phi \in \mathcal{L}^{\bigcirc\langle\cdot\rangle}$, $n|_\phi \neq \emptyset$ iff $n \vDash \phi$.*

Proof. We prove it by induction on the length of ϕ.

- $\phi := p$ or $\phi := \top$. It is obvious.
- $\phi := \psi \vee \chi$. Apparently, $len(\psi) < len(\psi \vee \chi)$ and $len(\chi) < len(\psi \vee \chi)$. By induction hypothesis, we get $n|_\psi \neq \emptyset$ iff $n \vDash \psi$ and $n|_\chi \neq \emptyset$ iff $n \vDash \chi$. And by definition in (1), $n|_{\psi\vee\chi} \neq \emptyset$ if and only if the case that $n|_\psi \neq \emptyset$ or $n|_\chi \neq \emptyset$, thus by IH we can get it's equivalent to the condition that $n \vDash \psi$ or $n \vDash \chi$, i.e. $n \vDash \psi \vee \chi$.
- $\phi := \bigcirc\psi$. Since $len(\psi) < len(\bigcirc\psi)$, by IH we can get $(n+1)|_\psi \neq \emptyset$ iff $n+1 \vDash \psi$, i.e. $n|_{\bigcirc\psi}$ iff $n \vDash \bigcirc\psi$.
- $\phi := \langle\psi\rangle\chi$. We have to show $n|_{\langle\psi\rangle\chi} = \bigcup_{n'\in n|_\psi} n'|_\chi \neq \emptyset$ iff $n \vDash \psi$ and $\exists n' \in n|_\psi, n' \vDash \chi$. Since $len(\psi) < len(\langle\psi\rangle\chi)$, by IH we can get $n|_\psi \neq \emptyset$ iff $n \vDash \psi$. First, assuming that $n|_\psi = \emptyset$, which directly implies $n|_{\langle\psi\rangle\chi} = \emptyset$, moreover by IH on ψ, it's if and only if the case that $n \nvDash \psi$, thus we obtain $n \nvDash \langle\psi\rangle\chi$. Then, for the case that $n|_\psi \neq \emptyset$. Consider the condition that $n'|_\chi = \emptyset$, samely by IH on χ, we can get $n' \vDash \neg\chi$ for all $n' \in n|_\psi$, thus $n \nvDash \langle\psi\rangle\chi$. Conversely, for $n'|_\chi \neq \emptyset$, by IH on χ we get $n' \vDash \chi$, and it's not hard to obtain $n \vDash \langle\psi\rangle\chi$.
- $\phi := \neg\psi$. We then have the following 5 cases:
  - $\phi := \neg p$ or $\phi := \neg\top$. It is obvious.
  - $\phi := \neg\neg\chi$. Since $len(\chi) < len(\neg\neg\chi)$, by IH we can get $n|_\chi \neq \emptyset$ iff $n \vDash \chi$, and easily obtaining $n|_{\neg\neg\chi} \neq \emptyset$ iff $n \vDash \neg\neg\chi$.
  - $\phi := \neg\bigcirc\chi$. Note the fact that $\neg\bigcirc\chi = \bigcirc\neg\chi$. Since $len(\neg\chi) < len(\bigcirc\neg\chi)$, by IH we can get $(n+1)|_{\neg\chi} \neq \emptyset$ iff $n+1 \vDash \neg\chi$, which by semantics means $n|_{\bigcirc\neg\chi}$ iff $n \vDash \bigcirc\neg\chi$,i.e. $n|_{\neg\bigcirc\chi}$ iff $n \vDash \neg\bigcirc\chi$.
  - $\phi := \neg(\chi \vee \theta) = \neg\chi \wedge \neg\theta$. Since $len(\neg\chi) < len(\neg(\chi \vee \theta))$ and $len(\neg\theta) < len(\neg(\chi \vee \theta))$. By IH we get $n|_{\neg\chi} \neq \emptyset$ iff $n \vDash \neg\chi$ and $n|_{\neg\theta} \neq \emptyset$ iff $n \vDash \neg\theta$. If $n|_{\neg(\chi\vee\neg\theta)} \neq \emptyset$, by definition in Table 1 and Definition 3, it forces both $n|_{\neg\chi}$ and $n|_{\neg\theta}$ to be non-empty, then by IH we can get $n \vDash \neg\chi$ and $n \vDash \neg\theta$, thus $n \vDash \neg(\chi \vee \theta)$. Conversely, if $n \vDash \neg\chi \wedge \neg\theta$, likewise by IH we can get $n|_{\neg\chi} \neq \emptyset$ and $n|_{\neg\theta} \neq \emptyset$, again by definition in Table 1 and Definition 3, there must be at least one element in $n|_{\neg\chi\wedge\neg\theta}$, i.e. $n|_{\neg\chi\wedge\neg\theta} \neq \emptyset$.
  - $\phi := \neg\langle\chi\rangle\theta$.We have to show $n|_{\neg\langle\chi\rangle\theta} = n|_{\neg\chi} \cup (\biguplus_{n'\in n|_\chi} n'|_{\neg\theta}) \neq \emptyset$ iff $n \nvDash \chi$ or $\forall n' \in n|_\chi, n' \vDash \neg\theta$. Since $len(\neg\chi) < len(\neg\langle\chi\rangle\theta)$, by IH we can

get $n|_{\neg\chi} \neq \emptyset$ iff $n \vDash \neg\chi$. Assuming that $n|_{\neg\chi} = \emptyset$, by IH we can get $n \vDash \chi$. Consider the case that $n|_{\chi} = \emptyset$, similarly by IH on χ we get $n \nvDash \chi$, which contradicts to the assumption, so it is required that $n|_{\chi} \neq \emptyset$. Then, if $n'|_{\neg\theta} = \emptyset$, it follows that $n|_{\neg\langle\chi\rangle\theta} = \emptyset$, and by IH on $\neg\theta$, we can get $n' \nvDash \theta$, thus obtaining $n \vDash \neg\langle\chi\rangle\theta$. Conversely, if $n'|_{\neg\theta} \neq \emptyset$. And for the case that $n'|_{\neg\theta} \neq \emptyset$, it follows that $n|_{\neg\langle\chi\rangle\theta} \neq \emptyset$, and by IH on $\neg\theta$ it forces $\forall n' \in n|_{\chi}, n' \vDash \neg\theta$, so we obtain $n \vDash \neg\langle\chi\rangle\theta$. Moreover, for the case that $n|_{\neg\chi} \neq \emptyset$, by IH we can easily obtain $n \vDash \neg\langle\chi\rangle\theta$.

Due to $\vDash p \wedge \bigcirc\top \leftrightarrow p$, the following proposition indicates that the operation of equivalence replacement might not preserve truth value if the operation takes place in the scope of the modality $\langle\cdot\rangle$.

Proposition 2. $\nvDash \langle p \wedge \bigcirc\top\rangle p \leftrightarrow \langle p\rangle p$

Proof. Construct a linear model σ. Let the valuation sets $\sigma(k) = \sigma(k+1) = \{p\}$ ($k \in \mathbb{N}$, $p \in \mathbf{P}$) and assigns $\emptyset$ to every other natural numbers.

Since $p \in \sigma(k)$ implies $k \vDash p$, similarly, $p \in \sigma(k+1)$ implies $k+1 \vDash p$, from which we can derive $k+1 \in k|_p$ and it follows that $k \vDash \langle p\rangle p$.

But it is not the case that $k \vDash \langle p \wedge \bigcirc\top\rangle p$. Suppose it's true otherwise, then we get: $k \vDash p \wedge \bigcirc\top$ and $\exists k' \in k|_{p\wedge\bigcirc\top}, k' \vDash p$. Note that $k|_p = \{k+1\}$ and $k|_{\bigcirc\top} = (k+1)|_{\top} = \{k+2\}$, so we get $k|_{p\wedge\bigcirc\top} = k|_p \Cap k|_{\bigcirc\top} = \{k+2\}$. Since $p \notin \sigma(k+2)$, by semantics we get $k \nvDash \langle p \wedge \bigcirc\top\rangle p$ which contradicts to the assumption. Thus, we can obtain $\nvDash \langle p \wedge \bigcirc\top\rangle p \leftrightarrow \langle p\rangle p$.

3 Expressivity

In this section, we will make a comparison of the expressivity of $\mathcal{L}^{\bigcirc\langle\cdot\rangle}$ and the expressivity of the basic linear temporal logic $\mathcal{L}^{\bigcirc}$ defined below. Since $\mathcal{L}^{\bigcirc}$ is a fragment of $\mathcal{L}^{\bigcirc\langle\cdot\rangle}$, it means that $\mathcal{L}^{\bigcirc\langle\cdot\rangle}$ is at least as expressive as $\mathcal{L}^{\bigcirc}$. Afterward, we will show that each $\phi \in \mathcal{L}^{\bigcirc\langle\cdot\rangle}$ can be equivalently reduced to a formula $\psi \in \mathcal{L}^{\bigcirc}$, which means that $\mathcal{L}^{\bigcirc}$ is at least as expressive as $\mathcal{L}^{\bigcirc\langle\cdot\rangle}$. Thus, it is proved that $\mathcal{L}^{\bigcirc\langle\cdot\rangle}$ and $\mathcal{L}^{\bigcirc}$ are equally expressive.

The language $\mathcal{L}^{\bigcirc}$ is the fragment of $\mathcal{L}^{\bigcirc\langle\cdot\rangle}$ without the modality $\langle\cdot\rangle$ is defined as:

$$\phi ::= \top \mid p \mid \neg\phi \mid (\phi \vee \phi) \mid \bigcirc\phi.$$

Our strategy is to define a translation function which will translate each $\mathcal{L}^{\bigcirc\langle\cdot\rangle}$-formula into an $\mathcal{L}^{\bigcirc}$-formula, and then show that each $\mathcal{L}^{\bigcirc\langle\cdot\rangle}$-formula is equivalent to its translation $\mathcal{L}^{\bigcirc}$-formula.

In the translation, we need to rewrite a $\mathcal{L}^{\bigcirc}$-formula into its disjunctive normal form, which is defined as follows. The intuition of disjunctive normal form is that the negation symbol $\neg$ will occurs only next to atom letters and that the temporal modality $\bigcirc$ will not occur outside the conjunction and disjunction symbols. For example, a disjunctive normal form of the formula $\neg\bigcirc(p \wedge \bigcirc q)$ is $\bigcirc\neg p \vee \bigcirc\bigcirc\neg q$.

Definition 6 (Disjunctive normal form). *Given a formula $\phi \in \mathcal{L}^{\bigcirc}$, we do the following steps:*

1. *do the following replacement until each negation symbol is next to a propositional letter:*
   (a) *replace each subformula $\neg(\psi \vee \chi)$ with $\neg\psi \wedge \neg\chi$;*
   (b) *replace $\neg \bigcirc \psi$ with $\bigcirc\neg\psi$;*
   (c) *replace $\neg\neg\psi$ with ψ;*
2. *do the following replacement until each $\bigcirc$ symbol is next to either a propositional letter or a negation symbol:*
   (a) *replace $\bigcirc(\psi \vee \chi)$ with $(\bigcirc\psi \vee \bigcirc\chi)$;*
   (b) *replace $\bigcirc(\psi \wedge \chi)$ with $(\bigcirc\psi \wedge \bigcirc\chi)$;*
3. *do the following replacement until no disjunctive symbols are in the scope of conjunctive symbols:*
   (a) *replace $\psi \wedge (\chi_1 \vee \chi_2)$ with $(\psi \wedge \chi_1) \vee (\psi \wedge \chi_2)$*
   (b) *replace $(\chi_1 \vee \chi_2) \wedge \psi$ with $(\chi_1 \wedge \psi) \vee (\chi_2 \wedge \psi)$.*

We call the final formula a disjunctive normal form of ϕ, denoted by $\phi^{\vee}$.

Let $\bigcirc^{n+1}\phi$ be the formula $\bigcirc \bigcirc^{n} \phi$ and $\bigcirc^{0}\phi = \phi$. We name those formulas of form $\bigcirc^{n}p$ or $\bigcirc^{n}\neg p$ as *temporal literals*, where $n \in \mathbb{N}$ and $p \in \mathbf{P}$. From the definition of disjunctive normal form, it can be seen that each formula $\phi^{\vee}$ is a disjunction of conjunctions of temporal literals.

Note that a disjunctive normal form of ϕ is not defined as a disjunction of conjunctions of temporal literals that is equivalent to ϕ. The reason is that we would like to avoid the case that $p \wedge \bigcirc\top$ is a disjunctive normal form of p.

Given a disjunctive normal form $\phi^{\vee}$, we use $\mathrm{Cs}(\phi)$ to denote the set of disjuncts of $\phi^{\vee}$. For example, for $\phi^{\vee} = (p \wedge \bigcirc^{2}p) \vee (\bigcirc p \wedge \bigcirc^{4}\neg q)$, we have $\mathrm{Cs}(\phi^{\vee}) = \{p \wedge \bigcirc^{2}p, \bigcirc p \wedge \bigcirc^{4}\neg p\}$.

Next, we are going to show that $n|_{\phi} = n|_{\phi^{\vee}}$ for each $\phi \in \mathcal{L}^{\bigcirc}$ (i.e. Proposition 5). Before that, we need the following auxiliary propositions.

Proposition 3. *Given a formula $\phi \in \mathcal{L}^{\bigcirc}$ and a pointed model (σ, n), we have that $n|_{(\bigcirc\phi)^{\vee}} = (n+1)|_{\phi^{\vee}}$.*

Proof. Let $\phi^{\vee}$ be the formula $\bigvee_{i \leq n}(\bigwedge_{j \leq i_m} l_j)$, where l_j is a temporal literal. It follows that $(\bigcirc\phi)^{\vee}$ is the formula $\bigvee_{i \leq n}(\bigwedge_{j \leq i_m} \bigcirc l_j)$. Then, it is easy to check that $n|_{(\bigcirc\phi)^{\vee}} = (n+1)|_{\phi^{\vee}}$.

Proposition 4. *Given a formula $\phi \wedge \psi \in \mathcal{L}^{\bigcirc}$ and a pointed model (σ, n), we have that $n|_{(\phi\wedge\psi)^{\vee}} = n|_{\phi^{\vee}} \Cap n|_{\psi^{\vee}}$.*

Proof. Note that $(\phi \wedge \psi)^{\vee} = \bigvee_{\phi' \in \mathrm{Cs}(\phi^{\vee})} \bigvee_{\psi' \in \mathrm{Cs}(\psi^{\vee})} (\phi' \wedge \psi')$. So that $n|_{(\phi\wedge\psi)^{\vee}} = \bigcup_{\phi' \in \mathrm{Cs}(\phi^{\vee})} \bigcup_{\psi' \in \mathrm{Cs}(\psi^{\vee})} n|_{\phi'\wedge\psi'} = \bigcup_{\phi' \in \mathrm{Cs}(\phi^{\vee})} \bigcup_{\psi' \in \mathrm{Cs}(\psi^{\vee})} (n|_{\phi'} \Cap n|_{\psi'})$. On the other hand, $n|_{\phi^{\vee}} \Cap n|_{\psi^{\vee}} = \bigcup_{\phi' \in \mathrm{Cs}(\phi^{\vee})} n|_{\phi'} \Cap \bigcup_{\psi' \in \mathrm{Cs}(\psi^{\vee})} n|_{\psi'}$. Since it is not hard to check $\bigcup_{\phi' \in \mathrm{Cs}(\phi^{\vee})} \bigcup_{\psi' \in \mathrm{Cs}(\psi^{\vee})} (n|_{\phi'} \Cap n|_{\psi'}) = \bigcup_{\phi' \in \mathrm{Cs}(\phi^{\vee})} n|_{\phi'} \Cap \bigcup_{\psi' \in \mathrm{Cs}(\psi^{\vee})} n|_{\psi'}$, so we are able to obtain $n|_{(\phi\wedge\psi)^{\vee}} = n|_{\phi^{\vee}} \Cap n|_{\psi^{\vee}}$.

Now we are ready to show that $n|_\phi = n|_{\phi^\vee}$ for each $\phi \in \mathcal{L}^{\bigcirc}$.

Proposition 5. *Given a pointed model* (σ, n), *we have that for each* $\phi \in \mathcal{L}^{\bigcirc}$, $n|_\phi = n|_{\phi^\vee}$.

Proof. We prove it by induction on the length of ϕ.

- $\phi := p$ or $\phi := \top$. It is obvious.
- $\phi := \psi \vee \chi$. Since $(\psi \vee \chi)^\vee = \psi^\vee \vee \chi^\vee$, along with the inductive hypothesis we can derive $n|_{(\psi\vee\chi)^\vee} = n|_{\psi^\vee\vee\chi^\vee} = n|_{\psi^\vee} \cup n|_{\chi^\vee} = n|_\psi \cup n|_\chi = n|_{\psi\vee\chi}$.
- $\phi := \bigcirc\psi$. By Proposition 3 and IH, we can easily get $n|_{(\bigcirc\psi)^\vee} = (n+1)|_{\psi^\vee} = (n+1)|_\psi = n|_{\bigcirc\psi}$.
- $\phi := \neg\psi$. We then have the following 5 cases:
  - $\phi := \neg p$ or $\phi := \neg\top$. It is obvious.
  - $\phi := \neg\neg\chi$. It is obvious due to the fact that $(\neg\neg\chi)^\vee = \chi^\vee$.
  - $\phi := \neg\bigcirc\chi$. Note that $(\neg\bigcirc\chi)^\vee = (\bigcirc\neg\chi)^\vee$, and again by Proposition 3 and IH, we can get $n|_{(\neg\bigcirc\chi)^\vee} = n|_{(\bigcirc\neg\chi)^\vee} = (n+1)|_{(\neg\chi)^\vee} = (n+1)|_{\neg\chi} = n|_{\bigcirc\neg\chi} = n|_{\neg\bigcirc\chi}$.
  - $\phi := \neg(\chi \vee \theta) = \neg\chi \wedge \neg\theta$. By Proposition 4 and IH, we can easily get $n|_{(\neg\chi\wedge\neg\theta)^\vee} = n|_{(\neg\chi)^\vee} \Cap n|_{(\neg\theta)^\vee} = n|_{\neg\chi} \Cap n|_{\neg\theta} = n|_{\neg\chi\wedge\neg\theta}$.

The following proposition states that each $\mathcal{L}^{\bigcirc}$-formula is equivalent to its disjunctive normal form.

Proposition 6. *For each* $\phi \in \mathcal{L}^{\bigcirc}$, $\vDash \phi \leftrightarrow \phi^\vee$.

Proof. This follows from Propositions 1 and 5.

Next we are going to define the translation function. Before that, we first introduce the notion of modal depth.

Definition 7 (Modal depth). *The function* $d : \mathcal{L}^{\bigcirc} \to \mathbb{N}$ *is defined as follows:*

$$
\begin{array}{rr}
d(\top) = 0 & d(p) = 0 \\
d(\neg\phi) = d(\phi) & d(\phi \vee \psi) = max(d(\phi), d(\psi)) \\
d(\bigcirc\phi) = d(\phi) + 1 &
\end{array}
$$

Definition 8 (Translation). *The translation function* $t : \mathcal{L}^{\bigcirc\langle\cdot\rangle} \to \mathcal{L}^{\bigcirc}$ *is defined in Table 3.*

The following proposition states that the translation function indeed translates each $\mathcal{L}^{\bigcirc\langle\cdot\rangle}$-formula into an $\mathcal{L}^{\bigcirc}$-formula.

Proposition 7. *For each* $\phi \in \mathcal{L}^{\bigcirc\langle\cdot\rangle}$, $t(\phi) \in \mathcal{L}^{\bigcirc}$.

Proof. This can be easily shown by induction on the length of ϕ.

Next we are going to show that each $\mathcal{L}^{\bigcirc\langle\cdot\rangle}$-formula is equivalent to its translation formula. The key is to show that $n|_\phi = n|_{t(\phi)}$ for each $\phi \in \mathcal{L}^{\bigcirc\langle\cdot\rangle}$ (i.e. Proposition 10). Before that, we need the following two auxiliary propositions.

Table 3. The translation function

$t(\top)$	$= \top$	$t(\neg\top)$	$= \neg\top$
$t(p)$	$= p$	$t(\neg p)$	$= \neg p$
		$t(\neg\neg\phi)$	$= t(\phi)$
$t(\phi \vee \psi)$	$= t(\phi) \vee t(\psi)$	$t(\neg(\phi \vee \psi))$	$= t(\neg\phi) \wedge t(\neg\psi)$
$t(\bigcirc\phi)$	$= \bigcirc t(\phi)$	$t(\neg\bigcirc\phi)$	$= \bigcirc t(\neg\phi)$
$t(\langle\phi\rangle\psi)$	$= \bigvee_{\phi_i \in \mathrm{Cs}(t(\phi)^{\vee})}(\phi_i \wedge \bigcirc^{d(\phi_i)+1} t(\psi))$		
$t(\neg\langle\phi\rangle\psi)$	$= t(\neg\phi) \vee \bigwedge_{\phi_i \in \mathrm{Cs}(t(\phi)^{\vee})}(\phi_i \wedge \bigcirc^{d(\phi_i)+1} \neg t(\psi))$		

Proposition 8. *Given $\phi, \psi \in \mathcal{L}^{\bigcirc}$ and a pointed model (σ, n), we have that* $\bigcup_{n' \in n|_{\phi^{\vee}}} n'|_{\psi} = \bigcup_{\phi' \in Cs(\phi^{\vee})} n|_{\phi' \wedge \bigcirc^{d(\phi')+1}\psi}$.

Proof. Note that $n' \in n|_{\phi^{\vee}}$ implies $n' \in n|_{\bigvee_{\phi' \in \mathrm{Cs}(\phi^{\vee})}\phi'}$, thus $n' \in \bigcup_{\phi' \in \mathrm{Cs}(\phi^{\vee})} n|_{\phi'}$. And, we have that $\bigcup_{n' \in n|_{\phi^{\vee}}} n'|_{\psi} = \bigcup_{\phi' \in \mathrm{Cs}(\phi^{\vee})} (n|_{\phi'})|_{\psi}$. On the other hand, we have $\bigcup_{\phi' \in \mathrm{Cs}(\phi^{\vee})} n|_{\phi' \wedge \bigcirc^{d(\phi')+1}\psi} = \bigcup_{\phi' \in \mathrm{Cs}(\phi^{\vee})} (n|_{\phi'} \Cap n|_{\bigcirc^{d(\phi')+1}\psi})$. Now we are ready to show that $\forall \phi' \in \mathrm{Cs}(\phi^{\vee}), (n|_{\phi'})|_{\psi} = (n|_{\phi'} \Cap n|_{\bigcirc^{d(\phi')+1}\psi})$:

Let $\phi' = \bigcirc^{d(\phi')} p_{\phi'}$, where $p_{\phi'} \in \mathbf{P}$. Then $n|_{\phi'} = n|_{\bigcirc^{d(\phi')} p_{\phi'}} = (n + d(\phi'))|_{p_{\phi'}} = \{n + d(\phi') + 1 \mid n + d(\phi') \vDash p_{\phi'}\}$. For $n|_{\phi'} \Cap n|_{\bigcirc^{d(\phi')+1}\psi}$, note the fact that it requires $n|_{\phi'} \neq \emptyset$ and $n|_{\bigcirc^{d(\phi')+1}\psi} \neq \emptyset$. By Proposition 1, we can derive $n + d(\phi') \vDash p_{\phi'}$ from $n|_{\bigcirc^{d(\phi')} p_{\phi'}} \neq \emptyset$. Moreover, $d(\phi') < d(\bigcirc^{d(\phi')+1}\psi)$ implies $n|_{\phi'} \Cap n|_{\bigcirc^{d(\phi')+1}\psi} = \{n + d(\phi') + 1 \mid n + d(\phi') \vDash p_{\phi'}$ and $n + d(\phi') + 1 \vDash \psi\}$. Again, by Proposition 1, we are able to obtain $(n|_{\phi'})|_{\psi} = n|_{\phi'} \Cap n|_{\bigcirc^{d(\phi')+1}\psi}$. Finally, it is suffice to show that $\bigcup_{n' \in n|_{\phi^{\vee}}} n'|_{\psi} = \bigcup_{\phi' \in \mathrm{Cs}(\phi^{\vee})} n|_{\phi' \wedge \bigcirc^{d(\phi')+1}\psi}$.

Proposition 9. *Given $\phi, \psi \in \mathcal{L}^{\bigcirc}$ and a pointed model (σ, n), we have that* $\Cap_{n' \in n|_{\phi^{\vee}}} n'|_{\psi} = \Cap_{\phi' \in Cs(\phi^{\vee})} n|_{\phi' \wedge \bigcirc^{d(\phi')+1}\psi}$.

Proof. Since $n' \in n|_{\phi^{\vee}}$ implies $n' \in \bigcup_{\phi' \in \mathrm{Cs}(\phi^{\vee})} n|_{\phi'}$, we can derive $\Cap_{n' \in n|_{\phi^{\vee}}} n'|_{\psi} = \Cap_{\phi' \in \mathrm{Cs}(\phi^{\vee})} (n|_{\phi'})|_{\psi}$. On the other hand, we have $\Cap_{\phi' \in \mathrm{Cs}(\phi^{\vee})} n|_{\phi' \wedge \bigcirc^{d(\phi')+1}\psi} = \Cap_{\phi' \in \mathrm{Cs}(\phi^{\vee})} (n|_{\phi'} \Cap n|_{\bigcirc^{d(\phi')+1}\psi})$. Following from the former proof of Proposition 8, we are capable to show that: $\forall \phi' \in \mathrm{Cs}(\phi^{\vee}), (n|_{\phi'})|_{\psi} = (n|_{\phi'} \Cap n|_{\bigcirc^{d(\phi')+1}\psi})$. So it's reasonable to reach the conclusion that $\Cap_{n' \in n|_{\phi^{\vee}}} n'|_{\psi} = \Cap_{\phi' \in \mathrm{Cs}(\phi^{\vee})} n|_{\phi' \wedge \bigcirc^{d(\phi')+1}\psi}$.

Now we are ready to show that $n|_{\phi} = n|_{t(\phi)}$ for each $\phi \in \mathcal{L}^{\bigcirc\langle\cdot\rangle}$.

Proposition 10. *Given a pointed model (σ, n), we have that for each $\phi \in \mathcal{L}^{\bigcirc\langle\cdot\rangle}$, $n|_{\phi} = n|_{t(\phi)}$.*

Proof. We prove it by induction on the length of ϕ.

- $\phi := p$ or $\phi := \top$. It is obvious.
- $\phi := \psi \vee \chi$. Since $n|_{t(\psi\vee\chi)} = n|_{t(\psi)\vee t(\chi)} = n|_{t(\psi)} \cup n|_{t(\chi)}$, and by IH we get $n|_{t(\psi)} \cup n|_{t(\chi)} = n|_{\psi} \cup n|_{\chi} = n|_{\psi\vee\chi}$.
- $\phi := \bigcirc\psi$. Since $n|_{t(\bigcirc\psi)} = n|_{\bigcirc t(\psi)} = (n+1)|_{t(\psi)}$ and $n|_{\bigcirc\psi} = (n+1)|_{\psi}$. By IH on ψ, we can get $(n+1)|_{t(\psi)} = (n+1)|_{\psi}$, thus $n|_{t(\bigcirc\psi)} = n|_{\bigcirc\psi}$.
- $\phi := \langle\psi\rangle\chi$. Following from the translation function in Table 3 along with the Proposition 8, we can get $n|_{t(\langle\psi\rangle\chi)} = n|_{\bigvee_{\psi_i\in \mathrm{Cs}(t(\psi)^{\vee})}(\psi_i\wedge\bigcirc^{d(\psi_i)+1}t(\chi))} = \bigcup_{\psi\in \mathrm{Cs}(t(\psi)^{\vee})} n|_{\psi_i\wedge\bigcirc^{d(\psi_i)+1}t(\chi)} = \bigcup_{n'\in n|_{t(\psi)^{\vee}}} n'|_{t(\chi)}$. Then by Proposition 5 and IH, we can obtain $n|_{\langle\psi\rangle\chi} = \bigcup_{n'\in n|_{\psi}} n'|_{\chi} = \bigcup_{n'\in n|_{t(\psi)^{\vee}}} n'|_{t(\chi)} = n|_{t(\langle\psi\rangle\chi)}$.
- $\phi := \neg\psi$. We then have the following 5 cases:
  - $\phi := \neg p$ or $\phi := \neg\top$. It is obvious.
  - $\phi := \neg\neg\chi$. It is obvious due to the fact that $t(\neg\neg\chi) = t(\chi)$.
  - $\phi := \neg\bigcirc\chi$. Note the fact that $\neg\bigcirc\chi = \bigcirc\neg\chi$, so we have $n|_{\neg\bigcirc\chi} = n|_{\bigcirc\neg\chi} = (n+1)|_{\neg\chi}$. Moreover, $n|_{t(\neg\bigcirc\chi)} = n|_{\bigcirc t(\neg\chi)} = (n+1)|_{t(\neg\chi)}$. By IH on χ, we can get $(n+1)|_{\neg\chi} = (n+1)|_{t(\neg\chi)}$, which means $n|_{\bigcirc\neg\chi} = n|_{t(\bigcirc\neg\chi)}$, i.e. $n|_{\neg\bigcirc\chi} = n|_{t(\neg\bigcirc\chi)}$.
  - $\phi := \neg(\chi \vee \theta) = \neg\chi \wedge \neg\theta$. By IH, it is not hard to get $n|_{t(\neg\chi\wedge\neg\theta)} = n|_{t(\neg\chi)\wedge t(\neg\theta)} = n|_{t(\neg\chi)} \Cap n|_{t(\neg\theta)} = n|_{\neg\chi} \Cap n|_{\neg\theta} = n|_{\neg\chi\wedge\neg\theta}$.
  - $\phi := \neg\langle\chi\rangle\theta$. Following from the translation function in Table 3 and Proposition 9, we can get $n|_{t(\langle\psi\rangle\chi)} = n|_{\bigvee_{\psi_i\in \mathrm{Cs}(t(\psi)^{\vee})}(\psi_i\wedge\bigcirc^{d(\psi_i)+1}t(\chi))} = \bigcup_{\psi\in \mathrm{Cs}(t(\psi)^{\vee})} n|_{\psi_i\wedge\bigcirc^{d(\psi_i)+1}t(\chi)} = \bigcup_{n'\in n|_{t(\psi)^{\vee}}} n'|_{t(\chi)}$. Then by Proposition 5 and IH, we get $n|_{\neg\langle\chi\rangle\theta} = n|_{\neg\chi} \cup \Cap_{n'\in n|_{\chi}} (n'|_{\neg\theta}) = n|_{t(\neg\langle\chi\rangle\theta)}$.

The following theorem states that each $\mathcal{L}^{\bigcirc\langle\cdot\rangle}$-formula is equivalent to its translation formula.

Theorem 1. *For each $\phi \in \mathcal{L}^{\bigcirc\langle\cdot\rangle}$, $\models \phi \leftrightarrow t(\phi)$.*

Proof. This follows from Propositions 1 and 10.

4 Discussion

4.1 Succinctness

It is known that public announcement logic has the same expressivity as epistemic logic, but public announcement logic is exponentially more succinct than epistemic logic (see [13]). In this section, we will discuss the succinctness of $\mathcal{L}^{\bigcirc\langle\cdot\rangle}$.

For each $i \in \mathbb{N}$, the formula $\phi_i \in \mathcal{L}^{\bigcirc\langle\cdot\rangle}$ is defined as follows:

$$\begin{aligned} \phi_0 &= (p \vee \bigcirc p) \\ \phi_{i+1} &= \langle\phi_i\rangle(p \vee \bigcirc p) \end{aligned}$$

Table 4. Models of ϕ_2

0	1	2	3	4	5
p	p	p			
p	p		p		
p		p	p		
p		p		p	
	p	p	p		
	p	p		p	
	p		p	p	
	p		p		p

It is obvious that the length ϕ_i is bounded by a linear function of i. However, each ϕ_i contains 2^i different cases. Take ϕ_2 as an example, which is the following formula:

$$\langle\langle(p \vee \bigcirc p)\rangle(p \vee \bigcirc p)\rangle(p \vee \bigcirc p).$$

It contains the following 8 different cases:

$$\begin{array}{llll} \langle\langle p\rangle p\rangle p, & \langle\langle p\rangle p\rangle \bigcirc p, & \langle\langle p\rangle \bigcirc p\rangle p, & \langle\langle p\rangle \bigcirc p\rangle \bigcirc p, \\ \langle\langle \bigcirc p\rangle p\rangle p, & \langle\langle \bigcirc p\rangle p\rangle \bigcirc p, & \langle\langle \bigcirc p\rangle \bigcirc p\rangle p, & \langle\langle \bigcirc p\rangle \bigcirc p\rangle \bigcirc p. \end{array}$$

These 8 cases correspond to the 8 models depicted in Table 4.

The 8 cases can also be expressed by the following $\mathcal{L}^{\bigcirc}$-formula:

$$\begin{array}{l} (p \wedge \bigcirc((p \wedge \bigcirc(p \vee \bigcirc p)) \vee \bigcirc(p \wedge \bigcirc(p \vee \bigcirc p)))) \\ \vee \bigcirc\ (p \wedge \bigcirc((p \wedge \bigcirc(p \vee \bigcirc p)) \vee \bigcirc(p \wedge \bigcirc(p \vee \bigcirc p)))), \end{array}$$

which is the formula ψ_2 defined below. For each $i \in \mathbb{N}$, ψ_i is defined as follows:

$$\begin{array}{ll} \psi_0 & = (p \vee \bigcirc p) \\ \psi_i' & = (p \wedge \bigcirc \psi_i) \\ \psi_{i+1} & = (\psi_i' \vee \bigcirc \psi_i') \end{array}$$

Moreover, it can be shown that $\vDash \phi_i \leftrightarrow \psi_i$ for each $i \in \mathbb{N}$. It is obvious that the length of ψ_i is no less than 2^i. We conjecture that ψ_i is the shortest $\mathcal{L}^{\bigcirc}$-formula that is equivalent to ψ_i.

4.2 Comparison with Interval-Based Temporal Logic

Interval-based temporal logic is interpreted on temporal models where the underlying temporal ontology is time intervals, instead of time instants (see [2,6,8]). There have been various proposals and developments of interval-based temporal logics in the philosophical logic literature, such as [1,3,8–12,14,20,22].

An interval on the natural numbers $\mathbb{N}$ is a set of continuous natural numbers. For example, the interval $[n, m]$ is the set $\{k \in \mathbb{N} \mid n \leq k \leq m\}$. In an influential early work on the formal study of interval-based temporal ontology, Allen in [2] considered the family of all binary relations that can arise between two intervals in a linear order, including the *meet* relation. The interval A meets the interval B if and only if the last number of A equals the first number of B. For example, the interval $[1,3]$ meets the interval $[3,5]$.

The *next* modality $\langle\cdot\rangle$ discussed in this paper can be interpreted on intervals. The interval relation corresponding the modality $\langle\cdot\rangle$ is not the relation *meet*. For example, if $p \wedge \bigcirc p$ is true on both the intervals $[1,2]$ and the intervals $[3,4]$, we can say that the successive events $\langle p \wedge \bigcirc p\rangle(p \wedge \bigcirc p)$ is true on the interval $[1,4]$. However, the interval $[1,2]$ does not meet the interval $[3,4]$.

Although we can say that the interval $[1,2]$ is *before* the interval $[3,4]$, the *before* relation is not the relation modeled by the modality $\langle\cdot\rangle$. The *before* relation discussed by Allen in [2] is that the interval A is before the interval B (in other words, B is after A) if and only if the last number of A is bigger than the first number of B. The relation modeled by $\langle\cdot\rangle$ is the *immediate after* relation on intervals. To the limitation of the author's knowledge, such relation is not discussed in literatures.

5 Conclusion

In this paper, we proposed a new temporal modality $\langle\cdot\rangle$ to capture successive events. We extended the basic linear temporal logic, which refers to the linear temporal logic with the only modality *next*, by including the new successive modality. In semantics, we interpreted this new modality in a dynamic way that is similar to the public announcement modality in public announcement logic. We studied the expressivity of this new modality and showed that this new modality can be defined in the basic linear temporal logic. Additionally, We also discussed the succinctness of this extended linear temporal logic.

Regarding future work, one natural question is whether this extended linear temporal logic is exponentially more succinct than the basic linear temporal logic. In Sect. 4, we merely discuss this question through an example, which does not provide precise proof. In the future, we will try to prove its succinctness in detail. Another interesting direction is to interpret this logic on interval-based temporal logic and find out whether the modality $\langle\cdot\rangle$ can be expressed by the interval-based temporal logic by [8].

Acknowledgement. The authors thank four anonymous reviewers for their useful comments, which helped us to improve the presentation. This work is supported by the Fundamental Research Funds for the Central Universities (No. 63233137).

References

1. Allen, J.F.: Towards a general theory of action and time. Artif. Intell. **23**(2), 123–154 (1984)
2. Allen, J.F., Ferguson, G.: Actions and events in interval temporal logic. In: Stock, O. (ed.) Spatial and Temporal Reasoning, pp. 205–245. Springer, Dordrecht (1997). https://doi.org/10.1007/978-0-585-28322-7_7
3. Burgess, J.P.: Axioms for tense logic. II. Time periods. Notre Dame J. Formal Logic **23**(4), 375–383 (1982)
4. Gabbay, D., Pnueli, A., Shelah, S., Stavi, J.: On the temporal analysis of fairness. In: Proceedings of the 7th ACM SIGPLAN-SIGACT Symposium on Principles of Programming Languages, pp. 163–173 (1980)
5. Gerbrandy, J., Groeneveld, W.: Reasoning about information change. J. Logic Lang. Inform. **6**(2), 147–169 (1997). https://doi.org/10.1023/A:1008222603071
6. Goranko, V., Montanari, A., Sciavicco, G.: A road map of interval temporal logics and duration calculi. J. Appl. Non-Classical Logics **14**(1–2), 9–54 (2004). https://doi.org/10.3166/jancl.14.9-54
7. Goranko, V., Rumberg, A.: Temporal logic. In: Zalta, E.N. (ed.) The Stanford Encyclopedia of Philosophy. Metaphysics Research Lab, Stanford University, Summer 2022 edn. (2022)
8. Halpern, J.Y., Shoham, Y.: A propositional modal logic of time intervals. J. ACM **38**(4), 935–962 (1991). https://doi.org/10.1145/115234.115351
9. Hamblin, C.L.: Instants and intervals. In: Fraser, J.T., Haber, F.C., Müller, G.H. (eds.) The Study of Time, pp. 324–331. Springer, Heidelberg (1972). https://doi.org/10.1007/978-3-642-65387-2_23
10. Hansen, M.R., Chaochen, Z.: Duration calculus: logical foundations. Formal Aspects Comput. **9**, 283–330 (1997)
11. Humberstone, I.: Interval semantics for tense logic: some remarks. J. Philos. Logic 171–196 (1979)
12. Kowalski, R., Sergot, M.: A logic-based calculus of events. N. Gener. Comput. **4**, 67–95 (1986)
13. Lutz, C.: Complexity and succinctness of public announcement logic. In: Proceedings of the Fifth International Joint Conference on Autonomous Agents and Multiagent Systems, AAMAS 2006, pp. 137–143. Association for Computing Machinery, New York (2006). https://doi.org/10.1145/1160633.1160657
14. Moszkowski, B.C.: Reasoning about digital circuits. Stanford University (1983)
15. Plaza, J.: Logics of public communications. Synthese **158**(2), 165–179 (2007)
16. Pnueli, A.: The temporal logic of programs. In: 18th Annual Symposium on Foundations of Computer Science (SFCS 1977), pp. 46–57. IEEE (1977)
17. Prior, A.N.: Time and Modality. Oxford University Press, Oxford (1957)
18. Prior, A.N.: Present and Future. Oxford University Press, Oxford (1967)
19. Prior, A.N.: Papers on Time and Tense. Oxford University Press, Oxford (1968)
20. Roper, P.: Intervals and tenses. J. Philos. Log. **9**, 451–469 (1980)
21. Van Ditmarsch, H., van Der Hoek, W., Kooi, B.: Dynamic Epistemic Logic, vol. 337. Springer, Heidelberg (2007)
22. Zhou, C., Hansen, M.R.: Duration Calculus: A Formal Approach to Real-Time Systems. Monographs in Theoretical Computer Science. an Eatcs Series, Springer, Heidelberg (2004)

Non-labelled Sequent Calculi of Public Announcement Expansions of K45 and S5

Sizhuo Liu[1(✉)] and Katsuhiko Sano[2]

[1] Graduate School of Humanities and Human Sciences, Hokkaido University, Sapporo, Japan
liuliusizhuo@outlook.com

[2] Faculty of Humanities and Human Sciences, Hokkaido University, Sapporo, Japan
v-sano@let.hokudai.ac.jp

Abstract. This paper proposes non-labelled sequent calculi, G(**K45PAL**) and G(**S5PAL**), for the public announcement expansions of modal logics **K45** and **S5**. We transform each of the recursion axioms of PAL into left and right rules for the sequent calculi. For G(**K45PAL**), the cut elimination theorem is shown using the complexity measure introduced by van Ditmarsch et al. (2007). This measure was originally employed to establish semantic completeness via recursion axioms. While the cut elimination theorem fails in G(**S5PAL**), we adopt Takano's strategy (1992) to establish that the cut formula in G(**S5PAL**) can be restricted to the set of suitably extended subformulas (i.e., closure) of the conclusion of the cut rule.

Keywords: Public Announcement Logic · Epistemic Logic · Cut Elimination · Sequent Calculus · Analytic Cut

1 Introduction

Sequent calculus, also called Gentzen system, is a logic system devised by Gerhard Gentzen in the 1930s, which opened a field of proof theory by structuring logical deductions [8,9], with the notion of sequent $\Gamma \Rightarrow \Delta$, which expresses the implication from premises Γ to conclusions Δ, i.e., if all formulas in Γ hold then some formula in Δ holds. A crucial aspect of sequent calculus is the concept of cut-elimination (Gentzen's *Hauptsatz*), which simplifies proofs by removing instances of the cut rule, thereby yielding more intuitive and transparent "cut-free" proofs, where the cut rule is of the following form:

$$\frac{\Gamma \Rightarrow \Delta, \varphi \quad \varphi, \Pi \Rightarrow \Sigma}{\Gamma, \Pi \Rightarrow \Delta, \Sigma}\ (Cut).$$

This process uncovers the constructive content of proofs and contributes to our understanding of proof normalization and proof search [30].

We would like to thank the anonymous referees for their valuable comments. The work of the first author was supported by Graduate Grant Program of Graduate School of Humanities and Human Sciences, Hokkaido University. The work of the second author was partially supported by JSPS KAKENHI Grant-in-Aid for Scientific Research (B) Grant Number JP 22H00597 and (C) Grant Number JP 19K12113.

N. Alechina et al. (Eds.): LORI 2023, LNCS 14329, pp. 190–206, 2023.
https://doi.org/10.1007/978-3-031-45558-2_15

When we turn our eyes on modal logics, sequent calculi were provided, e.g., by Ohnishi and Matsumoto [19]. For modal logics such as **KT** and **S4**, traditional sequent calculi are acknowledged to be sound, complete, and cut-free in [19]. On the other hand, for modal logic **S5**, the cut-elimination theorem fails for a sequent calculus G(**S5**) of modal logic **S5**, as explained in [20, p.116]. To recover the cut elimination theorem, it is necessary to either incorporate global side conditions, as in the work of Braüner [5], or extend the sequent format with additional structures such as hypersequents (for example, see Poggiolesi [23]), display calculus (for example, see Dosen [7]), or labelled sequents by Negri [15]. In particular, Negri [15] introduced a universal method for producing contraction- and cut-free *labelled* sequent calculi for a broad spectrum of normal modal logics. Instead of regaining the cut elimination, Takano [28] established that, while system G(**S5**) does not allow the elimination of cuts, it can be shown that every application of the cut rule in G(**S5**) can be replaced by an analytic application of the cut rule, where the cut rule is *analytic* if the cut formula φ is a subformula of the conclusion $\Gamma, \Pi \Rightarrow \Delta, \Sigma$ of the cut rule above.

Public announcement logic **PAL** serves as a theoretical framework for interpreting knowledge shifts within multi-agent systems [22]. It expands the multi-agent version of modal logic **S5** (epistemic logic) with a public announcement operator $[\varphi]\psi$ that reads as "if φ is true, then after the public announcement of φ, ψ holds." The semantic completeness of **PAL** is reduced to that of **S5** by an equivalence-preserving translation of a formula from the syntax of **PAL** to a formula in the syntax of **S5**. A key part of this reduction consists of the recursion axioms that push an occurrence of a public announcement operator inside the whole formula, along with the complexity measure that enables us to conduct induction properly.

The labelled formalism introduced by Negri [15] has subsequently been employed by several authors in the construction of labelled *cut-free* sequent calculi for **PAL**, as documented in the works of Maffezioli and Negri [14], Balbiani et al. [2], Nomura et al. [18], and Wu et al. [32]. Furthermore, sequent calculi have been defined for various logics in the field of dynamic epistemic logic, even if one insists on adhering to the conventional notion of sequents and avoiding labelled formalism. Hatano and Sano [10] introduced a cut-free non-labelled sequent calculus for a constructive variant of dynamic logic of relation changers by van Benthen and Liu [4]. They transformed each of the recursion axioms to left and right inference rules of the proposed sequent calculus and emphasized the importance of the complexity measure from [6] as part of a cut-elimination argument. Then Wirsing and Knapp [31] provided a cut-free non-labelled sequent calculus, denoted as $\mathbf{G4}_{P,A}$, for action model logic with **S4** as its base logic (since we cannot eliminate the cut rule for **S5**). They also transformed recursion axioms for action model logic to sequent rules and utilized the complexity measure defined in [6] for the cut-elimination argument. However, there have been no non-labelled sequent calculi for public announcement logic or action model logic with the multi-agent version of **S5** as the base logic. This paper focuses on public announcement logic.

In this work, we introduce sequent calculi named G(**K45PAL**) and G(**S5PAL**) for the public announcement expansions of multi-agent **K45** and **S5**, respectively. Unlike labelled sequent calculi, our proposed calculi do not internalize accessibility relations or incorporate labels. Instead, like in [10,31], our inference rules for public announcement operators are derived naturally from the recursion axioms specified in [6] (it is worth noting that our inference rules for public announcement operators have almost the same form as those in [32] when disregarding the labels).

To emphasize the proof-theoretic importance of the complexity measure introduced in [6], we first establish that G(**K45PAL**) enjoys the cut-elimination theorem. Although our sequent calculi do not enjoy the subformula property (i.e., a sequent is derivable only in terms of sequents consisting of subformulas of the original sequent), we naturally extend the notion of subformula to the notion of closure. With this notion of closure, we follow Takano's approach [28] to syntactically establish that every application of the cut rule in G(**S5PAL**) can be replaced by such an application of the cut rule that the cut formula can be taken from the closure of the conclusion of the cut rule. In other words, our calculus G(**S5PAL**) remains analytic in terms of the notion of closure. We also emphasize that the complexity measure from [6] plays an important role in this syntactic argument.

The structure of this paper is organized in the following manner: Section 2 reviews syntax and semantics of multi-agent version of **K45** and **S5**. In Sect. 3, we introduce the sequent calculi G(**K45PAL**) and G(**S5PAL**). Section 4 presents the proof of cut-elimination for G(**K45PAL**) in terms of the complexity measure introduced by van Ditmarsch et al. (2007). Section 5 adopts Takano's strategy (1992) to establish that the cut formula in G(**S5PAL**) can be restricted to the set of suitably extended subformulas (i.e., closure) of the conclusion of the cut rule. Finally, Sect. 6 concludes the paper.

2 Preliminaries

Let Prop be a countably infinite set of propositional variables and Ag a non-empty finite set of agents. The syntax L for epistemic logic is defined inductively as follows:

$$L \ni \varphi ::= p \mid \neg\varphi \mid \varphi \wedge \psi \mid \varphi \rightarrow \psi \mid K_a\varphi$$

where $p \in \mathsf{Prop}$ and $a \in \mathsf{Ag}$. Moreover, the syntax L^+ for public announcement logic is defined inductively as follows:

$$L^+ \ni \varphi ::= p \mid \neg\varphi \mid \varphi \wedge \psi \mid \varphi \rightarrow \psi \mid K_a\varphi \mid [\varphi]\psi$$

We use $\varphi \leftrightarrow \psi$ as an abbreviation for $(\varphi \rightarrow \psi) \wedge (\psi \rightarrow \varphi)$ and $\varphi \vee \psi$ as an abbreviation for $\neg\varphi \rightarrow \psi$. For a finite set Δ, $\bigwedge\Delta$ and $\bigvee\Delta$ are the conjunction or disjunction of all formulas in Δ (when Δ is empty, $\bigwedge\Delta := \top$ and $\bigvee\Delta := \bot$). A *K-formula* is a formula of the form $K_a\varphi$.

The following definition and lemma are from [6, Definition 7.21] and [6, Lemma 7.22], respectively and they are key ingredients for establishing semantic completeness of the public announcement logic in [6].

Definition 1. *The complexity* c *of a formula* φ *in* L^+ *is inductively defined as*

$$\begin{aligned}
\mathsf{c}(p) &:= 1,\\
\mathsf{c}(\neg\varphi) &:= 1 + \mathsf{c}(\varphi),\\
\mathsf{c}(\varphi_1 \bullet \varphi_2) &:= 1 + \max(\mathsf{c}(\varphi_1), \mathsf{c}(\varphi_2)) \quad (\bullet \in \{\rightarrow, \wedge\}),\\
\mathsf{c}(K_a\varphi) &:= 1 + \mathsf{c}(\varphi),\\
\mathsf{c}([\varphi]\psi) &:= (4 + \mathsf{c}(\varphi)) \cdot \mathsf{c}(\psi).
\end{aligned}$$

Lemma 1. *The following inequalities hold in* L^+*:*

(1) $\mathsf{c}(\psi) > \mathsf{c}(\varphi)$ *if* $\varphi \in \mathsf{Sub}(\psi)$ *and* φ *is distinct from* ψ *as a formula,*
(2) $\mathsf{c}([\varphi]\neg\psi) > \mathsf{c}([\varphi]\psi)$,
(3) $\mathsf{c}([\varphi](\psi_1 \bullet \psi_2)) > \mathsf{c}([\varphi]\psi_i) \quad (\bullet \in \{\to, \wedge\})$,
(4) $\mathsf{c}([\varphi]K_a\psi) > \mathsf{c}(K_a[\varphi]\psi)$,
(5) $\mathsf{c}([\varphi][\psi]\gamma) > \mathsf{c}([\varphi \wedge [\varphi]\psi]\gamma)$.

The set $\mathsf{Sub}(\varphi)$ of all subformulas is defined as usual. Moreover we define the extended notion $\mathsf{CL}(\varphi)$ of $\mathsf{Sub}(\varphi)$ as follows (our notion is a simplification of [6, Definition 7.27] in the setting without the common knowledge operators). Inference rules for public announcements in our sequent calculi do not satisfy the subformula property but the *extended* version of the subformula property in terms of $\mathsf{CL}(\varphi)$.

Definition 2. *The* closure $\mathsf{CL}(\varphi)$ *of an* L^+*-formula* φ *is defined inductively as follows:*

$$\begin{aligned}
\mathsf{CL}(p) &:= \{p\}\\
\mathsf{CL}(\neg\varphi) &:= \mathsf{CL}(\varphi) \cup \{\neg\varphi\}\\
\mathsf{CL}(\varphi \bullet \psi) &:= \mathsf{CL}(\varphi) \cup \mathsf{CL}(\psi) \cup \{\varphi \bullet \psi\} (\bullet \in \{\to, \wedge\})\\
\mathsf{CL}(K_a\varphi) &:= \mathsf{CL}(\varphi) \cup \{K_a\varphi\}\\
\mathsf{CL}([\varphi]p) &:= \mathsf{CL}(\varphi) \cup \{p, [\varphi]p\}\\
\mathsf{CL}([\varphi]\neg\psi) &:= \mathsf{CL}(\varphi) \cup \mathsf{CL}([\varphi]\psi) \cup \{[\varphi]\neg\psi\}\\
\mathsf{CL}([\varphi](\psi \bullet \gamma)) &:= \mathsf{CL}([\varphi]\psi) \cup \mathsf{CL}([\varphi]\gamma) \cup \{[\varphi](\psi \bullet \gamma)\} (\bullet \in \{\to, \wedge\})\\
\mathsf{CL}([\varphi]K_a\psi) &:= \mathsf{CL}(K_a[\varphi]\psi) \cup \mathsf{CL}(\varphi) \cup \{[\varphi]K_a\psi\}\\
\mathsf{CL}([\varphi][\psi]\gamma) &:= \mathsf{CL}([\varphi \wedge [\varphi]\psi]\gamma) \cup \{[\varphi][\psi]\gamma\}
\end{aligned}$$

For a set Ξ *of formulas, we define* $\mathsf{CL}(\Xi) := \bigcup_{\varphi \in \Xi} \mathsf{CL}(\varphi)$.

Note that $\mathsf{Sub}(\varphi) \subseteq \mathsf{CL}(\varphi)$ and $\mathsf{CL}(\varphi)$ is finite. Moreover, if a formula φ is from L, it follows that $\mathsf{CL}(\varphi) = \mathsf{Sub}(\varphi)$. By Lemma 1, we obtain the following.

Lemma 2. $c(\psi) > c(\varphi)$ *if* $\varphi \in \mathsf{CL}(\psi)$ *and* φ *is distinct from* ψ *as a formula.*

We move on to the semantics. A *frame* F is a tuple $(W, (R_a)_{a \in Ag})$ where W is a set of states and $R_a \subseteq W \times W$ is a binary relation on W for each $a \in \mathsf{Ag}$. A *model* M is a tuple $(W, (R_a)_{a \in Ag}, V)$ where $(W, (R_a)_{a \in Ag})$ is a frame and V is a function from Prop to $\mathcal{P}(W)$. Let $M = (W, (R_a)_{a \in Ag}, V)$ be a model and $w \in W$. The notion of φ *being true* at w in M (notation: $M, w \models \varphi$) is defined inductively as follows:

$$\begin{array}{lll}
M, w \models p & \text{iff} & w \in V(p),\\
M, w \models \neg\varphi & \text{iff} & M, w \not\models \varphi,\\
M, w \models \varphi \wedge \psi & \text{iff} & M, w \models \varphi \text{ and } M \models \psi,\\
M, w \models \varphi \to \psi & \text{iff} & M, w \not\models \varphi \text{ or } M \models \psi,\\
M, w \models K_a\varphi & \text{iff} & \text{for all } v \in W, wR_av \text{ implies } M, v \models \varphi,\\
M, w \models [\varphi]\psi & \text{iff} & M, w \models \varphi \text{ implies } M^\varphi, w \models \psi,
\end{array}$$

where $M^\varphi = (W^\varphi, (R^\varphi_a)_{a \in Ag}, V^\varphi)$, $W^\varphi = \{ w \in W \mid M, w \models \varphi \}$, $R^\varphi_a = R_a \cap (W^\varphi \times W^\varphi)$ and $V^\varphi(p) = V(p) \cap W^\varphi$.

A formula φ is *valid* in a model (notation: $M \models \varphi$) if $M, w \models \varphi$ for all $w \in W$. A formula φ is *valid* in a frame F (notation: $F \models \varphi$) if $(F, V) \models \varphi$ for all valuations V. Given a class $\mathbb{F}$ of frames, φ is *valid* in $\mathbb{F}$ if φ is valid in F for every $F \in \mathbb{F}$.

Definition 3. *We define $\mathbb{F}_{\mathbf{K45}}$ to be the class of all transitive and euclidean frames and $\mathbb{F}_{\mathbf{S5}}$ to be the class of all reflexive, transitive and euclidean frames. We define that* **K45** *and* **S5** *are the sets of all L-formulas that are valid in the class of $\mathbb{F}_{\mathbf{K45}}$ and $\mathbb{F}_{\mathbf{S5}}$, respectively. The public announcement expansion* **K45PAL** *of* **K45** *is the set of all L^+-formulas that are valid in the class $\mathbb{F}_{\mathbf{K45}}$. The public announcement expansion* **S5PAL** *of* **S5** *is the set of all L^+-formulas that are valid in the class $\mathbb{F}_{\mathbf{S5}}$.*

It is well-known that all logics in Definition 3 can be axiomatized in Hilbert systems by the axioms and inference rules as in Table 1 (see [6]).

Table 1. Hilbert Systems H(**K45**), H(**S5**), H(**K45PAL**) and H(**S5PAL**)

All the Axioms and Rules of H(**K45**)	
(Taut)	all instances of propositional tautologies
(K)	$K_a(\varphi \to \psi) \to (K_a\varphi \to K_a\psi)$
(4)	$K_a\varphi \to K_aK_a\varphi$
(5)	$\neg K_a\varphi \to K_a\neg K_a\varphi$
(MP)	From $\varphi \to \psi$ and φ, we may infer ψ
(Nec)	From φ, we may infer $K_a\varphi$
Additional Axiom for H(**S5**)	
(T)	$K_a\varphi \to \varphi$
Additional Recursion Axioms for H(**K45PAL**) and H(**S5PAL**)	
([]at)	$[\varphi]p \leftrightarrow (\varphi \to p)$
([]$\wedge$)	$[\varphi](\psi \wedge \gamma) \leftrightarrow ([\varphi]\psi \wedge [\varphi]\gamma)$
([] $\to$)	$[\varphi](\psi \to \gamma) \leftrightarrow ([\varphi]\psi \to [\varphi]\gamma)$
([]$\neg$)	$[\varphi]\neg\psi \leftrightarrow (\varphi \to \neg[\varphi]\psi)$
([]K)	$[\varphi]K_a\psi \leftrightarrow (\varphi \to K_a[\varphi]\psi)$
([][])	$[\varphi][\psi]\gamma \leftrightarrow [\varphi \wedge [\varphi]\psi]\gamma$

Proposition 1. *Hilbert systems* H(**K45**) *and* H(**S5**) *are sound and complete for $\mathbb{F}_{\mathbf{K45}}$ and $\mathbb{F}_{\mathbf{S5}}$, respectively. Hilbert systems* H(**K45PAL**) *and* H(**S5PAL**) *are sound and complete for $\mathbb{F}_{\mathbf{K45}}$ and $\mathbb{F}_{\mathbf{S5}}$, respectively.*

Proof. Since all the results are well-known, we review how the complexity $\mathtt{c}(\varphi)$ plays a role in the case of the semantic completeness proof of Hilbert system H(**S5PAL**) as did in [6]. We can define a translation $\mathtt{t}$ from L^+ to L that removes all the occurrences of the public announcement operators from outermost occurrences. Then we can prove by induction on the complexity $\mathtt{c}(\varphi)$ that $\mathtt{t}(\varphi) \leftrightarrow \varphi$ are both valid in $\mathbb{F}_{\mathbf{S5}}$ and a theorem of H(**S5PAL**) (see [6, Lemma 7.24]). Then we proceed as follows. Suppose that a formula φ in L^+ is valid in $\mathbb{F}_{\mathbf{S5}}$. It follows that $\mathtt{t}(\varphi) \in L$ is also valid in $\mathbb{F}_{\mathbf{S5}}$. Since H(**S5**) is semantically complete for $\mathbb{F}_{\mathbf{S5}}$, $\mathtt{t}(\varphi)$ is a theorem of H(**S5**) hence of H(**S5PAL**). By the equivalence $\mathtt{t}(\varphi) \leftrightarrow \varphi$ in H(**S5PAL**), we conclude that φ is a theorem of H(**S5PAL**). □

3 Sequent Calculi

In this section we introduce our sequent calculi G(**K45PAL**) and G(**S5PAL**).

3.1 Sequent Calculi G(K45) and G(S5)

A *sequent* $\Gamma \Rightarrow \Delta$ is a pair of finite multisets Γ and Δ of formulas and it read as: if all formulas in Γ hold, then at least one formula in Δ holds. This section reviews the known sequent calculi G(**K45**) and G(**S5**) for multi-agent **K45** and **S5**, respectively. Both calculi build on $\mathbf{LK_0}$ (see [28]), which is the propositional fragment of a sequent calculus **LK** [8, 9] of the first-order classical logic.

Definition 4. *A sequent calculus* $\mathbf{LK_0}$ *consists of the following.*

– *Axioms:*

$$\frac{}{\varphi \Rightarrow \varphi}\ (\mathtt{id})$$

– *Structural Rules:*

$$\frac{\Gamma \Rightarrow \Delta}{\Gamma \Rightarrow \Delta, \varphi}\ (\Rightarrow w) \qquad \frac{\Gamma \Rightarrow \Delta}{\varphi, \Gamma \Rightarrow \Delta}\ (w \Rightarrow) \qquad \frac{\Gamma \Rightarrow \Delta, \varphi, \varphi}{\Gamma \Rightarrow \Delta, \varphi}\ (\Rightarrow c) \qquad \frac{\varphi, \varphi, \Gamma \Rightarrow \Delta}{\varphi, \Gamma \Rightarrow \Delta}\ (c \Rightarrow)$$

– *Logical Rules:*

$$\frac{\Gamma \Rightarrow \Delta, \varphi_1 \quad \Gamma \Rightarrow \Delta, \varphi_2}{\Gamma \Rightarrow \Delta, \varphi_1 \wedge \varphi_2}\ (\Rightarrow \wedge) \qquad \frac{\varphi_i, \Gamma \Rightarrow \Delta}{\varphi_1 \wedge \varphi_2, \Gamma \Rightarrow \Delta}\ (\wedge \Rightarrow) \qquad \frac{\varphi, \Gamma \Rightarrow \Delta}{\Gamma \Rightarrow \Delta, \neg \varphi}\ (\Rightarrow \neg) \qquad \frac{\Gamma \Rightarrow \Delta, \varphi}{\neg \varphi, \Gamma \Rightarrow \Delta}\ (\neg \Rightarrow)$$

$$\frac{\varphi, \Gamma \Rightarrow \Delta, \psi}{\Gamma \Rightarrow \Delta, \varphi \to \psi}\ (\Rightarrow\to) \qquad \frac{\Gamma \Rightarrow \Delta, \varphi \quad \psi, \Pi \Rightarrow \Sigma}{\varphi \to \psi, \Gamma, \Pi \Rightarrow \Delta, \Sigma}\ (\to\Rightarrow)$$

– *Cut:*

$$\frac{\Gamma \Rightarrow \Delta, \varphi \quad \varphi, \Pi \Rightarrow \Sigma}{\Gamma, \Pi \Rightarrow \Delta, \Sigma}\ (Cut)_\varphi$$

The sequent calculus G(**K45**) *is* $\mathbf{LK_0}$ *expanded with the following rule:*

$$\frac{\Gamma, K_a\Gamma \Rightarrow K_a\Delta, \varphi}{K_a\Gamma \Rightarrow K_a\Delta, K_a\varphi}\ (K_{\mathbf{K45}}).$$

The sequent calculus G(**S5**) *expands* $\mathbf{LK_0}$ *with the following rules:*

$$\frac{K_a\Gamma \Rightarrow K_a\Delta, \varphi}{K_a\Gamma \Rightarrow K_a\Delta, K_a\varphi}\ (\Rightarrow K_{\mathbf{S5}}) \qquad \frac{\varphi, \Gamma \Rightarrow \Delta}{K_a\varphi, \Gamma \Rightarrow \Delta}\ (K \Rightarrow).$$

For each calculus, we define the notion of derivability of a sequent as a finite tree generated from axioms (id) *by inference rules specific to the calculus.*

Proposition 2. *Let* $\Lambda \in \{\mathbf{K45}, \mathbf{S5}\}$.

(1) If $\Gamma \Rightarrow \Delta$ *is derivable in* G(Λ) *then* $\bigwedge \Gamma \to \bigvee \Delta$ *is a theorem of* H(Λ).
(2) If φ *is a theorem of* H(Λ), *then* $\Rightarrow \varphi$ *is derivable in* G(Λ).

Proposition 3 (Shvarts [27]**).** *The cut rule is admissible in* G(**K45**)*, i.e., if* $\Gamma \Rightarrow \Delta$ *is derivable in* G(**K45**) *then the same sequent is also derivable in the sequent calculus* G(**K45**) *without the cut rule.*

A formula $p \to K_a\neg K_a\neg p$ is derivable in G(**S5**) with the cut rule as follows:

$$\dfrac{\dfrac{\dfrac{K_a\neg p \Rightarrow K_a\neg p}{\Rightarrow \neg K_a\neg p, K_a\neg p}(\Rightarrow\neg)}{\Rightarrow K_a\neg K_a\neg p, K_a\neg p}(\Rightarrow K_{S5}) \quad \dfrac{\dfrac{p \Rightarrow p}{\neg p, p \Rightarrow}(\neg\Rightarrow)}{K_a\neg p, p \Rightarrow}(K\Rightarrow)}{\dfrac{p \Rightarrow K_a\neg K_a\neg p}{\Rightarrow p \to K_a\neg K_a\neg p}(\Rightarrow\to)}(Cut)_{K_a\neg p}$$

As noted in [20], however, we cannot eliminate the above application of the cut rule. Moreover, Takano [28] established the following syntactically (a semantic argument can be found in [29]).

Proposition 4 (Takano [28]**).** *Every application of the cut rule of* G(**S5**) *can be replaced with an analytic one.*

3.2 Sequent Calculi G(K45PAL) and G(S5PAL)

Definition 5. *The sequent calculi* G(**K45PAL**) *and* G(**S5PAL**) *are the ones which are obtained from* G(**K45**) *and* G(**S5**)*, respectively, by supplementing the following rules on public announcements:*

$$\dfrac{\varphi, \Gamma \Rightarrow \Delta, p}{\Gamma \Rightarrow \Delta, [\varphi]p}(\Rightarrow[]\mathtt{at}) \qquad \dfrac{\Gamma \Rightarrow \Delta, \varphi \quad p, \Pi \Rightarrow \Sigma}{[\varphi]p, \Gamma, \Pi \Rightarrow \Delta, \Sigma}([]\mathtt{at}\Rightarrow)$$

$$\dfrac{\varphi, [\varphi]\psi, \Gamma \Rightarrow \Delta}{\Gamma \Rightarrow \Delta, [\varphi]\neg\psi}(\Rightarrow[]\neg) \qquad \dfrac{\Gamma \Rightarrow \varphi, \Delta \quad \Pi \Rightarrow \Sigma, [\varphi]\psi}{[\varphi]\neg\psi, \Gamma, \Pi \Rightarrow \Delta, \Sigma}([]\neg\Rightarrow)$$

$$\dfrac{\Gamma \Rightarrow \Delta, [\varphi]\psi_1 \quad \Gamma \Rightarrow \Delta, [\varphi]\psi_2}{\Gamma \Rightarrow \Delta, [\varphi](\psi_1 \wedge \psi_2)}(\Rightarrow[]\wedge) \qquad \dfrac{[\varphi]\psi_i, \Gamma \Rightarrow \Delta}{[\varphi](\psi_1 \wedge \psi_2), \Gamma \Rightarrow \Delta}([]\wedge\Rightarrow)$$

$$\dfrac{[\varphi]\psi, \Gamma \Rightarrow \Delta, [\varphi]\gamma}{\Gamma \Rightarrow \Delta, [\varphi](\psi \to \gamma)}(\Rightarrow[]\to) \qquad \dfrac{\Gamma \Rightarrow \Delta, [\varphi]\psi \quad [\varphi]\gamma, \Pi \Rightarrow \Sigma}{[\varphi](\psi \to \gamma), \Gamma, \Pi \Rightarrow \Delta, \Sigma}([]\to\Rightarrow)$$

$$\dfrac{\varphi, \Gamma \Rightarrow \Delta, K_a[\varphi]\psi}{\Gamma \Rightarrow \Delta, [\varphi]K_a\psi}(\Rightarrow[]K) \qquad \dfrac{\Gamma \Rightarrow \Delta, \varphi \quad K_a[\varphi]\psi, \Pi \Rightarrow \Sigma}{[\varphi]K_a\psi, \Gamma, \Pi \Rightarrow \Delta, \Sigma}([]K\Rightarrow)$$

$$\dfrac{\Gamma \Rightarrow \Delta, [\varphi \wedge [\varphi]\psi]\gamma}{\Gamma \Rightarrow \Delta, [\varphi][\psi]\gamma}(\Rightarrow[][]) \qquad \dfrac{[\varphi \wedge [\varphi]\psi]\gamma, \Gamma \Rightarrow \Delta}{[\varphi][\psi]\gamma, \Gamma \Rightarrow \Delta}([][]\Rightarrow)$$

As the reader may see, these new rules are naturally obtained from recursion axioms for public announcements in Table 1. For example, a formula $K_a[\varphi]\psi$ in the premise of the rule $(\Rightarrow []K)$ is *not* a subformula of the conclusion of the rule, but it is in the closure of the conclusion. For each of the new rules above, every formula in premises of the rule is an element of the closure of the conclusion of the rule. In this sense, our calculi enjoy the extended subformula property in terms of the notion of closure (Definition 2).

A derivation of the left-to-right direction of the axiom ($[]\,K$) is given as follows.

$$\dfrac{\dfrac{\dfrac{\varphi \Rightarrow \varphi \quad K_a[\varphi]\psi \Rightarrow K_a[\varphi]\psi}{\varphi, [\varphi]K_a\psi \Rightarrow K_a[\varphi]\psi}\,([]\,K_a \Rightarrow)}{[\varphi]K_a\psi \Rightarrow \varphi \to K_a[\varphi]\psi}\,(\Rightarrow\to)}{\Rightarrow [\varphi]K_a\psi \to (\varphi \to K_a[\varphi]\psi)}\,(\Rightarrow\to) \quad .$$

Proposition 5. *Let* $\Lambda \in \{\mathbf{K45PAL}, \mathbf{S5PAL}\}$.

(1) If $\Gamma \Rightarrow \Delta$ *is derivable in* $\mathsf{G}(\Lambda)$ *then* $\bigwedge \Gamma \to \bigvee \Delta$ *is a theorem of* $\mathsf{H}(\Lambda)$.
(2) If φ *is a theorem of* $\mathsf{H}(\Lambda)$*, then* $\Rightarrow \varphi$ *is derivable in* $\mathsf{G}(\Lambda)$.

4 Cut-Elimination for G(K45PAL)

This section establishes the cut-elimination theorem for G(**K45PAL**). The following is an extended version of the cut rule from [12], which plays the similar role essentially as Mix rule by Gentzen [8] to handle the difficulty caused by contraction rules.

Definition 6. *(Ecut) is the following rule:*

$$\dfrac{\Gamma \Rightarrow \Delta, \rho^m \quad \rho^n, \Pi \Rightarrow \Sigma}{\Gamma, \Pi \Rightarrow \Delta, \Sigma}\,(Ecut)_\rho \, ,$$

where $m, n \in \mathbb{N}$*, i.e.,* m *and* n *are possibly zero.*

It is noted that when $m = n = 1$ $(Ecut)$ is the same as (Cut).

Definition 7. *We define* G^-(**K45PAL**) *as the same calculus as* G(**K45PAL**) *except that the cut rule is excluded. The sequent calculus* G^*(**K45PAL**) *is the same calculus as* G(**K45PAL**) *except that the cut rule is replaced with* $(Ecut)$.

Definition 8. *The formulas that do not change in an inference rule except for* $(K_{\mathbf{K45}})$ *in Definition 4 and* $(\Rightarrow K_{\mathbf{S5}})$ *in Definition 4 are called* parameters. *For the other formulas, those occurring in the lower sequent of an inference rule are called* principal formulas, *and those occurring in the upper sequent are called* active formulas. *For* $(K_{\mathbf{K45}})$ *and* $(\Rightarrow K_{\mathbf{S5}})$*, the principal formulas are defined to be* $K_a\Gamma$*,* $K_a\Delta$ *and* $K_a\varphi$ *hence there are no parameters in these rules.*

For example, the principal formula of $([]\,[] \Rightarrow)$ is $[\varphi][\psi]\gamma$.

Theorem 1. *If* $\Gamma \Rightarrow \Delta$ *is derivable in* G^*(**K45PAL**)*, then it is also derivable in* G^-(**K45PAL**)*. Therefore, if* $\Gamma \Rightarrow \Delta$ *is derivable in* G(**K45PAL**) *then it is also derivable in* G^-(**K45PAL**).

Proof. It suffices to prove the first statement. For our purpose, we establish the following (*Hauptsatz*): If in the following derivation $\mathcal{D}$ of $\mathsf{G}^*(\mathbf{K45PAL})$:

$$\mathcal{D} \equiv \dfrac{\dfrac{\vdots\ \mathcal{L}}{\Gamma \Rightarrow \Delta, \varphi^m}\ \mathtt{rule}(\mathcal{L}) \quad \dfrac{\vdots\ \mathcal{R}}{\varphi^n, \Pi \Rightarrow \Sigma}\ \mathtt{rule}(\mathcal{R}),}{\Gamma, \Pi \Rightarrow \Delta, \Sigma}\ (Ecut)_\varphi$$

there is no other $(Ecut)$ in derivations of $\Gamma \Rightarrow \Delta, \varphi^m$ and $\varphi^n, \Pi \Rightarrow \Sigma$, then there is a derivation in $\mathsf{G}^-(\mathbf{K45PAL})$ of the sequent $\Gamma, \Pi \Rightarrow \Delta, \Sigma$. We prove this statement on a derivation $\mathcal{D}$ by double induction on the *complexity* $\mathsf{c}(\mathcal{D})$ and *weight* $\mathtt{w}(\mathcal{D})$ of the derivation $\mathcal{D}$ (or the lexicographic order of the complexity and the weight), which are defined as follows (see [12, pp.135–136]):

- The *complexity* $\mathsf{c}(\mathcal{D})$ of a derivation $\mathcal{D}$ is $\mathsf{c}(\varphi)$ of the cut-formula φ of $(Ecut)$, as defined in Definition 1.
- The *weight* $\mathtt{w}(\mathcal{D})$ of a derivation $\mathcal{D}$ is the number of all sequents in derivations $\mathcal{L}$ and $\mathcal{R}$.

If $m = 0$ or $n = 0$, it follows our goal can be obtained by weakening rules. Hence we suppose $m > 0$ and $n > 0$ below. We divide the argument into the following cases.

1. Let $\mathtt{rule}(\mathcal{L})$ or $\mathtt{rule}(\mathcal{R})$ be $(\mathtt{id})$.
2. Let $\mathtt{rule}(\mathcal{L})$ or $\mathtt{rule}(\mathcal{R})$ be structural rules.
3. Let $\mathtt{rule}(\mathcal{L})$ or $\mathtt{rule}(\mathcal{R})$ be logical rules or rules for K_a and public announcements where the cut-formula is not principal.
4. Let $\mathtt{rule}(\mathcal{L})$ and $\mathtt{rule}(\mathcal{R})$ be rules of the same connective and the cut-formula is principal in both rules.

Since cases (1) and (2) are handled similarly to the cut-elimination proof for $\mathbf{LK}_0$ (cf. [12]), we comment on the cases where our new public announcement rules are concerned with. In particular, we deal with one typical instance of case (4) where the complexity measure of Definition 1 plays a role.

Suppose $\mathtt{rule}(\mathcal{L})$ is $(\Rightarrow [\,][\,])$ and $\mathtt{rule}(\mathcal{R})$ is $([\,][\,] \Rightarrow)$ where the cut-formula is principal in both rules. In this case, the derivation $\mathcal{D}$ is of the following form:

$$\dfrac{\dfrac{\Gamma \Rightarrow \Delta, ([\varphi][\psi]\gamma)^{m-1}, [\varphi \wedge [\varphi]\psi]\gamma}{\Gamma \Rightarrow \Delta, ([\varphi][\psi]\gamma)^m}\ (\Rightarrow [\,][\,]) \quad \dfrac{[\varphi \wedge [\varphi]\psi]\gamma, ([\varphi][\psi]\gamma)^{n-1}, \Pi \Rightarrow \Sigma}{([\varphi][\psi]\gamma)^n, \Pi \Rightarrow \Sigma}\ ([\,][\,] \Rightarrow)}{\Gamma, \Pi \Rightarrow \Delta, \Sigma}\ (Ecut)_{[\varphi][\psi]\gamma}\ .$$

We transform this derivation into:

$$\dfrac{\Gamma \Rightarrow \Delta, ([\varphi][\psi]\gamma)^{m-1}, [\varphi \wedge [\varphi]\psi]\gamma \quad ([\varphi][\psi]\gamma)^n, \Pi \Rightarrow \Sigma}{\Gamma, \Pi \Rightarrow \Delta, \Sigma, [\varphi \wedge [\varphi]\psi]\gamma}\ (Ecut)^1_{[\varphi][\psi]\gamma}$$

$$\dfrac{\Gamma \Rightarrow \Delta, ([\varphi][\psi]\gamma)^m \quad [\varphi \wedge [\varphi]\psi]\gamma, ([\varphi][\psi]\gamma)^{n-1}, \Pi \Rightarrow \Sigma}{[\varphi \wedge [\varphi]\psi]\gamma, \Gamma, \Pi \Rightarrow \Delta, \Sigma}\ (Ecut)^2_{[\varphi][\psi]\gamma}$$

By induction hypothesis $(Ecut)^1$ and $(Ecut)^2$ can be eliminated since their complexities are the same and their weights are reduced. Then, we obtain the following:

$$\dfrac{\dfrac{\dfrac{\vdots\ \mathcal{L}}{\Gamma, \Pi \Rightarrow \Delta, \Sigma, [\varphi \wedge [\varphi]\psi]\gamma} \quad \dfrac{\vdots\ \mathcal{R}}{[\varphi \wedge [\varphi]\psi]\gamma, \Gamma, \Pi \Rightarrow \Delta, \Sigma}}{\Gamma, \Gamma, \Pi, \Pi \Rightarrow \Delta, \Delta, \Sigma. \Sigma}\ (Ecut)^3_{[\varphi \wedge [\varphi]\psi]\gamma}}{\Gamma, \Pi \Rightarrow \Delta, \Sigma}\ (c)\ ,$$

where $(Ecut)^3$ can be eliminated because its complexity is reduced by Lemma 1 (5): $\mathsf{c}([\varphi][\psi]\gamma) > \mathsf{c}([\varphi \wedge [\varphi]\psi]\gamma)$. □

As a corollary of Theorem 1, we can show that G(**K45PAL**) is a conservative extension of G(**K45**) as follows.

Corollary 1. *If a sequent $\Gamma \Rightarrow \Delta$ in L is derivable in* G(**K45PAL**) *then it is also derivable in* G(**K45**).

5 Extended Analytic Cut Property of G(S5PAL)

This section establishes that every application of the cut rule in G(**S5PAL**) can be replaced with an application of the cut rule such that the cut formula is in the *closure* of the conclusion of the cut.

Definition 9. *A cut*

$$\frac{\Gamma \Rightarrow \Delta, \varphi \quad \varphi, \Pi \Rightarrow \Sigma}{\Gamma, \Pi \Rightarrow \Delta, \Sigma}\ (Cut)_{\varphi}$$

is analytic *if $\varphi \in \mathsf{CL}(\Gamma, \Pi, \Delta, \Sigma)$, otherwise it is* non-analytic. *A derivation without non-analytic cuts is said to be an* analytic derivation, *or an* a-derivation *for short. The sequent calculus* G^{a}(**S5PAL**) *is the same calculus as* G(**S5PAL**) *except that (Cut) is always analytic.*

We say that an $(Ecut)_{\rho}$ of Definition 6 is *admissible* in G^{a}(**S5PAL**) if derivability in G^{a}(**S5PAL**) of both upper sequents $\Gamma \Rightarrow \Delta, \rho$ and $\rho, \Pi \Rightarrow \Sigma$ of the $(Ecut)_{\rho}$ implies derivability in G^{a}(**S5PAL**) of the lower sequent $\Gamma, \Pi \Rightarrow \Delta, \Sigma$. In order to establish the following main lemma regarding the extended analytic cut property of G(**S5PAL**), we employ Takano's argument for [28, Lemma 3.2] with necessary modifications to account for our public announcement operators. The proof of the lemma can be found in Appendix A.

Lemma 3. *Suppose that $(Ecut)_{\psi}$ is admissible in* G^{a}(**S5PAL**) *for all $\psi \in \mathsf{CL}(\varphi)$. If the sequent $K_a\Gamma \Rightarrow K_a\Delta$ is derivable in* G^{a}(**S5PAL**), *but with an application of $(\Rightarrow K_{\mathrm{S5}})$ for its lowest inference, then the sequent $K_a\Gamma \Rightarrow K_a(\Delta_{\varphi}), \varphi$ is derivable in* G^{a}(**S5PAL**), *where Δ_{φ} denotes the set which is obtained from Δ by deleting all the occurrences of φ.*

Definition 10. *We define* G^{*}(**S5PAL**) *as the same calculus as* G(**S5PAL**) *except that the cut rule is replaced with $(Ecut)$.*

Theorem 2. *If $\Gamma \Rightarrow \Delta$ is derivable in* G^{*}(**S5PAL**) *then it is also derivable in* G^{a}(**S5PAL**). *Therefore, if $\Gamma \Rightarrow \Delta$ is derivable in* G(**S5PAL**) *then it is also derivable in* G^{a}(**S5PAL**).

Proof. The following argument is also an improvement of Takano's argument in [28], in which we use $(Ecut)$ instead of Mix rule by Gentzen [8] to simplify the outline of the argument. It suffices to prove the first statement. For our purpose, we establish the following: If in the following derivation $\mathcal{D}$ of $\mathsf{G}^*(\mathbf{S5PAL})$:

$$\mathcal{D} \equiv \dfrac{\dfrac{\vdots\ \mathcal{L}}{\Gamma \Rightarrow \Delta, \varphi^m}\ \mathtt{rule}(\mathcal{L}) \quad \dfrac{\vdots\ \mathcal{R}}{\varphi^n, \Pi \Rightarrow \Sigma}\ \mathtt{rule}(\mathcal{R}),}{\Gamma, \Pi \Rightarrow \Delta, \Sigma}\ (Ecut)_\varphi$$

there is no other $(Ecut)$ in derivations of $\Gamma \Rightarrow \Delta, \varphi^m$ and $\varphi^n, \Pi \Rightarrow \Sigma$, then there is a derivation in $\mathsf{G}^a(\mathbf{S5PAL})$ (i.e., all cuts are analytic) of the sequent $\Gamma, \Pi \Rightarrow \Delta, \Sigma$. That is, we can assume without loss of generality that the last $(Ecut)$ is non-analytic.

We prove this statement on a derivation $\mathcal{D}$ by double induction on the complexity and weight (or the lexicographic order of the complexity and the weight) of a derivation $\mathcal{D}$, which are defined as the same as in the proof of Theorem 1. If $m = 0$ or $n = 0$, it follows that our goal can be obtained by weakening rules, hence we suppose $m > 0$ and $n > 0$ below. We divide the argument into the following cases determined by the rules applied last above the $(Ecut)$.

(1) Let $\mathtt{rule}(\mathcal{L})$ or $\mathtt{rule}(\mathcal{R})$ be $(\mathtt{id})$.
(2) Let $\mathtt{rule}(\mathcal{L})$ or $\mathtt{rule}(\mathcal{R})$ be a structural rule.
(3) Let $\mathtt{rule}(\mathcal{L})$ or $\mathtt{rule}(\mathcal{R})$ be an analytic cut.
(4) Let $\mathtt{rule}(\mathcal{L})$ or $\mathtt{rule}(\mathcal{R})$ be a logical rule or a rule for K_a and public announcements where the cut-formula is not principal.
(5) Let $\mathtt{rule}(\mathcal{L})$ and $\mathtt{rule}(\mathcal{R})$ be rules of the same connective and the cut-formula is principal in both rules.

Since cases (1) and (2) are handled similarly as in the proof of Theorem 1. We comment on the new case (3) and case (5) where we need to use Lemma 3.

- (3) We only deal with the case where $\mathtt{rule}(\mathcal{L})$ is an analytic cut (when $\mathtt{rule}(\mathcal{R})$ is an analytic cut, the argument is analogous). Our $\mathcal{D}$ looks as follows:

$$\dfrac{\dfrac{\Gamma' \Rightarrow \Delta', \psi, \rho^a \quad \psi, \Gamma'' \Rightarrow \Delta'', \rho^b}{\Gamma', \Gamma'' \Rightarrow \Delta', \Delta'', \rho^{a+b}}\ (Cut)^{\mathsf{a}}_\psi \quad \rho^n, \Pi \Rightarrow \Sigma}{\Gamma', \Gamma'', \Pi \Rightarrow \Delta', \Delta'', \Sigma}\ (Ecut)_\rho$$

where $\psi \in \mathsf{CL}(\Gamma', \Gamma'', \Delta', \Delta'', \rho^{a+b})$ since (Cut) is analytic. Here we divide the argument into the following two cases: $\psi \equiv \rho$ and $\psi \not\equiv \rho$. Suppose $\psi \equiv \rho$. We transform $\mathcal{D}$ into:

$$\dfrac{\dfrac{\Gamma' \Rightarrow \Delta', \psi, \rho^a \quad \rho^n, \Pi \Rightarrow \Sigma}{\Gamma', \Pi \Rightarrow \Delta', \Sigma}\ (Ecut)_\rho}{\Gamma', \Gamma'', \Pi \Rightarrow \Delta', \Delta'', \Sigma}\ (w)\ .$$

By induction hypothesis, this $(Ecut)$ can be eliminated since its weight is reduced. Suppose $\psi \not\equiv \rho$. We transform $\mathcal{D}$ into:

$$\dfrac{\dfrac{\dfrac{\Gamma' \Rightarrow \Delta', \psi, \rho^a \quad \rho^n, \Pi \Rightarrow \Sigma}{\Gamma', \Pi \Rightarrow \Delta', \psi, \Sigma}\ (Ecut)^1_\rho \quad \dfrac{\psi, \Gamma'' \Rightarrow \Delta'', \rho^b \quad \rho^n, \Pi \Rightarrow \Sigma}{\psi, \Gamma'', \Pi \Rightarrow \Delta'', \Sigma}\ (Ecut)^2_\rho}{\Gamma', \Gamma'', \Pi, \Pi \Rightarrow \Delta', \Delta'', \Sigma, \Sigma}\ (Ecut)^3_\psi}{\Gamma', \Gamma'', \Pi \Rightarrow \Delta', \Delta'', \Sigma}\ (c)\ .$$

In this derivation $(Ecut)^1$ and $(Ecut)^2$ can be eliminated by induction hypothesis since their weights are reduced. For $(Ecut)^3$, we divide the argument into the following two cases: $\psi \in \mathsf{CL}(\rho)$ and $\psi \notin \mathsf{CL}(\rho)$.

- Suppose $\psi \in \mathsf{CL}(\rho)$. Since $\psi \not\equiv \rho$, the complexity of $(Ecut)^3$ is reduced by Lemma 2: $\mathsf{c}(\rho) > \mathsf{c}(\psi)$. Hence it can be eliminated by induction hypothesis.
- Suppose $\psi \notin \mathsf{CL}(\rho)$. Since the original (Cut) is analytic, $\psi \in \mathsf{CL}(\Gamma', \Gamma'', \Delta', \Delta'', \rho^{a+b})$. Then we have $\psi \in \mathsf{CL}(\Gamma', \Gamma'', \Delta', \Delta'')$. Thus $(Ecut)^3$ becomes an analytic cut since $\psi \in \mathsf{CL}(\Gamma', \Gamma'', \Delta', \Delta'') \subseteq \mathsf{CL}(\Gamma', \Gamma'', \Pi, \Pi, \Delta', \Delta'', \Sigma, \Sigma)$.

– (5) Suppose `rule`$(\mathcal{L})$ is $(\Rightarrow K_{\mathrm{S5}})$ and `rule`$(\mathcal{R})$ is $(K \Rightarrow)$ where the cut-formula is the principal formula in both rules. In this case, the derivation $\mathcal{D}$ runs as follows:

$$\dfrac{\dfrac{K_a\Gamma \Rightarrow K_a\Delta, \varphi, (K_a\rho)^m}{K_a\Gamma \Rightarrow K_a\Delta, K_a\varphi, (K_a\rho)^m}(\Rightarrow K_{\mathrm{S5}}) \quad \dfrac{\rho, (K_a\rho)^{n-1}, \Pi \Rightarrow \Sigma}{(K_a\rho)^n, \Pi \Rightarrow \Sigma}(K \Rightarrow)}{K_a\Gamma, \Pi \Rightarrow K_a\Delta, K_a\varphi, \Sigma}(Ecut)_{K_a\rho}.$$

Here we divide the argument into the following two cases: $\varphi \equiv \rho$ and $\varphi \not\equiv \rho$. Since the standard argument can be applied to the case of $\varphi \equiv \rho$, we focus on the case of $\varphi \not\equiv \rho$. Suppose $\varphi \not\equiv \rho$. By Lemma 3, if $K_a\Gamma \Rightarrow K_a\Delta, K_a\varphi, (K_a\rho)^m$ is derivable in G^{a}(**S5PAL**), then $K_a\Gamma \Rightarrow K_a(\Delta_\rho), K_a\varphi, \rho$ is also derivable in G^{a}(**S5PAL**). Hence we transform the derivation into:

$$\dfrac{\dfrac{K_a\Gamma \Rightarrow K_a(\Delta_\rho), K_a\varphi, \rho \quad \dfrac{K_a\Gamma \Rightarrow K_a\Delta, K_a\varphi, (K_a\rho)^m \quad \rho, (K_a\rho)^{n-1}, \Pi \Rightarrow \Sigma}{K_a\Gamma, \Pi, \rho \Rightarrow K_a\Delta, K_a\varphi, \Sigma}(Ecut)^1_{K_a\rho}}{K_a\Gamma, K_a\Gamma, \Pi \Rightarrow K_a(\Delta_\rho), K_a\Delta, K_a\varphi, K_a\varphi, \Sigma}(Ecut)^2_\rho}{K_a\Gamma, \Pi \Rightarrow K_a\Delta, K_a\varphi, \Sigma}(c).$$

By induction hypotheses, $(Ecut)^1$ can be eliminated since its weight is reduced, and $(Ecut)^2$ can be eliminated since its complexity is reduced by Lemma 2 and $\rho \in \mathsf{Sub}(K_a\rho) \subseteq \mathsf{CL}(K_a\rho)$. □

As a corollary of Theorem 2, we can show that G(**S5PAL**) is a conservative extensiion of G(**S5**) as follows.

Corollary 2. *If a sequent from L is derivable in* G(**S5PAL**) *then it is also derivable in* G(**S5**).

Proof. Suppose a sequent from L is derivable in G(**S5PAL**). By Theorem 2, it follows that the sequent is derivable in G^{a}(**S5PAL**) by a derivation $\mathcal{D}$. To regard $\mathcal{D}$ as a derivation in G^{a}(**S5**) it suffices to show that each application of analytic cut in G(**S5PAL**) preserves the subformula property (*not* the extended subformula property in terms of CL) when the conclusion of the cut is in $\mathcal{L}$. Let the cut formula ρ be in the closure of the conclusion $\Gamma, \Pi \Rightarrow \Delta, \Sigma$ of the cut, i.e., $\rho \in \mathsf{CL}(\Gamma, \Pi, \Delta, \Sigma)$. Since the conclusion is from L, we get $\mathsf{CL}(\Gamma, \Pi, \Delta, \Sigma) = \mathsf{Sub}(\Gamma, \Pi, \Delta, \Sigma)$. So, $\rho \in \mathsf{Sub}(\Gamma, \Pi, \Delta, \Sigma)$ holds, as desired. □

6 Conclusion

We propose two non-labelled sequent calculi for the public announcement expansions of the multi-agent **K45** and **S5**. The calculus G(**K45PAL**) is demonstrated to be cut-free, while G(**S5PAL**) is proven to enjoy the extended subformula property in terms of closure. In both proofs, the complexity measure $\mathsf{c}(\varphi)$ from [6] plays a crucial role from a proof-theoretic viewpoint. This aspect was not emphasized in [32] due to the fact that the cut-elimination theorem of the labelled sequent calculus for public announcement expansion in [32] was *reduced* to that of multi-agent **S5** (similarly to semantic completeness, as seen in the proof of Proposition 1). By Theorems 1 and 2, any sequent $\Gamma \Rightarrow \Delta$ that is derivable in G(**K45PAL**) or G(**S5PAL**) has a derivation in which all the sequents are constructed from $\mathsf{CL}(\Gamma, \Delta)$ (it should be noted that $\mathsf{CL}(\Gamma, \Delta)$ is finite and that $\mathsf{CL}(\Gamma, \Delta) = \mathsf{Sub}(\Gamma, \Delta)$ when Γ and Δ are from L). By employing Gentzen's argument (cf. [21]) in terms of *reduced sequents*, we can establish the following:

Corollary 3. G(**K45PAL**) *and* G(**S5PAL**) *are decidable.*

This decidability argument for our non-labelled sequent calculi can be regarded as a merit compared to the existing labelled sequent calculi [2,18,32] for public announcement logic, because the notion of label makes [32]'s argument for the decidability more involved than ours ([2,18] did not discuss the decidability of the calculi).

There are three future directions of further research. First of all, there are other extensions of logics with public announcements that also incorporate recursion axioms, such as intuitionistic **PAL** [3,13,17], bilattice **PAL** [25] and relevant **PAL** [24]. We can endeavor to develop sequent calculi for those logics with the similar approach as in this paper.

Second, in light of the cut-free non-labelled sequent calculus in [31] for action model logic that is based on **S4**, it would be worthwhile to explore the development of a comparable calculus for that based on **S5** (for a cut-free labelled sequent calculus for action model logic, the reader can refer to [16]).

Finally, although cut-elimination theorem does not hold in G(**S5**), there are attempts for obtaining cut-free sequent calculi such as hypersequent calculus by [1], the structured system of [26], and bi-sequent calculus in [11]. Based on these calculi, it would be interesting to see if we can obtain a cut-free sequent calculus for **PAL** and how the complexity of Definition 1 will be used.

A Proof of Lemma 3

Recall that a *K-formula* is a formula of the form $K_a\varphi$ for some formula φ. Let φ be a formula and Ξ and Ω sets of formulas. Then, Ξ_φ (or Ξ_Ω) denotes the resulting set by deleting from Ξ all the occurrences of φ (or formulas in Ω, respectively). In what follows, when $\Gamma \Rightarrow \Delta$ is derivable in $\mathsf{G^a}$(**S5PAL**), we denote $\mathsf{G^a}$(**S5PAL**) $\vdash \Gamma \Rightarrow \Delta$. We can establish the following in the same way as in the proof of [28, Lemma 3.1].

Lemma 4. *Suppose* $\Xi \subseteq \mathsf{CL}(\Gamma, \Delta)$*. If the sequent* $\Phi, \Gamma \Rightarrow \Delta, \Psi$ *is derivable in* $\mathsf{G^a}$(**S5PAL**) *for every partition* $\langle \Phi; \Psi \rangle$ *of* Ξ*, then* $\Gamma \Rightarrow \Delta$ *has a derivation in* $\mathsf{G^a}$(**S5PAL**).

In what follows, we provide a proof of Lemma 3, whose statement is the following.

> Suppose that $(Ecut)_\psi$ is admissible in G^{a}(**S5PAL**) for all $\psi \in \mathsf{CL}(\varphi)$. If the sequent $K_a\Gamma \Rightarrow K_a\Delta$ is derivable in G^{a}(**S5PAL**) with an application of $(\Rightarrow K_{\mathrm{S5}})$ for its lowest inference, then the sequent $K_a\Gamma \Rightarrow K_a(\Delta_\varphi), \varphi$ is derivable in G^{a}(**S5PAL**), where Δ_φ denotes the set which is obtained from Δ by deleting all the occurrences of φ.

Proof. Let $\mathcal{D}$ be such derivation of $K_a\Gamma \Rightarrow K_a\Delta$ in G^{a}(**S5PAL**) with an application of $(\Rightarrow K_{\mathrm{S5}})$ for its lowest inference. We divide our argument into the following two cases: $K_a\varphi \in \mathsf{CL}(K_a\Gamma, K_a(\Delta_\varphi))$ and $K_a\varphi \notin \mathsf{CL}(K_a\Gamma, K_a(\Delta_\varphi))$. When $K_a\varphi \in \mathsf{CL}(K_a\Gamma, K_a(\Delta_\varphi))$ we have the following derivation:

$$\dfrac{\dfrac{\vdots\, \mathcal{D}}{K_a\Gamma \Rightarrow K_a\Delta} \quad \dfrac{\varphi \Rightarrow \varphi}{K_a\varphi \Rightarrow \varphi}\,(K\Rightarrow)}{K_a\Gamma \Rightarrow K_a(\Delta_\varphi), \varphi}\,(Cut)^a_{K_a\varphi},$$

where $(Cut)^a_{K_a\varphi}$ is analytic by assumption. So, in what follows, we always assume

(a) $K_a\varphi \notin \mathsf{CL}(K_a\Gamma, K_a(\Delta_\varphi))$.

Moreover, we replace any application of (Cut) in $\mathcal{D}$

$$\dfrac{\Theta \Rightarrow \Omega, \varphi \quad \varphi, \Pi \Rightarrow \Sigma}{\Theta, \Pi \Rightarrow \Omega, \Sigma}\,(Cut)_\varphi$$

such that $\varphi \in \Theta \cup \Pi \cup \Omega \cup \Sigma$ with applications of weakening rules. By (a) $K_a\varphi \notin \mathsf{CL}(K_a\Gamma, K_a(\Delta_\varphi))$, we can establish the following claim (as did in the proof of [28, Lemma 3.2]).

Claim 1 None of the sequents of $\mathcal{D}$ contain $K_a\varphi$ in its antecedent.

Define Ξ to be a sequence of all formulas except $K_a\varphi$ which occur in the lower sequents of some applications of $(\Rightarrow K_{\mathrm{S5}})$ in $\mathcal{D}$. It follows that Ξ only consist K-formulas. By the extended subformula property, we have $\Xi \subseteq \mathsf{CL}(K_a\Gamma, K_a(\Delta_\varphi), \varphi)$. Our goal is to show that $K_a\Gamma \Rightarrow K_a(\Delta_\varphi), \varphi$ is derivable in G^{a}(**S5PAL**). By Lemma 4, it suffices to show the following.

(2.1) For every partition $\langle K_a\Phi; K_a\Psi\rangle$ of Ξ, $K_a\Phi, K_a\Gamma \Rightarrow K_a(\Delta_\varphi), \varphi, K_a\Psi$ is derivable in G^{a}(**S5PAL**).

Fix any partition $\langle K_a\Phi; K_a\Psi\rangle$ of Ξ. Then, we divide our argument into the following two cases:

(i) $(K_a\Phi \cap K_a(\Delta_\varphi) \neq \varnothing$ or $K_a\Gamma \cap K_a\Psi \neq \varnothing)$ or
(ii) $(K_a\Phi \cap K_a(\Delta_\varphi) = \varnothing$ and $K_a\Gamma \cap K_a\Psi = \varnothing)$.

Suppose that (i) $K_a\Phi \cap K_a(\Delta_\varphi) \not\equiv \varnothing$ or $K_a\Gamma \cap K_a\Psi \not\equiv \varnothing$. Then, either $K_a\Phi \Rightarrow K_a(\Delta_\varphi)$ or $K_a\Gamma \Rightarrow K_a\Psi$ is derivable in G^{a}(**S5PAL**) by weakening rules. Hence $K_a\Phi, K_a\Gamma \Rightarrow K_a(\Delta_\varphi), \varphi, K_a\Psi$ is derivable in G^{a}(**S5PAL**) by weakening rules.

Therefore, we always suppose that (ii) $K_a\Phi \cap K_a(\Delta_\varphi) = \varnothing$ and $K_a\Gamma \cap K_a\Psi = \varnothing$ in what follows. By definition of Ξ, we have $K_a\Gamma \subseteq \Xi$ and $K_a(\Delta_\varphi) \subseteq \Xi$. Since $\langle K_a\Phi; K_a\Psi\rangle$ is a partition of Ξ, it follows from (ii) that (c) $K_a\Gamma \subseteq K_a\Phi$ and $K_a(\Delta_\varphi) \subseteq K_a\Psi$.

Let us consider the uppermost application $\mathtt{rule}(\mathcal{D}')$ of $(\Rightarrow K_{\mathrm{S5}})$ in $\mathcal{D}$:

$$\dfrac{\begin{array}{c}\mathcal{D}'\\ K_a\Pi \Rightarrow K_a\Sigma, \psi\end{array}}{K_a\Pi \Rightarrow K_a\Sigma, K_a\psi}\ (\Rightarrow K_{\mathrm{S5}}) = \mathtt{rule}(\mathcal{D}')$$

such that (d) $K_a\Pi \subseteq K_a\Phi$ and $K_a\Sigma \cup \{K_a\psi\} \subseteq K_a\Psi \cup \{K_a\varphi\}$. Such an application exists certainly, because, in the lowermost application $(\Rightarrow K_{\mathrm{S5}})$ in the derivation $\mathcal{D}$ (i.e., $\mathtt{rule}(\mathcal{D})$), by (c) $K_a\Gamma \subseteq K_a\Phi$ and $K_a(\Delta_\varphi) \subseteq K_a\Psi$ hence $K_a\Delta \subseteq K_a\Psi \cup \{K_a\varphi\}$. Moreover we have the following claim:

Claim 2 $\psi \not\equiv K_a\varphi$.

Proof. Now our goal is to show the following item:

(2.1′) $K_a\Phi, K_a\Gamma \Rightarrow K_a(\Delta_\varphi), \varphi, K_a\Psi$ is derivable in G^{a}(**S5PAL**).

It suffices to show the following.

(2.2) For every sequent $\Theta \Rightarrow \Omega$ which lies above the lower sequent of $\mathtt{rule}(\mathcal{D}')$, (i.e., in $\mathcal{D}'$), the sequent $K_a\Phi, \Theta \Rightarrow \Omega_{K_a\varphi}, K_a\Psi$ is derivable in G^{a}(**S5PAL**).

Suppose (2.2) holds. By applying (2.2) to the upper sequent $K_a\Pi \Rightarrow K_a\Sigma, \psi$ of $\mathtt{rule}(\mathcal{D}')$, it follows that $K_a\Phi, K_a\Pi \Rightarrow K_a(\Sigma_\varphi), \psi, K_a\Psi$ is derivable in G^{a}(**S5PAL**). To show (2.1′), we divide the argument into the following two cases: $\psi \equiv \varphi$ and $\psi \not\equiv \varphi$.

First, let $\psi \equiv \varphi$. Then, by (d) $K_a\Pi \subseteq K_a\Phi$ and $K_a\Sigma \cup \{K_a\psi\} \subseteq K_a\Psi \cup \{K_a\varphi\}$, we have $K_a(\Sigma_\varphi) \subseteq K_a\Psi$. Hence (2.1′) holds by structural rules:

$$\dfrac{K_a\Phi, K_a\Pi \Rightarrow K_a(\Sigma_\varphi), \varphi, K_a\Psi}{K_a\Phi, K_a\Gamma \Rightarrow K_a(\Delta_\varphi), \varphi, K_a\Psi}\ .$$

Second, let $\psi \not\equiv \varphi$. Then by (d) $K_a\Pi \subseteq K_a\Phi$ and $K_a\Sigma \cup \{K_a\psi\} \subseteq K_a\Psi \cup \{K_a\varphi\}$, we have $K_a(\Sigma_\varphi) \cup \{K_a\psi\} \subseteq K_a\Psi$. Hence (2.1′) holds by structural rules:

$$\dfrac{\dfrac{K_a\Phi, K_a\Pi \Rightarrow K_a(\Sigma_\varphi), \psi, K_a\Psi}{K_a\Phi, K_a\Pi \Rightarrow K_a(\Sigma_\varphi), K_a\psi, K_a\Psi}\ (\Rightarrow K_{\mathrm{S5}})}{K_a\Phi, K_a\Gamma \Rightarrow K_a(\Delta_\varphi), \varphi, K_a\Psi}\ .$$

Therefore we are going to establish (2.2). Fix any sequent $\Theta \Rightarrow \Omega$ in $\mathcal{D}'$. Let n be the height of derivation $\mathcal{D}'$. To show (2.2), it suffices to prove the following statement by induction on $k \leqslant n$:

$$\forall \Theta \Rightarrow \Omega \text{ in } \mathcal{D}' \text{ (if the height of } \Theta \Rightarrow \Omega \text{ is } k, \text{ then } K_a\Phi, \Theta \Rightarrow \Omega_{K_a\varphi}, K_a\Psi \text{ is derivable in } \mathsf{G}^{\mathsf{a}}(\mathbf{S5PAL})).$$

□

References

1. Avron, A.: The method of hypersequents in the proof theory of propositional non-classical logics. In: Logic: From Foundations to Applications: European Logic Colloquium, pp. 1–32. Clarendon Press, USA (1996)
2. Balbiani, P., Demange, V., Galmiche, D.: A sequent calculus with labels for public announcement logic. In: Conference on Advances in Modal Logic (AiML 2014), vol. 10 (2014)
3. Balbiani, P., Galmiche, D.: About intuitionistic public announcement logic. In: 11th Conference on Advances in Modal Logic (AiML 2016), pp. 97–116. College Publications (2016)
4. van Benthem, J., Liu, F.: Dynamic logic of preference upgrade. J. Appl. Non-Classical Logics **17**(2), 157–182 (2007)
5. Braüner, T.: A cut-free Gentzen formulation of the modal logic S5. Logic J. IGPL **8**(5), 629–643 (2000)
6. van Ditmarsch, H., van der Hoek, W., Kooi, B.P.: Dynamic Epistemic Logic, Synthese Library, vol. 337. Springer, Dordrecht (2007). https://doi.org/10.1007/978-1-4020-5839-4
7. Došen, K.: Sequent-systems and groupoid models. I. Studia Logica: Int. J. Symbolic Logic **47**(4), 353–385 (1988)
8. Gentzen, G.: Investigations into logical deduction. Am. Philos. Q. **1**(4), 288–306 (1964)
9. Gentzen, G.: Investigations into logical deduction: II. Am. Philos. Q. **2**(3), 204–218 (1965)
10. Hatano, R., Sano, K.: Three faces of recursion axioms: the case of constructive dynamic logic of relation changers. J. Log. Comput. (2022). https://doi.org/10.1093/logcom/exac013
11. Indrzejczak, A.: Sequents and Trees: An Introduction to the Theory and Applications of Propositional Sequent Calculi. Studies in Universal Logic, Springer, Switzerland (2021)
12. Kashima, R.: Mathematical Logic. Modern Foundamental mathematics, sakura Publishing Co., Ltd (2009). (In Japanese)
13. Ma, M., Palmigiano, A., Sadrzadeh, M.: Algebraic semantics and model completeness for intuitionistic public announcement logic. Ann. Pure Appl. Logic **165**(4), 963–995 (2014)
14. Maffezioli, P., Negri, S.: A proof-theoretical perspective on public announcement logic. Logic Philos. Sci. **IX**, 49–59 (2011)
15. Negri, S.: Proof analysis in modal logic. J. Philos. Log. **34**(5), 507–544 (2005)
16. Nomura, S., Ono, H., Sano, K.: A cut-free labelled sequent calculus for dynamic epistemic logic. J. Log. Comput. **30**(1), 321–348 (2020)
17. Nomura, S., Sano, K., Tojo, S.: A labelled sequent calculus for intuitionistic public announcement logic. In: Davis, M., Fehnker, A., McIver, A., Voronkov, A. (eds.) LPAR 2015. LNCS, pp. 187–202. Springer, Heidelberg (2015)
18. Nomura, S., Sano, K., Tojo, S.: Revising a labelled sequent calculus for public announcement logic. In: Yang, S.C.-M., Deng, D.-M., Lin, H. (eds.) Structural Analysis of Non-Classical Logics. LASLL, pp. 131–157. Springer, Heidelberg (2016). https://doi.org/10.1007/978-3-662-48357-2_7
19. Ohnishi, M., Matsumoto, K.: Gentzen method in modal calculi. Osaka Math. J. **9**(2), 113–130 (1957)
20. Ohnishi, M., Matsumoto, K.: Gentzen method in modal calculi. II. Osaka Math. J **11**(2), 115–120 (1959)
21. Ono, H.: Proof-theoretic methods in nonclassical logic -an introduction. In: Theories of Types and Proofs, vol. 2, pp. 207–255. Mathematical Society of Japan (1998)
22. Plaza, J.: Logics of public communications. Synthese **158**(2), 165–179 (2007)
23. Poggiolesi, F.: A cut-free simple sequent calculus for modal logic S5. Rev. Symbolic Logic **1**(1), 3–15 (2008)
24. PunČochář, V., Sedlár, I., Tedder, A.: Relevant epistemic logic with public announcements and common knowledge. J. Log. Comput. **33**(2), 436–461 (2023)

25. Rivieccio, U.: Bilattice public announcement logic. Adv. Modal Logic **10**, 459–477 (2014)
26. Sato, M.: A cut-free Gentzen-type system for the modal logic S5. J. Symbolic Logic **45**(1), 67–84 (1980)
27. Shvarts, G.F.: Gentzen style systems for K45 and K45D. In: Meyer, A.R., Taitslin, M.A. (eds.) Logic at Botik 1989. LNCS, vol. 363, pp. 245–256. Springer, Heidelberg (1989). https://doi.org/10.1007/3-540-51237-3_20
28. Takano, M.: Subformula property as a substitute for cut-elimination in modal propositional logics. Math. Japon. **37**(6), 1129–1145 (1992)
29. Takano, M.: A semantical analysis of cut-free calculi for modal logics. Rep. Math. Logic **53**, 43–65 (2018)
30. Troelstra, A.S., Schwichtenberg, H.: Basic Proof Theory, 2nd edn. Cambridge University Press, Cambridge (2000)
31. Wirsing, M., Knapp, A.: A reduction-based cut-free Gentzen calculus for dynamic epistemic logic. Logic J. IGPL (2022). https://doi.org/10.1093/jigpal/jzac078
32. Wu, H., van Ditmarsch, H., Chen, J.: A labelled sequent calculus for public announcement logic (2022). https://doi.org/10.48550/arXiv.2207.10262

On the Finite Model Property of Non-normal Modal Logics

Yu Peng(✉) and Yiheng Wang(✉)

Department of Philosophy, Institute of Logic and Cognition, Sun Yat-sen University, Guangzhou 510275, China
amberlogos@163.com, ianwang747@gmail.com

Abstract. In this paper, we consider the non-normal modal logics over monotonic modal logic M with extensions of any combination of (N), (P), (T), and (4) i.e. $\{\mathsf{M}, \mathsf{MN}, \ldots, \mathsf{MNPT4}\}$. We study the algebras corresponding to these logics and give some examples of them. We further introduce the Gentzen-style sequent calculi with soundness and completeness proved. Finally, we prove the FMP of these logics and thus decidability based on our systems by algebraic proof-theoretic methods.

Keywords: Non-normal modal logics · FMP · Decidability

1 Introduction

A modal logic is called normal if it contains the following axioms and rule (M): if $\alpha \to \beta$, then $\Diamond\alpha \to \Diamond\beta$, (C): $\Diamond(\alpha \vee \beta) \to \Diamond\alpha \vee \Diamond\beta$, and (N): $\Diamond\bot \to \bot$. The classical propositional logic (denoted by CPL) extended with (M), (C), and (N) is denoted by system K, of which the corresponding axiom is (K): $\Diamond(\alpha \vee \beta) \leftrightarrow \Diamond\alpha \vee \Diamond\beta$. Clearly, non-normal modal logics are modal logics that do not satisfy these conditions. For instance, Lewis's [20] systems $\mathsf{S1}$, $\mathsf{S2}$, $\mathsf{S3}$ and Lemmon's [19] systems $\mathsf{E2}$ and $\mathsf{E3}$ are all non-normal modal logics. Non-normal modal logics play a significant role in modern modal logic since different interpretations of normal modalities may lead to anti-intuitive and unacceptable conclusions in various fields of logic and philosophy. In epistemology and epistemic logic, the legitimacy of an equivalent form of axiom (K): $\Box(\alpha \to \beta) \to (\Box\alpha \to \Box\beta)$ (known as epistemic closure in epistemology) is controversial since many scholars claimed that knowledge is not closed under known implications, which may cause logical omniscience problem (cf. [11]). In ethics and deontic logic, Forrester's paradox shows that the acceptance of axiom (K) may result in contradictory consequences like moral dilemmas (cf. [24]). Similar motivations can be found in the logic of agency and ability, philosophy of action, majority logic, logic of high probability, and logic of group decision making as well (cf. [6,28]). Besides, when it comes to more exact disciplines such as mathematics, non-normal modal logics can be used to represent some classes of arithmetic formulas (cf. [1]).

Non-normal modal logics have been studied since the beginning of modern modal logic (cf. [3,16,19,20,25,29,30]). They are defined as CPL extended with (M), (C), (N) and (E). (E) means the congruence rule: if $\alpha \leftrightarrow \beta$, then $\Diamond\alpha \leftrightarrow \Diamond\beta$. A logic that satisfies (E) is called congruential. Taking system E (CPL extended

N. Alechina et al. (Eds.): LORI 2023, LNCS 14329, pp. 207–221, 2023.
https://doi.org/10.1007/978-3-031-45558-2_16

with (E)) as a minimal logic, then one gets the so-called "classical cube" (cf. [3]). Various investigations have been made for these structures. Vardi [32] proved that E and any its extension with (M), (C), (N), (P), (T), (4) are decidable. Lavendhomme and Lucas [17] introduced the sequent calculi for monotonic modal logic M and congruential modal logic E which admit cut elimination. They provided decision procedures for M, E, EC, MC and MN based on inversion of rules. Orlandelli [27] provided the first sequent calculi for non-normal logics EP and MP. Indrzejczak [12] further considered some extensions of M with (D), (T), (4), (B), (5), and gave decision procedures for MT, MD, M4, MT4, M5, MD5, M45 and MD45. Indrzejczak [12] also studied the MN-counterparts of M and showed that sequent calculi for MND, MNT, MN4, MN5, MNT4, MND4, MN45 and MND45 admit cut-elimination. Similar results were obtained on E as well (cf. [15]). Tableau, natural deduction, labeled sequent calculi, nested sequent calculi, hypersequent calculi and display calculi for some of these logics were studied as well (cf. [4,5,7–9,13,14,18,26]). From the algebraic perspective, Hansen [10] described the algebraization of monotonic modal logics. For the FMP and decidability result, Chellas [3] proved that the logics in classical cube extended with (T), (P), (D) enjoy FMP by the filtration technique. However, to the best of the authors' knowledge, there is no proof of FMP for these logics extended with (4) (cf. [6]). Shkatov and Alten [31] showed the universal theory of monotonic modal algebra is co-NP-complete by the theory of partial algebra.

We continue this line of research. We consider the non-normal modal logic M and its extensions with axioms from (N): $\Diamond\bot \to \bot$, (P): $\top \to \Diamond\top$, (T): $\alpha \to \Diamond\alpha$, and (4): $\Diamond\Diamond\alpha \to \Diamond\alpha$. These extensions are listed in the following Fig. 1, where the direction of an arrow represents the extension with another axiom:

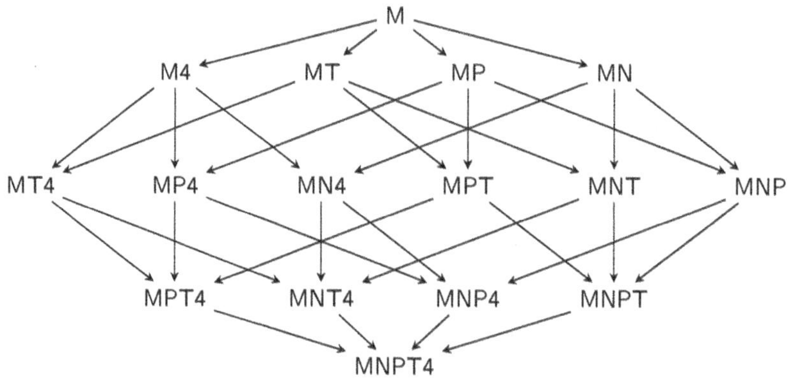

Fig. 1. The class of non-normal modal logic based on M

Clearly, some of the logics in the above figure are not independent. For example, the MNPT4 is equivalent to MNT4 since (P) is simply a special case of (T). Although some of these logics have been studied, there still lacks systematic research about them from the algebraic proof-theoretic approach. Since axioms

(T) and (4) are quite essential in representing many objects' properties, like knowledge and time, it is pretty natural to take these axioms into consideration.

In this paper, we will introduce a serial monotonic modal algebras $A_M = \{\text{M}, \ldots, \text{MNPT4}\}$ for all logics in $L_M = \{\mathsf{M}, \ldots, \mathsf{MNPT4}\}$. The examples of algebras M, MN, MP, M4, MNP, MN4, MPT, MP4, MNPT, MNP4, and MNPT4 are given. We prove the FMP of these logics and thus decidability. The FMP results of these logics are beneficial supplements to current research, especially given the fact that the lacking of FMP results of logics in the classical cube extended with (4). These results show that a further study on the context-freeness of these logics and thus a decidable algorithm is possible. Our method is inspired by [21–23].

The present paper is organized as follows. In Sect. 2, we introduce a class of algebras corresponding to $L_M = \{\mathsf{M}, \ldots, \mathsf{MNPT4}\}$. In Sect. 3, we show their Gentzen-style sequent calculi with soundness and completeness results. In Sect. 4, we prove the finite model property of these logics.

2 Algebra

In this section, we study a class of algebras $A_M = \{\text{M}, \ldots, \text{MNPT4}\}$ that correspond to the logics in previous Fig. 1. We first introduce the monotonic modal algebra M as the basis for the entire discussion. Further, we consider various extensions of M with any combinations of (N), (P), (T), (4), and give some examples of them.

Definition 1 ([10]). *A monotonic modal algebra* M *is a structure* $(A, \wedge, \vee, \neg, \Diamond, 0, 1)$, *where* $(A, \wedge, \vee, \neg, 0, 1)$ *is a Boolean algebra and* $\Diamond$ *is a unary operation on* A *satisfying the following condition: for all* $a, b \in A$,

$$\text{(M) If } a \leq b, \text{ then } \Diamond a \leq \Diamond b.$$

Definition 2. *Algebra* MX *is any extension of* M *satisfying* X *which is (possibly empty) any combinations of the following list of conditions: for all* $a \in A$,

$$\text{(N) } \Diamond 0 = 0; \;\; \text{(P) } \Diamond 1 = 1; \;\; \text{(T) } a \leq \Diamond a; \;\; \text{(4) } \Diamond\Diamond a \leq \Diamond a.$$

Clearly, MX is M when X is empty. Then we use MX to denote any algebras from the class of algebra M with its various possible extensions namely MX $\in A_M$. In what follows, we show some algebraic properties of the algebra structures we have defined so far, including the basic algebra structure M and various extensions of it. We denote $\Diamond^n(\alpha) = \Diamond(\Diamond^{n-1}(\alpha)), \Diamond^0(\alpha) = \alpha$.

Lemma 1. *For any* MX, *the following properties hold: for any* $a, b, c \in A$ *and* $n, m \geq 0$,

(1) $a \wedge b \leq 0$ *iff* $b \leq \neg a$;
(2) If $\Diamond^n a \leq b$, *then* $\Diamond^n (a \wedge c) \leq b$;
(3) If $\Diamond^n a \leq b$ *and* $\Diamond^n a \leq c$, *then* $\Diamond^n a \leq b \wedge c$;

(4) If $\Diamond^{n+1}a \leq b$*, then* $\Diamond^{n}\Diamond a \leq b$*;*
(5) If $\Diamond^{n}a \leq b$*, then* $\Diamond^{n+1}a \leq \Diamond b$*;*
(6) If $\Diamond^{n}a \leq b$ *and* $\Diamond^{m}b \leq c$*, then* $\Diamond^{n+m}a \leq c$*.*

Lemma 2. *The following holds for corresponding* MX*: for any* $a, b, c \in A$ *and* $n \geq 0$*,*

(1) $\Diamond^{n}0 \leq 0$ *holds for any* MX *where* (N) *is in* X*;*
(2) If $\Diamond^{n+1}1 \leq a$*, then* $\Diamond^{n}1 \leq a$ *holds for any* MX *where* (P) *is in* X*;*
(3) If $\Diamond^{n+1}a \leq b$*, then* $\Diamond^{n}a \leq b$ *holds for any* MX *where* (T) *is in* X*;*
(4) If $\Diamond^{n+1}a \leq b$*, then* $\Diamond^{n+2}a \leq b$ *holds for any* MX *where* (4) *is in* X*.*

Example 1. The lattices in Fig. 2, 3, 4, 5, 6, 7, 8, 9, 10, 11 and 12 are examples of M, MN, MP, M4, MNP, MN4, MPT, MP4, MNPT, MNP4, and MNPT4 respectively. Note that $\Diamond(a \vee b) \neq \Diamond a \vee \Diamond b$ for any $a, b \in A$ for all these examples. For instance, $\Diamond(a \vee c) = d \neq \Diamond a \vee \Diamond c = f$ in Fig. 2 and $\Diamond(a \vee c) = 1 \neq \Diamond a \vee \Diamond c = e$ in Fig. 12.

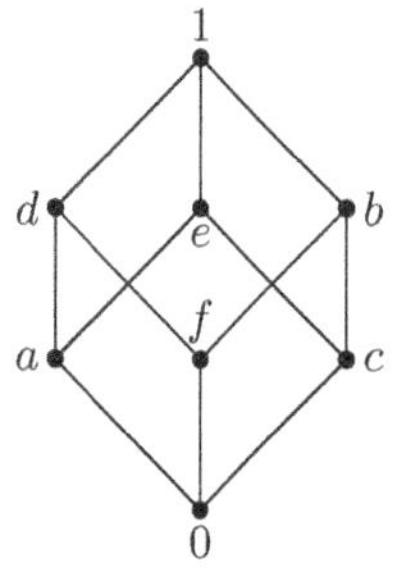

	1	0	a	b	c	d	e	f
$\Diamond$	d	f	f	d	f	d	d	d

Fig. 2. An example of M

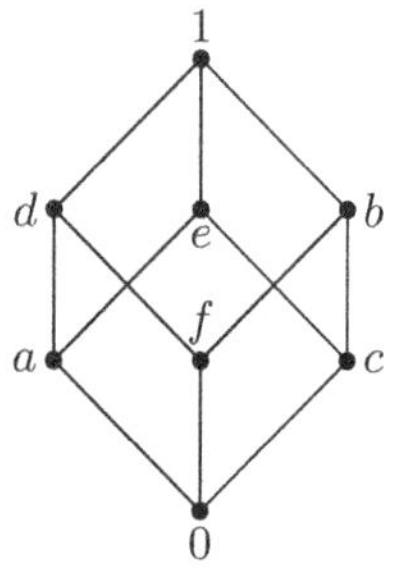

	1	0	a	b	c	d	e	f
$\Diamond$	d	0	d	a	0	d	d	0

Fig. 3. An example of MN

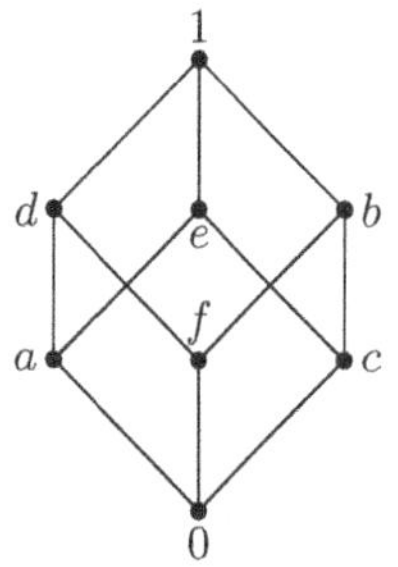

	1	0	a	b	c	d	e	f
$\Diamond$	1	c	c	1	1	1	1	c

Fig. 4. An example of MP

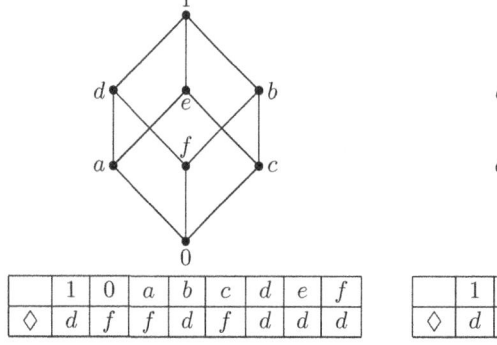

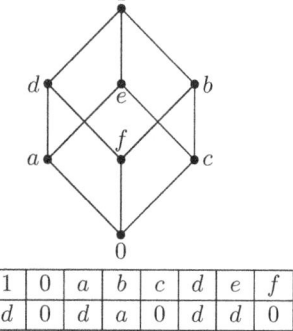

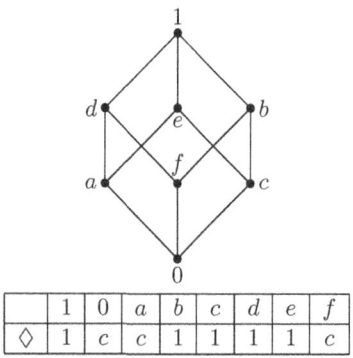

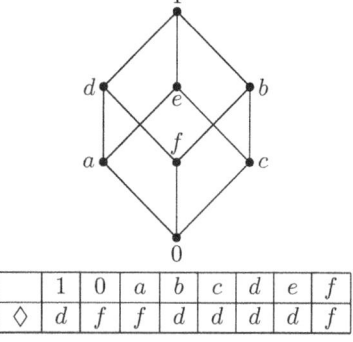

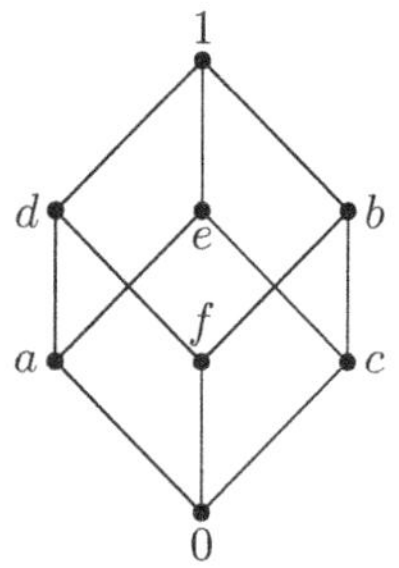

	1	0	a	b	c	d	e	f
$\Diamond$	d	f	f	d	d	d	d	f

Fig. 5. An example of M4

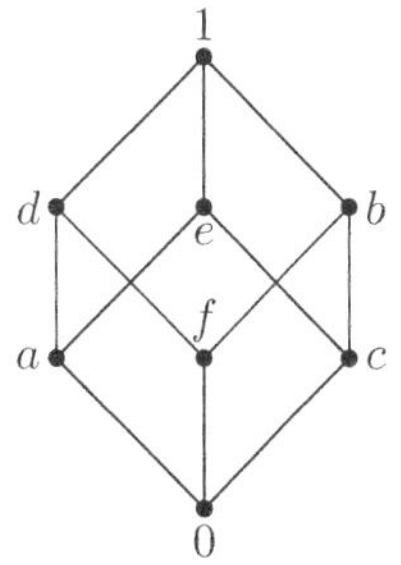

	1	0	a	b	c	d	e	f
$\Diamond$	1	0	0	1	e	c	1	0

Fig. 6. An example of MNP

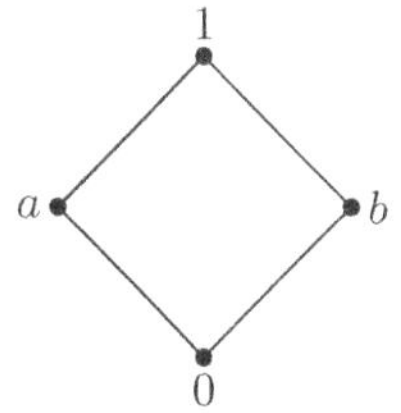

	1	0	a	b
$\Diamond$	a	0	0	0

Fig. 7. An example of MN4

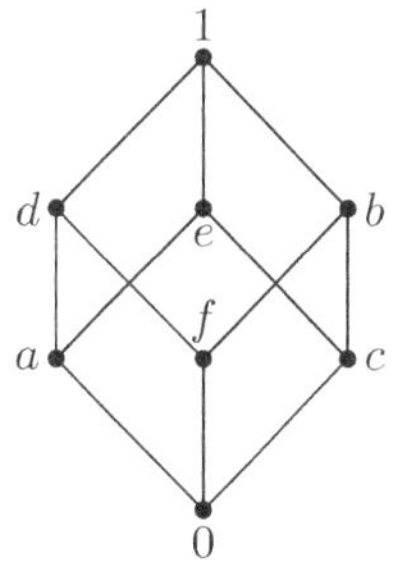

	1	0	a	b	c	d	e	f
$\Diamond$	1	d	d	1	1	1	1	d

Fig. 8. An example of MPT

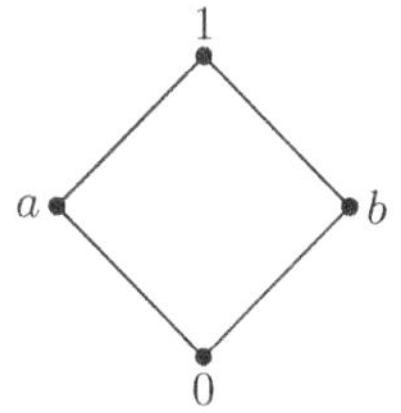

	1	0	a	b
$\Diamond$	1	a	a	a

Fig. 9. An example of MP4

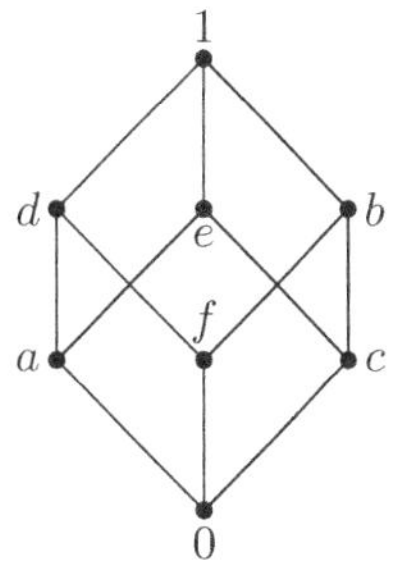

	1	0	a	b	c	d	e	f
$\Diamond$	1	0	d	1	c	1	1	f

Fig. 10. An example of MNPT

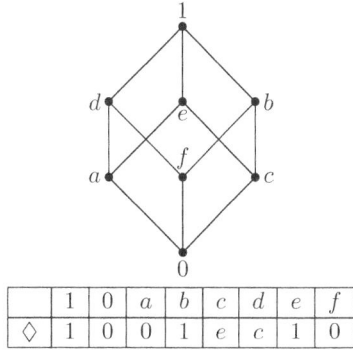

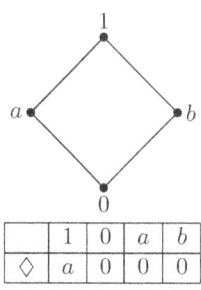

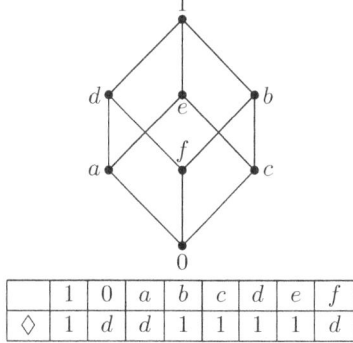

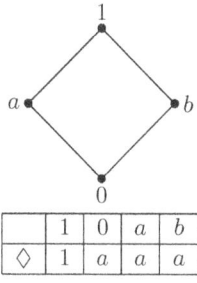

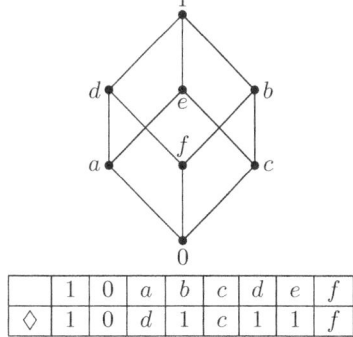

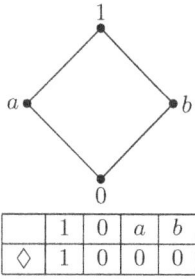

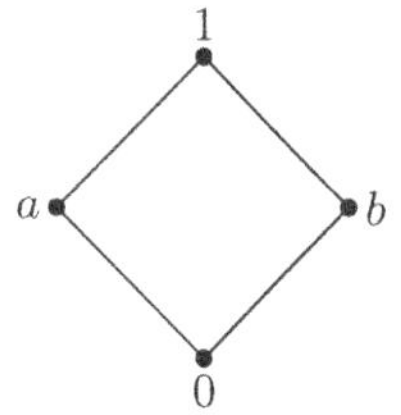

	1	0	a	b
$\Diamond$	1	0	0	0

Fig. 11. An example of MNP4

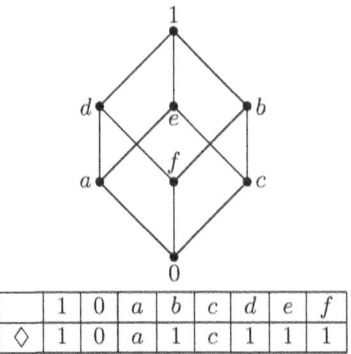

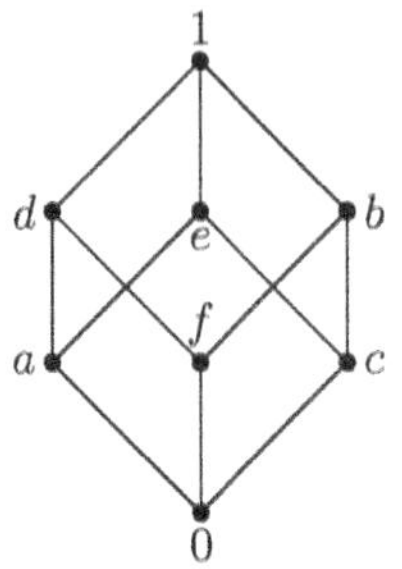

	1	0	a	b	c	d	e	f
$\Diamond$	1	0	a	1	c	1	1	1

Fig. 12. An example of MNPT4

Hereafter, we use **MX** and MX to denote the class of all MXs and the logic of **MX** respectively.

3 Sequent Calculus

In this section, we introduce the Gentzen-style sequent calculi for any $\mathsf{MX} \in L_M$, denoted by $\mathsf{GMX} \in G_M = \{\mathsf{GM}, \ldots, \mathsf{GMNPT4}\}$. We start with the basic system GM and present its various extensions by adding corresponding axioms or rules. Their soundness and completeness results will be proved as well.

Definition 3. *The set of formulas (terms) $\mathcal{F}$ is defined inductively as follows:*

$$\mathcal{F} \ni \alpha ::= p \mid \top \mid \bot \mid \alpha \wedge \beta \mid \neg\alpha \mid \Diamond\alpha$$

*where $p \in$ **Var**, a denumerable set of propositional variables. A formula α is called* atomic *if $\alpha \in \mathbf{Var} \cup \{\top, \bot\}$. The* complexity *of a formula α, denoted by $c(\alpha)$, is defined inductively as follows:*

$$c(\alpha) = 0 \text{ if } \alpha \text{ is atomic;}$$
$$c(\dagger\alpha) = c(\alpha) + 1 \text{ if } \dagger \text{ is a unary operator;}$$
$$c(\alpha \star \beta) = max\{c(\alpha), c(\beta)\} + 1 \text{ if } \star \text{ is a binary operator.}$$

Further, we use the following abbreviations:

$$\alpha \vee \beta := \neg(\neg\alpha \wedge \neg\beta) \quad \Box\alpha := \neg\Diamond\neg\alpha$$

Definition 4. *The set of all formula structures $\mathcal{FS}$ is defined inductively as follows:*

$$\mathcal{FS} \ni \circ^n(\alpha) ::= \alpha \mid \circ^n(\alpha)$$

where $\alpha \in \mathcal{F}$. Here we denote $\circ^n(\alpha) = \circ(\circ^{n-1}(\alpha)), \circ^0(\alpha) = \alpha$. A sequent *is an expression of the form $\circ^n(\alpha) \Rightarrow \beta$ where $\circ^n(\alpha)$ is a formula structure and β is a formula. Let $\alpha \Leftrightarrow \beta$ denote $\alpha \Rightarrow \beta$ and $\beta \Rightarrow \alpha$. A* context *is a formula structure*

$\circ^n(-)$ *with a designated* position $(-)$ *which can be filled with a formula structure. In particular, a single position* $(-)$ *is a context. Let* $\circ^n(\alpha)$ *be a formula structure obtained from* $\circ^n(-)$ *by substituting formula* α *for* $(-)$. *For any formula structure* $\circ^n(\alpha)$, *the formula* $f(\circ^n(\alpha))$ *means a formula obtained from the formula structure* $\circ^n(\alpha)$ *by replacing all structure operators* $\circ$ *with their corresponding formula operators.* $f(\circ^n(\alpha))$ *is defined inductively as follows:*

$$f(\alpha) = \alpha \quad f(\circ(\alpha)) = \Diamond(f(\alpha))$$

Clearly, $\circ$ *is the structure operator for* $\Diamond$ *and* $f(\circ^n(\alpha)) = \Diamond^n(f(\alpha))$. *Every nonempty formula structure* $\circ^n(\alpha)$ *has a parsing tree with formulas in* $\circ^n(\alpha)$ *as leaf nodes and structural operators in* $\circ^n(\alpha)$ *as non-leaf nodes.*

Example 2. The expression $\circ^{10}(-)$ is a context. If we replace the formula $\alpha = p\wedge q$ for the position $-$ in $\circ^n(-)$, we get the formula structure $\circ^{10}(\alpha) = \circ^{10}(p\wedge q)$.

Definition 5. *The Gentzen-style sequent calculus* GM *for non-normal modal logic* M *consists of the following axioms and rules: for* $n \geq 0$,

(1) Axioms:

$$\text{(Id)}\ \alpha \Rightarrow \alpha \quad \text{(Dis)}\ \alpha \wedge (\beta \vee \gamma) \Rightarrow (\alpha \wedge \beta) \vee (\alpha \wedge \gamma)$$

$$(\top)\ \circ^n(\alpha) \Rightarrow \top \quad (\bot)\ \bot \Rightarrow \alpha \quad \text{(DN)}\ \neg\neg\alpha \Rightarrow \alpha$$

(2) Logical rules:

$$\frac{\alpha \wedge \beta \Rightarrow \bot}{\beta \Rightarrow \neg\alpha}(\neg\text{I}) \quad \frac{\beta \Rightarrow \neg\alpha}{\alpha \wedge \beta \Rightarrow \bot}(\neg\text{E}) \quad \frac{\circ^n(\alpha_i) \Rightarrow \beta}{\circ^n(\alpha_1 \wedge \alpha_2) \Rightarrow \beta}(\wedge\text{L})(i = \{1, 2\})$$

$$\frac{\circ^n(\alpha) \Rightarrow \beta \quad \circ^n(\alpha) \Rightarrow \gamma}{\circ^n(\alpha) \Rightarrow \beta \wedge \gamma}(\wedge\text{R}) \quad \frac{\circ^{n+1}(\alpha) \Rightarrow \beta}{\circ^n(\Diamond\alpha) \Rightarrow \beta}(\Diamond\text{L}) \quad \frac{\circ^n(\alpha) \Rightarrow \beta}{\circ^{n+1}(\alpha) \Rightarrow \Diamond\beta}(\Diamond\text{R})$$

(3) Cut rule:

$$\frac{\circ^n(\alpha) \Rightarrow \beta \quad \circ^m(\beta) \Rightarrow \gamma}{\circ^{m+n}(\alpha) \Rightarrow \gamma}(\text{Cut})$$

The sequent calculus GMX for MX is obtained from GM by adding (possibly empty) X which is any combination of the following axiom or rules:

$$\text{(N)}\ \circ^{n+1}(\bot) \Rightarrow \bot \quad \frac{\circ^{n+1}(\top) \Rightarrow \beta}{\circ^n(\top) \Rightarrow \beta}(\text{P}) \quad \frac{\circ^{n+1}(\alpha) \Rightarrow \beta}{\circ^n(\alpha) \Rightarrow \beta}(\text{T}) \quad \frac{\circ^{n+1}(\alpha) \Rightarrow \beta}{\circ^{n+2}(\alpha) \Rightarrow \beta}(4)$$

A *derivation* of a sequent $\circ^n(\alpha) \Rightarrow \beta$ in GMX is a finite tree of sequents in which each node is either an instance of an axiom schema or derived from child node(s) by an inference rule and the root node is $\circ^n(\alpha) \Rightarrow \beta$. A sequent $\circ^n(\alpha) \Rightarrow \beta$ is *provable* in GMX, denoted by $\vdash_{\mathsf{GMX}} \circ^n(\alpha) \Rightarrow \beta$, if there is a derivation of $\circ^n(\alpha) \Rightarrow \beta$ in GMX. We write $\vdash_{\mathsf{GMX}} \alpha \Leftrightarrow \beta$ if $\vdash_{\mathsf{GMX}} \alpha \Rightarrow \beta$ and $\vdash_{\mathsf{GMX}} \beta \Rightarrow \alpha$. For the derivation tree $\mathcal{D}$, the *height* of the derivation tree (also called the *length* of proof), denoted by $|\mathcal{D}|$, is the maximal length of branches in $\mathcal{D}$. The height of a single node derivation is zero. The subscript GMX in $\vdash_{\mathsf{GMX}}$ is omitted if no confusion arises.

Lemma 3. *The following properties hold in any* GMX*:*

(1) $\vdash \alpha \vee \neg\alpha \Leftrightarrow \top$;
(2) $\vdash \alpha \wedge \neg\alpha \Leftrightarrow \bot$;
(3) $\vdash \alpha \Rightarrow \neg\neg\alpha$;
(4) If $\vdash \alpha \Rightarrow \beta$, *then* $\vdash \neg\beta \Rightarrow \neg\alpha$;
(5) $\vdash (\alpha \wedge \beta) \vee (\alpha \wedge \gamma) \Rightarrow \alpha \wedge (\beta \vee \gamma)$;
(6) If $\vdash \alpha \Rightarrow \beta$, *then* $\vdash \Diamond\alpha \Rightarrow \Diamond\beta$;

Proof. We only provide the proofs for (4), (5), (6). The derivations of (4), (6) are as follows:

$$\dfrac{\dfrac{\dfrac{\alpha \Rightarrow \beta \qquad \beta \Rightarrow \neg\neg\beta}{\alpha \Rightarrow \neg\neg\beta}\,(\mathrm{Cut})}{\alpha \wedge \neg\beta \Rightarrow \bot}\,(\neg\mathrm{E})}{\neg\beta \Rightarrow \neg\alpha}\,(\neg\mathrm{I}) \qquad\qquad \dfrac{\dfrac{\alpha \Rightarrow \beta}{\circ(\alpha) \Rightarrow \Diamond\beta}\,(\Diamond\mathrm{R})}{\Diamond\alpha \Rightarrow \Diamond\beta}\,(\Diamond\mathrm{L})$$

For (5), It suffices to show that $\alpha \wedge \beta \Rightarrow \alpha \wedge (\beta \vee \gamma)$:

$$\dfrac{\dfrac{\dfrac{\dfrac{\dfrac{\dfrac{\neg\beta \wedge \beta \Leftrightarrow \bot \qquad \bot \Rightarrow \neg\neg\gamma}{\neg\beta \wedge \beta \Rightarrow \neg\neg\gamma}\,(\mathrm{Cut})}{(\neg\beta \wedge \neg\gamma) \wedge \beta \Rightarrow \bot}\,(\neg\mathrm{E})}{\beta \Rightarrow \neg(\neg\beta \wedge \neg\gamma)}\,(\neg\mathrm{I})}{\alpha \wedge \beta \Rightarrow \neg(\neg\beta \wedge \neg\gamma)}\,(\wedge\mathrm{L}) \qquad \dfrac{\alpha \Rightarrow \alpha}{\alpha \wedge \beta \Rightarrow \alpha}\,(\wedge\mathrm{L})}{\alpha \wedge \beta \Rightarrow \alpha \wedge \neg(\neg\beta \wedge \neg\gamma)}\,(\wedge\mathrm{R})}{}$$

Lemma 4. *The following properties hold for corresponding* GMX*:*

(1) $\vdash \Diamond\bot \Rightarrow \bot$ *for any* GMX *containing* (N) *axiom;*
(2) $\vdash \top \Rightarrow \Diamond\top$ *for any* GMX *containing* (P) *rule;*
(3) $\vdash \alpha \Rightarrow \Diamond\alpha$ *for any* GMX *containing* (T) *rule;*
(4) $\vdash \Diamond\Diamond\alpha \Rightarrow \Diamond\alpha$ *for any* GMX *containing* (4) *rule.*

Proof. We only provide the proofs for (3) and (4).

$$\dfrac{\dfrac{\alpha \Rightarrow \alpha}{\circ(\alpha) \Rightarrow \Diamond\alpha}\,(\Diamond\mathrm{R})}{\alpha \Rightarrow \Diamond\alpha}\,(\mathrm{T}) \qquad\qquad \dfrac{\dfrac{\dfrac{\alpha \Rightarrow \alpha}{\circ(\alpha) \Rightarrow \Diamond\alpha}\,(\Diamond\mathrm{R})}{\circ\circ(\alpha) \Rightarrow \Diamond\alpha}\,(4)}{\Diamond\Diamond\alpha \Rightarrow \Diamond\alpha}\,(\Diamond\mathrm{L}) \times 2$$

Definition 6. *Given a* MX $\mathbb{A} = (A, \wedge, \vee, \neg, \Diamond, 0, 1)$, *an* assignment *in* $\mathbb{A}$ *is a function* $\theta : \mathbf{Var} \to A$. *Every assignment* σ *in* $\mathbb{A}$ *can be extended homomorphically. Let* $\hat{\sigma}(\alpha)$ *be the element in* A *denoted by* α. *An* algebraic model *is a pair* $(\mathbb{A}, \sigma)$ *where* $\mathbb{A}$ *is an algebraic structure, and* σ *is an assignment in* $\mathbb{A}$. *A sequent* $\circ^n(\alpha) \Rightarrow \beta$ *is* true *in an algebraic model* $(\mathbb{A}, \sigma)$, *notation* $\models_{\mathbb{A},\sigma} \circ^n(\alpha) \Rightarrow \beta$, *if* $\hat{\sigma}(f(\circ^n(\alpha))) \leq \hat{\sigma}(\beta)$. *A sequent* $\circ^n(\alpha) \Rightarrow \beta$ *is* true *in a class of algebraic structure* $\mathcal{K}$, *notation* $\models_{\mathcal{K}} \circ^n(\alpha) \Rightarrow \beta$, *if* $\models_{\mathbb{A},\sigma} \circ^n(\alpha) \Rightarrow \beta$ *for any algebraic model* $(\mathbb{A}, \sigma)$ *with* $\mathbb{A} \in \mathcal{K}$. *A sequent rule with premises* $\circ_1^n(\alpha_1) \Rightarrow \beta_1, \ldots, \circ_m^n(\alpha_m) \Rightarrow \beta_m$ *and conclusion* $\circ_0^n(\alpha_0) \Rightarrow \beta_0$ preserves truth *in* $\mathcal{K}$, *if* $\models_{\mathbb{A},\sigma} \circ_0^n(\alpha_0) \Rightarrow \beta_0$ *whenever* $\models_{\mathbb{A},\sigma} \circ_i^n(\alpha_i) \Rightarrow \beta_i$ *for* $1 \leq i \leq m$, *for any algebraic model* $(\mathbb{A}, \sigma)$ *with* $\mathbb{A} \in \mathcal{K}$.

Theorem 1 (Soundness and Completeness). *Any* GMX *is sound and complete with respect to its corresponding* **MX**.

Proof. The soundness can be obtained by the induction on the length of proof with Lemma 1 and Lemma 2, while the completeness can be obtained by the FMP result in the next section.

4 Finite Model Property

In this section, we prove the FMP results for any GMX, that is, we will show that if $\nvdash_{\mathsf{GMX}} \alpha \Rightarrow \beta$, then there exists a finite MX model $\mathcal{M}$ such that $\not\models_{\mathcal{M}} \alpha \Rightarrow \beta$. Then, we get the decidability for all $\mathsf{GM} \in G_M$. Our proof strategy is that we will focus on GMNPT4's FMP and decidability since it is the strongest system in Fig. 1. Then the FMP and decidability result of other GMX weaker than GMNPT4 can be obtained independently from GMNPT4's proof simply by deleting the redundant cases. Note that the $\vdash$ symbol in this section means $\vdash_{\mathsf{GMNPT4}}$ if no confusion arises.

Definition 7. *Let $\mathcal{T}$ be a set of formulas, a formula structure $\circ^n(\alpha)$ is a $\mathcal{T}$-formula structure if $\alpha \in \mathcal{T}$. Let $\mathcal{FS}(\mathcal{T})$ be the set of all $\mathcal{T}$-formula structures. Let the notation $c(\mathcal{T})$ denote the closure of $\mathcal{T}$ under $(\top, \bot, \wedge, \neg)$ and subformulas. A sequent $\circ^n(\alpha) \Rightarrow \beta$ is a $\mathcal{T}$-sequent if $\alpha, \beta \in \mathcal{T}$. We use $\vdash \circ^n(\alpha) \Rightarrow_{\mathcal{T}} \beta$ if there is a derivation of $\circ^n(\alpha) \Rightarrow \beta$ such that all sequents appearing in it are $\mathcal{T}$-sequents.*

Lemma 5 (Interpolation). *If $\vdash \circ^{m+n}(\alpha) \Rightarrow_{\mathcal{T}} \beta$ and $n \geq 1$, then there is a $\gamma \in \mathcal{T}$ such that $\vdash \circ^n(\alpha) \Rightarrow_{\mathcal{T}} \gamma$, $\vdash \circ^m(\gamma) \Rightarrow_{\mathcal{T}} \beta$ and $\vdash \circ(\gamma) \Rightarrow_{\mathcal{T}} \gamma$ for any m, n.*

Proof. We proceed by induction on the length of proof of $\circ^{m+n}(\alpha) \Rightarrow_{\mathcal{T}} \beta$. The proof for axioms is obvious. Take $(\top)$ as an example, that is $\circ^{m+n}(\alpha) \Rightarrow \top$. Since $\top \in \mathcal{T}$, then $\top$ is the required interpolant. For (N), $\bot$ is the required interpolant. Assume that the end sequent is obtained by rule (R). Let us consider the following cases, others can be treated similarly.

(R)=($\wedge$L). Assume the premise is $\vdash \circ^{m+n}(\varphi) \Rightarrow_{\mathcal{T}} \beta$ and the conclusion is $\vdash \circ^{m+n}(\varphi \wedge \psi) \Rightarrow_{\mathcal{T}} \beta$ where $\alpha = \varphi \wedge \psi$. By induction hypothesis, there is a $\gamma \in \mathcal{T}$ such that (1) $\vdash \circ^n(\varphi) \Rightarrow_{\mathcal{T}} \gamma$, $\vdash \circ^m(\gamma) \Rightarrow_{\mathcal{T}} \beta$ and $\vdash \circ(\gamma) \Rightarrow_{\mathcal{T}} \gamma$. Then from (1) by ($\wedge$L), one obtains $\vdash \circ^n(\varphi \wedge \psi) \Rightarrow_{\mathcal{T}} \gamma$. Therefore, γ is a required interpolant.

(R)=($\wedge$R). Assume the premises are $\vdash \circ^{m+n}(\alpha) \Rightarrow_{\mathcal{T}} \varphi$ and $\vdash \circ^{m+n}(\alpha) \Rightarrow_{\mathcal{T}} \psi$ where $\beta = \varphi \wedge \psi$. By induction hypothesis, there are $\gamma_1, \gamma_2 \in \mathcal{T}$ such that (1) $\vdash \circ^n(\alpha) \Rightarrow_{\mathcal{T}} \gamma_1$, (2) $\vdash \circ^n(\alpha) \Rightarrow_{\mathcal{T}} \gamma_2$, (3) $\vdash \circ^m(\gamma_1) \Rightarrow_{\mathcal{T}} \varphi$, (4) $\vdash \circ^m(\gamma_2) \Rightarrow_{\mathcal{T}} \psi$, (5) $\vdash \circ(\gamma_1) \Rightarrow_{\mathcal{T}} \gamma_1$, (6) $\vdash \circ(\gamma_2) \Rightarrow_{\mathcal{T}} \gamma_2$. By applying ($\wedge$L) to (3) and (4), one has (7) $\vdash \circ^m(\gamma_1 \wedge \gamma_2) \Rightarrow_{\mathcal{T}} \varphi$ and (8) $\vdash \circ^m(\gamma_1 \wedge \gamma_2) \Rightarrow_{\mathcal{T}} \psi$. Next, we apply ($\wedge$R) to (7) and (8), one has $\vdash \circ^m(\gamma_1 \wedge \gamma_2) \Rightarrow_{\mathcal{T}} \varphi \wedge \psi$. Again by applying ($\wedge$R) to (1) and (2), one has $\vdash \circ^n(\alpha) \Rightarrow_{\mathcal{T}} \gamma_1 \wedge \gamma_2$. Similarly, we apply ($\wedge$L) and ($\wedge$R) to (5) and (6), one has $\vdash \circ(\gamma_1 \wedge \gamma_2) \Rightarrow_{\mathcal{T}} \gamma_1 \wedge \gamma_2$. Therefore, $\gamma_1 \wedge \gamma_2$ is a required interpolant.

(R)=($\Diamond$R). Assume the premise is $\circ^k(\alpha) \Rightarrow_{\mathcal{T}} \beta$, then the conclusion is $\vdash \circ^{k+1}(\alpha) \Rightarrow_{\mathcal{T}} \Diamond\beta$ where $1 \leq n \leq k$. By induction hypothesis, there is a $\gamma \in \mathcal{T}$

such that (1) $\vdash \circ^n(\alpha) \Rightarrow_{\mathcal{T}} \gamma$, (2) $\vdash \circ^{k-n}(\gamma) \Rightarrow_{\mathcal{T}} \beta$ and (3) $\vdash \circ(\gamma) \Rightarrow_{\mathcal{T}} \gamma$. By applying ($\Diamond$R) to (2), one has $\vdash \circ^{k-n+1}(\gamma) \Rightarrow_{\mathcal{T}} \Diamond\beta$. Note that (1) and (3) still hold for the conclusion. Therefore, γ is a required interpolant. The ($\Diamond$L) case can be treated similarly.

(R)=(T). Assume the premise is $\vdash \circ^{k+1}(\alpha) \Rightarrow_{\mathcal{T}} \beta$, then the conclusion is $\vdash \circ^k(\alpha) \Rightarrow_{\mathcal{T}} \beta$. Assume $n + m = k$, then by induction hypothesis, there is a $\gamma \in \mathcal{T}$ such that (1) $\vdash \circ^n(\alpha) \Rightarrow_{\mathcal{T}} \gamma$, (2) $\vdash \circ^{k+1-n}(\gamma) \Rightarrow_{\mathcal{T}} \beta$ with $m = k+1-n$ and (3) $\vdash \circ(\gamma) \Rightarrow_{\mathcal{T}} \gamma$. Next, by applying (T) to (2), one has $\vdash \circ^{k-n}(\gamma) \Rightarrow_{\mathcal{T}} \beta$ with $m = k - n$. Note that (1) and (3) still hold for the conclusion. Therefore, γ is a required interpolant. The (P) case can be treated similarly.

(R)=(4). Assume the premise is $\vdash \circ^{k+1}(\alpha) \Rightarrow_{\mathcal{T}} \beta$, then the conclusion is $\vdash \circ^{k+2}(\alpha) \Rightarrow_{\mathcal{T}} \beta$. If $m + n = 2$ i.e. $k = 0$, then by induction hypothesis, there is a $\gamma \in \mathcal{T}$ such that $\vdash \circ(\alpha) \Rightarrow_{\mathcal{T}} \gamma$, (1) $\vdash \gamma \Rightarrow_{\mathcal{T}} \beta$ and (2) $\vdash \circ(\gamma) \Rightarrow_{\mathcal{T}} \gamma$. Then by applying (Cut) to (1) and (2), one obtains $\vdash \circ(\gamma) \Rightarrow_{\mathcal{T}} \beta$. Thus γ is a required interpolant. Otherwise, assume $m + n \geq 3$, by induction hypothesis, there is a $\gamma \in \mathcal{T}$ such that $\vdash \circ^n(\alpha) \Rightarrow_{\mathcal{T}} \gamma$, $\vdash \circ^{k+1-n}(\gamma) \Rightarrow_{\mathcal{T}} \beta$ and $\vdash \circ(\gamma) \Rightarrow_{\mathcal{T}} \gamma$. If $n \neq k+1$, then by rule (4) on $\vdash \circ^{k+1-n}(\gamma) \Rightarrow_{\mathcal{T}} \beta$ or $\vdash \circ^n(\alpha) \Rightarrow_{\mathcal{T}} \gamma$, one has $\vdash \circ^{k+2-n}(\gamma) \Rightarrow_{\mathcal{T}} \beta$ or $\vdash \circ^{k+2}(\alpha) \Rightarrow_{\mathcal{T}} \gamma$. Therefore, γ is a required interpolant. Assume $n = k + 1$, then this case can be treated by a similar method when $n + m = 2$. Therefore, γ is a required interpolant.

(R)=(Cut). Assume the premises are (1) $\vdash \circ^k(\alpha) \Rightarrow_{\mathcal{T}} \varphi$ and (2) $\vdash \circ^l(\varphi) \Rightarrow_{\mathcal{T}} \beta$, then the conclusion is $\vdash \circ^{k+l}(\alpha) \Rightarrow_{\mathcal{T}} \beta$. Obviously $m + n = k + l$. Further, assume $n \leq k$, then by induction hypothesis one has (3) $\vdash \circ^n(\alpha) \Rightarrow_{\mathcal{T}} \gamma$, (4) $\vdash \circ^{k-n}(\gamma) \Rightarrow_{\mathcal{T}} \varphi$ and (5) $\vdash \circ(\gamma) \Rightarrow_{\mathcal{T}} \gamma$. By applying (Cut) to (2) and (4), one has $\vdash \circ^{k+l-n}(\gamma) \Rightarrow_{\mathcal{T}} \beta$ with $m = k + l - n$. Note that (3) and (5) still hold for the conclusion. Therefore, γ is a required interpolant. Otherwise assume $n > k$, one has $m < l$. Then by induction hypothesis, there is a $\gamma \in \mathcal{T}$ such that (6) $\vdash \circ^{l-m}(\varphi) \Rightarrow_{\mathcal{T}} \gamma$, (7) $\vdash \circ^m(\gamma) \Rightarrow_{\mathcal{T}} \beta$ and (8) $\vdash \circ(\gamma) \Rightarrow_{\mathcal{T}} \gamma$. By applying (Cut) to (1) and (6), one has $\vdash \circ^{k+l-m}(\alpha) \Rightarrow_{\mathcal{T}} \gamma$ with $n = k + l - m$. Note that (7) and (8) still hold for the conclusion. Therefore, γ is a required interpolant.

Remark 1. The above interpolation lemma is not the standard form of Craig's interpolation. It is more like an inverse of the analytic cut. The additional content i.e. $\vdash \circ(\gamma) \Rightarrow_{\mathcal{T}} \gamma$ is specially designed for solving the case of (R)=(4) when $m + n = 2$. Such a lemma is rooted in Buszkowski's work [2] towards finite embeddability property and FMP of nonassociative Lambek calculus and its various lattice extensions. Lin [21,22] further studied this lemma to prove some non-classical modal logics' FMP or SFMP.

We define a kind of disjunctive normal form formula on the language. A formula α is called a letter if $\alpha \in \mathbf{Var}, \alpha = \top, \alpha = \bot$ or $\alpha = \Diamond\beta$ for some formulas β. Let $\mathcal{L}_e$ be the set of all letters in the language. A formula α is called a literal if $\alpha \in \mathcal{L}_e$ or $\alpha = \neg\beta$ for some $\beta \in \mathcal{L}_e$. We denote the set of literals under language $\mathcal{L}$ by $\mathcal{L}_i$. A formula in disjunctive normal form (DNF) is the disjunction of one or more disjuncts, each of which is the conjunction of one or more literals. Note that the DNF of a formula defined here is actually the equivalent expression obtained by the abbreviation of $\alpha \vee \beta := \neg(\neg\alpha \wedge \neg\beta)$.

Let $\mathcal{T}$ be a set of formulas such that $\mathcal{T} = c(\mathcal{T})$. Suppose that $\mathcal{T}_{li} \subseteq \mathcal{T}$ is the set of all literals in $\mathcal{T}$. We say $\mathcal{T}$ is finitely based if $\mathcal{T}_{li}$ is finite. For any formula $\alpha \in \mathcal{T}$, there is a DNF formula $\beta \in \mathcal{T}$ such that α is equivalent to β by axiom (Dis) and axiom (DN). If one omits the repetition of literals, then one has a unique formula in DNF which is equivalent to α. We denote the unique DNF formula corresponding to α by $df_{\mathcal{T}}(\alpha)$. Let $df(\mathcal{T}) = \{df_{\mathcal{T}}(\alpha) \mid \alpha \in \mathcal{T}\}$. If $\mathcal{T}_{li}$ is finite, then $df(\mathcal{T})$ is finite.

Corollary 1. *If* $\vdash \circ^{n+m}(\alpha) \Rightarrow_{\mathcal{T}} \beta$ *and* $n \geq 1$, *then there is a* $\gamma \in df(\mathcal{T})$ *such that* $\vdash \circ^n(\alpha) \Rightarrow_{\mathcal{T}} \gamma$ *and* $\vdash \circ^m(\gamma) \Rightarrow_{\mathcal{T}} \beta$ *for any* m, n.

Definition 8 (Order $\leq_{\mathcal{T}}$ on $\mathcal{FS}(\mathcal{T})$). *Let* $\circ^n(\alpha), \circ^m(\beta) \in \mathcal{FS}(\mathcal{T})$, *we define an order* $\leq_{\mathcal{T}}$ *on* $\mathcal{FS}(\mathcal{T})$ *as follows:* $\circ^n(\alpha) \leq_{\mathcal{T}} \circ^m(\beta)$ *iff for any context* $\circ^k(-)$ *and formula* $\varphi \in \mathcal{T}$, *if* $\circ^k(\circ^m(\beta)) \Rightarrow_{\mathcal{T}} \varphi$, *then* $\circ^k(\circ^n(\alpha)) \Rightarrow_{\mathcal{T}} \varphi$.

Let $\circ^n(\alpha) \approx_{\mathcal{T}} \circ^m(\beta)$ be $\circ^n(\alpha) \leq_{\mathcal{T}} \circ^m(\beta)$ and $\circ^m(\beta) \leq_{\mathcal{T}} \circ^n(\alpha)$, then $\approx_{\mathcal{T}}$ is an equivalence relation. Let $[\alpha]_{\mathcal{T}} = \{\circ^m(\beta) \mid \circ^m(\beta) \approx_{\mathcal{T}} \alpha \ \& \ \circ^m(\beta) \in \mathcal{FS}(\mathcal{T})\}$ for any $\alpha \in \mathcal{T}$. Let $[\mathcal{T}] = \{[\alpha]_{\mathcal{T}} \mid \alpha \in \mathcal{T})\}$. Since $[\alpha]_{\mathcal{T}} = [df_{\mathcal{T}}(\alpha)]_{\mathcal{T}}$ and the number of $[df_{\mathcal{T}}(\alpha)]_{\mathcal{T}}$ is finite, $[\mathcal{T}]$ is finite.

Lemma 6. *For any* $\circ(\alpha) \in \mathcal{FS}(\mathcal{T})$, *there is a* $\beta \in df(\mathcal{T})$ *such that* $\circ(\alpha) \approx_{\mathcal{T}} \beta$.

Proof. For any $\circ^k(-)$ and $\gamma_j \in \mathcal{T}$, assume that $\vdash \circ^k(\circ(\alpha)) \Rightarrow_{\mathcal{T}} \gamma_j$. By Corollary 1, there is a $\beta_j \in df(\mathcal{T})$ such that $\vdash \circ(\alpha) \Rightarrow_{\mathcal{T}} \beta_j$ and $\vdash \circ^k(\beta_j) \Rightarrow_{\mathcal{T}} \gamma_j$. Obviously, the number of γ_j is finite. Let δ be the conjunction of all β_j. Clearly $\delta \in \mathcal{T}$. By ($\wedge$R) and ($\wedge$L), one has (1) $\vdash \circ(\alpha) \Rightarrow_{\mathcal{T}} \delta$ and (2) $\vdash \circ^k(\delta) \Rightarrow_{\mathcal{T}} \gamma_j$. Then $df_{\mathcal{T}}(\delta) \in df(\mathcal{T})$. Let $\beta = df_{\mathcal{T}}(\delta)$, then one has (3) $\vdash \circ(\alpha) \Rightarrow_{\mathcal{T}} \beta$ and (4) $\vdash \circ^k(\beta) \Rightarrow_{\mathcal{T}} \gamma_j$. Thus, the assumption $\vdash \circ^k(\circ(\alpha)) \Rightarrow_{\mathcal{T}} \gamma_j$ implies (4) $\vdash \circ^k(\beta) \Rightarrow_{\mathcal{T}} \gamma_j$, then one has $\beta \leq_{\mathcal{T}} \circ(\alpha)$. Further, assume that (5) $\vdash \circ^l(\beta) \Rightarrow_{\mathcal{T}} \delta$ for some context $\circ^l(-)$ and formula $\delta \in \mathcal{T}$. By applying (Cut) to (3) and (5), one has $\vdash \circ^l(\circ(\alpha)) \Rightarrow_{\mathcal{T}} \delta$. Therefore, one has $\circ(\alpha) \leq_{\mathcal{T}} \beta$. Consequently, one has $\circ(\alpha) \approx_{\mathcal{T}} \beta$.

Definition 9 (Quotient Algebra). *The quotient algebra of* $[\mathcal{T}]$ *is a structure* $\mathbf{Qa} = ([\mathcal{T}], \wedge^*, \vee^*, \neg^*, \Diamond^*, \bot^*, \top^*)$ *where* $\top^*, \bot^*$ *and operations* $\wedge^*, \neg^*$ *and* $\Diamond^*$ *in* $[\mathcal{T}]$ *are defined as follows:*

(1) $\top^* = [\top]_{\mathcal{T}}$;
(2) $\bot^* = [\bot]_{\mathcal{T}}$;
(3) $\neg^*[\alpha]_{\mathcal{T}} = [\neg\alpha]_{\mathcal{T}}$;
(4) $[\alpha]_{\mathcal{T}} \wedge^* [\beta]_{\mathcal{T}} = [\alpha \wedge \beta]_{\mathcal{T}}$;
(5) $\Diamond^*[\alpha]_{\mathcal{T}} = [\gamma]_{\mathcal{T}}$ *s.t.* $\gamma \approx_{\mathcal{T}} \circ(\alpha)$.

We define $[\alpha]_{\mathcal{T}} \leq^* [\beta]_{\mathcal{T}}$ as $[\alpha]_{\mathcal{T}} \wedge^* [\beta]_{\mathcal{T}} = [\alpha]_{\mathcal{T}}$ and $[\alpha]_{\mathcal{T}} \vee^* [\beta]_{\mathcal{T}} = \neg^*(\neg^*[\alpha]_{\mathcal{T}} \wedge^* \neg^*[\beta]_{\mathcal{T}})$. Clearly by the definition of $\Diamond^*[\alpha]_{\mathcal{T}}$ and Lemma 6, $[\gamma]_{\mathcal{T}}$ exists and is unique. Further, let $[\alpha_1]_{\mathcal{T}} = [\alpha_2]_{\mathcal{T}}$, one can show that $\Diamond^*[\alpha_1]_{\mathcal{T}} = \Diamond^*[\alpha_2]_{\mathcal{T}}$. Since $\alpha_1 \in [\alpha_1]_{\mathcal{T}}$, then $\alpha_1 \in [\alpha_2]_{\mathcal{T}}$. By the definition of the equivalence class, one has $\alpha_1 \approx_{\mathcal{T}} \alpha_2$. Given any context $\circ^k(-)$ and formula $\beta \in \mathcal{T}$. Assume that $\vdash \circ^k(\circ(\alpha_1)) \Rightarrow_{\mathcal{T}} \beta$, then one has $\vdash \circ^k(\circ(\alpha_2)) \Rightarrow_{\mathcal{T}} \beta$. Therefore, one has

$\circ(\alpha_2) \leq_{\mathcal{T}} \circ(\alpha_1)$. By similar argument, one has $\circ(\alpha_1) \leq_{\mathcal{T}} \circ(\alpha_2)$. Consequently, one has $\circ(\alpha_1) \approx_{\mathcal{T}} \circ(\alpha_2)$. Assume $\lozenge^*[\alpha_1]_{\mathcal{T}} = [\gamma]_{\mathcal{T}}$ such that $\gamma \approx_{\mathcal{T}} \circ(\alpha_1)$, then $\gamma \approx_{\mathcal{T}} \circ(\alpha_2)$ and $[\gamma]_{\mathcal{T}} = \lozenge^*[\alpha_2]_{\mathcal{T}}$. Therefore $\lozenge^*[\alpha_1]_{\mathcal{T}} = \lozenge^*[\alpha_2]_{\mathcal{T}}$. Other operations in Definition 9 can be checked similarly. Consequently, all these operations are well-defined.

Lemma 7. *The following are equivalent: (1)* $\alpha \leq_{\mathcal{T}} \beta$*; (2)* $\vdash \alpha \Rightarrow_{\mathcal{T}} \beta$*; (3)* $[\alpha]_{\mathcal{T}} \leq^* [\beta]_{\mathcal{T}}$.

Proof. For (1) and (2), assume $\vdash \alpha \Rightarrow_{\mathcal{T}} \beta$. Given any context $\circ^k(-)$ and formula $\varphi \in \mathcal{T}$, assume that $\vdash \circ^k(\beta) \Rightarrow_{\mathcal{T}} \varphi$. By (Cut) one has $\vdash \circ^k(\alpha) \Rightarrow_{\mathcal{T}} \varphi$. Therefore, one has $\alpha \leq_{\mathcal{T}} \beta$. Conversely, assume $\alpha \leq_{\mathcal{T}} \beta$. Since $\vdash \beta \Rightarrow_{\mathcal{T}} \beta$, then one has $\vdash \alpha \Rightarrow_{\mathcal{T}} \beta$. For (2) and (3), Assume $[\alpha]_{\mathcal{T}} \leq^* [\beta]_{\mathcal{T}}$, then one has $[\alpha]_{\mathcal{T}} \wedge^* [\beta]_{\mathcal{T}} = [\alpha]_{\mathcal{T}}$. Since $[\alpha]_{\mathcal{T}} \wedge^* [\beta]_{\mathcal{T}} = [\alpha \wedge \beta]_{\mathcal{T}}$, then one has $[\alpha \wedge \beta]_{\mathcal{T}} = [\alpha]_{\mathcal{T}}$. By the definition of the equivalence class, one has $\alpha \wedge \beta \approx_{\mathcal{T}} \alpha$. Further, one has $\alpha \leq_{\mathcal{T}} \alpha \wedge \beta$ and $\alpha \wedge \beta \leq_{\mathcal{T}} \beta$. Therefore one has $\alpha \leq_{\mathcal{T}} \beta$. Conversely, assume $\alpha \leq_{\mathcal{T}} \beta$, then one has $\vdash \alpha \Rightarrow_{\mathcal{T}} \beta$. By ($\wedge$R), one has $\vdash \alpha \wedge \beta \Leftrightarrow_{\mathcal{T}} \alpha$. Therefore, one has $[\alpha \wedge \beta]_{\mathcal{T}} = [\alpha]_{\mathcal{T}} = [\alpha]_{\mathcal{T}} \wedge^* [\beta]_{\mathcal{T}}$. Therefore, $[\alpha]_{\mathcal{T}} \leq^* [\beta]_{\mathcal{T}}$. Consequently, $\alpha \leq_{\mathcal{T}} \beta$ iff $[\alpha]_{\mathcal{T}} \leq^* [\beta]_{\mathcal{T}}$.

Lemma 8. *The following conditions hold for* **Qa***: for any* $[\alpha]_{\mathcal{T}}, [\beta]_{\mathcal{T}}, [\gamma]_{\mathcal{T}} \in [\mathcal{T}]$,

(Inf) $[\gamma]_{\mathcal{T}} \leq^* [\alpha]_{\mathcal{T}} \wedge^* [\beta]_{\mathcal{T}}$ *iff* $[\gamma]_{\mathcal{T}} \leq^* [\alpha]_{\mathcal{T}}$ *and* $[\gamma]_{\mathcal{T}} \leq^* [\beta]_{\mathcal{T}}$;
(Sup) $[\alpha]_{\mathcal{T}} \vee^* [\beta]_{\mathcal{T}} \leq^* [\gamma]_{\mathcal{T}}$ *iff* $[\alpha]_{\mathcal{T}} \leq^* [\gamma]_{\mathcal{T}}$ *and* $[\beta]_{\mathcal{T}} \leq^* [\gamma]_{\mathcal{T}}$;
(Dis) $[\alpha]_{\mathcal{T}} \wedge^* ([\beta]_{\mathcal{T}} \vee^* [\gamma]_{\mathcal{T}}) = ([\alpha]_{\mathcal{T}} \wedge^* [\beta]_{\mathcal{T}}) \vee^* ([\alpha]_{\mathcal{T}} \wedge^* [\gamma]_{\mathcal{T}})$;
(Bound) $\bot^* \leq^* [\alpha]_{\mathcal{T}} \leq^* \top^*$;
(LC) $[\alpha]_{\mathcal{T}} \wedge^* \neg^*[\alpha]_{\mathcal{T}} = \bot^*$;
(LEM) $[\alpha]_{\mathcal{T}} \vee^* \neg^*[\alpha]_{\mathcal{T}} = \top^*$;
(M) *If* $[\alpha]_{\mathcal{T}} \leq^* [\beta]_{\mathcal{T}}$*, then* $\lozenge^*[\alpha]_{\mathcal{T}} \leq^* \lozenge^*[\beta]_{\mathcal{T}}$;
(N) $\lozenge^*\bot^* = \bot^*$;
(P) $\lozenge^*\top^* = \top^*$;
(T) $[\alpha]_{\mathcal{T}} \leq^* \lozenge^*[\alpha]_{\mathcal{T}}$;
(4) $\lozenge^*\lozenge^*[\alpha]_{\mathcal{T}} \leq^* \lozenge^*[\alpha]_{\mathcal{T}}$.

Proof. We only provide the proofs for (M), (N), (P), (T), and (4).

- (M) Assume $[\alpha]_{\mathcal{T}} \leq^* [\beta]_{\mathcal{T}}$. Let $\lozenge^*[\alpha]_{\mathcal{T}} = [\gamma_1]_{\mathcal{T}}$ and $\lozenge^*[\beta]_{\mathcal{T}} = [\gamma_2]_{\mathcal{T}}$ such that $\circ(\alpha) \approx_{\mathcal{T}} \gamma_1$ and $\circ(\beta) \approx_{\mathcal{T}} \gamma_2$. It suffices to prove that $\vdash \gamma_1 \Rightarrow_{\mathcal{T}} \gamma_2$. From the assumption by Lemma 7, one obtains $\vdash \alpha \Rightarrow_{\mathcal{T}} \beta$. By Lemma 7 and Definition 8, $\vdash \circ(\beta) \Rightarrow_{\mathcal{T}} \gamma_2$. Hence by (Cut), one obtains $\vdash \circ(\alpha) \Rightarrow_{\mathcal{T}} \gamma_2$. Therefore $\vdash \gamma_1 \Rightarrow_{\mathcal{T}} \gamma_2$.
- (N) Let $\lozenge^*[\bot]_{\mathcal{T}} = [\gamma]_{\mathcal{T}}$ such that $\gamma \approx_{\mathcal{T}} \circ(\bot)$. It suffices to prove $\gamma \approx_{\mathcal{T}} \bot$. Clearly $\vdash \bot \Rightarrow \gamma$ from ($\bot$). From axiom (N), one has $\vdash \circ(\bot) \Rightarrow \bot$. Thus $\vdash \gamma \Rightarrow \bot$.
- (P) Let $\lozenge^*[\top]_{\mathcal{T}} = [\gamma]_{\mathcal{T}}$ such that $\circ(\top) \approx_{\mathcal{T}} \gamma$, then one has $\top \Rightarrow_{\mathcal{T}} \gamma$ by Lemma 7 and (P). Obviously, one has $\circ(\top) \Rightarrow_{\mathcal{T}} \top$. Therefore, one has $\lozenge^*\top^* = \top^*$.
- (T) Let $\lozenge^*[\alpha]_{\mathcal{T}} = [\gamma]_{\mathcal{T}}$ such that $\gamma \approx_{\mathcal{T}} \circ(\alpha)$. It suffices to prove $\vdash \alpha \Rightarrow_{\mathcal{T}} \gamma$. Clearly, $\vdash \circ(\alpha) \Rightarrow_{\mathcal{T}} \gamma$. Then by (T), one obtains $\vdash \alpha \Rightarrow_{\mathcal{T}} \gamma$.

(4) Let $\Diamond^*[\alpha]_\mathcal{T} = [\theta_1]_\mathcal{T}$ and $\Diamond^*[\theta_1]_\mathcal{T} = [\theta_2]_\mathcal{T}$ such that $\theta_1 \approx_\mathcal{T} \circ(\alpha), \theta_2 \approx_\mathcal{T} \circ(\theta_1)$. It suffices to show $\theta_2 \leq_\mathcal{T} \theta_1$ by Lemma 7. Assume that $\vdash \circ^k(\theta_1) \Rightarrow_\mathcal{T} \varphi$ for some context $\circ^k(-)$ and $\varphi \in \mathcal{T}$. Hence $\vdash \circ^k(\circ(\alpha)) \Rightarrow_\mathcal{T} \varphi$. Then by rule (4), one obtains $\vdash \circ^k(\circ(\circ(\alpha))) \Rightarrow_\mathcal{T} \varphi$. Whence $\vdash \circ^k(\circ(\theta_1)) \Rightarrow_\mathcal{T} \varphi$. Therefore $\vdash \circ^k(\theta_2) \Rightarrow_\mathcal{T} \varphi$. Thus, $\theta_2 \leq_\mathcal{T} \theta_1$.

Theorem 2. $\mathbf{Qa} = ([\mathcal{T}], \wedge^*, \vee^*, \neg^*, \Diamond^*, \bot^*, \top^*)$ *is a finite* MNPT4.

Lemma 9. *If* $\Diamond\alpha \in \mathcal{T}$, *then* $\Diamond^*[\alpha]_\mathcal{T} = [\Diamond\alpha]_\mathcal{T}$.

Proof. Assume $\Diamond^*[\alpha]_\mathcal{T} = [\gamma]_\mathcal{T}$ such that $\gamma \in \mathcal{T}$ and $\circ(\alpha) \approx_\mathcal{T} \gamma$. It suffices to show $\gamma \approx_\mathcal{T} \Diamond\alpha$. Assume that $\vdash \circ^k(\gamma) \Rightarrow_\mathcal{T} \varphi$ for some context $\circ^k(-)$ and $\varphi \in \mathcal{T}$. Then $\vdash \circ^k(\circ(\alpha)) \Rightarrow_\mathcal{T} \varphi$. By $(\Diamond\mathrm{L})$, $\vdash \circ^k(\Diamond\alpha) \Rightarrow_\mathcal{T} \varphi$. Assume $\vdash \circ^k(\Diamond\alpha) \Rightarrow_\mathcal{T} \varphi$. Clearly $\vdash \circ(\alpha) \Rightarrow_\mathcal{T} \Diamond\alpha$. Then by (Cut), $\vdash \circ^k(\circ(\alpha)) \Rightarrow_\mathcal{T} \varphi$. Thus $\vdash \circ^k(\gamma) \Rightarrow_\mathcal{T} \varphi$.

Lemma 10. *If* $\nvdash_{\mathsf{GMNPT4}} \alpha \Rightarrow \beta$, *then* $\not\models_{\mathbf{Qa}} [\alpha]_\mathcal{T} \leq^* [\beta]_\mathcal{T}$.

Proof. Let $\mathcal{T}$ be the smallest set containing α, β such that $\mathcal{T} = c(\mathcal{T})$, assume that $\nvdash_{\mathsf{GMNPT4}} \alpha \Rightarrow \beta$, then $\nvdash_{\mathsf{GMNPT4}} \alpha \Rightarrow_\mathcal{T} \beta$. Construct $\mathbf{Qa} = ([\mathcal{T}], \wedge^*, \vee^*, \neg^*, \Diamond^*, \bot^*, \top^*)$ as above and an assignment $\sigma : \mathbf{Var} \longrightarrow [\mathcal{T}]$ such that $\sigma(p) = [p]_\mathcal{T}$. By induction on the complexity of the formula, one can easily prove that $\hat{\sigma}(\delta) = [\delta]_\mathcal{T}$ by Definition 9 and Lemma 9. Assume that $\models_{\mathbf{Qa},\sigma} [\alpha]_\mathcal{T} \leq^* [\beta]_\mathcal{T}$, then by Lemma 7, one has $\vdash \alpha \Rightarrow_\mathcal{T} \beta$, which contradicts to our initial assumption. Therefore, if $\nvdash_{\mathsf{GMNPT4}} \alpha \Rightarrow \beta$, then $\not\models_{\mathbf{Qa}} [\alpha]_\mathcal{T} \leq^* [\beta]_\mathcal{T}$.

Corollary 2. *If* $\nvdash_{\mathsf{GMNPT4}} \alpha \Rightarrow \beta$, *then* $\not\models_{\mathrm{MNPT4}} \mu(\alpha) \leq \mu(\beta)$ *where* μ *is a mapping from* **Var** *to a finite* MNPT4.

Theorem 3 (FMP). *All* $\mathsf{GMX} \in \{\mathsf{GM}, \ldots, \mathsf{GMNPT4}\}$ *have FMP.*

Proof. One can check that all the proofs of GMX can be obtained independently by deleting the redundant cases in the proof of $\mathsf{GMNPT4}$.

Theorem 4 (Decidability). *All* $\mathsf{GMX} \in \{\mathsf{GM}, \ldots, \mathsf{GMNPT4}\}$ *are decidable.*

5 Concluding Remark

We investigate the non-normal modal logics M and various extensions of it i.e. logics from $\{\mathsf{M}, \mathsf{MN}, \ldots, \mathsf{MNPT4}\}$ and obtain their FMP and thus decidability. Such results are new contributions to the current research of these logics, especially for those logics with axiom (4). Naturally, the method used in this paper can be certainly extended to those logics with axioms expressed by $\Diamond$ and $\wedge$. Another future work will be establishing decidable algorithms for these logics.

Acknowledgments. This work of both authors was supported by the Chinese National Funding of Social Sciences (Grant no. 18ZDA033).

References

1. Boolos, G.: The Logic of Provability. Cambridge University Press, Cambridge (1995)
2. Buszkowski, W.: Interpolation and FEP for logics of residuated algebras. Log. J. IGPL **19**(3), 437–454 (2011)
3. Chellas, B.F.: Modal Logic: An Introduction. Cambridge University Press, Cambridge (1980)
4. Chen, J., Greco, G., Palmigiano, A., Tzimoulis, A.: Non normal logics: semantic analysis and proof theory. In: Iemhoff, R., Moortgat, M., de Queiroz, R. (eds.) WoLLIC 2019. LNCS, vol. 11541, pp. 99–118. Springer, Heidelberg (2019). https://doi.org/10.1007/978-3-662-59533-6_7
5. Chen, J., Greco, G., Palmigiano, A., Tzimoulis, A.: Non-normal modal logics and conditional logics: semantic analysis and proof theory. Inf. Comput. **287**, 104756 (2022)
6. Dalmonte, T.: Non-normal modal logics: neighbourhood semantics and their calculi. Ph.D. thesis, Aix-Marseille (2020)
7. Dalmonte, T., Lellmann, B., Olivetti, N., Pimentel, E.: Hypersequent calculi for non-normal modal and deontic logics: countermodels and optimal complexity. J. Log. Comput. **31**(1), 67–111 (2021)
8. Dalmonte, T., Olivetti, N., Negri, S.: Non-normal modal logics: bi-neighbourhood semantics and its labelled calculi. In: Advances in Modal Logic 2018 (2018)
9. Governatori, G., Luppi, A.: Labelled tableaux for non-normal modal logics. In: Lamma, E., Mello, P. (eds.) AI*IA 1999. LNCS (LNAI), vol. 1792, pp. 119–130. Springer, Heidelberg (2000). https://doi.org/10.1007/3-540-46238-4_11
10. Hansen, H.H.: Monotonic modal logics. Master's thesis, Institute for Logic, Language and Computation (ILLC), University of Amsterdam (2003)
11. Holliday, W.H.: Epistemic closure and epistemic logic I: relevant alternatives and subjunctivism. J. Philos. Log. **44**, 1–62 (2015)
12. Indrzejczak, A.: Sequent calculi for monotonic modal logics. Bull. Section Logic **34**(3), 151–164 (2005)
13. Indrzejczak, A.: Labelled tableau calculi for weak modal logics. Bull. Section Logic **36**(3–4), 159–173 (2007)
14. Indrzejczak, A.: Natural Deduction, Hybrid Systems and Modal Logics, vol. 30. Springer, Heidelberg (2010)
15. Indrzejczak, A.: Admissibility of cut in congruent modal logics. Logic Log. Philos. **20**(3), 189–203 (2011)
16. Kripke, S.A.: Semantical analysis of modal logic II. Non-normal modal propositional calculi. In: The Theory of Models, pp. 206–220. Elsevier (2014)
17. Lavendhomme, R., Lucas, T.: Sequent calculi and decision procedures for weak modal systems. Stud. Logica **66**, 121–145 (2000)
18. Lellmann, B., Pimentel, E.: Modularisation of sequent calculi for normal and non-normal modalities. ACM Trans. Comput. Logic (TOCL) **20**(2), 1–46 (2019)
19. Lemmon, E., Scott, D.: The 'Lemmon Notes': An Introduction to Modal Logic. American Philosophical Quarterly Monograph Series. Basil Black Well, Oxford (1977)
20. Lewis, C.I., Langford, C.H., Lamprecht, P.: Symbolic Logic, vol. 170. Dover Publications, New York (1959)
21. Lin, Z.: Non-associative Lambek calculus with modalities: interpolation, complexity and FEP. Logic J. IGPL **22**(3), 494–512 (2014)

22. Lin, Z., Chakraborty, M.K., Ma, M.: Residuated algebraic structures in the vicinity of pre-rough algebra and decidability. Fund. Inform. **179**(3), 239–274 (2021)
23. Lin, Z., Ma, M.: Gentzen sequent calculi for some intuitionistic modal logics. Logic J. IGPL **27**(4), 596–623 (2019)
24. McNamara, P.: Deontic logic. In: Handbook of the History of Logic, vol. 7, pp. 197–288. Elsevier (2006)
25. Montague, R.: Universal grammar. Theoria **36**(3) (1970)
26. Negri, S.: Proof theory for non-normal modal logics: the neighbourhood formalism and basic results. IfCoLog J. Logics Appl. **4**(4), 1241–1286 (2017)
27. Orlandelli, E.: Proof analysis in deontic logics. In: Cariani, F., Grossi, D., Meheus, J., Parent, X. (eds.) DEON 2014. LNCS (LNAI), vol. 8554, pp. 139–148. Springer, Cham (2014). https://doi.org/10.1007/978-3-319-08615-6_11
28. Pacuit, E.: Neighborhood Semantics for Modal Logic. Springer, Heidelberg (2017)
29. Scott, D.: Advice on modal logic. In: Philosophical Problems in Logic: Some Recent Developments, pp. 143–173 (1970)
30. Segerberg, K.K.: An essay in classical modal logic. Stanford University (1971)
31. Shkatov, D., Van Alten, C.J.: Complexity of the universal theory of modal algebras. Stud. Logica **108**, 221–237 (2020)
32. Vardi, M.Y.: On the complexity of epistemic reasoning. IBM Thomas J. Watson Research Division (1989)

Belief Base: A Minimal Logic of Fine-Grained Information Dynamics

Pengfei Song(✉)

College of Philosophy, Law & Political Science, Shanghai Normal University, Shanghai 200234, China
pfsong@shnu.edu.cn

Abstract. This paper proposes a minimal logic of fine-grained information dynamics via belief bases. The framework is shown to be able to accommodate explicit belief, implicit belief, awareness of and awareness that, where awareness of agents is not treated as a tacit premise. A sound and complete axiomatization of static logic is established, upon which a series of dynamic operations are defined. It is argued that these dynamics adapt different scenarios. Our logic is minimal because to each agent we only attach two databases, from which a variety of epistemic attitudes are generated.

Keywords: Belief base · Awareness · Dynamic epistemic logic

1 Introduction

Since the first recognition of the logical omniscience problem in epistemic logic [12], awareness has long been a well-accepted concept for formalizing realistic agents. The earliest work on awareness by Fagin & Halpern [5] establishes a solid foundation for researchers, who mainly follow two approaches: the *semantic* approach [2,11,19,20,25], where awareness is generated by atomic propositions, and the *syntactic* approach [1,22,26], where the set of formulas of which an agent is aware can simply be any given set of formulas.

Grossi & Velázquez-Quesada (G&V) [9,10] combine the two approaches and present a formal analysis of *awareness of* that concerns the atomic propositions an agent has available or can resort to, and *awareness that* that concerns the formulas an agent acknowledges as true, where the two notions technically correspond to the two approaches, respectively. To motivate their work, the authors probe into a scenario from the classic movie "12 Angry Men", where agents become aware of previously unnoticed details (some atomic propositions) and make inherent knowledge (being true in every accessible state and acknowledged as true) explicit.

However, it is possible that agents cannot remember the specific piece of information while having available all the relevant atomic propositions. Consider the following scenario.

N. Alechina et al. (Eds.): LORI 2023, LNCS 14329, pp. 222–237, 2023.
https://doi.org/10.1007/978-3-031-45558-2_17

> Lucy is typesetting an article. She has made a mistake and wants to undo her previous action. But she cannot recall the keyboard shortcut of "undo" although she is facing the keyboard and aware of every button on it.

As we all know, the "undo" shortcut is CTRL+Z. Let p indicate "'CTRL' is pressed", q indicate "'Z' is pressed", r indicate "the previous action is undone". Thus, Lucy is aware of p, q and r, but cannot remember $p \wedge q \rightarrow r$. And this situation is not covered by G&V's theory. Yet someone may argue that Lucy simply doesn't know this shortcut. To respond to this challenge, let's consider the following scenario which explains Lucy indeed having some information about the shortcut.

> Now Lucy is attending an exam on basic computer skills. On the test paper there is a choice question "Choose the correct key combination of the shortcut of 'undo the previous action'." The four answers to this question are "ALT+A", "ALT+Z", "CTRL+A" and "CTRL+Z". Staring at the four options, Lucy suddenly recollects what she has learned in class and is confident that the correct answer is "CTRL+Z".

This example is quite familiar to us as the options effectively reminds Lucy of what she has acquired before. To make this happen, an agent needs to notice the exact piece of information rather than its elements.

To tackle this problem, we argue that agents can remember[1] complex formulas as well as atomic propositions. In contrast with G&V's work, here we adopt *belief bases* [15,16] rather than Dynamic Epistemic Logic (DEL, [4]) to represent beliefs or knowledges for two reasons. One is that it follows Levesque [14], "...a sentence is explicitly believed when it is actively held to be true by an agent and implicitly believed when it follows from what is believed" (p. 198). We argue that it better conforms to the actual process of reasoning compared with Fagin & Halpern (F&H)'s logic of general awareness which defines explicit belief as a formula implicitly believed by an agent and of which the agent is aware. The other is that, in light of Lorini & Song [17,18], an agent's belief base is a rough approximation of his working memory, and an agent's awareness is not a primitive but is directly computed from the agent's belief base. As such, belief bases well captures the previous scenario, where $p \wedge q \rightarrow r$ is known by Lucy somehow but not inside her working memory, then the option "CTRL+Z" reminds her and put $p \wedge q \rightarrow r$ into her working memory, and Lucy becomes aware of $p \wedge q \rightarrow r$ and other formulas formed by p, q and r at the same time.

This work is an extension of [18] by incorporating the notion of awareness that and awareness of agents, where the latter is introduced by van Ditmarsch and French [2] but is not involved in [18]. We follow G&V [9,10] and let awareness that concern the formulas an agent acknowledges as true but not necessarily explicitly known i.e., not necessarily within her working memory. After

[1] G&V argue that becoming aware of is different with remembering. However, their framework does not deal with the notion of remembering. In our work, we process the two notions at the same time, and simply treat the awareness-of set as the set of atomic propositions occurring in the working memory.

establishing the static framework, we study a series of dynamics including learning, becoming aware of, recalling, forgetting, deductive inference, etc. From this perspective, we offer a minimalistic logic approach to explicit, implicit belief, awareness of and awareness that.

The remainder of this paper is organized as follows. In Sect. 2 we define the language for implicit belief, explicit belief, awareness of and awareness that. Section 3 proposes the belief base semantics for our language. Section 4 offers an axiomatization and sketches the proof of soundness and completeness for it. Section 5 is the principal part of this paper. We define a series of actions including learning, forgetting, explicit announcement, etc. Several examples formalized by these dynamic operations are presented. Section 6 concludes the paper and points out future works.

2 Language

This section presents the language of the Logic of Fine-grained Belief (LFB). It extends the language in [18] with modalities of awareness that and awareness of agents. Let $Atm = \{p, q, \ldots\}$ be a countably infinite set of atomic propositions and let $Agt = \{1, \ldots, n\}$ be a finite set of agents. Let $Var = Atm \cup Agt$, where Var is the set of variables. The language $\mathcal{L}_0(Atm, Agt)$ is defined as follows:

$$\alpha ::= p \mid \neg\alpha \mid \alpha_1 \wedge \alpha_2 \mid \triangle_i\alpha \mid \text{☆}_i\alpha \mid \bigcirc_i\alpha \mid \bigcirc_{i,j},$$

where p ranges over Atm and i, j ranges over Agt.

The language $\mathcal{L}_{\mathsf{LFB}}(Atm, Agt)$ extends $\mathcal{L}_0(Atm, Agt)$ by building a new level of language with implicit belief operators and is defined as follows:

$$\varphi ::= \alpha \mid \neg\varphi \mid \varphi_1 \wedge \varphi_2 \mid \Box_i\varphi \mid \bigcirc_i\varphi,$$

where α ranges over $\mathcal{L}_0(Atm, Agt)$ and i ranges over Agt.

When the context is unambiguous, we write $\mathcal{L}_0$ instead of $\mathcal{L}_0(Atm, Agt)$ and $\mathcal{L}_{\mathsf{LFB}}$ instead of $\mathcal{L}_{\mathsf{LFB}}(Atm, Agt)$. The other Boolean connectives $\vee$, $\rightarrow$ and $\leftrightarrow$ are defined from $\neg$ and $\wedge$ in the standard way. We make $\top$ a primitive symbol and treat it as vacant truth. The formula $\triangle_i\alpha$ is read "agent i explicitly believes that α is true", $\bigcirc_i\varphi$ is read "agent i is aware of φ", $\text{☆}_i\alpha$ is read "agent i is aware that α is true", $\bigcirc_{i,j}$ is read "agent i is aware of agent j". Note that $\bigcirc_{i,j}$ can be treated as an atom, and other operators from the first level can be iterated or nested, such as $\triangle_i\triangle_i\alpha$, $\triangle_i\text{☆}_i\alpha$ and $\text{☆}_i\bigcirc_i(\alpha \wedge \bigcirc_{i,j})$.

The formula $\Box_i\varphi$ is read "agent i implicitly believes that φ is true". The dual operator $\Diamond_i$ is defined as follows:

$$\Diamond_i\varphi \stackrel{\mathsf{def}}{=} \neg\Box_i\neg\varphi,$$

where $\Diamond_i\varphi$ is read "φ is consistent with agent i's explicit beliefs". Note that the awareness-of operator $\bigcirc_i$ appears at both levels of the language, but the modalities $\triangle_i$ and ☆_i only appears at the first level. As a result, we can have awareness operators in the scope of $\triangle_i$ or $\star_i$, but not implicit belief operators in the scope of $\triangle_i$ or $\star_i$.

The following function $Var : \mathcal{L}_{\mathsf{LFB}} \longrightarrow 2^{Var}$ specifies the variables occurring in a formula:

$$\begin{aligned}
Var(\top) &= \emptyset, \\
Var(p) &= \{p\}, \text{ for } p \in Atm, \\
Var(\bigcirc_{i,j}) &= \{i, j\}, \\
Var(\neg\varphi) &= Var(\varphi), \\
Var(\varphi_1 \wedge \varphi_2) &= Var(\varphi_1) \cup Var(\varphi_2), \\
Var(Y_i\alpha) &= \{i\} \cup Var(\alpha), \text{ for } Y \in \{\triangle, \bigcirc, \star\}, \\
Var(Y_i\varphi) &= \{i\} \cup Var(\varphi), \text{ for } Y \in \{\Box, \bigcirc\}.
\end{aligned}$$

Let $\Gamma \subseteq \mathcal{L}_{\mathsf{LFB}}$ be finite, we define $Var(\Gamma) := \bigcup_{\varphi\in\Gamma} Var(\varphi)$. For simplicity, let$Atm(\varphi) = Var(\varphi) \cap Atm$, $Atm(\Gamma) = Var(\Gamma) \cap Atm$, $Agt(\varphi) = Var(\varphi) \cap Agt$, $Agt(\Gamma) = Var(\Gamma) \cap Agt$.

3 Semantics

In this section, we present the belief base semantics for $\mathcal{L}_{\mathsf{LFB}}$, where an agent's set of doxastic alternatives are not primitive but computed from them. The basic constituent of our semantics is the following notion of state.

Definition 1. *A state is a tuple $S = (B, At, Ao, V)$, where*

- $B = \{B_1, \ldots, B_n\}$, *where $B_i \subseteq \mathcal{L}_0$ is agent i's belief base for every $i \in Agt$,*
- $At = \{At_1, \ldots, At_n\}$, *where $B_i \subseteq At_i \subseteq \mathcal{L}_0$ is agent i's awareness-that set for every $i \in Agt$,*
- $Ao = \{Ao_1, \ldots, Ao_n\}$, *where $Ao_i = Var(B_i)$ is agent i's awareness-of set for every $i \in Agt$,*
- $V \subseteq Atm$ *is the actual environment.*

Compared with [18], this definition is enriched with an awareness-that set for every agent. And we let each agent's belief base be a subset of her awareness-that set. It follows the intuition that each agent acknowledges some formulas as true, and part of them are active in her working memory and form her belief base. The set of all states is denoted by $\mathbf{S}$. With the definition of state, we have the following interpretations for the formulas in $\mathcal{L}_0$.

Definition 2. *For any $S = (B, At, Ao, V) \in \mathbf{S}$:*

$$\begin{aligned}
S \models p &\iff p \in V, \\
S \models \neg\alpha &\iff S \not\models \alpha, \\
S \models \alpha_1 \wedge \alpha_2 &\iff S \models \alpha_1 \text{ and } S \models \alpha_2, \\
S \models \triangle_i\alpha &\iff \alpha \in B_i, \\
S \models \star_i\alpha &\iff \alpha \in At_i, \\
S \models \bigcirc_i\alpha &\iff Var(\alpha) \subseteq Ao_i, \\
S \models \bigcirc_{i,j} &\iff j \in Ao_i.
\end{aligned}$$

The only caveat is that, for an agent being aware of a formula, we require her to be aware of every variable (atomic propositions and agents) occurring in the formula. This is different with [18] that assumes an agent is aware of all agents in *Agt*. The following definition builds multi-agent fine-grained belief model from states.

Definition 3. *A multi-agent fine-grained belief model* (MFBM) *is a pair* (S, Cxt)*, where* $S \in \mathbf{S}$ *and* $Cxt \subseteq \mathbf{S}$.

Following [16], in the following definition we compute the agents' epistemic accessibility relations from their belief bases.

Definition 4. *For any* $i \in Agt$, $\mathcal{R}_i$ *is the binary relation on* $\mathbf{S}$ *such that for any* $S = (B, At, Ao, V), S' = (B', At', Ao', V') \in \mathbf{S}$,

$$(S, S') \in \mathcal{R}_i \text{ if and only if } \forall \alpha \in B_i,\ S' \models \alpha^2.$$

The following definition gives interpretations for formulas in the second level of $\mathcal{L}_{\mathsf{LFB}}$. The boolean cases are omitted.

Definition 5. *Let* (S, Cxt) *be a* MFBM *with* $S = (B, At, Ao, V)$. *Then,*

$$\begin{aligned}
(S, Cxt) \models \alpha &\Longleftrightarrow S \models \alpha,\\
(S, Cxt) \models \Box_i\varphi &\Longleftrightarrow \forall S' \in Cxt,\ \text{if } S\mathcal{R}_iS'\\
&\qquad\quad \text{then } (S', Cxt) \models \varphi,\\
(S, Cxt) \models \bigcirc_i\varphi &\Longleftrightarrow Var(\varphi) \subseteq Ao_i.
\end{aligned}$$

The following two definitions specify two properties of MFBMs, that correspond to consistency and reflexivity in Kripke model.

Definition 6. *The* MFBM (S, Cxt) *satisfies global consistency* (GC) *if and only if, for every* $i \in Agt$ *and for every* $S' \in (\{S\} \cup Cxt)$, *there exists* $S'' \in Cxt$ *such that* $(S', S'') \in \mathcal{R}_i$.

Definition 7. *The* MFBM (S, Cxt) *satisfies belief correctness* (BC) *if and only if* $S \in Cxt$ *and, for every* $i \in Agt$ *and for every* $S' \in Cxt$, $(S', S') \in \mathcal{R}_i$.

For $\mathrm{X} \subseteq \{\mathrm{GC}, \mathrm{BC}\}$, $\mathbf{MFBM}_{\mathrm{X}}$ is the class of MFBMs satisfying all the conditions in X. It is easy to see that $\mathbf{MFBM}_{\{\mathrm{GC},\mathrm{BC}\}} = \mathbf{MFBM}_{\{\mathrm{BC}\}}$.

---

[2] An alternative is to define the epistemic alternatives by all formulas in the awareness-that set of agent i, so that everything the agent acknowledges as true holds in all her epistemic possibilities. Such a definition gives the model an entirely different meaning. We draw the arrows based on belief bases because an agent makes inferences from what she has in her working memory. If the arrows are generated from her awareness-that set, then her epistemic possibilities denote what she can "potentially" infer from what she acknowledges as true.

4 Axiomatics

In this section, we define some variants of the LFB logics and prove their soundness and completeness for their corresponding model classes.

We define the base logic LFB to be the extension of classical propositional logic given by the following axioms and rule of inference:

$$\big(\Box_i\varphi \wedge \Box_i(\varphi \rightarrow \psi)\big) \rightarrow \Box_i\psi \tag{$\mathbf{K}_{\Box_i}$}$$

$$\triangle_i\alpha \rightarrow \Box_i\alpha \tag{$\mathbf{Int}_{\triangle_i,\Box_i}$}$$

$$\triangle_i\alpha \rightarrow \medwhitestar_i\alpha \tag{$\mathbf{Int}_{\triangle_i,\medwhitestar_i}$}$$

$$\triangle_i\alpha \rightarrow \bigcirc_i\alpha \tag{$\mathbf{Int}_{\triangle_i,\bigcirc_i}$}$$

$$\bigcirc_i\,\varphi \leftrightarrow \bigwedge_{p\in Atm(\varphi)} \bigcirc_i p \wedge \bigwedge_{j\in Agt(\varphi)} \bigcirc_{i,j} \tag{$\mathbf{AGPA}$}$$

$$\frac{\varphi}{\Box_i\varphi} \tag{$\mathbf{Nec}_{\Box_i}$}$$

For $\mathrm{X} \subseteq \{\mathbf{D}_{\Box_i}, \mathbf{T}_{\Box_i}\}$, let LFB_X be the extension of logic LFB by every axiom in X, where,

$$\neg(\Box_i\varphi \wedge \Box_i\neg\varphi) \tag{$\mathbf{D}_{\Box_i}$}$$

$$\Box_i\varphi \rightarrow \varphi \tag{$\mathbf{T}_{\Box_i}$}$$

To prove completeness of each logic LFB_X, let us define the following correspondence function between axioms and semantic properties:

- $cf(\mathbf{D}_{\Box_i}) = \mathrm{GC}$,
- $cf(\mathbf{T}_{\Box_i}) = \mathrm{BC}$.

Theorem 1. *Let* $\mathrm{X} \subseteq \{\mathbf{D}_{\Box_i}, \mathbf{T}_{\Box_i}\}$. *Then, the logic* LFB_X *is sound and complete for the class* $\mathbf{MFBM}_{\{cf(x):x\in\mathrm{X}\}}$.

Proof. As the proof procedure is similar with that of [18], here we only sketch the main techniques. Above all, we need to define two Kripke-style semantics. One is called notional model semantics, the model of which satisfies certain properties corresponding to MFBMs. The other is called quasi-model semantics, where the model relaxes the properties of notional model semantics. Then we prove that the three semantics are equivalent in terms of satisfiability with respect to a finite set of formulas. Following that, we employ the canonical model method to prove that LFB_X is complete for the corresponding class of quasi-models. Subsequently, by the equivalence result, it is straightforward that LFB_X is complete for the class $\mathbf{MFBM}_{\{cf(x):x\in\mathrm{X}\}}$. For soundness, the proof is standard.

5 Dynamics of Belief and Awareness

This section accommodates a variety of dynamics involving awareness and belief. According to Lorini [16], belief base models have distinct advantage on private dynamics, as operations only modify the belief bases of some agents but not of all agents, and the accessibility relations are recomputed after that. This leads to a "parsimonious" account of private informative actions. MFBMs inherits this merit and adapts to even richer notions of private dynamics.

5.1 Operation Definitions

Firstly, we provide a formal definition of the action of learning. In fact, it is almost identical with "Private Belief Expansion" in [16]. The only difference is that we have to attend to the awareness-of and awareness-that set when adding a new formula to an agent's belief base. One may notice an agent can learn something already in her belief base, or in her awareness-that set but not in her belief base, or in neither of the two sets.

Definition 8. *Let* (S, Cxt) *be a* MFBM *with* $S = (B, At, Ao, V)$. *Then the learning operation works as follows.*

$$(S, Cxt) \models [+_i\alpha]\varphi \Longleftrightarrow (S^{+_i\alpha}, Cxt) \models \varphi,$$

with $S^{+_i\alpha} = (B^{+_i\alpha}, At^{+_i\alpha}, Ao^{+_i\alpha}, V)$, *where for all* $j \in Agt$:

$$\begin{array}{lll} B_j^{+_i\alpha} = B_j \cup \{\alpha\} & At_j^{+_i\alpha} = At_j \cup \{\alpha\} & \text{if } i = j, \\ B_j^{+_i\alpha} = B_j & At_j^{+_i\alpha} = At_j & \text{otherwise}, \\ Ao_j^{+_i\alpha} = Var(B_j^{+_i\alpha}). & & \end{array}$$

Secondly, we move to the action of recalling. Being different with the learning action, an agent can only recall what is already in her awareness-that set.

Definition 9. *Let* (S, Cxt) *be a* MFBM *with* $S = (B, At, Ao, V)$. *Then the recalling operation works as follows.*

$$(S, Cxt) \models [\oplus_i\alpha]\varphi \Longleftrightarrow (S^{\oplus_i\alpha}, Cxt) \models \varphi,$$

with $S^{\oplus_i\alpha} = (B^{\oplus_i\alpha}, At, Ao^{\oplus_i\alpha}, V)$, *where for all* $j \in Agt$:

$$\begin{array}{ll} B_j^{\oplus_i\alpha} = B_j \cup \{\alpha\} & \text{if } i = j \text{ and } \alpha \in At_i, \\ B_j^{\oplus_i\alpha} = B_j & \text{otherwise}, \\ Ao_j^{\oplus_i\alpha} = Var(B_j^{\oplus_i\alpha}). & \end{array}$$

Thirdly, we define the action of becoming aware of. This operation is a little tricky, as we want to make the agent become aware of some propositions or agents without changing what she believes or acknowledges as true. To make this happen, we stipulate formulas of the form $\varphi \vee \top$ as vacant formulas that only affect an agent's awareness-of set without forcing her to believe anything.

Definition 10. *Let* (S, Cxt) *be a* MFBM *with* $S = (B, At, Ao, V)$. *Then the becoming aware of operation works as follows.*

$$(S, Cxt) \models [\odot_i \psi]\varphi \iff (S^{\odot_i \psi}, Cxt) \models \varphi,$$

with $S^{\odot_i \psi} = (B^{\odot_i \psi}, At^{\odot_i \psi}, Ao^{\odot_i \psi}, V)$, *where for all* $j \in Agt$:

$$\begin{aligned}
B_j^{\odot_i \psi} &= B_j \cup \{\psi \vee \top\} \quad At_j^{\odot_i \psi} = At_j \cup \{\psi \vee \top\} && \textit{if } i = j \\
B_j^{\odot_i \psi} &= B_j \quad\quad\quad\quad\quad\; At_j^{\odot_i \psi} = At_j && \textit{otherwise,} \\
Ao_j^{\odot_i \psi} &= Var(B_j^{\odot_i \psi}).
\end{aligned}$$

To make the semantics simple, we don't have separate operators for becoming aware of propositions and agents. As a result, when capturing an action of becoming aware of an agent j, simply make j occur in the formula ψ, for instance, let ψ be the form $\triangle_j \top$[3]. It makes sense because when we are aware of some agent, we believe that she has the ability to believe something i.e., to believe truth.

It's worth mentioning that a similar notion "attention" is used interchangeably with awareness in some literature [13]. According to the latter, inattention concerns "...concepts that the agent in principle understands, but has not thought to apply to the case at hand", which seems another depiction of working memory. So our notion of awareness can be regarded as attention, and it is different with another notion of "conceptual grasp" awareness.

Fourthly, we enter into the actions of reducing information and talk about forgetting. To forget is to completely lose one piece of information, and one needs to relearn it if she wants to explicitly believe or know it again. It roots in the experience shared by us that, when we believe that we are touching something new, someone tells us that we have learned it before, and such a reminder may frustrate us a little bit.

Definition 11. *Let* (S, Cxt) *be a* MFBM *with* $S = (B, At, Ao, V)$. *Then the forgetting operation works as follows.*

$$(S, Cxt) \models [-_i \alpha]\varphi \iff (S^{-_i \alpha}, Cxt) \models \varphi,$$

[3] We are using "dummy" formulas to formalize the actions of becoming aware. One can argue that this forces the agent to know/believe something, regardless of how simple that might be. We have to acknowledge that this is the shortcoming or defect of our approach of awareness dynamics. However, we would like to put more emphasis on the minimality and simplicity of our structure. The purpose is to propose a compatible method to incorporate awareness dynamics into belief base structure by making the minimal change on the original model. Compared with Fagin & Halpern's awareness logic, belief base semantics models explicit and implicit belief without invoking the notion of awareness and it is good thing [16]. We want to show that our semantics supports awareness and its dynamics anyway. Another defense for this treatment is that it conforms to the way of computers processing data. Belief bases are analogous to computer memories. Raw data are stored in them and are "translated" into different kinds of messages based on certain rules.

with $S^{-_i\alpha} = (B^{-_i\alpha}, At^{-_i\alpha}, Ao^{-_i\alpha}, V)$, *where for all* $j \in Agt$:

$$B_j^{-_i\alpha} = B_j \setminus \{\alpha\} \qquad At_j^{-_i\alpha} = At_j \setminus \{\alpha\} \qquad \textit{if } i = j,$$
$$B_j^{-_i\alpha} = B_j \qquad At_j^{-_i\alpha} = At_j \qquad \textit{otherwise},$$
$$Ao_j^{-_i\alpha} = Var(B_j^{-_i\alpha}).$$

From this definition, we can find out two forms of forgetting. One is some information disappearing from the agent's working memory and inactive storage all at once, maybe her head is impacted by blunt force unfortunately, causing her to lose track of what she is thinking about and lose part of her memory. The other is some information fading from the agent's awareness-that set unconsciously, which frequently occurs in our life. For simplicity, we don't differentiate these two kinds of forgetting and capture them by only one type of action.

There are proposals for representing actions for forgetting [3,6] working within the DEL setting. Van Ditmarsch et al. [3] formalize forgetting by means of a dynamic operator, the latter corresponds to an event model that transforms the original Kripke-style model into a new one containing copies of states with different valuations, so that the agent becomes ignorant of certain atomic propositions (for instance, from being certain of p to uncertainty of p and $\neg p$). Fernández-Duque et al. [6] extend [3] to capture the forgetting of complex propositional formulas using a slightly more complicated technique that applies to a Kripke-style model. Compared with their works, our approach provides a "parsimonious" account of actions of forgetting, since it does not require to duplicate epistemic models and to make them exponentially larger in a series of actions. Moreover, our approach easily supports multi-agent dynamics of forgetting as one agent's belief change does not influence others' belief bases. And it captures agents forgetting modal formulas as well as propositional formulas without complicating the technique.

Fifthly, we would like to introduce a notion indicating information being dropped from working memory while still stored in the agent's awareness-that set. We simply call the action "dropping".

Definition 12. *Let* (S, Cxt) *be a* MFBM *with* $S = (B, At, Ao, V)$. *Then the dropping operation works as follows.*

$$(S, Cxt) \models [\ominus_i\alpha]\varphi \Longleftrightarrow (S^{\ominus_i\alpha}, Cxt) \models \varphi,$$

with $S^{\ominus_i\alpha} = (B^{\ominus_i\alpha}, At, Ao^{\ominus_i\alpha}, V)$, *where for all* $j \in Agt$:

$$B_j^{\ominus_i\alpha} = B_j \setminus \{\alpha\} \qquad \textit{if } i = j,$$
$$B_j^{\ominus_i\alpha} = B_j \qquad \textit{otherwise},$$
$$Ao_j^{\ominus_i\alpha} = Var(B_j^{\ominus_i\alpha}).$$

Intuitively, we can only drop some information currently active in our working memory[4].

[4] A reviewer puts forward an interesting question: couldn't an agent have (in reality) an explicit belief that is not in her working memory? For example, a belief that she

Sixthly, we define a new form of announcement quite different from *public announcement* introduced by Plaza [21]. As we all know, DEL interprets knowledge by means of uncertainty i.e., an agent not knowing whether p is equivalent to she being uncertain about p (p does not have a uniform value in all her accessible states). As a result, to formalize public announcement, we need to shrink the DEL model and make agents certain about what is announced. However, in our model, explicit knowledge is captured by formulas included in agents' belief bases. Then, if we want to formalize public announcement in an explicit manner, we have to add the particular formula into the belief bases of all agents.

Definition 13. *Let* (S, Cxt) *be a* MFBM *with* $S = (B, At, Ao, V)$*. Then the operation of explicit announcement works as follows.*

$$(S, Cxt) \models [\uplus_i \alpha]\varphi \iff (S^{\uplus_i \alpha}, Cxt) \models \varphi,$$

with $S^{\uplus_i \alpha} = (B^{\uplus_i \alpha}, At^{\uplus_i \alpha}, Ao^{\uplus_i \alpha}, V)$*, where for all* $j \in Agt$*:*

$$\begin{array}{lll} B_j^{\uplus_i \alpha} = B_j \cup \{\alpha\} & At_j^{\uplus_i \alpha} = At_j \cup \{\alpha\} & \text{if } \alpha \in B_i, \\ B_j^{\uplus_i \alpha} = B_j & At_j^{\uplus_i \alpha} = At_j & \text{otherwise}, \\ Ao_j^{\uplus_i \alpha} = Var(B_j^{\uplus_i \alpha}). & & \end{array}$$

There are two more features of the announcement operation that deserved to be noted. Compared with [21], an explicit announcement is made by a specific agent rather than an external source, which follows the announcement operation in [10]. Not only that, it is quite a different notion of public announcements in [16]. In fact, it can be viewed as a group version of private belief base expansion in [16], except that the formula being added should be already included in agent i's belief base.

Finally, following G&V [9,10], the deductive inference operation also turns out to be a model transformation, which is not regarded by Lorini [16].

Definition 14. *Let* (S, Cxt) *be a* MFBM *with* $S = (B, At, Ao, V)$*. Then the operation of deductive inference works as follows.*

$$(S, Cxt) \models \left[\xhookrightarrow{\alpha \to \beta}_i\right]\varphi \iff (S^{\xhookrightarrow{\alpha \to \beta}_i}, Cxt) \models \varphi,$$

with $S^{\xhookrightarrow{\alpha \to \beta}_i} = (B^{\xhookrightarrow{\alpha \to \beta}_i}, At^{\xhookrightarrow{\alpha \to \beta}_i}, Ao, V)$*, where for all* $j \in Agt$*:*

$$\begin{array}{lll} B_j^{\xhookrightarrow{\alpha \to \beta}_i} = B_j \cup \{\beta\} & At_j^{\xhookrightarrow{\alpha \to \beta}_i} = At_j \cup \{\beta\} & \text{if } i = j \text{ and } \{\alpha, \alpha \to \beta\} \subseteq B_i, \\ B_j^{\xhookrightarrow{\alpha \to \beta}_i} = B_j & At_j^{\xhookrightarrow{\alpha \to \beta}_i} = At_j & \text{otherwise}. \end{array}$$

holds explicitly (say that she uses it as justifications of her actions) but that she is not currently explicitly considering. But it is still there, in her mind somewhere, not dropped. Here is my answer. An agent needs certain explicit beliefs to guide or justify her behavior. In the beginning, she has to keep them in her working memory. After a while, when such behavior turns into a habit, she does not need to hold such beliefs in her working memory anymore. Then, the habit becomes independent of the beliefs that used to guide or justify the behavior.

By the definition, in order to make a deduction, an agent should explicitly believe the premises in the first place.

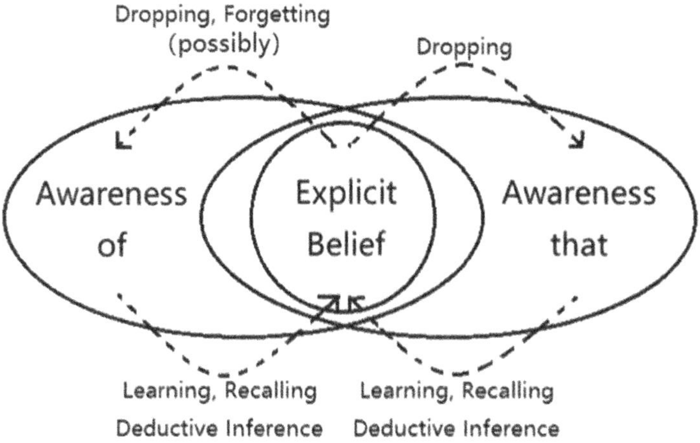

Fig. 1. Logical relationships among the notions of: awareness of, awareness that and explicit knowledge. The dashed arrows denote the actions by which we can move between different attitudes. Note that, by dropping or forgetting the belief of α, the agent possibly remains aware of atomic propositions occurring in α (if there remain other pieces of beliefs involving them), and possibly ceases to be aware of them (if no other belief involves them).

Figure 1 provides an overview on how these three notions are related by the private actions defined above. Through learning, recalling and deductive inference, an agent can get explicit beliefs from what she is aware of or aware that. Among the three actions, learning is the strongest that helps agents obtaining totally unfamiliar information. However, from explicit belief, we could move to awareness that by dropping, and possibly move to awareness of by dropping and forgetting. It is worth noting that, explicit belief is not equal to awareness of plus awareness that, i.e., the latter two notions are just a necessary condition for the former.

5.2 Scenarios

Example 1. Now we are ready to account for the scenario in Introduction. Assume Lucy is agent i at state $S = (B, At, Ao, V)$ included in a MFBM (S, Cxt). In the beginning, Lucy is aware that $p \wedge q \to r$, but does not explicitly know it[5].

$$(S, Cxt) \models \text{☆}_i(p \wedge q \to r) \wedge \neg \triangle_i(p \wedge q \to r)$$

[5] In this case, since $S \models p \wedge q \to r$, we use knowing instead of believing.

Then she stares at the keyboard and becomes aware of p and q. Besides, as she is thinking about undoing the previous action, she becomes aware of r. However, these actions together do not lead her to recall $p \wedge q \rightarrow r$.

$$(S, Cxt) \models [\odot_i p]\,[\odot_i q]\,[\odot_i r]((\bigcirc_i p \wedge \bigcirc_i q \wedge \bigcirc_i r) \wedge \neg\triangle_i(p \wedge q \rightarrow r))$$

When Lucy is attending the exam, she notices the correct answer of "undo the previous action". She recalls what she has learned before and becomes explicitly knowing $p \wedge q \rightarrow r$.

$$(S, Cxt) \models [\oplus_i(p \wedge q \rightarrow r)]\triangle_i(p \wedge q \rightarrow r)$$

Need to add that, there is a substantial difference between learning and recalling. To learn something, we need to get clear-cut information, not like the choice question that does not inform us of which one is the correct answer. In this case, it is the forth option as a reminder about what Lucy is aware that makes Lucy recalls $p \wedge q \rightarrow r$ and become explicitly know it.

Example 2. This part is devoted to capture the famous scenario from "12 Angry Men" brought up by G&V [9,10]. Assume agent A is at state $S = (B, At, Ao, V)$ included in a MFBM (S, Cxt). In the beginning, the jury explicitly know nothing[6] but are aware that $mkns \rightarrow gls$, $gls \rightarrow esq$ and $esq \rightarrow \neg glt$. We takes

$$(S, Cxt) \models [+_A mkns]\text{☆}_A(mkns \rightarrow gls) \wedge \text{☆}_A(gls \rightarrow esq) \wedge \text{☆}_A(esq \rightarrow \neg glt)$$

The relevant atomic propositions are defined as follows.

gls : the woman wears glasses $\quad$ $mkns$: she has marks in the nose

esq : her eyesight is in question $\quad$ glt : the accused is guilty

In court, A learns $mkns$ and then drops it.

$$(S, Cxt) \models [+_A mkns]\triangle_A mkns \qquad (S^{+_A mkns}, Cxt) \models [\ominus_A mkns]\text{☆}_A mkns$$

During the jury discussion, agent H's action of scratching his nose makes A recall $mkns$. After that, A recalls the three relevant implications from his personal experience.

$$((S^{+_A mkns})^{\ominus_i \alpha}, Cxt) \models [\oplus_A mkns]\triangle_A mkns$$
$$((S^{+_A mkns})^{\ominus_i \alpha}, Cxt) \models [\oplus_A(mkns \rightarrow gls)]\triangle_A(mkns \rightarrow gls)$$
$$((S^{+_A mkns})^{\ominus_i \alpha}, Cxt) \models [\oplus_A(gls \rightarrow esq)]\triangle_A(gls \rightarrow esq)$$
$$((S^{+_A mkns})^{\ominus_i \alpha}, Cxt) \models [\oplus_A(esq \rightarrow \neg glt)]\triangle_A(esq \rightarrow \neg glt)$$

Let $S_1 = (((((S^{+_A mkns})^{\ominus_i \alpha})^{\oplus_A mkns})^{\oplus_A(mkns \rightarrow gls)})^{\oplus_A(gls \rightarrow esq)})^{\oplus_A(esq \rightarrow \neg glt)}$ be the state after A recalls all the relevant information. Then A makes an explicit announcement about $mkns$.

$$\text{For all } j \in Agt,\ (S_1, Cxt) \models [\uplus_A mkns]\triangle_j mkns.$$

[6] Again, since $S \models mkns \wedge gls \wedge esq \wedge \neg glt$, we use knowing instead of believing.

A can also make deductive inferences to reach the conclusion that the accused is not guilty[7].

$$(S_1, Cxt) \models \left[\xrightarrow{mkns \rightarrow gls}_A\right] \left[\xrightarrow{gls \rightarrow esq}_A\right] \left[\xrightarrow{esq \rightarrow \neg glt}_A\right] \triangle_A \neg glt$$

Example 3. There are two robots i and j that never met each other before, where i is a new type, and j is an old type. Given the condition that every robot is coded with the information of all robots previously manufactured, i is aware of j, while j is unaware of i. Now i and j are passing through a corridor in opposite directions. Because they are sending identifiers by radio, they notice each other and write the information into their databases. In this case, j receives relatively more information than i. It is for the reason that, j not only becomes knowing that i is passing through the corridor, but also becomes aware of i.

Let i and j be at state $S = (B, At, Ao, V)$ included in a MFBM (S, Cxt). Initially, we have

$$(S, Cxt) \models \bigcirc_{i,j} \wedge \neg \bigcirc_{j,i} .$$

Let α indicate "i passes the corridor", let β be "j passes the corridor". When the two robots meet each other, both learn that α and β.

$$(S, Cxt) \models [+_i\alpha]\triangle_i\alpha \wedge [+_i\beta]\triangle_i\beta \wedge [+_j\alpha]\triangle_j\alpha \wedge [+_j\beta]\triangle_j\beta$$

Not only that, both recognize each other and learn that the other robot becomes explicitly knowing $\alpha \wedge \beta$, and j becomes aware of i

$$(S, Cxt) \models [+_i(\triangle_j(\alpha \wedge \beta))]\triangle_i\triangle_j(\alpha \wedge \beta)$$
$$(S, Cxt) \models [+_j(\triangle_i(\alpha \wedge \beta))]\triangle_j\triangle_i(\alpha \wedge \beta) \wedge \bigcirc_{j,i}$$

6 Conclusion and Perspectives

This paper has proposed an investigation of notions of explicit and implicit belief, awareness of and awareness that. The novelty of our work rests with two aspects: one is that we do not assume awareness of all agents as a primitive, the other comes from the belief bases model we are using, that renders a minimal logic supporting fine-grained information dynamics. The latter involves a series of actions including learning, recalling, forgetting, dropping, becoming aware of, explicit announcement and deductive inference. These actions are defined as operations on agents' belief bases and awareness-that sets, highlighting the computational advantages of belief base model. Through these actions, a variety of examples are formalized. Besides, we have provided an axiomatization of the static logic, of which the soundness and completeness are proved by a similar process as that in [18].

An extended version of this paper will deliberate about a detailed comparison between our structure and the framework by G&V [9,10]. Specifically, our

[7] We have simplified the scenario and omitted the agents other than A and H.

definition of awareness that is slightly different with theirs, and we have different opinions on the notions of becoming aware of and recalling. Moreover, the reduction axioms of the dynamic operators are also part of the future work. Since our work is an extension of the logic in [18], and several embedding results are proved in the latter, it is reasonable to conjecture that analogous embeddings are to be held with respect to the logic in this paper.

Due to the technically concise nature of our model, it's worth to consider incorporating more notions into the belief base structure. One candidate is *aboutness* [7,8,27]. The idea simple: though a proposition is false generally, it can be true *about* certain issue. So a proposition may be partial truth. Here we quote a simple example from Russell [23].

> Certain philosophers, he says, having "arrived at results incompatible with the existence of error, . . . have then had to add a postscript explaining that what we call error is really partial truth. If we think it is Tuesday when it is really Wednesday, we are at least right in thinking that it is a day of the week. If we think America was discovered in 1066, we are at least right in thinking that something important happened in that year".

In my opinion, partial truth reveals certain connections underlying atomic propositions. Let p, q and r indicate "It happened on Tuesday", "It happened on Wednesday" and "It happened on a week day", respectively. Then, if p is true, r must be true. But the reverse does not hold. If a witness gives her testimony that p, but the truth is q, it needs the attorney to make more inquiries, such as "Do you have any proof that it was a week day?", to discover the truth of r. In fact, our previous paper [24] involves a similar notion concerning the connections between different atomic propositions leading to agents becoming aware of new propositions. A feasible way to tackle this problem is making such connections rules explicitly known by certain agents, i.e., let the rules be included in their belief bases, who can make inferences based on these rules and make inquiries according to consequence.

References

1. van Benthem, J., Velázquez-Quesada, F.R.: The dynamics of awareness. Synthese **177**, 5–27 (2010)
2. van Ditmarsch, H., French, T.: Becoming aware of propositional variables. In: Banerjee, M., Seth, A. (eds.) ICLA 2011. LNCS (LNAI), vol. 6521, pp. 204–218. Springer, Heidelberg (2011). https://doi.org/10.1007/978-3-642-18026-2_17
3. van Ditmarsch, H., Herzig, A., Lang, J., Marquis, P.: Introspective forgetting. In: Wobcke, W., Zhang, M. (eds.) AI 2008. LNCS (LNAI), vol. 5360, pp. 18–29. Springer, Heidelberg (2008). https://doi.org/10.1007/978-3-540-89378-3_2
4. van Ditmarsch, H., van der Hoak, W., Kooi, B.: Dynamic Epistemic Logic. Synthese Library, vol. 337. Springer, Dordrecht (2007). https://doi.org/10.1007/978-1-4020-5839-4
5. Fagin, R., Halpern, J.Y.: Belief, awareness, and limited reasoning. Artif. Intell. **34**(1), 39–76 (1987). https://doi.org/10.1016/0004-3702(87)90003-8

6. Fernández-Duque, D., Nepomuceno-Fernández, Á., Sarrión-Morrillo, E., Soler-Toscano, F., Velázquez-Quesada, F.R.: Forgetting complex propositions. Logic J. IGPL **23**(6), 942–965 (2015). https://doi.org/10.1093/jigpal/jzv049
7. Fine, K.: Truth-maker semantics for intuitionistic logic. J. Philos. Log. **43**(2), 549–577 (2014). https://doi.org/10.1007/s10992-013-9281-7
8. Fine, K.: Angellic content. J. Philos. Log. **45**(2), 199–226 (2015). https://doi.org/10.1007/s10992-015-9371-9
9. Grossi, D., Velázquez-Quesada, F.R.: *Twelve angry men*: a study on the fine-grain of announcements. In: He, X., Horty, J., Pacuit, E. (eds.) LORI 2009. LNCS (LNAI), vol. 5834, pp. 147–160. Springer, Heidelberg (2009). https://doi.org/10.1007/978-3-642-04893-7_12
10. Grossi, D., Velázquez-Quesada, F.R.: Syntactic awareness in logical dynamics. Synthese **192**(12), 4071–4105 (2015). https://doi.org/10.1007/s11229-015-0733-1
11. Heifetz, A., Meier, M., Schipper, B.C.: Interactive unawareness. J. Econ. Theory **130**(1), 78–94 (2006). https://doi.org/10.1016/j.jet.2005.02.007
12. Hintikka, J.: Knowledge and belief. Cornell University Press, Ithaca (1962)
13. de Jager, S.: "Now that you mention it, I wonder..." : awareness, attention, assumption. PhD dissertation, Institute for Logic, Language and Computation, University of Amsterdam (2009)
14. Levesque, H.J.: A logic of implicit and explicit belief. In: Proceedings of the Fourth AAAI Conference on Artificial Intelligence, pp. 198–202. AAAI 1984, AAAI Press (1984)
15. Lorini, E.: In praise of belief bases: doing epistemic logic without possible worlds. In: McIlraith, S.A., Weinberger, K.Q. (eds.) Proceedings of the Thirty-Second AAAI Conference on Artificial Intelligence (AAAI-18), pp. 1915–1922. AAAI Press (2018)
16. Lorini, E.: Rethinking epistemic logic with belief bases. Artif. Intell. **282**, 103233 (2020). https://doi.org/10.1016/j.artint.2020.103233
17. Lorini, E., Song, P.: Grounding awareness on belief bases. In: Martins, M.A., Sedlár, I. (eds.) DaLi 2020. LNCS, vol. 12569, pp. 170–186. Springer, Cham (2020). https://doi.org/10.1007/978-3-030-65840-3_11
18. Lorini, E., Song, P.: A computationally grounded logic of awareness. J. Logic Comput. **33**, 1463–1496 (2022). https://doi.org/10.1093/logcom/exac035
19. Modica, S., Rustichini, A.: Awareness and partitional information structures. Theor. Decis. **37**(1), 107–124 (1994). https://doi.org/10.1007/BF01079207
20. Modica, S., Rustichini, A.: Unawareness and partitional information structures. Games Econ. Behav. **27**(2), 265–298 (1999). https://doi.org/10.1006/game.1998.0666
21. Plaza, J.: Logics of public communications. In: Emrich, M., Pfeifer, M., Hadzikadic, M., Ras, Z. (eds.) Proceedings of the 4th International Symposium on Methodologies for Intelligent Systems, pp. 201–216 (1989)
22. Ågotnes, T., Alechina, N.: Full and relative awareness: a decidable logic for reasoning about knowledge of unawareness. In: Proceedings of the 11th Conference on Theoretical Aspects of Rationality and Knowledge, pp. 6–14. TARK 2007, Association for Computing Machinery, New York, NY, USA (2007). https://doi.org/10.1145/1324249.1324255
23. Russell, B.: Pragmatism. Philosophical Essays. Longmans, Green (1910)
24. Song, P., Xiong, W.: Awareness as potential for knowledge. J. Philos. Log. **52**(2), 669–703 (2023). https://doi.org/10.1007/s10992-022-09684-2

25. van Ditmarsch, H., French, T., Velázquez-Quesada, F.R., Wáng, Y.N.: Implicit, explicit and speculative knowledge. Artif. Intell. **256**, 35–67 (2018). https://doi.org/10.1016/j.artint.2017.11.004
26. Velázquez-Quesada, F.R.: Inference and update. Synthese **169**(2), 283–300 (2009). https://doi.org/10.1007/s11229-009-9556-2
27. Yablo, S.: Aboutness. In: Hempel, C.G. (ed.) Lecture Series. Princeton University Press (2014)

Hyperintensionality in Relevant Logics

Shawn Standefer(✉)

Department of Philosophy, National Taiwan University, Taipei City, Taiwan
standefer@ntu.edu.tw

Abstract. In this article, we present a definition of a hyperintensionality appropriate to relevant logics. We then show that relevant logics are hyperintensional in this sense, drawing consequences for other non-classical logics, including HYPE and some substructural logics. We further prove results concerning extensionality in relevant logics. We close by discussing related concepts for classifying formula contexts and potential applications of these results.

1 Introduction

Hyperintensionality, being able to distinguish necessarily equivalent formulas, has become an important topic in philosophical logic.[1] The growing importance of hyperintensionality for philosophical concepts has been highlighted by Nolan [24], calling it the "hyperintensional revolution." One can, of course, extend classical logic with hyperintensional operators,[2] but one might wonder whether other logics could offer something distinctive with respect to hyperintensional operators. Recently, Leitgeb [19] defended the non-classical logic HYPE as exhibiting a distinctive combination of simplicity and strength. Among its claimed features is providing a kind of hyperintensionality, a claim disputed by Odintsov and Wansing [25]. We will offer some support to Leitgeb's claim, proceeding via a discussion of relevant logics. Given some of the distinctions that relevant logics draw, such as distinguishing logical truths, it is natural to suspect that relevant logics build in a kind of hyperintensionality. We will argue that this suspicion is borne out by providing some hyperintensional contexts in relevant logics. In so doing, we will draw out some consequences for HYPE and other substructural logics.

In the remainder of this section, we will supply some brief background on relevant logics, in particular the logic R. Then, we will precisely define some concepts

[1] See Berto and Nolan [4].

[2] Some of the standard examples of hyperintensional operators added to classical logic, often though not always modeled using impossible worlds, include belief operators, knowledge operators, and conditional operators. See Wansing [40], Alechina and Logan [1], and Berto et al. [5], among others, for recent examples, and see Berto and Jago [6, ch. 7] for an overview of the work on epistemic logics. For a general approach to hyperintensional operators, see Sedlár [31].

N. Alechina et al. (Eds.): LORI 2023, LNCS 14329, pp. 238–250, 2023.
https://doi.org/10.1007/978-3-031-45558-2_18

to classify formula contexts in Sect. 2, notably extensionality and hyperintensionality. In Sect. 3, we will present our main results concerning hyperintensional contexts in relevant logics, drawing out a consequence for HYPE. Finally, in Sect. 4, we will look at two further definitions for classifying formula contexts and discuss some upshots of our results.

Relevant logics are a family of non-classical logics with a distinctive conditional, or implication, connective.[3] One of the important ways in which the relevant conditional is distinctive can be found in Belnap's variable sharing criterion: If $A \to B$ is valid, then A and B share a propositional variable. The variable sharing criterion is typically taken as a necessary condition on being a relevant logic. We will focus on the standard logical vocabulary of $\{\to, \wedge, \vee, \neg\}$, considering the addition of a modal operator $\Box$, below. The biconditional, $A \leftrightarrow B$, will be defined as $(A \to B) \wedge (B \to A)$. To contrast the relevant conditional and biconditional with the classical material ones, we will use $\supset$ and $\equiv$ for the latter connectives, defining $A \supset B$ as $\neg A \vee B$ and $A \equiv B$ as $(A \supset B) \wedge (B \supset A)$. In the context of relevant logics, and generally any non-classical logic, $A \supset B$ and $A \equiv B$ will be defined as in classical logic.

While there are many relevant logics, our focus will be on the logic R. R is a relatively strong logic. We will present the axioms and rules for it, where $\Rightarrow$ is used to demarcate premises from conclusion in the rules.

(1) $A \to A$
(2) $(A \wedge B) \to A$, $(A \wedge B) \to A$
(3) $((A \to B) \wedge (A \to C)) \to (A \to (B \wedge C))$
(4) $A \to (A \vee B)$, $B \to (A \vee B)$
(5) $((A \to C) \wedge (B \to C)) \to ((A \vee B) \to C)$
(6) $(A \wedge (B \vee C)) \to ((A \wedge B) \vee (A \wedge C))$
(7) $\neg\neg A \to A$
(8) $(A \to \neg B) \to (B \to \neg A)$
(9) $(A \to B) \to ((B \to C) \to (A \to C))$
(10) $A \to ((A \to B) \to B)$
(11) $(A \to (A \to B)) \to (A \to B)$
(12) $A, A \to B \Rightarrow B$
(13) $A, B \Rightarrow A \wedge B$

The logic R is the least set of formulas containing all the axioms and closed under the rules. Other relevant logics can be obtained by variation of axioms (8)–(11), dropping those axioms or possibly adding others, and by addition of other rules. The focus will be on R, although we will briefly consider some weaker relevant logics towards the end of Sect. 3. Let us now turn to some concepts for classifying formula contexts.

2 Classifying Contexts

Let us begin with some definitions. Following Williamson [41], define a formula context as a pair (C, p), of a formula and an atom. Given a context (C, p), the formula $C(A)$ is what results by replacing every occurrence of p in C with the formula A.

[3] See Dunn and Restall [10], Bimbó [7], or Mares [20] for overviews of the area. See Anderson and Belnap [2] and Routley et al [27] for broader discussions.

Definition 1 (Extensionality). *A formula context* (C, p) *is* extensional *iff for all formulas* A *and* B*,*

- $\models (A \equiv B) \supset (C(A) \equiv C(B))$

This is a fine definition of extensionality for classical logic and its extensions. It is not, however, appropriate for all non-classical logics. The reason is that in many non-classical logics, including relevant logics, the interest is on the primitive conditional connective, and the associated biconditional, rather than the material conditional of the logic, and the associated material biconditional.[4] Therefore, we will replace the definition of extensional context with one that uses the appropriate conditional and biconditional of the logic.

Definition 2 (Extensionality in L). *A formula context* (C, p) *is* extensional *in the logic* L *iff for all formulas* A *and* B*,*

- $\models_{\mathsf{L}} (A \leftrightarrow B) \rightarrow (C(A) \leftrightarrow C(B))$,

where $\models_{\mathsf{L}}$ *is the consequence relation of* L.

This is a natural adaptation of Williamson's definition to a non-classical context. For a more general study of extensionality and related concepts, we would need to make the relativity to the chosen conditional and biconditional explicit, so that the two options above would be $(\supset, \equiv)$-extensionality and $(\rightarrow, \leftrightarrow)$-extensionality, respectively. There are alternative definitions of extensionality using different combinations of $\rightarrow$, $\supset$, $\leftrightarrow$, and $\equiv$, but we won't explore those further here.[5] Our interest is not on extensional contexts per se, although we will return to them at the end of the next section. Our interest is, rather, in their use in the definition of non-hyperintensional contexts.

Definition 3 (Non-hyperintensionality, hyperintensionality). *A formula context* (C, p) *is* non-hyperintensional *in* L *iff for all formulas* A *and* B*,*

- $\models_{\mathsf{L}} \Box(A \leftrightarrow B) \rightarrow \Box(C(A) \leftrightarrow C(B))$.

A formula context is hyperintensional *in* L *iff it is not non-hyperintensional.*

A logic L *is hyperintensional iff there is a formula context* (C, p) *that is hyperintensional in* L.

Unpacking the definitions, a formula context (C, p) is *hyperintensional* iff there are formulas A and B such that $\not\models_{\mathsf{L}} \Box(A \leftrightarrow B) \rightarrow \Box(C(A) \leftrightarrow C(B))$. An immediate consequence of the definitions is the following proposition.

Proposition 4. *Let* M *be a sublogic of* L*. If* L *is hyperintensional, then so is* M.

[4] In the context of relevant logics, many of the contraction-free logics lack any theorems not containing '$\rightarrow$', for which see Slaney [32]; so $(\supset, \equiv)$-extensionality will be a less useful concept there. Yet, it still seems sensible to say that those logics have some extensional contexts made up only of the vocabulary $\{\wedge, \vee, \neg\}$. Thanks to an anonymous referee for raising this point.

[5] See Humberstone [15,16] and [17, 455] for more on extensionality of connectives.

Hyperintensionality is preserved downwards to sublogics. This will be important for our main result.

Before we proceed, it is worth noting an important intermediate category of formula contexts that we will not discuss below, namely the *intensional contexts*. These are contexts that are not extensional but are non-hyperintensional. Investigation of intensional contexts will be left for future work.

3 Hyperintensionality

Although there are many relevant logics, we will focus on the logic R, which is the strongest of the standard relevant logics.[6] The definitions of extensional, non-hyperintensional, and hyperintensional contexts should be understood as indexed to R, and its modal extensions, with the displayed conditional and biconditional being those of R. One could obtain versions of the definitions for other logics by changing the index.

Once we have settled the question of the base logic, there is a further question concerning which necessity to use in the statement of non-hyperintensionality. For a general study of hyperintensionality, care needs to be taken regarding what modal axioms, if any, should be required to ensure that the non-hyperintensionality definition yields satisfactory results. Williamson uses the necessity of S5 in stating his definition. The necessity of S5 would be a fine necessity for our purposes, but we can obtain stronger results with a different necessity.[7] A logic being hyperintensional is a matter of the invalidity of an instance of the non-hyperintensionality scheme, and, since invalidity is preserved from stronger logics down to weaker logics, using stronger modal principles will give stronger results concerning hyperintensionality. To motivate the appropriate modal principles, we will take a detour through logical necessity.

Anderson and Belnap showed how to define logical necessity in their logic E, a close relative of R, obtained by changing axiom (10) to its rule form, $A \Rightarrow (A \to B) \to B$, and adding a reductio axiom, $(A \to \neg A) \to \neg A$. Anderson and Belnap define $\Box A$ as $(A \to A) \to A$.[8] This can be understood as saying that logic implies A, which is a fair definition of logical necessity. In the context of E, $\Box$, so defined, has an S4-ish logic, and in the context of weaker relevant logics, it obeys weaker principles. In the context of R, however, the defined connective $\Box$ is trivial in the sense that $A \leftrightarrow \Box A$ is a logical truth. Taking this biconditional as a logic's modal axioms gives the modal logic known as TRIV. We will call the extension of R with the TRIV biconditional the logic R.TRIV. While the necessity of R.TRIV is not plausible as a kind of logical necessity, it is useful for the sort of

[6] See Mares [23] for defense of R.

[7] The concept of S5 necessity exhibits some subtleties in the context of relevant logics, for which see Standefer [36].

[8] One can obtain an alternative definition by using the Ackermann truth constant, t, which is glossed as the conjunction of all logical truths. Using the Ackermann constant, $\Box A$ can be defined as $t \to A$. The equivalence of the two definitions is demonstrated by Mares and Standefer [21], among others.

negative results we are after, so we will use it as the necessity in the definitions of non-hyperintensionality and hyperintensionality.

To obtain our main result, namely that many plausible modal extensions of R are hyperintensional, we first prove a lemma using matrix methods. A matrix has a set V of semantic values, with a subset of designated values $D \subseteq V$, and operations on V for interpreting each connective of the language. A valuation v is a function from atoms to V that is extending to the whole language using the operations of the matrix. A valuation v on a matrix is a counterexample to a formula A iff $v(A) \notin D$.

Lemma 5. *The formula* $(p \leftrightarrow q) \rightarrow ((p \wedge r) \leftrightarrow (q \wedge r))$ *is not a theorem of* R.

Proof. We will use a three-valued matrix. For the set of values, V, we take $\{0, \frac{1}{2}, 1\}$, with $D = \{\frac{1}{2}, 1\}$. The value of complex formulas is computed using the following tables.

$\rightarrow$	0	$\frac{1}{2}$	1	$\neg$
0	1	1	1	1
$\frac{1}{2}$	0	$\frac{1}{2}$	1	$\frac{1}{2}$
1	0	0	1	0

$\wedge$	0	$\frac{1}{2}$	1
0	0	0	0
$\frac{1}{2}$	0	$\frac{1}{2}$	$\frac{1}{2}$
1	0	$\frac{1}{2}$	1

$\vee$	0	$\frac{1}{2}$	1
0	0	$\frac{1}{2}$	1
$\frac{1}{2}$	$\frac{1}{2}$	$\frac{1}{2}$	1
1	1	1	1

A valuation v is a countermodel for a formula A iff $v(A) = 0$, which is to say that $v(A)$ is not designated.

Every axiom of R is designated on every valuation and the rules preserve designation.[9] By an inductive argument, this implies that every theorem of R receives a designated value. To show that a formula is not a theorem of R, it suffices to provide a valuation that assigns it 0. In the case of interest, $v(p) = 1$, $v(q) = 1$, and $v(r) = \frac{1}{2}$ will work.[10] This valuation gives $v(p \leftrightarrow q) = 1$, while $v((p \wedge r) \leftrightarrow (q \wedge r)) = \frac{1}{2}$. As $1 \rightarrow \frac{1}{2} = 0$,

$$v((p \leftrightarrow q) \rightarrow ((p \wedge r) \leftrightarrow (q \wedge r))) = 0,$$

as desired.

The formula scheme $(A \leftrightarrow B) \rightarrow ((A \wedge C) \leftrightarrow (B \wedge C))$ is not a theorem of R.[11] With this result in hand, we can turn to our main result.

Theorem 6. *The logic* R.TRIV *is hyperintensional.*

Proof. To show that R.TRIV is hyperintensional, we need a formula context which is hyperintensional. Take the formula context $(s \wedge r, s)$. The formula

$$\Box(p \leftrightarrow q) \rightarrow \Box((p \wedge r) \leftrightarrow (q \wedge r))$$

[9] This was shown by Robert Meyer. See Anderson and Belnap [2, 470].

[10] This countermodel was found using John Slaney's program MaGIC. See https://users.cecs.anu.edu.au/~jks/magic.html.

[11] Axioms of this form were studied by Routley et al [27, 345] and by Urbas and Sylvan [38]. Thanks to Andrew Tedder for drawing my attention to these citations.

is not valid in R.TRIV. This is because we can use the fact that $A \leftrightarrow \Box A$ to focus on the equivalent

$$(p \leftrightarrow q) \rightarrow ((p \wedge r) \leftrightarrow (q \wedge r)),$$

which was shown not to be a theorem of R in Lemma 5.

Thus, we have demonstrated that R.TRIV is hyperintensional. It is worth noting that, for similar reasons, $(p \vee r, p)$ is a hyperintensional context as well. As an immediate corollary, we have the following result.

Corollary 7. *Let* L *be any sublogic of* R.TRIV. *Then* L *is also hyperintensional.*

The sublogics of R.TRIV include all the well-known relevant logics, such as T, E, and B, as well as (multiplicative, additive) linear logic, and further it includes many of their extensions with well-known modal principles. We can extend a base logic L with a non-trivial, primitive necessity operator, $\boxdot$, rather than a defined one. However, as long as L is a sublogic of R, we can, in many cases of interest, embed the result into R.TRIV using the embedding $\tau(\boxdot A) = \Box\tau(A)$, i.e. $\tau(\boxdot A) = (\tau(A) \rightarrow \tau(A)) \rightarrow \tau(A)$, provided the modal principles for $\boxdot$ are among those of TRIV. For such logics, the countermodel above will suffice to demonstrate hyperintensionality, setting $v(\boxdot A) = v(A)$.

There are modal logics that are not sublogics of TRIV, although the majority of the philosophically significant ones are sublogics of TRIV. Perhaps the most prominent modal logics that are not sublogics of TRIV are provability logics, logics that include the axiom $\Box(\Box A \rightarrow A) \rightarrow \Box A$.[12] These have not been studied much in the context of relevant logics, although Mares [22] is an exception, studying a provability logic extension of R. Although the above countermodel does not work for Mares's provability logic, the same invalid formula demonstrates that the logic is hyperintensional. For other modal logics that are not sublogics of R.TRIV, it is left open whether they are hyperintensional or not.

As noted above, in relevant logics, one can define a logical necessity operator: $\Box A$ is $(A \rightarrow A) \rightarrow A$. For the logic R, this necessity obeys the TRIV principles, although for weaker base logics, the defined necessity is more like a familiar kind of necessity. Using this definition, we can view relevant logics as modal logics and use the defined necessity in the definition of hyperintensionality. In this sense, R and its sublogics are hyperintensional.

We will observe one additional corollary of Lemma 5.

Corollary 8. *There are contexts that fail to be extensional in* R.

Proof. As the lemma shows, $(s \wedge r, s)$ fails to be extensional in R.

For similar reasons, $(s \vee r, s)$ also fails to be extensional in R. While it is perhaps not surprising that R, and all of its sublogics, contain non-extensional contexts, it is worth noting that the particular non-extensional contexts provided involve

[12] See Boolos [8] and Verbrugge [39] for more on provability logics.

only conjunction or only disjunction, both typically thought of as extensional.[13] In the context of R, at least, Williamson's definition of extensional context, with $\supset$ and $\equiv$, would say that $(s \wedge r, s)$ is an extensional context, an $(\supset, \equiv)$-extensional context in the nomenclature of the previous section. This is not the case for many of the weaker relevant logics, which is a consequence of the results of Slaney [32].

By contrast, if we consider the set of connectives often described as intensional, or non-extensional, $\{\rightarrow, \neg, \circ\}$, where $\circ$ is the fusion connective, we find that they are all extensional.

Proposition 9. *Let (C, p) be a context built from atoms and only the connectives $\rightarrow, \neg$, and $\circ$. Then (C, p) is extensional in* R.

Proof. The connective $\circ$ is definable in R as $A \circ B =_{Df} \neg(A \rightarrow \neg B)$. The result is then proved by induction on structure of C, which is straightforward using axioms (8) and (11). The inductive hypothesis is that $\models_{\mathsf{R}} (A \leftrightarrow B) \rightarrow (D(A) \leftrightarrow D(B))$, for less complex contexts (D, p).

For the conditional case, the context is $(D \rightarrow E, p)$. As $(D(A) \rightarrow E(A)) \rightarrow (D(A) \rightarrow E(A))$ is provable by axiom (1), we can prove

$$(A \leftrightarrow B) \rightarrow ((A \leftrightarrow B) \rightarrow ((D(A) \rightarrow E(A)) \rightarrow (D(B) \rightarrow E(B))))$$

with the two appeals to the inductive hypothesis and some simple transitivity moves available in R. An appeal to axiom (11) then yields half of the desired result. The other half is obtained similarly.

For the negation cases, we use (8) and the desired result follows immediately.

For logics that lack axioms (8), (10), or (11), the proposition may fail. In weaker logics, some contexts built from the connectives $\{\rightarrow, \neg, \circ\}$ can fail to be extensional. As we will see shortly, all the standard relevant logics include the rule form of axiom (9) used in the proof. Let us look at some examples of failures of extensionality in logics lacking axioms (8), (10), or (11). The logic RW is obtained from the axiomatization of R by dropping axiom (11).

Proposition 10. *In* RW, *the context $(r \rightarrow r, r)$ is not extensional.*

Proof. We leave it to the reader to find a countermodel using MaGIC.

The logic T is obtained from the axiomatization of R by removing (10) and adding $(A \rightarrow \neg A) \rightarrow \neg A$. In it, fusion is not definable in terms of negation and conditional. Contexts built from fusion fail to be extensional.

Proposition 11. *In* T, *$(p \circ r, p)$ is not extensional.*

Proof. We leave it to the reader to find a countermodel using MaGIC.

Although fusion fails to be extensional, T still enjoys some extensionality similar to that of R.

[13] Cf. Gabbay [13] corollary 21.

Proposition 12. *In* T, *all contexts constructed from the vocabulary* $\{\rightarrow, \neg\}$ *are extensional.*

Proof. The negation and conditional cases from the proof from Proposition 9 can be reproduced here, omitting fusion.

It is worth looking at an example of a failure of extensionality for contexts built from negation that can be found in the logic B. The logic B is the weakest relevant logic that is standardly discussed, and it is obtained from R by dropping axioms (8)–(11) and adding the following rules.

- $A \rightarrow \neg B \Rightarrow B \rightarrow \neg A$
- $A \rightarrow B \Rightarrow (C \rightarrow A) \rightarrow (C \rightarrow B)$
- $A \rightarrow B \Rightarrow (B \rightarrow C) \rightarrow (A \rightarrow C)$

Some formulas in the basic vocabulary fail to be extensional in B, beyond the examples provided above.

Lemma 13. *In* B, *the formula context* $(\neg p, p)$ *is not extensional.*

Proof. In B,

$$(p \leftrightarrow q) \rightarrow (\neg p \leftrightarrow \neg q)$$

is invalid. We can adapt the matrix from the proof of Lemma 5 to show this. We change the set of designated values to $\{1\}$, replace the conditional table with

$\rightarrow$	0	$\frac{1}{2}$	1
0	1	1	1
$\frac{1}{2}$	0	1	1
1	0	$\frac{1}{2}$	1

and all valuations on the resulting matrix assign all the theorems of B designated values.[14] The valuation v where $v(p) = 1$ and $v(q) = \frac{1}{2}$ is a counterexample to the target formula.

To obtain HYPE, or at least its logical truths, from R, we add $A \rightarrow (B \rightarrow A)$ and trade axiom (8) for its rule form, $A \rightarrow \neg B \Rightarrow B \rightarrow \neg A$. It follows that we can obtain HYPE by adding some axioms to B. B shares with HYPE the feature of having contraposition as a rule but, crucially, *not* as an axiom, which results in the failure of the pertinent instance of the extensionality scheme above. In fact, this example extends to HYPE as well. This provides an example of hyperintensionality in B.TRIV, as the context $(\neg p, p)$ is also hyperintensional in B.TRIV, and so in all sublogics. A similar point holds for HYPE, and in fact, the same matrix demonstrates the failure of extensionality. Thus, HYPE exhibits hyperintensionality in the same sense as relevant logics. With these results in hand, let us turn to some further concepts for classifying formula contexts and some discussion.

[14] This countermodel was found using John Slaney's program MaGIC.

4 Discussion

Odintsov and Wansing [25] adopt an alternative notion of hyperintensionality, using *self-extensionality*,[15] also known as *congruentiality*,[16] which they argue is closer to the suggestions of Cresswell [9]. Adapting their definition to the present setting, a formula context (C, p) is *congruential* (in L) iff for all formulas A and B,

- if $\models_{\mathsf{L}} A \leftrightarrow B$, then $\models_{\mathsf{L}} C(A) \leftrightarrow C(B)$.

A logic is *congruential* iff all formula contexts are congruential in that logic. Relevant logics and their usual modal extensions are congruential, although there are modal extensions which are not congruential.[17]

It is worth distinguishing congruentiality and hyperintensionality for two reasons. First, it is natural to maintain the distinction between (i) claims that are necessarily equivalent but not logically equivalent and (ii) claims that are both necessarily and logically equivalent. One might think that certain truths of mathematics or metaphysics are necessarily, but not logically, equivalent. Second, hyperintensionality builds in a modal element that is absent in congruentiality in the sense that the former, but not the latter requires a modal operator be used in its definition. Third, and relatedly, congruentiality can be given an alternative definition that does not involve object language biconditionals, instead using mutual entailments, but hyperintensionality cannot be given such definition. Both hyperintensionality and congruentiality are important and interesting classifications of formula contexts, so it is worth distinguishing them.

Let us consider one further concept that could be considered for hyperintensionality in the present context. Although the discussion so far has proceeded at the level of *logics*, independent of any models, one could introduce models for relevant logics, such as the ternary relational models,[18] enabling us to talk about the sets of worlds where formulas hold, using $[A]_M$ for the set of worlds where A holds in a model M. We could introduce a singulary modality, the universal modality $\mathbb{U}$, such that $\mathbb{U}A$ holds at a world iff $[A]_M$ is the set of all worlds in the model M. One could then say a formula context (C, p) is $\mathbb{U}$-*hyperintensional* in L iff for some formulas A and B,

- $\not\models_{\mathsf{L}} \mathbb{U}(A \leftrightarrow B) \rightarrow \mathbb{U}(C(A) \leftrightarrow C(B))$.

We'll say a logic is $\mathbb{U}$-*hyperintensional* iff it has a formula context (C, p) that is $\mathbb{U}$-hyperintensional. The logic R is not $\mathbb{U}$-hyperintensional. Since every sublogic of R.TRIV is hyperintensional, this tells us that the modal principles of $\mathbb{U}$ are not contained in TRIV, which puts it well outside the usual modal logics. Proponents of relevant logics, however, have a reason not to accept $\mathbb{U}$, as it leads to violations

[15] See Wójcicki [42, 342], who uses the term 'selfextensional', Font [12, ch. 7], Avron [3], for example. Thanks to Rohan French and Andrew Tedder for references.

[16] See Humberstone [18, 19], among others.

[17] See Savić and Studer [28] and Standefer [34] for examples.

[18] See Restall [26, ch. 11] for a good introduction to ternary relational models.

of Belnap's variable sharing criterion.[19] Proponents of relevant logics have reason not to accept that connective and to reject this sense of hyperintensionality. It is not the salient sense of hyperintensionality for the proponent of relevant logics.

It is worth pointing out that relevant logics have a feature that is, in some ways, similar in spirit to hyperintensionality. Classical logic is *monothetic* in the sense that for any two logical truths A and B, $A \leftrightarrow B$ is a logical truth.[20] From the point of view of classical logic, there is only a single logical truth. HYPE is also monothetic, replacing the classical biconditional with the biconditional of HYPE, and similarly for intuitionistic logic. Relevant logics are *polythetic* meaning that there are non-equivalent logical truths, that is, there are logical truths A and B such that $A \leftrightarrow B$ is not a logical truth.[21] In relevant logics, one can draw distinctions between logical truths, much as (classical) hyperintensionality allows one to draw distinctions among necessary truths. One can use logical truths, such as $p \to p$ and $q \to q$, to show that the formula context $(s \wedge r, s)$ is hyperintensional. By contrast, any logic that contains the weakening axiom, $A \to (B \to A)$, and where the conditional obeys *modus ponens* will be monothetic.

The results of this paper show that almost all the common modal extensions of relevant logics have hyperintensional contexts. This result extends to HYPE, although the range of such contexts appears more limited there than in R. As one weakens the logic, the range of hyperintensional contexts grows, a feature that extends to HYPE and other substructural logics as well. Hyperintensionality is of interest in a wide range of philosophical applications of logic, such logics of belief and epistemic logics. There is further work to do to see the extent to which the sorts of hyperintensionality identified here has natural application to, say, logics of belief or epistemic logics. A promising direction for future work is to precisely characterize the range of hyperintensional contexts in the different relevant and substructural logics. This will be useful in better understanding the ways in which logical omniscience can fail in non-classical settings.[22] One can, of course, appeal to various modeling techniques used to obtain hyperintensional contexts over classical logic to obtain such contexts in relevant logics. These modeling techniques will likely interact with the natural hyperintensionality of relevant logics in surprising ways.

To summarize, relevant logics are hyperintensional when considering many natural kinds of necessity, in at least one important sense. One can extend relevant logics with a singulary modal operator for universal necessity, $\mathbb{U}$, to obtain another sense of hyperintensionality. Relevant logics fail to be hyperintensional in that sense, although the relevant logician has antecedent reason not to accept

[19] See Standefer [35,36] for discussion.

[20] See Humberstone [17, 231].

[21] This point was also made by Standefer [33], albeit in discussion of justification logics.

[22] See, for example, Sedlár [29,30], Standefer, Shear, and French [37], and Ferenz [11], among others, for some discussion of logical omniscience in non-classical settings. For a contrasting recent discussion of omniscience in the setting of classical logic, see Hawke, Özgün, and Berto [14].

that modality and not to be interested in that sense of hyperintensionality. Relevant logics and their modal extensions are, generally, congruential, so they are not hyperintensional in the sense preferred by Odintsov and Wansing. Nonetheless, we do agree with Odintsov and Wansing's closing suggestion to study non-self-extensional, or non-congruential, operators, as non-classical logics likely have much to contribute in those areas. Despite being congruent, relevant logics are polythetic, which allows them to draw distinctions in ways reminiscent of hyperintensionality.

Acknowledgments. I would like to thank Greg Restall, Rohan French, Lloyd Humberstone, Andrew Tedder, and Heinrich Wansing, and the anonymous referees of LORI for comments and discussion that greatly improved this paper. This research was supported by the National Science and Technology Council of Taiwan grant 111-2410-H-002-006-MY3.

References

1. Alechina, N., Logan, B.: Belief ascription under bounded resources. Synthese **173**(2), 179–197 (2010). https://doi.org/10.1007/s11229-009-9706-6
2. Anderson, A.R., Belnap, N.D.: Entailment: The Logic of Relevance and Necessity, vol. I. Princeton University Press, Princeton (1975)
3. Avron, A.: Self-extensional three-valued paraconsistent logics. Log. Univers. **11**(3), 297–315 (2017). https://doi.org/10.1007/s11787-017-0173-4
4. Berto, F., Nolan, D.: Hyperintensionality. In: Zalta, E.N. (ed.) The Stanford Encyclopedia of Philosophy. Metaphysics Research Lab, Stanford University, Spring 2021 edn. (2021)
5. Berto, F., French, R., Priest, G., Ripley, D.: Williamson on counterpossibles. J. Philos. Log. **47**(4), 693–713 (2017). https://doi.org/10.1007/s10992-017-9446-x
6. Berto, F., Jago, M.: Impossible Worlds. Oxford University Press, Oxford (2019)
7. Bimbó, K.: Relevance logics. In: Jacquette, D. (ed.) Philosophy of Logic, Handbook of the Philosophy of Science, vol. 5, pp. 723–789. Elsevier (2007)
8. Boolos, G.: The Logic of Provability. Cambridge University Press, Cambridge (1993)
9. Cresswell, M.J.: Hyperintensional logic. Stud. Logica. **34**(1), 25–38 (1975). https://doi.org/10.1007/bf02314421
10. Dunn, J.M., Restall, G.: Relevance logic. In: Gabbay, D.M., Guenthner, F. (eds.) Handbook of Philosophical Logic, vol. 6, 2nd edn., pp. 1–136. Kluwer, Alphen aan den Rijn (2002)
11. Ferenz, N.: First-order relevant reasoners in classical worlds. Rev. Symbolic Logic, 1–26 (2023). https://doi.org/10.1017/s1755020323000096. Forthcoming
12. Font, J.M.: Abstract Algebraic Logic. An Introductory Textbook. College Publications (2016)
13. Gabbay, D.: What is a classical connective? Math. Log. Q. **24**(1–6), 37–44 (1978). https://doi.org/10.1002/malq.19780240106
14. Hawke, P., Özgün, A., Berto, F.: The fundamental problem of logical omniscience. J. Philos. Log. **49**(4), 727–766 (2019). https://doi.org/10.1007/s10992-019-09536-6
15. Humberstone, L.: Extensionality in sentence position. J. Philos. Log. **15**(1), 27–54 (1986). https://doi.org/10.1007/bf00250548

16. Humberstone, L.: Singulary extensional connectives: a closer look. J. Philos. Log. **26**(3), 341–356 (1997). https://doi.org/10.1023/a:1004240612163
17. Humberstone, L.: The Connectives. MIT Press, Cambridge (2011)
18. Humberstone, L.: Philosophical Applications of Modal Logic. College Publications, London (2016)
19. Leitgeb, H.: HYPE: a system of hyperintensional logic. J. Philos. Logic **48**(2), 305–405 (2019). https://doi.org/10.1007/s10992-018-9467-0
20. Mares, E.: Relevance logic. In: Zalta, E.N., Nodelman, U. (eds.) The Stanford Encyclopedia of Philosophy. Metaphysics Research Lab, Stanford University, Fall 2022 edn. (2022)
21. Mares, E., Standefer, S.: The relevant logic E and some close neighbours: a reinterpretation. IfCoLog J. Logics Appl. **4**(3), 695–730 (2017)
22. Mares, E.D.: The incompleteness of RGL. Stud. Logica. **65**(3), 315–322 (2000). https://doi.org/10.1023/A:1005283629842
23. Mares, E.D.: Relevant Logic: A Philosophical Interpretation. Cambridge University Press, Cambridge (2004)
24. Nolan, D.: Hyperintensional metaphysics. Philos. Stud. **171**(1), 149–160 (2013). https://doi.org/10.1007/s11098-013-0251-2
25. Odintsov, S., Wansing, H.: Routley star and hyperintensionality. J. Philos. Log. **50**(1), 33–56 (2020). https://doi.org/10.1007/s10992-020-09558-5
26. Restall, G.: An Introduction to Substructural Logics. Routledge, Milton Park (2000)
27. Routley, R., Plumwood, V., Meyer, R.K., Brady, R.T.: Relevant Logics and Their Rivals, vol. 1. Ridgeview, Atascadero (1982)
28. Savić, N., Studer, T.: Relevant justification logic. J. Appl. Logics **6**(2), 395–410 (2019)
29. Sedlár, I.: Substructural epistemic logics. J. Appl. Non-Classical Logics **25**(3), 256–285 (2015). https://doi.org/10.1080/11663081.2015.1094313
30. Sedlár, I.: Epistemic extensions of modal distributive substructural logics. J. Logic Comput. **26**(6), 1787–1813 (2016). https://doi.org/10.1093/logcom/exu034
31. Sedlár, I.: Hyperintensional logics for everyone. Synthese **198**(2), 933–956 (2019). https://doi.org/10.1007/s11229-018-02076-7
32. Slaney, J.K.: A metacompleteness theorem for contraction-free relevant logics. Stud. Logica. **43**(1–2), 159–168 (1984). https://doi.org/10.1007/BF00935747
33. Standefer, S.: Tracking reasons with extensions of relevant logics. Logic J. IGPL **27**(4), 543–569 (2019). https://doi.org/10.1093/jigpal/jzz018
34. Standefer, S.: Weak relevant justification logics. J. Logic Comput. (2022). https://doi.org/10.1093/logcom/exac057. Forthcoming
35. Standefer, S.: What is a relevant connective? J. Philos. Logic **51**(4), 919–950 (2022). https://doi.org/10.1007/s10992-022-09655-7
36. Standefer, S.: Varieties of relevant S5. Logic Logical Philos. **32**(1), 53–80 (2023). https://doi.org/10.12775/LLP.2022.011
37. Standefer, S., Shear, T., French, R.: Getting some (non-classical) closure with justification logic. Asian J. Philos. **2**(2), 1–25 (2023). https://doi.org/10.1007/s44204-023-00065-3
38. Urbas, I., Sylvan, R.: Prospects for decent relevant factorisation logics. J. Non-classical Logic **6**(1), 63–79 (1989)
39. Verbrugge, R.L.: Provability Logic. In: Zalta, E.N. (ed.) The Stanford Encyclopedia of Philosophy. Metaphysics Research Lab, Stanford University, Fall 2017 edn. (2017)

40. Wansing, H.: A general possible worlds framework for reasoning about knowledge and belief. Stud. Logica. **49**(4), 523–539 (1990). https://doi.org/10.1007/bf00370163
41. Williamson, T.: Indicative versus subjunctive conditionals, congruential versus non-hyperintensional contexts. Philos. Issues **16**(1), 310–333 (2006). https://doi.org/10.1111/j.1533-6077.2006.00116.x
42. Wójcicki, R.: Theory of Logical Calculi: Basic Theory of Consequence Operations. Kluwer Academic Publishers, Dordrecht (1988)

Reasons in Weighted Argumentation Graphs

David Streit(✉), Vincent de Wit, and Aleks Knoks

Department of Computer Science, University of Luxembourg, Esch-Sur-Alzette, Luxembourg
{david.streit,vincent.dewit,aleks.knoks}@uni.lu

Abstract. The philosophical literature that tackles foundational questions about normativity often appeals to normative reasons—or considerations that count in favor of or against actions—and their interaction. The interaction between normative reasons is usually made sense of by appealing to the metaphor of (normative) weight scales. This paper substitutes an argumentation-theoretic model for this metaphor. The upshot is a general and precise model that is faithful to the philosophical ideas.

Keywords: Argumentation theory · Normative reasons · Weighing

1 Introduction

Philosophers who explore normative matters often appeal to *normative (practical) reasons*, understanding them as considerations that count in favor of or against actions. When discussing the interaction between reasons, they often use such phrases as "the action supported on the balance of reasons" and "reasons in favor outweigh the reasons against", inviting an image of *weight scales for reasons*. The simplest model of these (normative) weight scales works, roughly, as follows. Reasons speaking in favor of φ-ing go in one pan of the scales, and reasons against go in the other. If the weight of the reasons in the first pan is greater, φ ought to be carried out. If the weight of the reasons in the second pan is greater, φ ought not to be carried out.

While philosophers have explored various ideas about the exact workings of the weight scales and also looked at some alternatives, their investigations have mostly been carried out in informal terms. The goal of this paper is to develop a *formal* model of (normative) weight scales, drawing on formal argumentation. Instead of starting from scratch, we repurpose Gordon and Walton's model of "balancing arguments" [4].

This paper is structured as follows. Section 2 sketches the philosophical ideas on weighing reasons. Sections 3 and 4 set up the model. Section 5 discusses our main results and some of the work that is most closely related to ours.

N. Alechina et al. (Eds.): LORI 2023, LNCS 14329, pp. 251–259, 2023.
https://doi.org/10.1007/978-3-031-45558-2_19

2 (Normative) Weight Scales

This section provides a bird's-eye view summary of the main ideas from the philosophical literature on weighing reasons. Note that it is an opinionated sketch: we simplify where possible and bracket a whole plethora of important and complex questions. (For more thorough overviews, see, e.g., [6] and [9, pp. 1–7].)

Normative reasons are typically taken to be *facts* that are not subject to debate. Thus, the fact that the person next door is in need of help is a reason for you to help them, regardless of your values and preferences, as well as your views on ethics and metaethics. Reasons are always reasons for someone: they favor or speak against someone's action. They are also intimately tied to their *weights*, which are comprised of a *magnitude* and a *polarity*. Magnitude has to do with the relative importance of the reason; polarity with whether it is a reason for or against. Reasons against an action count (either directly, or indirectly) as favoring alternative actions.

Reasons play a core role in determining the deontic statuses of actions. Thus, whether some action is permitted/required/ought to be taken depends on the reasons that count for/against it and their interaction. In staying with the weight scales metaphor, we say that one is permitted to φ just in case the net weight of the reasons for φ-ing is at least as high as the net weight of the reasons for the alternatives. (For a discussion of subtle changes one could make to this definition, see, e.g., [9].)

An important and hotly debated question concerns the effects of context on the weights of reasons. Positions range from extreme *atomist* views on which a reason's weight is context-independent to extreme *holist* views on which a fact that is a reason for φ-ing in one context can be a reason against φ-ing in a different one, or cease to be a reason at all. Most philosophers find positions at both ends of the spectrum implausible, preferring views on which there is both (some) stability in reasons' weights and that allow for (some) context-sensitivity. A common move here is to appeal to what we might call *normatively-relevant considerations that aren't reasons*. Such considerations don't qualify as reasons because they don't count for/against actions. However, they can affect the weights of reasons, and so have an (indirect) effect on an action's deontic status. It's common to distinguish between two types of such considerations: *undercutters* and *modifiers*. An undercutter nullifies the weight of a reason, effectively making it cease to be a reason. Modifiers are of two types: *attenuators* and *amplifiers*. An attenuator reduces the magnitude of a reason, making it less weighty. An amplifier amplifies the magnitude of a reason, making it more weighty. Undercutters and modifiers suggest the view that every reason has a context-independent *default weight* and a context-specific *final weight*, and that any difference between the two can be accounted for by appeal to undercutters and modifiers. This view is common, and it seems up to debate whether it is closer to atomism or holism.

3 Normative Graphs

In this section, we explain how to represent the structural relations that can obtain between normative reasons, other normatively-relevant considerations, and alternative actions (or options). (In the next one, we focus on modeling the normative effects that these can exert on each other.)

As background, we assume a propositional language ($\mathcal{L}$) with the standard connectives. We use the term 'normatively-relevant consideration' (or simply 'consideration') to refer to reasons, undercutters, and modifiers. Following the philosophical literature, we assume that every consideration (a) has a default weight; (b) can be undercut; and (c) can have its weight changed by modifiers. Jointly, the relevant considerations are meant to determine the deontic status of actions available to an agent. Adapting the notion of *issue* from [4], we think of options as a finite subset of $\mathcal{L}$ representing a set of mutually exclusive and exhaustive actions available to the agent. Adapting the notion of *argument* from [4], we define considerations as follows:

Definition 1 (Normatively-relevant consideration). *A* consideration *is a tuple* $C = (p, c, u, a^-, a^+, w)$, *where the first five elements are formulas of* $\mathcal{L}$ *called, respectively,* premise, conclusion, undercutter, attenuator, *and* amplifier, *while the sixth element is a positive real number called* default weight.

We will use the following scenario to illustrate this and future definitions—note that the expressions in brackets are atomic sentences of $\mathcal{L}$:

Example 1. You are to choose between two options: to go to the movies with me ($Movies$), or to have dinner with your mom at her favorite restaurant ($Dinner$). You have made a promise to me ($Promise$). What's more, you were very insistent when making the promise: you said that you would keep it no matter what ($Insist$). Dining with mom would make her happy ($MomHappy$). The restaurant also happens to serve your favorite cake ($Cake$). Also, it is Mother's Day ($MothersDay$) and you haven't seen your mom in a while ($LongTime$).

Notice that the options here are $\{Movies, Dinner\}$. The intuitive idea that your promise is a reason to go to the movies is captured by the consideration $C_1 = (Promise, Movies, u_1, a_1^-, a_1^+, w_1)$. Similarly, the idea that going to the restaurant will make your mom happy and that the restaurant serves your favorite cake are reasons to have dinner with mom is represented by $C_2 = (MomHappy, Dinner, u_2, a_2^-, a_2^+, w_2)$ and $C_3 = (Cake, Dinner, u_3, a_3^-, a_3^+, w_3)$ respectively. The idea that your being very insistent when making the promise *amplifies* C_1 is captured by consideration $C_4 = (Insist, a_1^+, u_4, a_4^-, a_4^+, w_4)$. Notice that the conclusion of C_4 (a_1^+) corresponds to the amplifier of C_1, which means that C_4 *amplifies* C_1 or that C_4 is an amplifier of C_1. The rest of the example is captured by considerations $C_5 = (MothersDay, a_2^+, u_5, a_5^-, a_5^+, w_5)$, $C_6 = (LongTime, a_2^+, u_6, a_6^-, a_6^+, w_6)$, and $C_7 = (Release, u_1, a_7^-, a_7^+, w_7)$.

The considerations that are in force in a given context form a graph structure. This is captured by our next definition—given a consideration $C = (p, c, u, a^-, a^+, w)$, we let $p_C = p$; $c_C = c$; $u_C = u$; $a_C^- = a^-$; $a_C^+ = a^+$, and $w_C = w$:[1]

[1] We follow [4] in calling the structures specified in Definition 2 *graphs* as they can be mapped straightforwardly to directed graphs.

Definition 2 (Normative graph). *A* normative graph *is a triple of the form* $N = (S, O, R)$, *where* S *is a finite subset of* $\mathcal{L}$; $O \subseteq S$ *called the* set of options*;* *and* R *is a finite set of considerations, where for every* $C \in R$, p_C, c_C, u_C, a_C^-, a_C^+ *are all members of* S *and* $w_C \in \mathbb{R}_{>0}$.

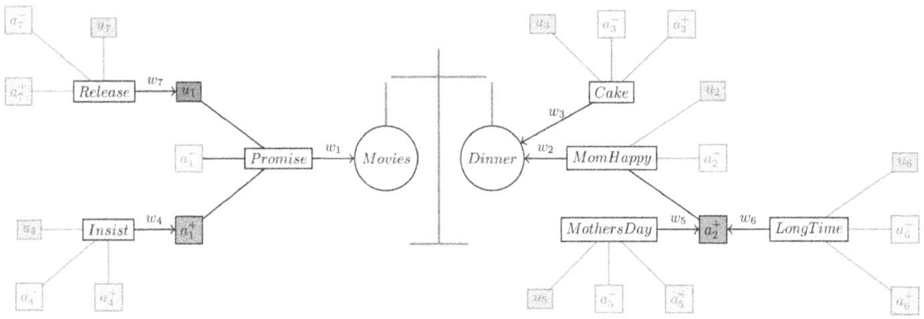

Fig. 1. Movies or Mom's favorite restaurant

Figure 1 depicts the full graph comprising our example visually. The two options are represented as circles in the middle. Each consideration is depicted as comprised of five nodes, some of which are shared between considerations. For example, the node $Insist$ is the premise of $C_4 = (Insist, a_1^+, u_4, a_4^-, a_4^+, w_4)$, and the node a_1^+ is the conclusion of C_4 and the amplifier of C_4. The nodes u_4, a_4^-, and a_4^+ stand for the remaining elements of C_4. They are grayed out, since no other considerations affect C_4. The default weight of C_4 is presented on the edge between premise and conclusion.

Notice that the graph structure in Fig. 1 is finite, directed, and acyclic. This is no coincidence. Following the philosophical literature, we allow that graphs representing the structural relations between normatively-relevant considerations and options are complex. However, we also require that they are finite and never contain any cycles. In particular, this rules out intra-consideration circularity, where e. g. the premise and conclusion are identical.

Before moving on, we introduce some terminology. Given a graph $N = (S, O, R)$, we call a consideration $C \in R$ a *reason* (for $o \in O$) if $c_C = o$. We say that C attenuates C' if c_C is the attenuator of some consideration C', we say that it amplifies C' if c_C is an amplifier of C' and that it undercuts C' if c_C is the uncercutter of C'. Overloading the terms, we sometimes call a consideration that undercuts or modifies another reason an undercutter or modifier respectively. Sometimes we may want to add, remove, or replace considerations in a graph. While the addition and removal are straightforward to define, one needs to be careful with replacement. Given a graph $N = (S, O, R)$ and a consideration C, we let $N + C$ denote the graph that results from adding C to N and $N - C$ the graph that results from removing C from N. A graph that results from replacing C by another consideration C' in N is denoted by $N_{-C}^{+C'}$.

It is defined as $(N - C) + C'$ only in case either both C and C' are reasons, or both C and C' modify or undercut the same consideration. This ensures that the replaced consideration occupies the same place in the graph.[2]

4 Weighing Functions

This section explains how we model the effects of normatively-relevant considerations on each other, along with their effects on options.

Suppose that we had a function f^N which, given any normative graph N, would output the *final weights* of all considerations in it. Such a function would put us in a position to determine which options are permitted and which are forbidden in a rather straightforward way, where $\mathcal{C}_s = \{C \in R \mid c_C = s, s \in S\}$:

Definition 3 (Resolution function). *Given a normative graph* $N = (S, O, R)$*, an option* $o \in O$*, and a function* f^N*, let* $r^N(o) =$ `permitted` *if, for every* $o^* \in O$, $\sum_{C \in \mathcal{C}_o} f^N(C) \geq \sum_{C \in \mathcal{C}_{o^*}} f^N(C)$; $r^N(o) =$ `forbidden` *otherwise.*

And if there is a unique permitted option, we say that it *ought* to be carried out.

Of course, we still need to specify f^N. The way we calculate the final weight of a consideration C is via a two-step process using an additional function g^N. In the first step, we aggregate the (final) weights of C's amplifiers, attenuators, and undercutters. In the second step, we obtain the final weight of C on the basis of these aggregated weights and the default weight of C. This will be the job of f^N. (For readability, we will often omit the superscript where the context makes clear which graph we are talking about.) Note that even if we have a concrete aggregation function g at our disposal, there are many choices we could make for how f calculates the final weight from the output of g and C's default weight. So, before we define any concrete functions, it will be useful to think about some plausible *constraints* or *principles* that they should satisfy. (Many of the principles of weighted argumentation analyzed in [2] can be translated to our setting and make sense here as well.)

Definition 4 (Principles). *Let* $N = (S, O, R)$ *be an arbitrary normative graph,* C *some consideration, and* C' *and* C^* *a modifier and an undercutter of* C*, then:*

---

[2] It's worth noting two features of our model that might turn out to be either advantages or drawbacks. First, we represent reasons with negative polarity—reasons that speak against actions—only indirectly. In our model, any reason is always a reason *for* an option. So, it is a reason *against* an option only in so far as it adds to the final weight of an alternative option. Second, it is sometimes claimed that reasons can switch their polarity when combined [8]. Thus, in an (in)famous example, Prakken and Sartor [8] describe the effects of heat and rain on your going jogging: taken by themselves, the facts that it is raining and that it is hot constitute reasons for you not to go jogging, but, taken in combination, they speak in favor of going jogging. If these cases exist, our model cannot account for them. However, given that their existence is disputed (see e.g., [3,7,9]), our model may well gives the correct verdict here.

1. ***No Effects from Spurious Reasons:*** *If C is a reason in $N + C$, then if $f^{N+C}(C) = 0$, then for all options o, $r^N(o) = r^{N+C}(o)$.*
2. ***No Effects from Spurious Modifiers:*** *For any modifier $C \notin R$ of a consideration $C' \in R$ with $f^{N+C}(C) = 0$, we have $f^N(C') = f^{N+C}(C')$.*
3. ***No Valence Flips for Modifiers:*** *Given an amplifier C of some consideration $C' \in R$, $f^N(C') \geq f^{N-C}(C')$. Analogously, for an attenuator C of some consideration C', $f^N(C') \leq f^{N-C}(C')$.*
4. ***Modifier Reciprocity:*** *For any $C \in R$, if $g(a_C^+) = g(a_C^-)$ and $g(u_C) = 0$, then we have $f(C) = w_C$.*
5. ***Modeler's Delight:*** *For any consideration $C \in R$, we have $f(C) \geq 0$. If $g(u_C) > 0$, then $f(C) = 0$.*
6. ***Normative Parsimony*** *The function governing the weight of different considerations is uniform (and not gerrymandered) for reasons, undercutters, and modifiers.*
7. ***Relativity*** *Given a consideration $C = (p, c, u, a^-, a^+, w)$ and $x > 0$, let $x \times C$ be $(p, c, u, a^-, a^+, x \times w)$. Given a set of considerations R, let $x \times R = \{x \times C : C \in R\}$ and $x \times N = (S, O, x \times R)$. Then $r^N(o) =$ `forbidden` iff $r^{x \times N}(o) =$ `forbidden`.*
8. ***Distinct Roles*** *If $C' \in R$, $C^* \notin R$, $f^{N+C^*}(C^*) \neq 0$ and $f^N(C) \neq 0$, then $f^{N+C^*}_{-C'}(C) \neq f^N(C)$. And if $C^* \in R$, $C' \notin R$, $f^N(C^*) \neq 0$ and $f^{N-C^*}(C) \neq 0$, then $f^{N+C'}_{-C^*}(C) \neq f^N(C)$.*

Principles 1 and 2 state that the addition of both reasons and modifiers with the (final) weight of 0 should have no effects on which options are permissible and the final values of other specific considerations. Principle 3 states that no attenuator (no matter the rest of the graph) should ever help strengthen the weight of the consideration it modifies and that no amplifier should help weaken the weight of the consideration it modifies. Similarly, Principle 4 states that if the (aggregated) weight of attenuators and amplifiers is the same, their contributions cancel out and the final weight of the consideration they modify (if it is not undercut) is simply its default weight. Principles 5 and 6 are meant to ensure that weights get calculated in accordance with the scales metaphor. Principle 5 says that the minimal weight a consideration can have is 0 and that 0 is the weight it has if it is undercut. Principle 6 requires that the final weights of different kinds of normatively-relevant considerations are computed in the same way. Principle 7 states that weights have no meaning outside of the ratio scale they constitute. Principle 8, which one may or may not accept depending on one's metaethical inclinations, states that the roles of modifiers and undercutters in the economy of reasons should be distinct. (In particular, it states that an attenuator is never strong enough to entirely remove the weight of a consideration.)

With these principles in mind, we turn to concrete examples. Recall that we are looking for two functions g and f. The simplest thing to do for g is to add the (final) values of the nodes that "feed into" a consideration:

$$g_+(x) = \sum_{C' \in C_x} f(C')$$

Notice that g_+ treats undercutters in an intuitive way, returning 0 just in case there either are no undercutters or all of them are themselves undercut. The leaves of the graph get assigned the weight 0. So, given a function f, we can compute the final weight of a consideration, working from the leaves up the tree.

Turning to the function f, we define two concrete functions here. The first of these, f_+, assigns a final weight of zero if either the consideration is undercut or if the combined weight of its attenuators is greater than that of its amplifiers and its default weight. Otherwise, it returns the default weight plus the difference between the combined weight of its amplifiers and attenuators. (Note that $\mathbb{1}$ is the indicator function.)

$$f_+(C) = \mathbb{1}_{\{0\}}(g(u_C)) * max(0, g(a_C^+) - g(a_C^-) + w_C)$$

The second function $f_\times$ uses modifiers as multipliers: amplifiers increase the weight of a consideration by a factor, and attenuators lower it by a factor.

$$f_\times(C) = \mathbb{1}_{\{0\}}(g(u_C)) * \frac{1 + g(a_C^+)}{1 + g(a_C^-)} * w_C$$

5 Results and Related Work

We take it that Principles 1–7 state conditions that should be satisfied by all functions, while Principle 8 (Distinct Roles) is up for debate. Our first result runs thus—its proof is straightforward and omitted for reasons of space.

Theorem 1. *Functions g_+ and $f_\times$ satisfy all principles stated in Definition 4. Functions g_+ and f_+ satisfy all principles but* Distinct Roles.

While ($f_\times$ and g_+) allow for attenuators to be as strong as undercutters and (f_+ and g_+) doesn't, a plausible result about the relationship between attenuators and undercutters can be established for both. Intuitively, it states that undercutters can be seen as a limit case of attenuators:

Theorem 2. *Given the functions ($f_\times$ and g_+) or (f_+ and g_+) and any consideration C_0, there is a series of attenuators C_i such that each C_i attenuates C_0 that can be added to the normative graph N, such that* $\lim_{i \to \infty} f^{N + \cup_{j=1}^{j=i} C_j}(C_0) = 0$.

Notice that Definition 4 is only the first step towards mapping out the space of normative principles that weighing functions can (should) satisfy. We leave fully-fledged principle-based analysis for future work. In the remainder of this section, we briefly compare our model to the work that comes the closest to it: Gordon and Walton's model of "balancing arguments" [4], the work on weighted argumentation by Amgoud et al. [1,2], and Horty's model of reasons [5].

We repurposed the model from [4] to a particular domain. As a result, we obtained a model that is simpler in a number of respects. For instance, the graphs that we work with are acyclic, and we have no need for labeling. The way we interpret weights is also different: where [4] assign values from [0, 1] to arguments, interpreting them as the "level of acceptance", we allow all positive reals. As a result, we have also dropped the notion of "proof standard", which gets at the idea of a threshold of evidence for accepting a conclusion. No such threshold seems to exist in the context of practical normative reasons.

Amgoud et al. [1,2] work with the interval [0, 1], representing argument acceptance in their principle-based analysis of weighted argumentation. Their model assigns both default and final weights to arguments: attacks on arguments lower the final weight, while support increases it. The bulk of their principles can be restated in our framework, and many of them apply to reasons for action. In fact, our Principles 1–6 can be seen as translations from [2]. There are two main differences between the approach of Amgoud et al. and ours. First, the target of Amgoud et al. is the acceptability of arguments and not the interaction between normative reasons. Second, due to the interval scale used in their mode, an argument can be fully accepted simply due to the arguments "feeding into it". In our model, an issue is essentially contrastive: an option's normative status always depends on all reasons for all options. It also means that some of our principles (e.g., Principle 8) make little sense in the framework of Amgoud et al., while other principles aren't applicable in our model. Nevertheless, our principles both serve to show the (dis)similarities between normative reasons and arguments, we believe that there is a whole host of principles in the style of [2] unique to normative reasons that are still to be explored in future work.

Lastly, there is the model of Horty [5], which is meant to play a similar role to ours. The main advantage of our model over Horty's is its closer alignment to the idea of weight scales: our model associates magnitudes with reasons and lets us model combination or aggregation of weights in a straightforward way.

Acknowledgement. Vincent de Wit and Aleks Knoks acknowledge financial support from the Luxembourg National Research Fund (FNR). Knoks was supported through the project OPEN O20/14776480, de Wit through the project PRIDE19/14268506. We also thank our three anonymous referees for their generous comments.

References

1. Amgoud, L., Ben-Naim, J.: Evaluation of arguments in weighted bipolar graphs. Int. J. Approximate Reasoning **99**, 39–55 (2018)
2. Amgoud, L., Doder, D., Vesic, S.: Evaluation of argument strength in attack graphs: foundations and semantics. Artif. Intell. **302**, 1–61 (2022)
3. Bader, R.: Conditions, modifiers and holism. In: Lord, E., Maguire, B. (eds.) Weighing Reasons, pp. 27–55. Oxford University Press (2016)
4. Gordon, T., Walton, D.: Formalizing balancing arguments. In: Baroni, P., Gordon, T., Scheffler, T., Stede, M. (eds.) Proceedings of the 2016 Conference on Computational Models of Argument (COMMA 2016), pp. 327–38. IOS Press (2016)
5. Horty, J.: Reasons as Defaults. Oxford University Press, Oxford (2012)
6. Lord, E., Maguire, B.: An opinionated guide to the weight of reasons. In: Lord, E., Maguire, B. (eds.) Weighing Reasons, pp. 3–24. Oxford University Press (2016)

7. Maguire, B., Snedegar, J.: Normative metaphysics for accountants. Philos. Stud. **178**, 363–84 (2018)
8. Prakken, H., Sartor, G.: Modelling reasoning with precedents in a formal dialogue game. Artif. Intell. Law **6**, 231–87 (1998)
9. Tucker, C.: A holist balance scale. J. Am. Philos. Assoc. First View **9**, 1–21 (2022)

Making Norms and Following Norms

Xingchi Su(✉)

Zhejiang Lab, Hangzhou, China
x.su@zhejianglab.com

Abstract. A normative sentence contains information which is used for either describing some deontic situations or prescribing a new norm. Considering the differences between them, we introduced the notion of relativized conditional obligations based on ideality sequences. On the one hand, each ideality sequence can be treated as a normative system which *prescribes* the relative ideality of states of affairs. On the other hand, a bare structure provides a factual background. Once these are done, every betterness structure based on a given ideality sequence and a bare structure *describes* the conditional obligations. Deletion and postfixing are two updates on the normative system which can bring about corresponding obligations successfully or not. Jørgensen's dilemma can be conceptualized by using the notion of successful updates. A sound and strongly complete axiom system for the logic of relativized conditional obligations $\mathbb{PCDL}$ is established.

Keywords: normative system · conditional obligation · ideality sequence · update normative system · successful update · Jørgensen's dilemma · axiomatization

1 Introduction

A normative sentence contains information which is used for either describing some deontic situations or prescribing a new norm. The descriptive use is normally shown in an indicative mood. The prescriptive use usually appears in an imperative mood. These different uses of normative sentences inspire different deontic logics.

In the *descriptive* sense, normative propositions describe agents' deontic states, thereby having truth values. For example, 'Pieter ought to drive on the right' has a certain truth value under some circumstances. In the Netherlands, it is true, but it is false in England. Varieties of deontic logic were developed following this approach, such as SDL [14], dyadic deontic logics [3,4], and deontic Stit logic [2,5], etc.

In the *prescriptive* sense, we pay more attention to norms rather than normative propositions since norms are used for prescribing new rules and hence do not have truth values. For example, a norm can be a command which *changes* agents' deontic states. As a consequence, a puzzling question arises: if imperatives have no truth values, is there a logic of imperatives? The question is known as *Jørgensen's dilemma*. The basic strategy to resolve the problem is introducing an independent set of norms which can be updated by adding new norms or subtracting norms from it (see [8,10,12]). Following the idea, this paper also splits off normative systems from deontic models. However, our normative systems will be characterized by ideality sequences where norms are not independent with each other. Rather, they are structural. Moreover, the updates on

N. Alechina et al. (Eds.): LORI 2023, LNCS 14329, pp. 260–268, 2023.
https://doi.org/10.1007/978-3-031-45558-2_20

ideality sequences would make effects on conditional obligations by changing betterness relations in deontic models. Accordingly, we give the notion of successful updates and the Jørgensen's dilemma can be resolved.

2 Preliminaries

We first introduce Hansson's dyadic deontic logic. The language is $\mathcal{L}_{\mathrm{CDL}}$ (CDL stands for 'conditional deontic logic'). Let **P** be a countable set of propositional atoms.

Definition 1 (Language $\mathcal{L}_{\mathrm{CDL}}$). *The language $\mathcal{L}_{\mathrm{CDL}}$ is given by the following BNF:*

$$\varphi ::= p \mid \neg\varphi \mid (\varphi \wedge \varphi) \mid U\varphi \mid \bigcirc(\varphi|\varphi)$$

where $p \in \boldsymbol{P}$.

The formula $U\varphi$ stands for "φ is necessary"; $\bigcirc(\varphi|\psi)$ can be read as "if ψ is the case, φ ought to be the case". $\hat{U}$ is the dual of U.

Definition 2. *A betterness structure is a tuple $M = \langle S, \leqslant, V\rangle$ where*

- *S is a nonempty set of states;*
- *$\leqslant \subset S \times S$ is a partial order (transitive and antisymmetric);*
- *$V \subset \boldsymbol{P} \to \mathcal{P}(S)$.*

The set of states S provides a factual background for a model. The betterness relation shows which states are better or worse. It is worth noting that in Hansson's tradition, betterness relations are given a priori.

Definition 3 (Semantics of $\mathcal{L}_{\mathrm{CDL}}$ [4]). *Let $M = \langle S, \leqslant, V\rangle$ be a betterness structure. The semantics of $\mathcal{L}_{\mathrm{CDL}}$ is defined as follows (only the non-trivial cases are shown):*

- *$M, s \models U\varphi \Longleftrightarrow$ for each $t \in S$, $M, t \models \varphi$;*
- *$M, s \models \bigcirc(\varphi|\psi) \Longleftrightarrow \max_{\leqslant} ||\psi||_M \subseteq ||\varphi||_M$.*

Here, $||\varphi||_M = \{t \mid M, t \models \varphi\}$ and $\max_{\leqslant} T = \{s \mid \forall t \in T(s \leqslant t \Rightarrow t \leqslant s)\}$

The set $\max_{\leqslant} ||\psi||_M$ represents these best ψ-states in $||\psi||_M$ with respect to $\leqslant$. An agent has a conditional obligation $\bigcirc(\varphi|\psi)$ in a possible world s if and only if all the best ψ-states also satisfy φ. Thus, $\bigcirc(\_|\_)$ is a global operator whose semantics are not related to which possible world that we are standing on. It *describes* the ideal proposition (φ) under some certain condition (ψ).

3 Betterness Structures Based on Ideality Sequences

Betterness structures provide a basis for the description aspect of our research. They are used for describing obligations that an agent has. In contrast, the notion of ideality sequence to be introduced is the normative system in this paper. It prescribes what propositions are better or worse.

Definition 4. *A pair* $\mathcal{I} = (I \cup \{\epsilon\}, \ll)$ *is an ideality sequence where*

- $I \subset \mathcal{L}_{\mathrm{PL}}$ *is a finite (can be empty) set of propositional formulas;*
- $\ll: I \times (I \cup \{\epsilon\})$ *is a strict linear order such that for each* $\varphi \in I$, $\varphi \ll \epsilon$.

An ideality sequence provides us an ordering on propositional formulas. For example, if $\varphi \ll \psi$, it means that ψ is *dominantly* better than φ. The constant ϵ is used for representing the empty sequence when $I = \varnothing$.

Notations: given an ideality sequence $\mathcal{I} = (I \cup \{\epsilon\}, \ll)$ and $I \neq \varnothing$, for each $\varphi \in I$,

- $\mathcal{I}_\varphi = \{\varphi' \in I \mid \varphi \ll \varphi' \text{ or } \varphi' = \varphi\}$;
- if $\varphi \in I$ and φ is not the maximal element in I, then $\bigvee I^+_\varphi = \bigvee\{\varphi' \in I \mid \varphi \ll \varphi'\}$;
- if $\varphi \in I$ and φ is the maximal element in I, then $\bigvee I^+_\varphi = \bot$.

Definition 5 (Bare structure). *A pair* $M = (W, V)$ *is a bare structure where*

- W *is a non-empty set of states;*
- $V : \boldsymbol{P} \to \mathcal{P}(W)$ *is a valuation.*

A bare structure only provides a factual background of a model. But given an ideality sequence, a betterness relation can be derived from a bare structure and thereafter, a betterness structure is constructed.

Definition 6 (Betterness structure based on ideality sequence). *Given an ideality sequence* $\mathcal{I} = (I \cup \{\epsilon\}, \ll)$ *and a bare structure* $M = (W, V)$*, the betterness structure based on* $\mathcal{I}$ *and* M *is* $M_\mathcal{I} = (W, \leqslant_\mathcal{I}, V)$ *where* $\leqslant_\mathcal{I}: W \times W$ *is the betterness relation between states satisfying the following condition:*

$$s \leqslant_\mathcal{I} t \iff \text{either (i) or (ii), where}$$
$$\text{(i) } \exists\varphi \in I : (M, t \models \varphi \text{ and } \forall\psi \in I(\varphi \ll \psi \to M, s \not\models \psi))$$
$$\text{(ii) } \neg\exists\psi \in I : (M, s \not\models \psi \text{ or } M, t \not\models \psi)$$

The definition of $s \leqslant_\mathcal{I} t$ intuitively means that the best formula from $\mathcal{I}$ that t satisfies is not worse than the best formula from $\mathcal{I}$ that s satisfies. The betterness relation derived in this way is a total preorder (transitive and strongly connected) which is different from original Hansson's definition. But it still shows which states are better or worse in a given bare structure.

Proposition 1. *Given an ideality sequence* $\mathcal{I} = (I \cup \{\epsilon\}, \ll)$ *and the betterness structure based on* $\mathcal{I}$*, i.e.,* $M_\mathcal{I} = (W, \leqslant_\mathcal{I}, V)$*, if* $T \subseteq W$ *is non-empty, then* $\max_{\leqslant_\mathcal{I}} T \neq \varnothing$.

Since I is finite, the proof is trivial. The idea of ideality sequence comes from the notion of priority structure and priority sequence in [11] and [7].

4 Making Norms: Generating Normative Systems

There have witnessed amount of research about the updates on obligations in deontic logic (see [9, 12, 16]). Here we will show several ways of updating a given ideality sequence. We treat ideality sequences as synonyms for 'normative systems'. Given an ideality sequence $\mathcal{I} = (I \cup \{\epsilon\}, \ll)$, we use $\mathcal{I} = \langle\varphi_1, \varphi_2, \cdots, \varphi_n, \epsilon\rangle$ to denote the sequence of formulas in $\mathcal{I}$ with respect to $\ll$, where $\varphi_i \ll \varphi_j$ if $i \leq j$.

The first fundamental update is *deletion*.

Definition 7 (Deletion). *Given an ideality sequence $\mathcal{I} = (I \cup \{\epsilon\}, \ll)$ and φ_1 is the least formula in $\mathcal{I}$, deleting φ_1 from $\mathcal{I}$ yields the ideality sequence $\mathcal{I} - \varphi_1$ where φ_1 in $\mathcal{I}$ is removed and the betterness order $\ll$ for the remaining formulas is preserved.*

$$\varphi_1 \dashrightarrow \varphi_2 \dashrightarrow \varphi_3 \quad \xrightarrow{\text{delete } \varphi_1} \quad \varphi_2 \dashrightarrow \varphi_3$$

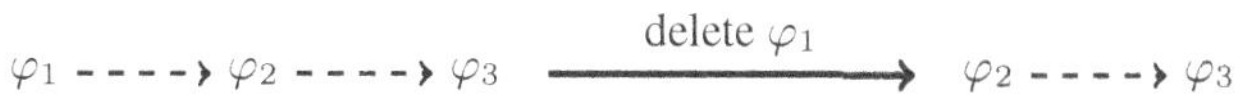

The deletion update captures abolishing a norm. It is related to the notion of repeal or annulment in law. The above figure is an example where the arrows from φ_i to φ_j represent $\varphi_i \ll \varphi_j$.

The second update is *postfixing* which was originally proposed in [7].

Definition 8 (Postfixing [7]). *Let $\mathcal{I} = (I \cup \{\epsilon\}, \ll)$ and $\varphi \notin I$.*

- *If $I \neq \emptyset$ and $\mathcal{I} = \langle \varphi_1, \cdots, \varphi_n, \epsilon \rangle$, then postfixing φ to $\mathcal{I}$ is $\mathcal{I} \triangleleft \varphi = \langle \varphi, \varphi_1, \cdots, \varphi_n, \epsilon \rangle$.*
- *If $I = \emptyset$, then postfixing φ to $\mathcal{I}$ is $\mathcal{I} \triangleleft \varphi = \langle \varphi, \epsilon \rangle$.*

Postfixing introduces a sub-ideal proposition to the original normative system. It is relevant to the notion of 'derogation' in law. An example of postfixing is shown as the following figure:

$$\varphi_1 \dashrightarrow \varphi_2 \quad \xrightarrow{\text{postfix } \varphi_0} \quad \varphi_0 \dashrightarrow \varphi_1 \dashrightarrow \varphi_2$$

The third update is *prefixing* which can be originally found in [11]. It is used for adding a better proposition than the original best proposition. The last update is insertion. It is employed to refine our ideality sequence. In the technical level, not all types of updates are elementary since we can construct any ideality sequence from a given ideality sequence by only deletion and postfixing. The idea is straightforward: delete all the formulas in the original one and then postfix the formulas one by one according to the order of the target sequence.

5 The Logic $\mathbb{PCDL}$

In this section, we will establish the logic of relativized conditional obligations $\mathbb{PCDL}$. The term $\mathbb{PCDL}$ is an acronym of 'Prescriptive Conditional Deontic Logic'.

5.1 Language and Semantics

Definition 9. *The language $\mathcal{L}_{\text{PCDL}}$ is given by the following BNF:*

$$\varphi ::= p \mid \neg\varphi \mid (\varphi \wedge \varphi) \mid U\varphi \mid \bigcirc_{\mathcal{I}}(\varphi|\varphi)$$
$$\mathcal{I} ::= \epsilon \mid \chi; \mathcal{I}$$

where $p \in \boldsymbol{P}$ and $\chi \in \mathcal{L}_{\text{PL}}$.

$\bigcirc_{\mathcal{I}}(\varphi|\psi)$ means that based on the ideality sequence $\mathcal{I}$, it ought to be φ given ψ.

Definition 10 (Semantics of $\mathcal{L}_{\text{PCDL}}$). *Let $M = \langle W, V\rangle$ be an arbitrary bare structure. The semantics of $\mathcal{L}_{\text{PCDL}}$ is given as follows (only nontrivial cases):*

$$\begin{aligned} M, s &\models U\varphi && \textit{iff } \|\varphi\|_M = W; \\ M, s &\models \bigcirc_{\mathcal{I}}(\varphi|\psi) && \textit{iff } \max_{\leqslant_{\mathcal{I}}} \|\psi\|_M \subseteq \|\varphi\|_M; \end{aligned}$$

The semantics of $\bigcirc_{\mathcal{I}}(\varphi|\psi)$ is also equivalent to $M_{\mathcal{I}}, s \models \bigcirc(\varphi|\psi)$ where $\bigcirc(\varphi|\psi)$ is Hansson's conditional obligations (see Definition 3). And we know that if $\|\psi\|_M \neq \varnothing$, $\max_{\leqslant_{\mathcal{I}}} \|\psi\|_M \neq \varnothing$ by Proposition 1. The truth value of $\bigcirc_{\mathcal{I}}(\varphi|\psi)$ is decided by the ideality sequence $\mathcal{I}$ and the bare structure we are concerning about.

Fact 1. *The following two formulas are valid:*

$$\begin{aligned} &(1)\ \bigcirc_{\varphi_1;\mathcal{I}}(\varphi_1|\neg\bigvee I) \wedge \hat{U}(\neg\bigvee I \wedge \neg\varphi_1) \rightarrow \neg\bigcirc_{\mathcal{I}}(\varphi_1|\neg\bigvee I); \\ &(2)\ \neg\bigcirc_{\mathcal{I}}(\varphi_1|\neg\bigvee I) \wedge \hat{U}(\neg\bigvee I \wedge \varphi_1) \rightarrow \bigcirc_{\mathcal{I}\lhd\varphi_1}(\varphi_1|\neg\bigvee I); \end{aligned}$$

The proof details are omitted due to page limitation. Formula (1) represents that deleting the norm φ_1 defeats the obligation to see to it that φ_1 when $\neg\varphi_1$ is possible under the condition $\neg\bigvee I$. Formula (2) means that postfixing the norm φ_1 brings about the new obligation to see to it that φ_1 when φ_1 is possible under the condition $\neg\bigvee I$.

5.2 Axiomatization

Our axiom system relies on the following formula schema which captures the best states in a bare structure under some condition. Given an ideality sequence $\mathcal{I} = (I \cup \{\epsilon\}, \ll)$ and an arbitrary $\mathcal{L}_{\text{PCDL}}$-formula ψ, the formula schema is:

$$\theta_\psi^{\mathcal{I}} : \bigvee_{\chi\in I}((\chi \wedge \psi) \wedge U(\bigvee I_\chi^+ \rightarrow \neg\psi)) \vee (\psi \wedge U(\bigvee I \rightarrow \neg\psi))$$

Lemma 1. *Given an ideality sequence $\mathcal{I} = \langle I \cup \{\epsilon\}, \ll\rangle$ and a bare structure (M, s),*

$$M, s \models \theta_\psi^{\mathcal{I}} \textit{ iff } s \in \max_{\leqslant_{\mathcal{I}}} \|\psi\|_M$$

The proof is omitted due to page limitation. Lemma 1 indicates that $\theta_\psi^{\mathcal{I}}$ captures the best ψ-states with respect to $\leqslant_{\mathcal{I}}$. In the light of $\theta_\psi^{\mathcal{I}}$, we can give the Kangerian-Andersonian reduction (KA-reduction) for the relativized conditional obligations and therefore the proof system of the logic $\mathbb{PCDL}$.

Proposition 2 (KA-reduction). *Given a bare structure (M, s) and an ideality sequence $\mathcal{I}$,*

$$M, s \models \bigcirc_{\mathcal{I}}(\varphi|\psi) \leftrightarrow U(\theta_\psi^{\mathcal{I}} \rightarrow \varphi)$$

It is easy to prove Proposition 2 by the semantics of $\bigcirc_{\mathcal{I}}(\varphi|\psi)$. The KA-reduction reduces formula $\bigcirc_{\mathcal{I}}(\varphi|\psi)$ to a $\mathcal{L}_{\text{ML}}$-formula[1] without any deontic operator. Therefore we can provide the proof system of the logic $\mathbb{PCDL}$.

[1] $\mathcal{L}_{\text{ML}}$ represents the language of the classical modal logic $S5$.

Definition 11. *The proof system* $\mathbb{PCDL}$ *consists of the following axiom schemas and inference rules:*

(TAUT)	*All instances of tautologies*
(K)	$U(\varphi \to \psi) \to (U\varphi \to U\psi)$
(T)	$U\varphi \to \varphi$
(4)	$U\varphi \to UU\varphi$
(5)	$\neg U\varphi \to U\neg U\varphi$
(KA)	$\bigcirc_{\mathcal{I}}(\varphi\vert\psi) \leftrightarrow U(\theta^{\mathcal{I}}_{\psi} \to \varphi)$
(MP)	*From* φ *and* $\varphi \to \psi$*, infer* ψ
(N)	*From* φ*, infer* $U\varphi$
(RE)	*From* $\varphi \leftrightarrow \psi$*, infer* $\chi \leftrightarrow \chi[\varphi/\psi]$

The soundness of the axiom (KA) has been given in Proposition 2. And it is easy to prove that the inference rule (RE) is sound since we can prove that 'from $\vdash \theta \leftrightarrow \chi$, infer $\vdash \bigcirc_{\mathcal{I}}(\theta|\psi) \leftrightarrow \bigcirc_{\mathcal{I}}(\chi|\psi)$ and $\vdash \bigcirc_{\mathcal{I}}(\varphi|\theta) \leftrightarrow \bigcirc_{\mathcal{I}}(\varphi|\chi)$' is valid. The remaining axioms and inference rules are also sound since they are classical axioms or rules from modal logic $S5$. By the axiom (KA), we can reduce every formula with the operator $\bigcirc_{\mathcal{I}}(\_|\_)$ to a formula without the operator. Therefore, completeness of the axiom system $\mathbb{PCDL}$ with respect to the semantics can be proved by translating $\mathcal{L}_{\text{PCDL}}$-formulas to $\mathcal{L}_{\text{ML}}$-formulas via reduction axioms and induction on the complexity of the formulas (see Chap. 7.4 in [13]).

Theorem 1. *The logic* $\mathbb{PCDL}$ *is sound and strongly complete with respect to the class of bare structures.*

6 Successful Updates and Jørgensen's Dilemma

Making a norm does not always bring the corresponding obligation to agents. We call it 'Moorean phenomena' in deontic context. In this section, we first provide some philosophical investigations on the agent's conditional obligations. Notations: If $\mathcal{I}_1 = \varphi_i; \mathcal{I}_2$, let $\mathcal{I}_1 - \varphi_i = \mathcal{I}_2$ and let $\mathcal{I}_2 \lhd \varphi_i = \mathcal{I}_1$.

6.1 Successful Updates

$\bigcirc_{\mathcal{I}}(\varphi|\psi)$ is a normative proposition which is satisfied or not in a bare structure. However, sometimes, an agent does not have an obligation to achieve φ even though φ has already been a norm in the concerning normative system. This happens since norms exist *outside* the bare structures. This suggests that obligations are not only decided by normative systems, but also by these facts embedded in the bare structures.

In order to clarify the discrepancy between norms and obligations, we first elaborate on the well-known Kantian principle that "*ought implies can*". How should we interpret the term 'can'? Many deontic logicians have been attempting to introduce the ability or agency into the deontic logics [1,6]. We, however, propose to interpret the 'can' in a more straightforward sense – possibility. We say φ is possible in a bare structure (M, s) if and only if $M, s \models \hat{U}\varphi$.

Fact 2.
$(a) \models \bigcirc_{\mathcal{I}}(\varphi|\top) \rightarrow \hat{U}\varphi$
$(b) \models (\bigcirc_{\mathcal{I}}(\varphi|\psi) \wedge \hat{U}\psi) \rightarrow \hat{U}\varphi$

The two validities illustrate that if an agent ought to achieve some situation, then it must be possible. In other words, they *can* be done.

Definition 12. *(Successful updates) Let $\mathcal{I}_1 = (I_1 \cup \{\epsilon\}, \ll)$ be an ideality sequence and let $\mathcal{I}_1 = \varphi_1; \mathcal{I}_2$.*

$-\varphi_1$ *is a successful update on $\mathcal{I}_1$ in (M, s) iff $M, s \models \neg \bigcirc_{\mathcal{I}_1 - \varphi_1} (\varphi_1 | \neg \bigvee I_2)$*
$\lhd\varphi_1$ *is a successful update on $\mathcal{I}_2$ in (M, s) iff $M, s \models \bigcirc_{\mathcal{I}_2 \lhd \varphi_1}(\varphi_1 | \neg \bigvee I^+_{1\varphi_1})$*

Briefly speaking, a successful command to delete a norm is supposed to release some obligations and a successful command to add a norm is meant to assign some new obligations. The issues of successful updates are closely related to CUGO principle ('Commands Usually Generate Obligations') put forward by Yamada [15]: $[!_{(i,j)}\varphi] \bigcirc_{(i,j)} \varphi$, which can be read as "after a command to i given by an authority j to see to it that φ, i has an obligation to j to see to it that φ". In our framework, we can distinguish between the commands which do generate obligations and those commands which do not. We therefore can rename the CUGO principle as SCGO which means that 'Successful Commands Generate Obligations'.

6.2 Resolving Jørgensen's Dilemma

Jørgensen's dilemma is know as the following inference:

Example 1.
(1) Let the door be open!
(2) It is impossible that the door is open but it is not unlocked.
---
(3) Let the door be unlocked!

Sentences (1) and (3) are commands given by a commander. Sentence (2) suggests that the content of norm (3) is implied by the content of norm (1). Our intuition is that the set of obligations brought about by (1) includes the obligations brought about by (3). We conceptualize Example 1 as follows.

Given an ideality sequence $\mathcal{I} = (I \cup \{\epsilon\}, \ll)$ and a bare structure (M, s), let $O^{\mathcal{I}}_{\psi} = \{\varphi \mid M, s \models \bigcirc_{\mathcal{I}}(\varphi|\psi)\}$. In the following part, proposition o represents that the door is open and u represents that the door is unlocked. Assume that, the original ideality sequence is $\mathcal{I} = (\epsilon, \ll)$ and the bare structure is (M, s). Sentence (1) is a speech act postfixing a new norm o to $\mathcal{I}$. It yields a new ideality sequence $\mathcal{I} \lhd o$. Sentence (2) can be interpreted as the formula $\neg\hat{U}(o \wedge \neg u)$ which is logically equivalent to $U(o \rightarrow u)$. Sentence (3) is a different speech act which postfixes a new norm u to $\mathcal{I}$ forming $\mathcal{I} \lhd u$. The set of obligations brought about by (1) is $O^{\mathcal{I}\lhd o}_{\neg \bigvee I} = \{\varphi \mid M, s \models \bigcirc_{\mathcal{I}\lhd o}(\varphi|\neg \bigvee I)\}$. The set of obligations brought about by (3) is $O^{\mathcal{I}\lhd u}_{\neg \bigvee I} = \{\varphi \mid M, s \models \bigcirc_{\mathcal{I}\lhd u}(\varphi|\neg \bigvee I)\}$. According to our intuition, it should be the case that $O^{\mathcal{I}\lhd u}_{\neg \bigvee I} \subseteq O^{\mathcal{I}\lhd o}_{\neg \bigvee I}$.

Proposition 3. *Let (M, s) be an arbitrary bare structure and $\mathcal{I} = (I \cup \{\epsilon\}, \ll)$ be an arbitrary ideality sequence. If $\lhd\psi$ is a successful update on $\mathcal{I}$ in (M, s) and $M, s \models U(\psi \rightarrow \chi)$, then $\lhd\chi$ is also a successful update on $\mathcal{I}$ in (M, s) and $O^{\mathcal{I}\lhd\chi}_{\neg \bigvee I} \subseteq O^{\mathcal{I}\lhd\psi}_{\neg \bigvee I}$.*

Proposition 3 indicates that the success of (1) and information (2) implies that (3) is successful and the obligations triggered by (3) are also triggered by (1).

Example 2.

(1*) There is no longer need to let the door be open!
(2) The door cannot be opened unless it is unlocked.
____
(3*) There is no longer need to let the door be unlocked!

Example 2 involves different updates on ideality sequences. Sentence (1*) deletes a norm o from the ideality sequence $o;\mathcal{I}$. Sentence (3*) deletes u from the ideality sequence $u;\mathcal{I}$. We consider it as a valid inference since we intuitively think deleting o defeats more obligations than deleting u.

Proposition 4. *Let (M, s) be an arbitrary bare structure and $\mathcal{I} = (I \cup \{\epsilon\}, \ll)$ be an arbitrary ideality sequence. If $-\psi$ is a successful updates on $\psi;\mathcal{I}$ in (M, s) and $M, s \models U(\psi \rightarrow \chi)$, then $-\chi$ is also a successful updates on $\chi;\mathcal{I}$ in (M, s) and $(O^{\chi;\mathcal{I}}_{\neg\bigvee I} - O^{(\chi;\mathcal{I})-\chi}_{\neg\bigvee I}) \subseteq (O^{\psi;\mathcal{I}}_{\neg\bigvee I} - O^{(\psi;\mathcal{I})-\psi}_{\neg\bigvee I})$.*

Proposition 4 indicates that the success of (1*) and information (2) implies that (3*) is successful and the obligations defeated by (3*) are also defeated by (1*).

7 Related Work and Conclusion

This paper studies the notion of relativized conditional obligations based on ideality sequences. Each ideality sequence is a normative system which provides a criterion on the relative ideality of propositions. Every betterness structure based on an ideality sequence describes the conditional obligations. Deletion and postfixing are two elementary updates on the normative system which can bring about corresponding obligations successfully or not. Jørgensen's dilemma is conceptualized properly. Furthermore, a sound and strongly complete axiom system for the logic of relativized conditional obligations $\mathbb{PCDL}$ is established.

Acknowledgement. The author is greatly indebted to Barteld Kooi, Rineke Verbrugge and Davide Grossi for many insightful discussions on the related topics of this work and helpful comments on earlier versions of this paper.

References

1. Broersen, J.: Action negation and alternative reductions for dynamic deontic logics. J. Appl. Log. **2**(1), 153–168 (2004)
2. Broersen, J.: Deontic epistemic stit logic distinguishing modes of mens rea. J. Appl. Log. **9**(2), 137–152 (2011)
3. Grossi, D., Kooi, B., Su, X., Verbrugge, R.: How knowledge triggers obligation. In: Ghosh, S., Icard, T. (eds.) LORI 2021. LNCS, vol. 13039, pp. 201–215. Springer, Cham (2021). https://doi.org/10.1007/978-3-030-88708-7_17
4. Hansson, B.: An analysis of some deontic logics. Nous, 373–398 (1969)
5. Horty, J.F.: Agency and Deontic Logic. Oxford University Press, Oxford (2001)

6. Horty, J.F., Pacuit, E.: Action types in stit semantics. Rev. Symbolic Logic **10**(4), 617–637 (2017)
7. Liu, F.: Changing for the better: preference dynamics and agent diversity. PhD Thesis, University of Amsterdam (2008)
8. Makinson, D., van der Torre, L.: What is input/output logic? input/output logic, constraints, permissions. In: Dagstuhl Seminar Proceedings. Schloss Dagstuhl-Leibniz-Zentrum für Informatik (2007)
9. Mastop, R.: Norm performatives and deontic logic. Eur. J. Anal. Philos. **7**(2), 83–105 (2011)
10. Meyer, J.J.C.: A different approach to deontic logic: deontic logic viewed as a variant of dynamic logic. Notre Dame J. Formal Logic **29**(1), 109–136 (1988)
11. van Benthem, J., Grossi, D., Liu, F.: Priority structures in deontic logic. Theoria **80**(2), 116–152 (2014)
12. van der Torre, L., Tan, Y.-H.: An update semantics for deontic reasoning. In: Mcnamara, P., Prakken, H. (eds.) Norms, Logics and Information Systems, pp. 73–90. IOS Press (1998)
13. van Ditmarsch, H., van der Hoek, W., Kooi, B.: Dynamic epistemic logic. Springer, Dordrecht (2007)
14. Wright, G.H.V.: Deontic logic. Mind **60**(237), 1–15 (1951)
15. Yamada, T.: Logical dynamics of some speech acts that affect obligations and preferences. Synthese **165**(2), 295–315 (2008)
16. Yamada, T.: Acts of requesting in dynamic logic of knowledge and obligation. Eur. J. Anal. Philos. **7**(2), 59–82 (2011)

Aggregating Credences into Beliefs: Threshold-Based Approaches

Minkyung Wang(✉)

CONCEPT, University of Cologne, Cologne, Germany
minkyungwang@gmail.com

Abstract. Binarizing belief aggregation tackles the problem of aggregating individuals' probabilistic beliefs on logically connected propositions into the group's binary beliefs. One common approach to associating probabilistic beliefs with binary beliefs would be applying thresholds to probabilities. This paper aims to introduce and classify a range of threshold-based binarizing belief aggregation rules while characterizing them based on different forms of monotonicity and other properties.

Keywords: Binarizing Belief Aggregation · Credence and Belief · Judgement Aggregation · Probabilistic Opinion Pooling · Threshold-based Approaches

1 Introduction: Binarizing Belief Aggregation

This paper addresses the challenge of aggregating individual probabilistic beliefs on logically connected propositions into collective binary beliefs, which we call binarizing belief aggregation. In [18,19], we showed impossibility results demonstrating that binarizing belief aggregation procedures cannot simultaneously satisfy proposition-wise independence and logical closure of collective belief, under certain reasonable conditions. In this study, we examine specific aggregation procedures, with a particular focus on threshold-based approaches.

Let me start with an example. Let A and B be two propositions. The agenda comprises A,B, $A \wedge B$, and their negations. A group consists of three individuals, and each individual provides probabilistic beliefs. The objective is to derive the group's binary beliefs based on individual credences. Table 1 illustrates various aggregation methods based on thresholds. The first method forms Group's beliefs 1 using the following procedure: the group believes a proposition X if and only if every individual's probability of X is at least 0.6. Note that disbelief in a proposition X means belief in its negation, while suspending judgment on a proposition X says neither believing X nor disbelieving X. The second method gives Group's beliefs 2 as follows: the group believes a proposition X if and only if the average of individuals' probabilities of X exceeds 0.6. While group beliefs are typically expected to be rational—consistent and deductively closed—, one might observe that the second collective beliefs do not meet this standard.

N. Alechina et al. (Eds.): LORI 2023, LNCS 14329, pp. 269–283, 2023.
https://doi.org/10.1007/978-3-031-45558-2_21

Indeed, the group believes two propositions A and $\neg(A \wedge B)$, which entails, by logical closure, that the group believes the proposition $\neg B$. To address this issue, we can devise belief binarization methods that guarantee rational binary beliefs. For instance, the third method demonstrates that selecting a threshold value of 0.8 for the average of individuals' probabilities of a proposition results in the group acquiring rational binary beliefs as shown in Group's beliefs 3.

Table 1. An Example of Event-wise Threshold-based Binarizing Belief Aggregation

Propositions	A	B	$A \wedge B$
Agent 1	0.9	0.7	0.6
Agent 2	0.8	0.4	0.2
Agent 3	0.4	0.7	0.1
Average	0.76	0.5	0.3
Group's Beliefs 1	Belief	Suspension	Suspension
Group's Beliefs 2	Belief	Suspension	Disbelief
Group's Beliefs 3	Suspension	Suspension	Suspension

Consistent and deductively closed binary beliefs can be conceived as a collection of doxastically plausible possible worlds or scenarios, which we call a belief core. A proposition is believed if and only if it is true at all doxastically plausible possible worlds, i.e., it is a superset of a belief core. In this context, applying a threshold to the probabilities of possible worlds to identify all plausible possible worlds would be a natural method. Table 2 illustrates this. The individuals' probabilistic beliefs in the previous table are now presented as probabilities of all possible worlds associated with the same agenda as before. The first method forms Group's beliefs 4 using the following procedure: the group's belief core includes possible world w if and only if every individual's probability of w is at least 0.3. As a result, only the possible world where $A \wedge \neg B$ is true constitutes the belief core. The second method yields Group's beliefs 5 as follows: the group's belief core includes possible world w if and only if the average of every individual's probability of w is at least 0.3. It is worth noting that these world-threshold procedures can always achieve consistent and deductively closed beliefs, unless the belief core is empty, as the group is supposed to believe all and only the supersets of the belief core.

Table 2. An Example of World-wise Threshold-based Binarizing Belief Aggregation

Possible Worlds	$A \wedge B$	$A \wedge \neg B$	$\neg A \wedge B$	$\neg A \wedge \neg B$
Agent 1	0.6	0.3	0.1	0
Agent 2	0.2	0.6	0.2	0
Agent 3	0.1	0.3	0.6	0
Average	0.3	0.4	0.3	0
Group's Beliefs 4	X	O	X	X
Group's Beliefs 5	O	O	O	X

As demonstrated in the examples above, threshold-based binarizing belief aggregation methods are closely intertwined with the problem of belief-credence connection, often referred to as belief binarization. Let's delve deeper into this aspect with broader theoretical interconnections. Binarizing belief aggregation can be viewed as a generalization of judgment aggregation dealing with binary beliefs, or as a generalization of belief binarization determining binary beliefs from probabilistic beliefs.[1] In judgment aggregation (e.g., [3]) or belief binarization, binary beliefs in a proposition are associated with a high quota of individuals believing the proposition or a high probability of the proposition. These rules are based on some sorts of thresholds to identify a high probability and are termed as *threshold-based approaches*. The most well-known approach is the *Lockean thesis* (LT^t), which posits that an agent (in the context of aggregation, the group) should believe a proposition if and only if its probability exceeds a given threshold t. However, the lottery paradox demonstrates that LT^t does not guarantee consistency and logical closure unless t is set at 1. Numerous resolutions have been put forth. Some challenge closure under conjunction [10,12], while others offer alternatives of probabilism [17], or modifications to the Lockean thesis [15,16]. Similar challenges are highlighted in the discursive dilemma and the aforementioned impossibility results in binarizing belief aggregation. In addressing these problems, our approach retains probabilism while developing a taxonomy to find ways to relax LT^t. To achieve this, we will systematically introduce and classify various types of threshold-based binarizing belief aggregators and characterize them. Threshold-based rules are typically characterized by some kind of monotonicity. For example, quota rules for judgment aggregation can be characterized using a specific type of monotonicity [3]. In belief binarization, Leitgeb's stability theory of belief [15] and Lin & Kelly's camera shutter rule [16] can also be interpreted as threshold-based rules that adhere to certain types of monotonicity. We will investigate exactly which types of monotonicity and what other properties of aggregators are needed to fully characterize threshold-based rules.

2 Classification of Threshold-Based Rules

We first set the notation and terminology that will be needed throughout this paper. Let W be a finite non-empty set of all possible worlds. We assume the set of all events to be the powerset $\mathcal{P}(W)$ of W, so that probabilities of the singleton set of each world are well-defined, which will be used for world-threshold rules. Further, let $N := \{1, ..., n\}$ be a set of individuals ($n \geq 2$). For each individual $i \in N$, P_i denotes i's probability function on $\mathcal{P}(W)$, and we write $\boldsymbol{P}$ for a profile $(P_1, ..., P_n)$ of probability functions. An *opinion pooling function (OP)* f is defined to be a function taking a profile $\boldsymbol{P}$ of probability functions and returning a probability function $f(\boldsymbol{P})$ on $\mathcal{P}(W)$; a *binarization rule (BR)* G is defined as a function mapping a probability function P to a binary belief $G(P)$,

[1] Indeed, various articles have pointed out the structural parallels between judgment aggregation and belief binarization [2,7–9].

i.e., a function from $\mathcal{P}(W)$ to $\{0,1\}$; we call F a *binarizing belief aggregator (BA)* if F is a function assigning to a profile $\boldsymbol{P}$ of probability functions a binary belief $F(\boldsymbol{P}) : \mathcal{P}(W) \to \{0,1\}$.

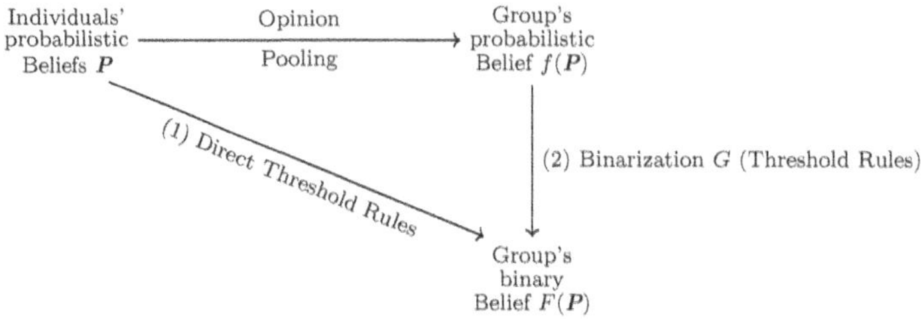

Fig. 1. Threshold-based BAs

Now we introduce and classify some relevant *threshold-based binarizing belief aggregators*. First of all, we can categorize BAs in general, including threshold-based ones, into two groups according to whether they can be represented by a combination of OP and BR (Fig. 1). The BAs in the first group do not go through an opinion pooling procedure and thus do not form the group's probability. They are called *direct threshold rules*. The other group is called *pooling + threshold-based binarization*. If a BA F belongs to the second group, then we have $F = G \circ f$ for some OP f and some BR G, and we write $F = f + G$.[2]

The next two criteria we propose for classifying threshold-based BAs pertain to types of thresholds employed. One is whether a threshold is applied to probabilities of events (in this case, we call it an *event-th.*) or probabilities of worlds (in this case, we call it a *world-th.*); we might have to believe all and only the events whose probability exceeds the event-th. or we might have to believe the set of all worlds with a probability above the world-th. and its supersets. The other is whether a threshold depends on inputs, i.e., profiles of individual probability functions. A threshold depending on inputs is called a *local* (event- or world-) th., and a threshold independent of inputs is called a *global* (event- or world-) th. On the basis of these three criteria, we can systematically present eight distinct classes of threshold-based BAs. Let us formulate the four classes of direct threshold rules first.

Definition 1 (Direct Threshold Rules). *Let F be a BA with the domain $\mathbb{P}^n$, where $\mathbb{P}$ denotes the set of all probability functions on $\mathcal{P}(W)$.*
(i) F is called a direct threshold rule with global event-th. if for each $A \in \mathcal{P}(W)$ there exist $(\rhd_{A,i})_{i \in N} \in \{>, \geq\}^N$ and $(t_{A,i})_{i \in N} \in [0,1]^N$ such that for all $\boldsymbol{P} \in \mathbb{P}^n$ it holds that $F(\boldsymbol{P})(A) = 1$ iff $P_i(A) \rhd_{A,i} t_{A,i}$ for all $i \in N$;
(ii) F is called the one with global world-th. if there exist $(\rhd_{w,i})_{(w,i) \in W \times N} \in \{>, \geq\}^{W \times N}$ and $(s_{w,i})_{(w,i) \in W \times N} \in [0,1]^{W \times N}$ such that for all $\boldsymbol{P} \in \mathbb{P}^n$ and all $A \in$

[2] For different kinds of probabilistic opinion pooling methods, see [4,5], and [6].

$\mathcal{P}(W)$ it holds that $F(\boldsymbol{P})(A) = 1$ iff $A \supseteq \{w \in W | P_i(w) \rhd_{w,i} s_{w,i}$ for all $i \in N\}$;
(iii) F is called the one with local event-th. if for each $\boldsymbol{P} \in \mathbb{P}^n$ there exist $(\rhd_{\boldsymbol{P},i})_{i \in N} \in \{>, \geq\}^N$ and $(t_{\boldsymbol{P},i})_{i \in N} \in [0,1]^N$ such that for all $A \in \mathcal{P}(W)$ it holds that $F(\boldsymbol{P})(A) = 1$ iff $P_i(A) \rhd_{\boldsymbol{P},i} t_{\boldsymbol{P},i}$ for all $i \in N$;
(iv) F is called the one with local world-th. if for each $\boldsymbol{P} \in \mathbb{P}^n$ there exist $(\rhd_{\boldsymbol{P},i})_{i \in N} \in \{>, \geq\}^N$ and $(s_{\boldsymbol{P},i})_{i \in N} \in [0,1]^N$ such that for all $A \in \mathcal{P}(W)$ it holds that $F(\boldsymbol{P})(A) = 1$ iff $A \supseteq \{w \in W | P_i(w) \rhd_{\boldsymbol{P},i} s_{\boldsymbol{P},i}$ for all $i \in N\}$.

Notice that in the definition of the rules with *event-ths.*($t_{A,i}$ and $t_{\boldsymbol{P},i}$), thresholds are applied to probabilities of events and used to determine the *belief set* $F(\boldsymbol{P})^{-1}(1)(:= \{A \in \mathcal{P}(W) | F(\boldsymbol{P})(A) = 1\})$. By contrast, *world-ths.*($s_{A,i}$ and $s_{\boldsymbol{P},i}$) are applied to obtain the *belief core* of the binary belief $F(\boldsymbol{P})$, defined as usual as the event of which supersets are all and the only believed events. Simply put, by the event-th. rules, the events with a probability being above (either greater than or not less than) the event-ths. form the belief set. And by the world-th. rules, the worlds with a probability being above the world-ths. constitute the belief core.

Both event-threshold and world-threshold approaches can be either global or local. *Local* thresholds ($t_{\boldsymbol{P},i}$ or $s_{\boldsymbol{P},i}$) are thresholds that vary with probability profiles $\boldsymbol{P}$. Conversely, *global* thresholds ($t_{A,i}$ or $s_{w,i}$) do not depend on probability profiles, but might differ for different events A or worlds w. We term thresholds *uniform*, if they are the same for all events or worlds. Notice that in our definition, local thresholds are all uniform by design since otherwise, the notion of *local* threshold rule would be empty in the sense that every rule can be seen as a local non-uniform threshold rule. It is important to notice one more dependency of thresholds in direct threshold rules. We encompass general cases where individuals may possess different threshold values, which is why we add the subscript i to all types of thresholds in direct rules.

Now consider an inequality symbol $\rhd$ in each definition, which designates either $\geq$ or $>$. The distinction between the two is not relevant when it comes to local thresholds. The reason is that for each $t_{\boldsymbol{P},i} \neq 0$ there exists $t'_{\boldsymbol{P},i} \neq 1$, and for each $t'_{\boldsymbol{P},i} \neq 1$ there exists $t_{\boldsymbol{P},i} \neq 0$ such that $\{A \in \mathcal{P}(W) | P_i(A) \geq t_{\boldsymbol{P},i}\} = \{A \in \mathcal{P}(W) | P_i(A) > t'_{\boldsymbol{P},i}\}$ because $\mathcal{P}(W)$ is finite. The same reasoning applies to local world-th. rules because W is finite. As for global threshold rules, an inequality with $\geq$ and one with $>$ cannot be represented by each other. For example, there exists no $t'_{A,i}$ satisfying $\{P_i \in \mathbb{P} | P_i(A) \geq t_{A,i}\} = \{P_i \in \mathbb{P} | P_i(A) > t'_{A,i}\}$, with A, i and $t_{A,i}$ being fixed, because $\mathbb{P}$ is not discrete. Therefore, to deal with global thresholds, the distinction is not superfluous. One more important feature regarding (strict) inequalities is that each individual i, each event A/world w (in the case of global event-/world- th.) and each profile $\boldsymbol{P}$ (in the case of local th.) can have a different kind of inequality—either strict or not—, just as each of them can have a different value of threshold. (The subscripts represent the dependency.) This enables us to study the most general cases.

Lastly, note that the inequalities in each definition should be satisfied for all individuals. This can be generalized by relaxing "all individuals" to certain proportion of individuals, but in this research we will focus on the basic case of unanimously exceeding each one's threshold, which is easy to characterize. Summarizing, we define direct threshold-based rules using the following inequalities to determine the belief set in the case of event-th. and the belief core in the case of world-th.

	event-th. for belief set	world-th. for belief core
global th	$P_i(A) \rhd_{A,i} \ t_{A,i}$ for all i	$P_i(w) \rhd_{w,i} \ s_{w,i}$ for all i
local th	$P_i(A) \rhd_{\boldsymbol{P},i} \ t_{\boldsymbol{P},i}$ for all i	$P_i(w) \rhd_{\boldsymbol{P},i} \ s_{\boldsymbol{P},i}$ for all i

We next turn to *pooling + threshold-based binarization*. Let f be an OP with the universal domain $\mathbb{P}^n$. Now we first form the group's probability $f(\boldsymbol{P}) \in \mathbb{P}$ and then use it as an input of a BR G, which outputs a binary belief $G(f(\boldsymbol{P}))$. In this way, the composition of an OP and a BR can be used as a method of binarizing belief aggregation. Since individuals' opinions are collected into the group's probability, we do not need to evaluate whether each individual's probability is above some threshold. Instead, we evaluate the group's probability $f(\boldsymbol{P}) \in \mathbb{P}$. Thus, on substituting $P_i(A)$ and $P_i(w)$ with $f(\boldsymbol{P})(A)$ and $f(\boldsymbol{P})(w)$, respectively, we obtain the definition of the four classes of *pooling + threshold-based Binarization*.

	event-th. for belief set	world-th. for belief core
global th	$f(\boldsymbol{P})(A) \rhd_A \ t_A$	$f(\boldsymbol{P})(w) \rhd_w \ s_w$
local th	$f(\boldsymbol{P})(A) \rhd_{\boldsymbol{P}} \ t_{\boldsymbol{P}}$	$f(\boldsymbol{P})(w) \rhd_{\boldsymbol{P}} \ s_{\boldsymbol{P}}$

The definition can be stated in full detail as follows.

Definition 2 (Pooling(f) + Threshold-based Binarization). *Let f be an OP with the domain $\mathbb{P}^n$ and F be a BA with the domain $\mathbb{P}^n$.*
(i) F is called a pooling(f) + threshold-based binarization with global event-th. if for each $A \in \mathcal{P}(W)$ there exist $\rhd_A \in \{>, \geq\}$ and $t_A \in [0,1]$ such that for all $\boldsymbol{P} \in \mathbb{P}(W)^n$ it holds that $F(\boldsymbol{P})(A) = 1$ iff $f(\boldsymbol{P})(A) \rhd_A \ t_A$;
(ii) F is called the one with global world-th. if there exist $(\rhd_w)_{w \in W} \in \{>, \geq\}^W$ and $(s_w)_{w \in W} \in [0,1]^W$ such that for all $\boldsymbol{P} \in \mathbb{P}(W)^n$ and all $A \in \mathcal{P}(W)$ it holds that $F(\boldsymbol{P})(A) = 1$ iff $A \supseteq \{w \in W | f(\boldsymbol{P})(w) \rhd_w \ s_w\}$;
(iii) F is called the one with local event-th. if for each $\boldsymbol{P} \in \mathbb{P}(W)^n$ there exist $\rhd_{\boldsymbol{P}} \in \{>, \geq\}$ and $t_{\boldsymbol{P}} \in [0,1]$ such that for all $A \in \mathcal{P}(W)$ it holds that $F(\boldsymbol{P})(A) = 1$ iff $f(\boldsymbol{P})(A) \rhd_{\boldsymbol{P}} \ t_{\boldsymbol{P}}$;
(iv) F is called the one with local world-th. if for each $\boldsymbol{P} \in \mathbb{P}(W)^n$ there exist $\rhd_{\boldsymbol{P}} \in \{>, \geq\}$ and $s_{\boldsymbol{P}} \in [0,1]$ such that for all $A \in \mathcal{P}(W)$ it holds that $F(\boldsymbol{P})(A) = 1$ iff $A \supseteq \{w \in W | f(\boldsymbol{P})(w) \rhd_{\boldsymbol{P}} \ s_{\boldsymbol{P}}\}$.

The contrast between event-ths. and world-ths., and the one between global and local thresholds can be made in the same way as in the direct threshold rules: (i) and (iii) utilize event-ths. so that the group believes events with a high group probability, and (ii) and (iv) apply world ths. so that the belief core consists of worlds with a high group probability. Thresholds and types of the inequalities might be different for each event and for each world in (i) and (ii), whereas they might be different for each probability profile in (iii) and (iv). The point that allowing two types of inequality is not redundant can be applied here as well.

If we focus only on the relation between the group's probability and the resulting binary belief, we can see that this framework embraces a variety of theories in the literature on belief binarization. The famous *Lockean thesis* (LT^t) combined with an OP f is no more than the rules with global even-th. t. Among various attempts to weaken LT^t, we note two of the most well-studied threshold-based approaches. One is Leitgeb's *stability theory of belief* [13–15]. According to the theory, an event should be believed if and only of it has a stably high probability, in the sense that its conditional probability on every not disbelieved event is above r, which he calls the Humean thesis of belief (HT^r, $r \in [\frac{1}{2}, 1)$). According to the theory, this relation between probability functions and binary beliefs ensures deductive closure, and so any binary belief satisfying HT^r has a belief core. For the stability condition to hold, the belief core must consist of worlds with high probability and the threshold identifying high probability depends on the probability function. This can be said in our framework as the following: the belief core is determined by a local world-th. Moreover, the theory shows that any event excluding any world in the belief core has lower probability than the belief core, so all and only the events with probability above a local event-th. are believed. So, the stability theory generates the rules with local event-ths., which can be seen as the rules with local world-ths. at the same time. And the other approach to note is Lin & Kelly's *Camera Shutter rules* (CS^s, $s > 1$), which collect the worlds whose probability ratio to the maximal probability is above $\frac{1}{s}$ as the elements of the belief core [16]. These rules utilize a special kind of local world-th., which is the maximal probability divided by s.[3]

3 Properties of Threshold-Based Rules

To characterize the eight classes of threshold-based BAs in Sect. 2, we introduce properties of BAs. The first property concerns the notion of deductive closure of binary beliefs, which are outputs of BAs. A BA F is called collectively deductively-closed (CDC) if $F(\boldsymbol{P})$ is deductively-closed, i.e., the belief set $F(\boldsymbol{P})^{-1}(1)$ is non-empty and closed under intersection and superset for all $\boldsymbol{P}$ in the domain of F. CDC will be used especially to characterize the rules with world-ths. because these rules presuppose that the resulting binary belief $F(\boldsymbol{P})$ have a belief core, which is equivalent to CDC.

Next we turn to *independence* and *neutrality*. In [18] and [19], we investigated *event-wise independence* and *event-neutrality*, whose tension with CDC leads to impossibility results. In this paper, we introduce, in addition, six different kinds of independence and neutrality. Roughly speaking, independence means that in order to decide whether an event/a world belongs to the belief set/belief core,

[3] For other types of belief binarization rules based on local world-th. see [1] and [11].

only the probability assigned to that event/world matters. By contrast, neutrality means that every event/world is determined to be included in the belief set/belief core by the same rule, and thereby, all events/worlds are considered equally. We can use this notion to characterize uniform thresholds.

Before providing the formal definition, let us mention some points needed to understand the definition: recall that world-ths. are used to determine the belief core, and bear in mind that $F(\boldsymbol{P})(\overline{w}) = 0$, where $\overline{w}$ is the complement of $\{w\}$, implies that w is in the belief core of $F(\boldsymbol{P})$, if $F(\boldsymbol{P})$ has the belief core. Here are the definitions of independence and neutrality to characterize the direct rules and the ones of *f-independence and f-neutrality* for pooling(f) + binarization.

Definition 3 (Independence and Neutrality). *A BA F is called*
(i) (event-wise) independent(IND) if for every $A \in \mathcal{P}(W)$, there is a function G_A such that $F(\boldsymbol{P})(A) = G_A(\boldsymbol{P}(A))$ for all $\boldsymbol{P}$ in the domain of F;
(ii) world-wise independent(IND^w) if for every $w \in W$, there is a function G_w such that $F(\boldsymbol{P})(\overline{w}) = G_w(\boldsymbol{P}(w))$ for all $\boldsymbol{P}$ in the domain of F;
(iii) event-neutral(eNEU) if for every $\boldsymbol{P} \in \mathbb{P}^n$, there is a function $G_{\boldsymbol{P}}$ such that $F(\boldsymbol{P})(A) = G_{\boldsymbol{P}}(\boldsymbol{P}(A))$ for all $A \in \mathcal{P}(W)$;
(iv) world-neutral(wNEU) if for every $\boldsymbol{P} \in \mathbb{P}^n$, there is a function $G_{\boldsymbol{P}}$ such that $F(\boldsymbol{P})(\overline{w}) = G_{\boldsymbol{P}}(\boldsymbol{P}(w))$ for all $w \in W$.

Definition 4 (f-Independence and f-Neutrality). *Let f be an OP. A BA F is called*
(i) f-(event-wise) independent(f-IND) if for every $A \in \mathcal{P}(W)$, there is a function G_A such that $F(\boldsymbol{P})(A) = G_A(f(\boldsymbol{P})(A))$ for all $\boldsymbol{P}$ in the domain;
(ii) f-world-wise independent(f-IND^w) if for every $w \in W$, there is a function G_w such that $F(\boldsymbol{P})(\overline{w}) = G_w(f(\boldsymbol{P})(w))$ for all $\boldsymbol{P}$ in the domain;
(iii) f-event-neutral(f-eNEU) if for every $f(\boldsymbol{P})(\in \mathbb{P})$, there is a function $G_{f(\boldsymbol{P})}$ such that $F(\boldsymbol{P})(A) = G_{f(\boldsymbol{P})}(f(\boldsymbol{P})(A))$ for all $A \in \mathcal{P}(W)$;
(iv) f-world-neutral(f-wNEU) if for every $f(\boldsymbol{P})(\in \mathbb{P})$, there is a function $G_{f(\boldsymbol{P})}$ such that $F(\boldsymbol{P})(\overline{w}) = G_{f(\boldsymbol{P})}(f(\boldsymbol{P})(w))$ for all $w \in W$.

Alternatively, F is defined to be IND/IND^w/eNEU/wNEU if
(i′) for every $A \in \mathcal{P}(W)$, if $\boldsymbol{P}(A) = \boldsymbol{P'}(A)$, then $F(\boldsymbol{P})(A) = F(\boldsymbol{P'})(A)$ for all $\boldsymbol{P}, \boldsymbol{P'}$ in the domain/
(ii′) for every $w \in W$, if $\boldsymbol{P}(w) = \boldsymbol{P'}(w)$, then $F(\boldsymbol{P})(\overline{w}) = F(\boldsymbol{P'})(\overline{w})$ for all $\boldsymbol{P}, \boldsymbol{P'}$ in the domain/
(iii′) for every $\boldsymbol{P}$, if $\boldsymbol{P}(A) = \boldsymbol{P}(B)$, then $F(\boldsymbol{P})(A) = F(\boldsymbol{P})(B)$ for all $A, B \in \mathcal{P}(W)$/
(iv′) for every $\boldsymbol{P}$, if $\boldsymbol{P}(w) = \boldsymbol{P}(v)$, then $F(\boldsymbol{P})(\overline{w}) = F(\boldsymbol{P})(\overline{v})$ for all $w, v \in W$.
While (i′) and(ii′) state that the same individual probabilities of an event/a world yield the same collective belief in the event/world, (iii′) and (iv′) assert that if two events/worlds have the same individual probabilities, then the collective belief in them should be the same. Similarly, equivalent definitions can be formulated for being f-IND/f-IND^w/f-eNEU/f-wNEU.

Now, we formalize various kinds of *monotonicity* that play a central role in characterizing threshold-based rules. Since we defined independence and neutrality in the above separately, here we define *strict* monotonicity (It will be shown that strict monotonicity taken together with independence and neutrality amounts to *monotonicity*). Informally, strict monotonicity means the following. Assume that an event/a world is in the belief set/belief core of the resulting collective binary belief of a probability profile. The first two kinds ((i) and (ii) in Definition 5) of strict monotonicity will imply that all other probability profiles with greater probability values of the event/the world should yield the same result. In contrast, the other two ones ((iii) and (iv)) will entail that all other events/worlds with greater probability values in the probability profile should also be in the belief set/belief core. *f-strict-monotonicity* in Definition 6 can be explained in a similar way if we replace a probability profile with a collective probability, which is the output of an OP f.

Definition 5 (Strict-Monotonicity). *A BA F is called*
(i) strict-monotone(SMON) if for every $A \in \mathcal{P}(W)$, if for some $i \in N$, $P_i(A) < P_i'(A)$ and for all $j \neq i$ $P_j(A) = P_j'(A)$, and if $F(\boldsymbol{P})(A) = 1$, then $F(\boldsymbol{P'})(A) = 1$ for all $\boldsymbol{P}, \boldsymbol{P'}$ in the domain;
(ii) worldwise strict-monotone($SMON^w$) if for every $w \in W$, if for some $i \in N$, $P_i(w) < P_i'(w)$ and for all $j \neq i$ $P_j(w) = P_j'(w)$, and if $F(\boldsymbol{P})(\overline{w}) = 0$, then $F(\boldsymbol{P'})(\overline{w}) = 0$ for all $\boldsymbol{P}, \boldsymbol{P'}$ in the domain;
(iii) event-strict-monotone(eSMON) if for every $\boldsymbol{P}$ in the domain, for some $i \in N$, $P_i(A) < P_i(B)$ and for all $j \neq i$ $P_j(A) = P_j(B)$, and if $F(\boldsymbol{P})(A) = 1$, then $F(\boldsymbol{P})(B) = 1$ for all $A, B \in \mathcal{P}(W)$;
(iv) world-strict-monotone(wSMON) if for every $\boldsymbol{P}$ in the domain, for some $i \in N$, $P_i(v) < P_i(w)$ and for all $j \neq i$ $P_j(v) = P_j(w)$, and if $F(\boldsymbol{P})(\overline{v}) = 0$, then $F(\boldsymbol{P})(\overline{w}) = 0$ for all $w, v \in W$.

Definition 6 (f-Strict-Monotonicity). *Let f be a OP and F be a HA on $(W, \mathcal{P}(W))$. F is called*
(i) f-strict-monotone(f-SMON) if for every $A \in \mathcal{P}(W)$, $f(\boldsymbol{P})(A) < f(\boldsymbol{P'})(A)$ and if $F(\boldsymbol{P})(A) = 1$, then $F(\boldsymbol{P'})(A) = 1$ for all $\boldsymbol{P}, \boldsymbol{P'}$ in the domain;
(ii) f-world-wise strict-monotone(f-$SMON^w$) if for every $w \in W$, $f(\boldsymbol{P})(w) < f(\boldsymbol{P'})(w)$ and if $F(\boldsymbol{P})(\overline{w}) = 0$, then $F(\boldsymbol{P'})(\overline{w}) = 0$ for all $\boldsymbol{P}, \boldsymbol{P'}$ in the domain;
(iii) f-event-strict-monotone(f-eSMON) if for every $\boldsymbol{P}$ in the domain, $f(\boldsymbol{P})(A) < f(\boldsymbol{P})(B)$ and if $F(\boldsymbol{P})(A) = 1$, then $F(\boldsymbol{P})(B) = 1$ for all $A, B \in \mathcal{P}(W)$;
(iv) f-world-strict-monotone(f-wSMON) if for every $\boldsymbol{P}$ in the domain, $f(\boldsymbol{P})(v) < f(\boldsymbol{P})(w)$ and if $F(\boldsymbol{P})(\overline{v}) = 0$, then $F(\boldsymbol{P})(\overline{w}) = 0$ for all $w, v \in W$.

Alternatively, in (i) of Definition 5 the condition that "for some $i \in N$, $P_i(A) < P_i'(A)$ and for all $j \neq i$ $P_j(A) = P_j'(A)$" can be replaced by "$\boldsymbol{P}(A) \leq \boldsymbol{P'}(A)$ and $\boldsymbol{P}(A) \neq \boldsymbol{P'}(A)$" where $\leq$ and $\neq$ between two vectors are understood as component-wise comparison. This is because the condition of (i) can be applied iteratively. The same can be said for (ii)-(iv) of Definition 5 as well. This indicates that combining independence or neutrality with strict-monotonicity yields monotonicity—e.g., IND plus SMON amounts to the statement that for every $A \in \mathcal{A}$, if $\boldsymbol{P}(A) \leq \boldsymbol{P'}(A)$, and if $F(\boldsymbol{P})(A) = 1$, then $F(\boldsymbol{P'})(A) = 1$ for all $\boldsymbol{P}, \boldsymbol{P'}$ in the domain, which we call *monotonicity*(MON). For other cases, the same reasoning can be applied.

Finally, we present the notion of *conjunctiveness*, which will prove pivotal in characterizing direct threshold rules. In essence, conjunctiveness can be elucidated as follows: Assume that two probability profiles ($\boldsymbol{P}$ and $\boldsymbol{P}'$) generate beliefs in a given event. And consider any probability profile ($\boldsymbol{P}''$) of which values of the event consist of the individual-wise minimum probabilities of the event out of the two profiles. Then it also should generate a belief in the event. For example, if $\boldsymbol{P}(A), \boldsymbol{P}'(A)$, and $\boldsymbol{P}''(A)$ are given by the following table and $F(\boldsymbol{P})(A) = F(\boldsymbol{P}')(A) = 1$, then $F(\boldsymbol{P}'')(A) = 1$. In the table, $minR$ signifies the minimum value in R for any $R \subseteq \mathbb{R}$.

	$\boldsymbol{P}(A)$	$\boldsymbol{P}'(A)$	$\boldsymbol{P}''(A) = (min\{P_i(A), P_i'(A)\})_i$
Agent 1	0.8	0.7	0.7
Agent 2	0.6	0.9	0.6

An analogous notion can also be applied to worlds in a belief core. There can be other kinds of conjunctiveness: if two events/worlds are believed/in the belief core, then so are any events/worlds whose probabilities are individual-wise minimum values out of probabilities of the first two events/worlds. These notions are needed, because, in the context of direct threshold rules, we demand that *all of the individuals' probabilities, not just a portion* should surpass their respective thresholds. This can be seen as the requirement that each individual-wise minimum of $P_i(A)$ in $\{\boldsymbol{P}(A) \in [0,1]^n | F(\boldsymbol{P})(A) = 1\}$ should also exceed each individual's threshold. Here is the formal definition.

Definition 7 (Conjunctiveness). *Let F be a HA on $(W, \mathcal{P}(W))$. F is called (i) conjunctive(Conj) if $F(\boldsymbol{P})(A) = 1$ and $F(\boldsymbol{P}')(A) = 1$, then $F(\boldsymbol{P}'')(A) = 1$ for any $\boldsymbol{P}''$ such that $P_i''(A) = min\{P_i(A), P_i'(A)\}$ for all i;*
(ii) world-wise conjunctive(Conj^w) if $F(\boldsymbol{P})(\overline{w}) = 0$ and $F(\boldsymbol{P}')(\overline{w}) = 0$, then $F(\boldsymbol{P}'')(\overline{w}) = 0$ for any $\boldsymbol{P}''$ such that $P_i''(w) = min\{P_i(w), P_i'(w)\}$ for all i;
(iii) event-conjunctive(eConj) if $F(\boldsymbol{P})(A) = 1$ and $F(\boldsymbol{P})(B) = 1$, then $F(\boldsymbol{P})(C) = 1$ for any C such that $P_i(C) = min\{P_i(A), P_i(B)\}$ for all i;
(iv) world-conjunctive(wConj) if $F(\boldsymbol{P})(\overline{v}) = 0$ and $F(\boldsymbol{P})(\overline{w}) = 0$, then $F(\boldsymbol{P})(\overline{u}) = 0$ for any u such that $P_i(u) = min\{P_i(v), P_i(w)\}$ for all i.

4 Characterizations of Threshold-Based Rules

We are now ready to characterize eight classes of threshold rules introduced in Sect. 2 in terms of properties in Sect. 3. Assume that a BA F and an OP f have the domain $\mathbb{P}^n$. Our results can be presented in Table 3.

As depicted in the table, every threshold rule satisfies some kind of strict monotonicity combined with independence in the case of global thresholds and with neutrality in the case of local thresholds, which we call monotonicity. For example, in case (2)(i), the rules are characterized by f-MON, which is defined

Table 3. Characterizations of Threshold Rules

	(1) Direct Threshold Rules	(2) Pooling(f)+Th.Binarization
(i) global event-th	IND, SMON, Conj	f-IND, f-SMON
(ii) global world-th	CDC, IND^w, SMON^w, Conj^w	CDC, f-IND^w, f-SMON^w
(iii) local event-th	eNEU, eSMON, eConj	f-eNEU, f-eSMON
(iv) local world-th	CDC, wNEU, wSMON, wConj	CDC, f-wNEU, f-wSMON

as f-IND plus f-SMON, and in case (2)(iii) by f-eMON, which is f-eNEU plus f-eSMON. To characterize the rules involving world threshold rules we additionally need CDC because they presuppose the existence of a belief core. Lastly, we need to add a kind of conjunctiveness to characterize direct threshold rules. The subsequent two theorems make this formally precise where the requirement of universal domain (UD) asserts that a BA F has the domain of $\mathbb{P}^n$.

Theorem 1 (Characterization of Direct Threshold Rules). *(i) The direct threshold rules with global event-ths. are fully characterized by UD, IND, SMON and CONJ;*
(ii) so are the ones with global world-ths. by UD, CDC, IND^w, SMON^w and CONJ^w;
(iii) so are the ones with local event-ths. by UD, eNEU, eSMON and eCONJ;
(iv) so are the ones with local world-ths. by UD, CDC, wNEU, wSMON and wCONJ

Proof. (i) It is obvious that the rule satisfies the properties. For the other direction, we need to find $t_{A,i}$ and $\rhd_{A,i}$ for each $i \in N$, with A being fixed. By UD and IND, we can let $F(\boldsymbol{P})(A) = G_A(\boldsymbol{P}(A))$ for all $\boldsymbol{P} \in \mathbb{P}(W)^n$. In the case of ${G_A}^{-1}(1) = \emptyset$, let $t_{A,i} := 1$ and $\rhd_{A,i} :=>$ for all $i \in N$. Otherwise, let $t_{Ai} := \inf\{a_i | \boldsymbol{a} \in {G_A}^{-1}(1)\}$ for each $i \in N$. We divide N into two subgroups N_1 and N_2 where N_1 is the set of individuals j such that the set $\{a_j | \boldsymbol{a} \in {G_A}^{-1}(1)\}$ has the infimum and N_2 is the set of the rest individuals. For every individual $j \in N_1$, set $\rhd_{A,j} :=\geq$ and for other individuals $k \in N_2$ define $\rhd_{A,k} :=>$. First observe that if for some $j \in N_1$, $x_j < t_{A,j}$ or for some $k \in N_2$, $x_k \leq t_{A,k}$, then $G_A(\boldsymbol{x}) = 0$ by the definition of infimum. What is left is to show that $G_A(\boldsymbol{y}) = 1$ for all $\boldsymbol{y}$ such that for every $j \in N_1$ and $k \in N_2$, $y_j \geq t_{A,j}$ and $y_k > t_{A,k}$. Since $t_{A,i}$ is the infimum of the i-th components of the vectors in ${G_A}^{-1}(1)$ and we have SMON and IND, it follows that for every i there is a vector $\boldsymbol{a}^i$ in ${G_A}^{-1}(1)$ such that the i-th component is y_i. Note that $\boldsymbol{a}^i$ has the following form where $n := |N|$:

$$\begin{aligned}
\boldsymbol{a}^1 &= (y_1, \quad a_2^1, \quad a_3^1, \quad \dots \quad, \quad a_n^1) \\
\boldsymbol{a}^2 &= (a_1^2, \quad y_2, \quad a_3^2, \quad \dots \quad, \quad a_n^2) \\
&\dots \\
\boldsymbol{a}^n &= (a_1^n, \quad a_2^n, \quad a_3^n, \quad \dots \quad, \quad y_n)
\end{aligned}$$

By iterated application of CONJ, we have $G_A((\min\{a_l^i | i \in N\})_{l \in N}) = 1$. Since we have $\min\{a_l^i | i \in N\} \leq y_l$ for all $l \in N$, by SMON and IND, we get $G_A(\boldsymbol{y}) = 1$, as desired.

(ii) We can prove this in much the same way, the only differences being (a) $F(\boldsymbol{P})(\overline{w}) = G_w(\boldsymbol{P}(w))$ where $F(\boldsymbol{P})(\overline{w}) = 0$ iff $w \in B$ by CDC, with B being the belief core of $F(\boldsymbol{P})$, (b) $t_{A,i}/\rhd_{A,i}/G_A/G_A^{-1}(1)/G_A(\boldsymbol{x}) = 0/G_A(\boldsymbol{y}) = 1$ replaced by $s_{w,i}/\rhd_{w,i}/G_w/G_w^{-1}(0)/G_w(\boldsymbol{x}) = 1/G_A(\boldsymbol{y}) = 0$ and (c) IND/SMON/CONJ replaced by IND^w/SMON^w/CONJ^w.

(iii) Similarly, (a) let $F(\boldsymbol{P})(A) = G_{\boldsymbol{P}}(\boldsymbol{P}(A))$ and replace (b) $t_{A,i}/\rhd_{A,i}/G_A/\ G_A^{-1}(1)$ by $t_{\boldsymbol{P},i}/\rhd_{\boldsymbol{P},i}/G_{\boldsymbol{P}}/G_{\boldsymbol{P}}^{-1}(1)$ and (c) IND/SMON/CONJ by eIND/eSMON/ eCONJ. Note that in this case $N_1 = N$(thereby $N_2 = \emptyset$), because given $\boldsymbol{P}$, $G_{\boldsymbol{P}}^{-1}(1)(\subseteq \{\boldsymbol{P}(A) | A \in \mathcal{P}(W)\})$ is finite since $\mathcal{P}(W)$ is finite.

(iv) Likewise, (a) let $F(\boldsymbol{P})(\overline{w}) = G_{\boldsymbol{P}}(\boldsymbol{P}(w))$ and replace (b) $t_{A,i}/\rhd_{A,i}/G_A/\ G_A^{-1}(1)\ /G_A(\boldsymbol{x}) = 0/G_A(\boldsymbol{y}) = 1$ by $s_{\boldsymbol{P},i}/\rhd_{\boldsymbol{P},i}/G_{\boldsymbol{P}}/G_{\boldsymbol{P}}^{-1}(0)/\ G_{\boldsymbol{P}}(\boldsymbol{x}) = 1/G_{\boldsymbol{P}}(\boldsymbol{y}) = 0$ and (c) IND/SMON/CONJ by wIND/wSMON/wCONJ. Note that in this case $N_1 = N$ as in (iii) since W is finite.

Theorem 2 (Characterization of Pooling(f) + Threshold-based Binarization). *Let f be an OP with the universal domain $\mathbb{P}(W)^n$.*
(i) The Pooling(f) + Threshold Binarization rules with global event-ths. are fully characterized by UD, f-IND and f-SMON;
(ii) so are the ones with global world-ths. by UD, CDC, f-IND^w and f-SMON^w;
(iii) so are the ones with local event-ths. by UD, f-eNEU and f-eSMON;
(iv) so are the ones with local world-ths. by UD, CDC, f-wNEU and f-wSMON

Proof. (i) It is clear that the rule satisfies the properties. For the other direction, by UD and f-IND we can let $F(\boldsymbol{P})(A) = G_A(f(\boldsymbol{P})(A))$ for all $\boldsymbol{P} \in \mathbb{P}(W)^n$. In the case of ${G_A}^{-1}(1) = \emptyset$, let $t_A := 1$ and $\rhd_A :=>$. Otherwise, let $t_A := \inf {G_A}^{-1}(1)$. If ${G_A}^{-1}(1)$ has the infimum, then let $\rhd_A :=\geq$, and otherwise let $\rhd_A :=>$. Our claim follows by f-SMON and f-IND.

(ii) This follows in the same manner with (a) $F(\boldsymbol{P})(\overline{w}) := G_w(f(\boldsymbol{P})(w))$ where $F(\boldsymbol{P})(\overline{w}) = 0$ iff $w \in B$ by CDC, where B is the belief core of $F(\boldsymbol{P})$. (b) $t_A/\rhd_A/G_A/G_A^{-1}(1)$ replaced by $s_w/\rhd_w/G_w/G_w^{-1}(0)$ and (c) f-IND/f-SMON replaced by f-IND^w/f-SMON^w.

(iii) Similarly, (a) let $F(\boldsymbol{P})(A) = G_{f(\boldsymbol{P})}(f(\boldsymbol{P})(A))$ and replace (b) $t_A/\rhd_A/G_A/\ G_A^{-1}(1)$ by $t_{\boldsymbol{P}}/\rhd_{\boldsymbol{P}}/G_{f(\boldsymbol{P})}/G_{f(\boldsymbol{P})}^{-1}(1)$ and (c) f-IND/f-SMON by f-eIND/f-eSMON. Note that when ${G_{f(\boldsymbol{P})}}^{-1}(1) \neq \emptyset$, ${G_{f(\boldsymbol{P})}}^{-1}(1)$ always has the infimum and thereby we can set $\rhd_{f(\boldsymbol{P})} :=\geq$, because given $\boldsymbol{P}$, $G_{f(\boldsymbol{P})}^{-1}(1)(\subseteq \{f(\boldsymbol{P})(A) | A \in \mathcal{P}(W)\})$ is finite since $\mathcal{P}(W)$ is finite.

(iv) Likewise, (a) let $F(\boldsymbol{P})(\overline{w}) = G_{f(\boldsymbol{P})}(f(\boldsymbol{P})(w))$ and replace (b) $t_A/\rhd_A/G_A/\ G_A^{-1}(1)$ by $s_{\boldsymbol{P}}/\rhd_{\boldsymbol{P}}/G_{f(\boldsymbol{P})}/G_{f(\boldsymbol{P})}^{-1}(0)$ and (c) f-IND/f-SMON by f-wIND/f-wSMON. when ${G_{f(\boldsymbol{P})}}^{-1}(0) \neq \emptyset$, we can set $\rhd_{f(\boldsymbol{P})} :=\geq$, because W is finite.

These two characterization theorems furnish a valuable framework for the analysis and comparison of the eight categories of threshold-based rules. We begin by examining part (i) of Theorem 1. It shows that the *direct rules with global event-thresholds* satisfy *IND*, which indicates that they are vulnerable to *the oligarchy result* described in [18] and [19]: any BA with CDC and IND leads to the oligarchy procedure under certain minor conditions unless the logical connection of the agenda is too simple. To circumvent this problem, alternative approaches become necessary, e.g., procedures reducing the complexity of the agenda such as premise-based rules, methods with some inconsistency management with minimal change, or other kinds of threshold rules without IND.

Global event-ths. might cause the same problem in *pooling(f)+threshold-based binarization* as well. As seen in part (i) of Theorem 2, it satisfies *f-IND*. If the OP f satisfies independence as well—e.g., linear pooling—in the sense that for every $A \in \mathcal{P}(W)$, if $\boldsymbol{P}(A) = \boldsymbol{P}'(A)$, then $f(\boldsymbol{P})(A) = f(\boldsymbol{P}')(A)$ for all $\boldsymbol{P}, \boldsymbol{P}'$, the entire process complies with *IND*. This, in turn, leads to the oligarchy result, akin to the scenario discussed earlier. It is also worth pointing out that employing binarization with global event-ths. does not guarantee collective deductive closure (CDC), as the lottery paradox shows.

Next, let us move to *direct threshold rules.* For the scope of this paper, we narrow our exploration to the simplest variants of direct threshold rules, wherein each individual's probability is required to *unanimously* surpass their individual threshold. This requirement aligns with *Conj.* While this norm might find justification in certain scenarios where every individual's opinion should be respected, it might seem less reasonable in various other contexts.

Now we turn to *pooling + threshold based binarization with world-ths. or local event-ths.* in parts (ii)-(iv) of Theorem 2. As highlighted in Sect. 2, binarization with local thresholds involves the rules satisfying *Humean thesis* (HT^r) and the *Camera Shutter* (CS^s) rules. In contrast to global event-ths., not only rules with world-ths. but also certain rules with local event-ths. elude the above problem of rationality. Firstly, the rules with world-ths. like the CS^s rules guarantee CDC, as demonstrated in parts (ii) and (iv). Secondly, not every local event-ths.-based binarization ensures rationality. However, it's noteworthy that HT^r can give rise to a special kind of local event-ths.-based rules. By integrating our findings with the stability theory of belief, we uncover the following observation:

a BA F satisfies f-eMON (f-eNEU plus f-eSMON) and CDC if and only if $(f(\boldsymbol{P}), F(\boldsymbol{P}))$ satisfies $\mathrm{HT}^{\frac{1}{2}}$ for all $\boldsymbol{P}$ in the domain.

It also deserves special mention that according to the stability theory of belief, the rules generated by HT^r can be seen as local world-threshold rules as well, which implies that the rules satisfy f-wMON (f-wNEU plus f-wSMON). Accordingly, the following also holds:

a BA F satisfies f-eMON, f-wMON and CDC if and only if $(f(\boldsymbol{P}), F(\boldsymbol{P}))$ satisfies $\mathrm{HT}^{\frac{1}{2}}$ for all $\boldsymbol{P}$ within the domain.

5 Conclusion

This paper introduced binarizing belief aggregation based on various kinds of thresholds. We systematically examined a range of potential threshold rules and described the characteristics associated with these rules. Within the literature on belief binarization, there have been previous approaches that rely on thresholds, and this paper proposed combining these approaches with opinion pooling.

However, alternative methods exist for both belief binarization [21] and binarizing belief aggregation [20], which involve strategies such as minimizing distance from probability functions or maximizing expected accuracy. These avenues offer promising directions for future research.

Furthermore, it's worth noting that several assumptions underpinning threshold-based rules could be relaxed to yield a more nuanced exploration. For instance, in direct threshold-based rules, the requirement of unanimity could be eased to accommodate a super-majority consensus. Additionally, the condition of deductive closure could potentially be replaced by consistency alone, or the universal domain could be relaxed, akin to what's observed in premise-based judgment aggregation approaches. These adjustments provide rich ground for further investigations and extensions.

In addition to the methods investigated in this paper, it would be valuable to explore alternative binarizing belief aggregation procedures. One potential approach to consider is the combination of threshold-based belief binarization with judgment aggregation methods. This involves applying a belief binarization method to individual probabilistic beliefs first, and then utilizing a judgment aggregation method. Further investigation into these approaches would allow for a comparison with the various binarizing belief aggregation rules examined in this paper.

Acknowledgements. I am deeply grateful to Hannes Leitgeb, Christian List, and Chisu Kim for the excellent and insightful feedback. This research is based on work supported by the German Academic Scholarship Foundation and the Alexander von Humboldt Foundation.

References

1. Cantwell, J., Rott, H.: Probability, coherent belief and coherent belief changes. Ann. Math. Artif. Intell. **87**(3), 259–291 (2019). https://doi.org/10.1007/s10472-019-09649-3
2. Cariani, F.: Local supermajorities. Erkenntnis **81**, 391–406 (2016)
3. Dietrich, F., List, C.: Judgment aggregation by quota rules: majority voting generalized. J. Theor. Polit. **19**(4), 391–424 (2007)
4. Dietrich, F., List, C.: Probabilistic opinion pooling. In: Hitchcock, C., Hajek, A. (eds.) Oxford handbook of probability and philosophy. Oxford University Press, Oxford (2016)
5. Dietrich, F., List, C.: Probabilistic opinion pooling generalized. Part one: general agendas. Soc. Choice Welfare **48**, 747–786 (2017a)
6. Dietrich, F., List, C.: Probabilistic opinion pooling generalized. Part two: the premise-based approach. Soc. Choice Welfare **48**, 787–814 (2017b)
7. Dietrich, F., List, C.: From degrees of belief to binary beliefs: lessons from judgement-aggregation theory. J. Philos. **115**(5), 787–814 (2018)
8. Dietrich, F., List, C.: The relation between degrees of belief and binary beliefs: a general impossibility theorem. In: Douven, I. (eds) Lotteries, Knowledge, and Rational Belief. Essays on the Lottery Paradox, pp. 223–254. CUP (2021)
9. Douven, I., Romeijn, J.-W.: The discursive dilemma as a lottery paradox. Econ. Philos. **23**(3), 301–319 (2007)

10. Easwaren, K., Fitelson, B.: Accuracy, Coherence and Evidence. In Oxford Studies in Epistemology, vol. 5, pp. 61–96. OUP (2015)
11. Goodman, J., Salow, B.: Epistemology normalized. Philos. Rev. **132**(1), 89–145 (2023)
12. Kyburg, H.E., Jr.: Probability and the Logic of Rational Belief. Wesleyan University Press, Middletown (1961)
13. Leitgeb, H.: Reducing belief simpliciter to degrees of belief. Ann. Pure Appl. Logic **164**, 1338–1389 (2013)
14. Leitgeb, H.: The stability theory of belief. Philos. Rev. **123**(2), 131–171 (2014)
15. Leitgeb, H.: The Stability of Belief: How Rational Belief Coheres with Probability. Oxford University Press, Oxford (2017)
16. Lin, H., Kelly, K.T.: A geo-logical solution to the lottery paradox. Synthese **186**(2), 531–575 (2012)
17. Spohn, W.: The Laws of Belief: Ranking Theory and Its Philosophical Applications. Oxford University Press, Oxford (2012)
18. Wang, M.: Aggregating Individual Credences into Collective Binary Beliefs: An Impossibility Result. Unpublished Manuscript (MS)
19. Wang, M., Kim, C.: Aggregating credences into beliefs: agenda conditions for impossibility results (extended abstract). In: Proceedings TARK 2023. EPTCS, vol. 379, pp. 518–526 (2023)
20. Wang, M., Kim, C.: Aggregating Credences into Beliefs: Distance- and Utility-based Approaches. Unpublished Manuscript (MSa)
21. Wang, M., Kim, C.: Distance- and Utility-based Belief Binarization. Unpublished Manuscript (MSb)

Cooperation Mechanisms for the Prisoner's Dilemma with Bayesian Games

Wei Xiong(✉)

Institute of Logic and Cognition, Department of Philosophy, Sun Yat-sen University, Guangzhou, China
hssxwei@mail.sysu.edu.cn

Abstract. This paper explores the cooperation mechanisms for the prisoner's dilemma game, a canonical example for studying cooperation mechanisms, with Bayesian games. By the approach allowing simultaneous moves with the assumption that the players might be self-interested or norm-following, we establish four possible Bayesian game models, all of which are cooperation mechanisms for the prisoner's dilemma game except for the model in which one of the two players must be self-interested.

Keywords: Cooperation · Prisoner's Dilemma · Norm-follower · Bayesian Game

1 Introduction

In game theory players are assumed to be rational individuals who represent preferences by their own interests when making decisions [13]. Based on this assumption, we can identify Nash equilibria as solutions for a given game. Nevertheless, we will face a social dilemma that the results obtained by the Nash equilibrium are inferior to the Pareto optimal outcomes. In particular, the prisoner's dilemma is a well-known example of social dilemmas.

Specifically, the prisoner's dilemma game (the PD game for short) can be illustrated by Table 1, where C stands for the action "cooperation", and D denotes the action "defection". It is clear that, whatever the other does, each player is better off defecting than cooperating. Consequently, for each player, the cooperation action is strictly dominated by the defection action, and (D, D) is the unique Nash equilibrium in the PD game. However, the outcome yielded by the equilibrium is worse for each than that they would have obtained when both cooperate.

N. Alechina et al. (Eds.): LORI 2023, LNCS 14329, pp. 284–291, 2023.
https://doi.org/10.1007/978-3-031-45558-2_22

Table 1. Prisoner's Dilemma

	C	D
C	$2, 2$	$0, 3$
D	$3, 0$	$1, 1$

There has been an increasing research focus on social cooperation involving the PD game in the literature. With which, researchers offer some valuable approaches to exploring the problem of social dilemmas, for example, norms for social cooperation under the concept of collective rationality [7,9,15], the repeated prisoner's deilemma [5,8,10], the evolutionary perspective [1,6,11,12], the mutually beneficial perspective [2,3], the reward and punishment approach [4,14,16], among many others.

The PD game is a strategic game, which assumes that each player must know the structure of the game. In many interactive situations the players, however, are not completely informed about the other players' characteristics or types. Consequently, the players might not know the other players' preferences, and thus the structure of the game cannot be determined.

In her seminal book [3], Bicchieri introduces some possible types for the players in the PD game, which is worthy of consideration here to guide for studying the cooperation mechanisms. In particular, she mentions a norm-following type who desires to cooperate, which means that for a norm-follower in the PD game, the action profile (C, C) is ranked first in the preference ordering. With the aid of this notion, this paper investigates some cooperation mechanisms for the PD game. More specifically, we find the cooperation mechanisms by simultaneous moves with two possible types. Since a Bayesian game is a generalized version of strategic games with incomplete information, we will construct some possible game models based on Bayesian games to deal with this problem.

2 Bayesian Games

To analyze a strategic game, we shall assume that each player must know the structure of the game she is playing. We need to establish a model under incomplete information to relax the assumption of strategic games. Such a model is called a Bayesian game.

Definition 1. *A Bayesian game is a tuple* $\mathcal{G} = (N, (A_i), \Omega, (T_i), P, (u_i))$, *where* N *is a set of players,* A_i *is the set of actions of player* i, Ω *is a set of states,* T_i *is the set of types of player* i, P *is a common belief (a probability distribution over* Ω*), and* u_i *is the payoff function of player* i *representing her preference.*

A strategy of player i in a Bayesian game s_i is a function that assigns to each of his type an action in the set of actions available to him. Formally, $s_i : T_i \to A_i$.

Let S_i be the set of strategies of player i. By the definition of strategy in an extensive game, we have

$$S_i = \underbrace{A_i \times \cdots \times A_i}_{\text{cardinality of the set } T_i}.$$

Before defining the notion of equilibrium for Bayesian games, let us present how to obtain the expected payoff to a strategy profile. According to Definition 1, a Bayesian game consists of some strategic games, in which each strategic game is determined by one of states. Since there is a common belief over the set of states, we can identify player i's expected payoff to any strategy profile $s = (s_i, s_{-i})$ as follows:

$$U_i(s) = \sum_{\omega \in \Omega} P(\omega) u_i^{\omega}(s_i(t_i), s_{-i}(t_{-i})), \tag{1}$$

where u_i^{ω} is player i's payoff function over the set $\prod_{i \in N} A_i$ in the strategic game determined by the state ω.

Using Formula (1), we can obtain the strategic form $(N, (S_i), (U_i))$ corresponding to a Bayesian game, and define Nash equilibria of the strategic game as the solutions for the Bayesian game, which are called Bayesian equilibria.

3 Mechanisms with Bayesian Games

In the PD game, each player is perfectly informed about the opponent's characteristics. In particular, the game is assumed that each player knows that the opponent is self-interested. As such each player knows the game she is playing. Now we relax the assumption: the player might be another type rather than a self-interested one. More specifically, in the PD game, each player is a self-interested type whose preference ranking is $(D, C) \succ (C, C) \succ (D, D) \succ (C, D)$. In a variant of the traditional PD game instead, each player might be a norm-follower whose preference ranking is $(C, C) \succ (D, D) \succ (D, C) \succ (C, D)$.

Next, we investigate the cooperation mechanisms based on this idea by verifying four possible cases. Let us first consider the case that each player has an indeterminate type: self-interested or norm-following. In this game there are two states, and neither player knows the state.

Definition 2. *A variant of the PD game in which neither player knows the state is defined as a Bayesian game* $\mathcal{G}_1 = (N, (A_i), \Omega, (T_i), P, (u_i))$*, where* $N = \{1, 2\}$*,* $A_1 = A_2 = \{C, D\}$*;* $\Omega = \{\omega_1, \omega_2\}$*, where* ω_1 *stands for the state that each player is the self-interested type denoted as* t_1*, and* ω_2 *stands for the state that each player is a norm-follower denoted as* t_2*;* $T_1 = T_2 = \{t\}$*, where* $t = t_1$ *or* $t = t_2$*;* $P = (p, 1-p)$ *is a probability distribution over* Ω*,* $p \in (0, 1)$*, and* u_1 *and* u_2 *are the players' payoff functions.*

The game $\mathcal{G}_1$ can be illustrated as Fig. 1. The frames labeled 1 and 2 enclosing both states indicate that player 1 and player 2 do not know the relevant state, which represents the players' cognitive conditions.

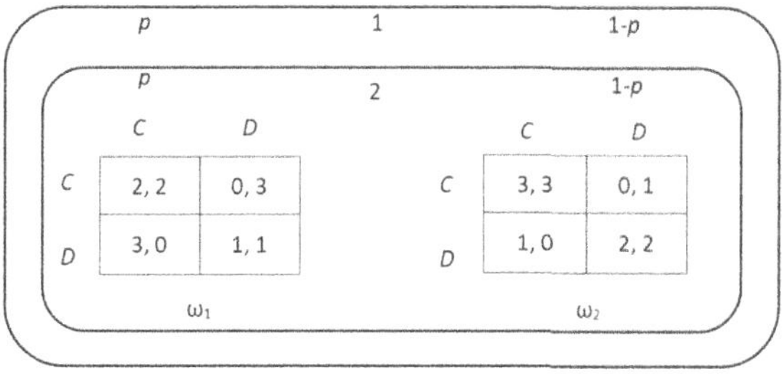

Fig. 1. $\mathcal{G}_1$: neither player knows the state

Remark 1. *The strategy sets of the players in the game* $\mathcal{G}_1$ *are* $S_1 = A_1 = \{C, D\}$ *and* $S_2 = A_2 = \{C, D\}$*, respectively.*

Proposition 1. *The strategy profile* (C, C) *is a Bayesian equilibrium of* $\mathcal{G}_1$ *if* $p \leq \frac{2}{3}$.

Proof. We first determine the strategic form of $\mathcal{G}_1$. As we shown, the strategy set of each player is $\{C, D\}$. To construct the strategic form of $\mathcal{G}_1$, we shall identify the expected payoff to each strategy profile. Consider the strategy profile (C, C). By Formula 1, we have

$$\begin{aligned} U_1(C, C) &= P(\omega_1)u_1^{\omega_1}(C, C) + P(\omega_2)u_1^{\omega_2}(C, C) \\ &= 2p + 3(1 - p) = 3 - p. \end{aligned}$$

We can similarly obtain the expected payoffs of each player to other strategy profiles. Thus, we have the strategic form of $\mathcal{G}_1$ as Table 2.

Table 2. Strategic form of $\mathcal{G}_1$

	C	D
C	$3 - p, 3 - p$	$0, 1 + 2p$
D	$1 + 2p, 0$	$2 - p, 2 - p$

By this game table, if $p \leq \frac{2}{3}$, then the best response of player 1 to player 2's strategy C is C, and the best response of player 2 to player 1's strategy C is also C. Therefore, the strategy profile (C, C) is a Bayesian equilibrium of $\mathcal{G}_1$ if $p \leq \frac{2}{3}$.

Proposition 1 shows that if the probability of each player being self-interested type ranges in $[0, \frac{2}{3}]$, then for each player C is an equilibrium strategy. As such, the proposition establishes that the Bayesian game $\mathcal{G}_1$ is a cooperation mechanism for PD game. Note that the strategy profile (D, D) is also a Bayesian equilibrium of $\mathcal{G}_1$, since for any p, the best response of player 1 to player 2's strategy D is D, and the best response of player 2 to player 1's strategy D is also D. Nevertheless, the two players have a motivation to choose the action C, as each player's expected payoff to the equilibrium (C, C) is bigger than that to the equilibrium (D, D).

Now consider the case that player 1, who has two possible types, is informed of the state: she knows the state ω_1 if she is self-interested, while the state ω_2 if she is a norm-follower; instead, player 2, a self-interested person, is not informed the state: she does not know player 1's type. In this case, a game modeling such a situation can be defined as follows.

Definition 3. *A variant of the PD game in which only player 1 knows the state and player 2 is self-interested is defined as a Bayesian game* $\mathcal{G}_2 = (N, (A_i), \Omega, (T_i), P, (u_i))$, *where* $T_1 = \{t_1, t_2\}$, *and* $T_2 = \{t_1\}$.

The game $\mathcal{G}_2$ can be illustrated as Fig. 2. The strategy sets of player 1 and player 2 in the game $\mathcal{G}_2$ are $S_1 = A_1 \times A_2 = \{C, D\} \times \{C, D\} = \{CC, CD, DC, DD\}$ and $S_2 = \{C, D\}$, respectively.

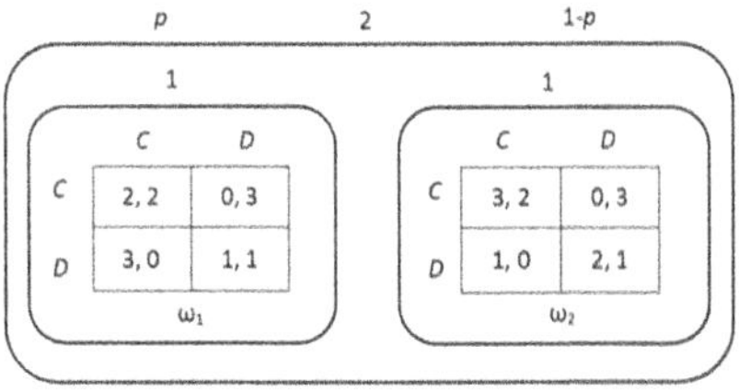

Fig. 2. $\mathcal{G}_2$: player 1 is informed of the state and player 2 is self-interested

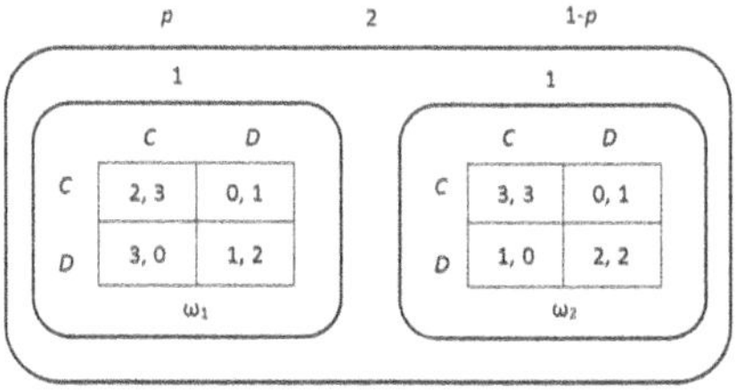

Fig. 3. $\mathcal{G}_3$: player 1 is informed of the state and player 2 is a norm-follower

Proposition 2. *The strategy profile* (DD, D) *is the unique Bayesian equilibrium of* $\mathcal{G}_2$.

Proof. We can identify the expected payoff of each player to each strategy profile by Formula 1. Thus, we have the strategic form of $\mathcal{G}_2$ as Table 3. According to the game Table 3, for player 2 the strategy D is strictly dominates his other strategies, and the best response of player 1 to player 2's strategy D is DD. Hence the strategy profile (DD, D) is the unique Bayesian equilibrium of $\mathcal{G}_2$.

Proposition 2 shows that the interactive situation designed by the game $\mathcal{G}_2$ is not a proper cooperation mechanism for the PD game, since the action "cooperation" is not a component of the Bayesian equilibrium of the game $\mathcal{G}_2$. Note that player 2 in this game is a self-interested person. It is, therefore, natural to investigate the question: what will happen if the game $\mathcal{G}_2$ remains the same expect that player 2 is a norm-follower?

Table 3. Strategic form of $\mathcal{G}_2$

	C	D
CC	$3-p, 2$	$0, 3$
CD	$1+p, 2p$	$2-2p, 1+2p$
DC	$3, 2-2p$	$p, 3-2p$
DD	$1+2p, 0$	$2-p, 1$

Table 4. Strategic form of $\mathcal{G}_3$

	C	D
CC	$3-p, 3$	$0, 1$
CD	$1+p, 3p$	$2-2p, 2-p$
DC	$3, 3-3p$	$p, 1+p$
DD	$1+2p, 0$	$2-p, 2$

Definition 4. *A variant of the PD game in which only player 1 knows the state and player 2 is a norm-follower is defined as a Bayesian game* $\mathcal{G}_3 = (N, (A_i), \Omega, (T_i), P, (u_i))$, *where* $T_1 = \{t_1, t_2\}$, *and* $T_2 = \{t_2\}$.

Similarly, we can set Fig. 3 to illustrate the game $\mathcal{G}_3$. Note that player 2 is norm-following in this case. As the game $\mathcal{G}_2$, the strategy sets of player 1 and player 2 in the game $\mathcal{G}_3$ are $S_1 = A_1 \times A_2 = \{C, D\} \times \{C, D\} = \{CC, CD, DC, DD\}$ and $S_2 = \{C, D\}$, respectively.

Proposition 3. *The strategy profile* (DC, C) *is a Bayesian equilibrium of* $\mathcal{G}_3$ *if* $p \leq \frac{1}{2}$.

Proof. By Formula 1, we can obtain the expected payoff of each player to each strategy profile, by which we can establish the strategic form of $\mathcal{G}_3$ as Table 4. According to this table, if $p \leq \frac{1}{2}$, then we have $3 - 3p \geq 1 + p$. In this case, the best response of player 1 to player 2's strategy C is DC, and the best response of player 2 to player 1's strategy DC is C. Therefore, the strategy profile (DC, C) is a Bayesian equilibrium of $\mathcal{G}_3$ if $p \leq \frac{1}{2}$.

Proposition 3 establishes that if the probability of player 1 being self-interested tpye ranges in $[0, \frac{1}{2}]$, then player 1's strategy DC and player 2's strategy C are combined to be a Bayesian equilibrium. As a result, the Bayesian game $\mathcal{G}_3$ can be regarded as a cooperation mechanism for the PD game.

Finally, we consider a more complicated case that neither player knows the state and each player has two possible types. Hence, there are four elements in the set of states.

Definition 5. *A variant of the PD game in which neither player knows the state and each player has two possible types is defined as a Bayesian game* $\mathcal{G}_4 = (N, (A_i), \Omega, (T_i), P, (u_i))$ *that can be illustrated by Fig. 4, where* $\Omega = \{\omega_1, \omega_2, \omega_3, \omega_4\}$, $T_1 = T_2 = \{t_1, t_2\}$, *and* $P = (p_1, p_2, p_3, p_4)$ *is a probability distribution over* Ω, $p_4 = 1 - p_1 - p_2 - p_3$.

Remark 2. *In the game* $\mathcal{G}_4$ *the strategy sets of players are* $S_1 = S_2 = \{C, D\} \times \{C, D\} = \{CC, CD, DC, DD\}$.

Proposition 4. *The strategy profile* (DC, DC) *is a Bayesian equilibrium of* $\mathcal{G}_4$ *if* $p_4 \geq p_2$ *and* $p_4 \geq p_3$.

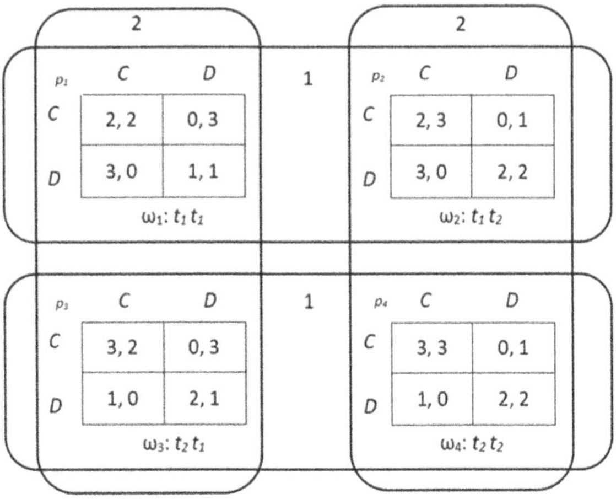

Fig. 4. $\mathcal{G}_4$: neither player knows the state and each player has two possible types

Table 5. Strategic form of $\mathcal{G}_4$

	CC	CD	DC	DD
CC	$3 - p_1 - p_2$, $3 - p_1 - p_3$	$2p_1 + 3p_3$, $1 + p_1 + p_3$	$3 - 3p_1 - p_2 - 3p_3$, 3	0, $1 + 2p_1 + 2p_3$
CD	$1 + p_1 + p_2$, $2p_1 + 3p_2$	$2 - 2p_2 - p_3$, $2 - p_2 - 2p_3$	$1 - p_1 + p_2 + p_3$, $3p_1 + 3p_2 + p_3$	$2 - 2p_1 - 2p_2$, $2 + p_1 - p_2 - p_3$
DC	3, $3 - 3p_1 - 3p_2$	$3p_1 + p_2 + 3p_3$, $1 - p_1 + p_2 + p_3$	$3 - 2p_1 - 3p_3$, $3 - 2p_1 - 3p_2$	$p_1 + p_2$, $1 + p_2 + 2p_3$
DD	$1 + 2p_1 + 2p_2$, 0	$2 + p_1 - p_2 - p_3$, $2 - 2p_1 - 2p_3$	$1 + 2p_2 + p_3$, $p_1 + p_3$	$2 - p_1 - p_2$, $2 - p_1 - p_3$

Proof. As before, we can identify the expected payoff of each player to each strategy profile by Formula 1. Hence, we have the strategic form of $\mathcal{G}_4$ as Table 5, where the upper and lower parts in the payoff matrix are the payoffs of player 1 and player 2, respectively.

It can be verified by Table 5 that if $p_4 \geq p_2$ and $p_4 \geq p_3$, then the best response of player 1 to player 2's strategy DC is DC, and the best response of player 2 to player 1's strategy DC is also DC. Therefore, the strategy profile (DC, DC) is a Bayesian equilibrium of $\mathcal{G}_4$.

By Proposition 4, the strategy DC is an equilibrium strategy for each player when the probability that they are of norm-following type is not less than the probability that they are of different types. As such, the game set by Definition 5 also provides a cooperation mechanism for the PD game.

4 Conclusion

The PD game is a canonical example for studying cooperation mechanisms. In this paper we have explored the cooperation mechanisms for the PD game with

Bayesian games. Under the assumption that the players in the PD game might be self-interested or norm-following, we have established four possible Bayesian game models: the game where there are two states, and neither player knows the state and players are self-interested or norm-following; the game where player 1, who has two possible types, is informed of the state, and player 2, a self-interested person, is not informed the state; the game where player 1, who has two possible types, is informed of the state, and player 2, a norm-follower, is not informed the state; and the game where neither player knows the state and each player has two possible types. We have further demonstrated that they are cooperation mechanisms for the PD game except for the model in which one of the two players must be self-interested.

Acknowledgements. This work was supported by the MOE Project of Key Research Institute of Humanities and Social Sciences at Universities 22JJD520001.

References

1. Axelrod, R., Hamilton, W.: The evolution of cooperation. Science **211**(4489), 1390–1396 (1981)
2. Bacharach, M.: Beyond Individual Choice: Teams and Frames in Game Theory. Princeton University Press, Princeton (2018)
3. Bicchieri, C.: The Grammar of Society: The Nature and Dynamics of Social Norms. Cambridge University Press, Cambridge (2005)
4. Brandt, H., Hauert, C., Sigmund, K.: Cooperation, punishment and reputation in spatial games. Proc. R. Soc. Lond. B **2701**, 1099–1104 (2003)
5. Dal Bó, P., Fréchette, G.: Strategy choice in the infinitely repeated Prisoner's Dilemma. Am. Econ. Rev. **109**(11), 3929–3952 (2019)
6. Doebeli, M., Hauert, C., Killingback, T.: The evolutionary origin of cooperators and defectors. Science **306**(5697), 859–862 (2004)
7. Gilbert, M.: Rationality in collective action. Philos. Soc. Sci. **36**(1), 3–17 (2006)
8. Grim, P.: The undecidability of the spatialized prisoner's dilemma. Theor. Decis. **42**, 53–80 (1997)
9. Huang, Y., Xiong, W.: Reasoning with Rawls' Maximin Criterion. Stud. Logic **12**(6), 96–107 (2019)
10. Kendall, G., Yao, X., Chong, S.: The Iterated Prisoners' Dilemma: 20 Years On. World Scientific, Singapore (2007)
11. McKenzie, A.: The Structural Evolution of Morality. Cambridge University Press, Cambridge (2007)
12. Nowak, M., Sasaki, A., Taylor, C., Fudenberg, D.: Emergence of cooperation and evolutionary stability in finite populations. Nature **4281**(6983), 646–650 (2004)
13. Osborne, M.J., Rubinstein, A.: A Course in Game Theory. MIT press, Massachusetts (1994)
14. Sigmund, K., Hauert, C., Nowak, M.: Reward and punishment. Proc. Natl. Acad. Sci. **98**(19), 10757–10762 (2001)
15. Sugden, R.: Team preferences. Econ. Philos. **16**(2), 175–204 (2000)
16. Villatoro, D., Sen, S., Sabater-Mir, J.: Of social norms and sanctioning: a game theoretical overview. Int. J. Agent Technol. Syst. **2**(1), 1–15 (2010)

Belief Revision with Satisfaction Measure

Wei Zhao(✉)

Institute of Logic and Cognition, Department of Philosophy, Sun Yat-sen University, Guangzhou, China
zhaow3516@126.com

Abstract. In this paper we propose a satisfaction measure for the theory of belief revision, which should be an alternative to the classical distance measure for defining revision operators. We present the idea of satisfaction measure as a monotonic condition, and characterize how it affects the revision operation by proving an extension of the *AGM* representation theorem (H. Katsuno and A. Mendelzon, 1991). A unique syntactical form, the disjunction of all prime implicants, is used to compile the belief bases. Then we develop a method to generate revision operators meet that monotonic condition. Lastly, we compare such an operator to four operators of other kinds with an example, then highlight its novelty and advantages.

Keywords: Belief revision · Satisfaction measure · Prime implicant

1 Introduction

The theory of belief revision [17] studies the change of beliefs when an agent receives new information. If the new information is contradictory with one's beliefs, there is no obvious way to combine them together into a consistent set of beliefs. C. Alchourrón, P. Gärdenfors and D. Makinson [1] established a theory where they proposed several rational postulates for this kind of change of beliefs. Known as *AGM* theory, their pioneering work has inspired plenty of studies in this literature [6].

A basic principle of *AGM* theory is that the new information is believed to be true and should be accepted unconditionally, and therefore the revision operation, if necessary, is imposed only on one's old beliefs. Others argued that in some scenarios we should give equal credence to both old beliefs and new information, and they defined revision operators that are non-prioritized [18] or commutative [13]. H. Katsuno and A. Mendelzon [9] concerned the case where beliefs are formalized by propositional formulae, and represented *AGM* postulates elegantly with pre-orders over possible worlds. From that perspective, an *AGM* revision operator chooses from the models of new information that are closest to the old beliefs, which, in this sense, are revised carefully by minimal change.

The (pseudo-)distance measure [12] is a natural way to evaluate the closeness from possible worlds to formulae. For instance, M. Dalal [5] proposed to take

N. Alechina et al. (Eds.): LORI 2023, LNCS 14329, pp. 292–305, 2023.
https://doi.org/10.1007/978-3-031-45558-2_23

each propositional atom as a distance unit, and as shown in [9], Dalal's revision operator turned out to be better than many others in terms of *AGM* postulates. However, according to its definition [12], a distance function, which measures the distance from a possible world to another, may or may not be relevant to the contents of possible worlds themselves. So we can see that the notion of "distance" is just used there in an abstract and metaphorical way.

This paper aims to present the measure of satisfaction as an alternative to the distance. The satisfaction measure firstly concerns the closeness from a possible world to a term, a formula of special form, and captures an intuition that the number of atoms on which they differ plays a critical role for evaluating that closeness. This idea is similar to Dalal's but expressed as a weaker condition. Furthermore, the measurement is extended to evaluate the closeness from a possible world to an arbitrary formula by taking as a bridge the set of all prime implicants (of that formula), which not only has the merit of syntactical uniqueness but also fits the satisfaction measure semantically.

Thus there are two pillars of our work: the notion of satisfaction and the syntax compilation towards prime implicants. Although neither one is completely new for this literature, we apply them in a different way. P. Pozos-Parra et al. have studied both belief merging [3] and belief revision [4] with a notion similar to satisfaction, which they called partial satisfiability. While they put forward specific operators for merging or revision, we intend to develop a general framework that characterizes a category of revision operators. We present the satisfaction measure as a monotonic condition in addition to *AGM* theory, and prove an extension of the *AGM* representation theorem [9]. Moreover, J. Marchi et al. [15] have proposed revision operators induced by prime implicants as well as prime implicates. While they combined prime forms with distance measure mostly for a better complexity performance, we take prime implicants as a necessary complement for satisfaction measure. We also give reasons for choosing prime implicants over prime implicates which was taken in [15] as the basis for their new notion of minimal change.

The rest of this paper is organized as follows. In Sect. 2, we introduce the notations, then define the prime forms and discuss the property of prime implicants in particular. A brief review of *AGM* theory and distance measure is given in Sect. 3. We characterize the satisfaction measure for revision operators in Sect. 4.1, then develop a method to generate operators of that kind in Sect. 4.2. We also in Sect. 5 compare such an operator to others of [4,5,15] with an example followed by some remarks. We conclude in Sect. 6 with a discussion of future work.

2 Preliminaries

2.1 Language

We consider a propositional language $\mathcal{L}$ over a finite set P of atoms and the connectives $(\neg, \vee, \wedge, \rightarrow, \bot, \top)$ are defined as usual. An *interpretation* or *possible world* I is a function from P to $\{\mathrm{T}, \mathrm{F}\}$, and it is a *model* of a formula $\varphi \in \mathcal{L}$ if and only if $I \models \varphi$ in the classical truth functional way. Let $[\![\varphi]\!]$ denote the set of all models of φ and $\mathcal{W} = [\![\top]\!]$. φ is *consistent* if $[\![\varphi]\!] \neq \emptyset$. φ_1 is *equivalent* to φ_2, denoted by $\varphi_1 \equiv \varphi_2$, if $[\![\varphi_1]\!] = [\![\varphi_2]\!]$. φ_2 is a *logical consequence* of φ_1, denoted by $\varphi_1 \models \varphi_2$, if $[\![\varphi_1]\!] \subseteq [\![\varphi_2]\!]$. $var(\varphi)$ is the set of atoms occurring in φ. For any set N, $|N|$ denotes the cardinality of it.

We say l is a *literal* if $l = p$ or $l = \neg p$ where $p \in P$. $C = \bigvee X$ is a *clause* and $D = \bigwedge X$ is a *term* if X is a set of literals. By this definition, there is no repetition of literals in a clause or term. In the case of $X = \emptyset$, we have $\bigvee \emptyset \equiv \bot$ and $\bigwedge \emptyset \equiv \top$. A *disjunctive normal form* (DNF) is a formula of the form $\bigvee Y$ where Y is a set of terms. A *conjunctive normal form* (CNF) is a formula of the form $\bigwedge Y$ where Y is a set of clauses. For any $\varphi \in \mathcal{L}$, there are algorithms [8] to equivalently transform it to a DNF or CNF. As usual, we identify two terms (clauses) as the same if they only differ in the order of literals within them, the same principle also applies to DNF and CNF.

By abuse of notation, a possible world I, regarded as a conjunction of all literals l s.t. $I(l) = \mathrm{T}$, is also a term. Moreover, we write $\rho \in \varphi$ for ρ being a conjunct (disjunct) of φ when φ is a term (clause). Then notations from set theory can be applied in the obvious way. Let D_1, D_2 be two terms, $D_1 \subseteq D_2$ means that $l \in D_1 \implies l \in D_2$, while $D_1 \cup D_2$, $D_1 - D_2$ and $D_1 \cap D_2$ denote respectively the terms obtained from D_1 by adding, omitting, and selecting all the literals in D_2. $\overline{D}$ is a term identified by flipping all the literals of D, e.g., if $D = p_1 \wedge \neg p_2 \wedge p_3$, then $\overline{D} = \neg p_1 \wedge p_2 \wedge \neg p_3$.

Let $\leq$ be a pre-order over $\mathcal{W}$. $I \simeq J$ denotes the case where $I \leq J$ and $J \leq I$, while $I < J$ means $I \leq J$ and $I \not\simeq J$. For any $\mathcal{A} \subseteq \mathcal{W}$, $\min(\mathcal{A}, \leq) = \{I \in \mathcal{A} \mid \forall J \in \mathcal{A}, J \not< I\}$ and $\max(\mathcal{A}, \leq) = \{I \in \mathcal{A} \mid \forall J \in \mathcal{A}, I \not< J\}$.

A *belief base* K is a finite set of formulae, which represents the current beliefs of one agent. For simplicity, a base K can also be regarded as the formula $\varphi = \bigwedge K$.

2.2 Prime Implicates and Prime Implicants

The dual notions of implicate and implicant [10] are closely related to CNF and DNF respectively.

Definition 1 (Implicate and implicant).

- *A clause C is an* implicate *of a formula φ if $\varphi \models C$.*
- *A term D is an* implicant *of a formula φ if $D \models \varphi$.*

Definition 2 (Prime implicate and prime implicant).

- *Let $\varphi \not\equiv \top$, C is a* prime implicate *of φ if $\varphi \models C$ and there is no other implicate C' of φ s.t. $C' \models C$.*
- *Let $\varphi \not\equiv \bot$, D is a* prime implicant *of φ if $D \models \varphi$ and there is no other implicant D' of φ s.t. $D \models D'$.*

Let dnf_φ be an arbitrary DNF s.t. $dnf_\varphi \equiv \varphi \not\equiv \bot$, it is easy to see that each disjunct of dnf_φ is an implicant of φ. We can convert dnf_φ to the set of all prime implicants of φ, denoted by IP_φ, and we have $\varphi \equiv \bigvee IP_\varphi$. The set of all prime implicates of φ, denoted by PI_φ, can be converted from an arbitrary cnf_φ in the same manner and we also have $\varphi \equiv \bigwedge PI_\varphi$. This kind of conversion could be achieved by the Tison's methods in [19] or the quantum notation in [15]; for more methods of generating prime implicates/implicants, see [10].

Example 3. Let φ be a formula such that $var(\varphi) = \{p_1, p_2, p_3, p_4\}$ and all of its models are listed as follows:

	p_1	p_2	p_3	p_4
I_1	T	F	T	T
I_2	T	F	F	T
I_3	T	F	F	F
I_4	F	T	F	T
I_5	F	F	T	T
I_6	F	F	F	T
I_7	F	F	F	F

Clearly, $\varphi \equiv I_1 \vee \cdots \vee I_7$. Notice that I_5 and I_6 only differ on the atom p_3, we can make an implicant (of φ) $D' = \neg p_1 \wedge \neg p_2 \wedge p_4$ by eliminating $p_3/\neg p_3$ from I_5/I_6 since $D' \equiv I_5 \vee I_6$ and $D' \models \varphi$. Later, $\neg p_1$ can also be eliminated form D' once we noticed another implicant (of φ) $D'' = p_1 \wedge \neg p_2 \wedge p_4$ generated from I_1 and I_2. This procedure ends up with $D_1 = \neg p_2 \wedge p_4$, $D_2 = \neg p_2 \wedge \neg p_3$ and $D_3 = \neg p_1 \wedge \neg p_3 \wedge p_4$, from which no more literals could be eliminated, and we get $IP_\varphi = \{D_1, D_2, D_3\}$.

An implicant D of φ is *non-trivial* if $D \not\equiv \bot$. Let $\mathcal{D}_\varphi$ be the set of all non-trivial implicants of φ. Intuitively, an implicant $D \in \mathcal{D}_\varphi$ is a partial description of worlds that can guarantee the truth of φ, i.e., for any world I, $I \supseteq D \implies I \models \varphi$, and a prime implicant D' of φ is such a description with no redundant information, as defined in [16]: $D' \in \mathcal{D}_\varphi$ and for all literals l, $D' - \{l\} \not\models \varphi$. So we can prove that $IP_\varphi = \min(\mathcal{D}_\varphi, \subseteq)$, and $[\![\varphi]\!] = \max(\mathcal{D}_\varphi, \subseteq)$. From this perspective, IP_φ describes all distinct minimal ways for making the truth of φ, while each $I \in [\![\varphi]\!]$ is maximal.

Taking the view of that IP_φ is the essential description for the truth-being of φ, we in this paper measure how close a world $J \not\models \varphi$ could be a model of φ by checking its relation with each $D \in IP_\varphi$, while the classical way of this literature is checking J's relation with each $I \in [\![\varphi]\!]$.

3 A Brief Review of Belief Revision

3.1 *AGM* Postulates

Suppose φ is a belief base which is going to accept a piece of new information μ, the revised belief base is denoted by $\varphi \circ \mu$. The prevailing *AGM* theory [1] proposed eight postulates which a revision operator should comply with, then they are rephrased as six postulates (R1)–(R6) [9] for the case of propositional language:

(R1) $\varphi \circ \mu \models \mu$.
(R2) If $\varphi \wedge \mu \not\models \bot$, then $\varphi \circ \mu \equiv \varphi \wedge \mu$.
(R3) If $\mu \not\models \bot$, then $\varphi \circ \mu \not\models \bot$.
(R4) If $\varphi_1 \equiv \varphi_2$ and $\mu_1 \equiv \mu_2$, then $\varphi_1 \circ \mu_1 \equiv \varphi_2 \circ \mu_2$.
(R5) $(\varphi \circ \mu) \wedge \psi \models \varphi \circ (\mu \wedge \psi)$.
(R6) If $(\varphi \circ \mu) \wedge \psi \not\models \bot$, then $\varphi \circ (\mu \wedge \psi) \models (\varphi \circ \mu) \wedge \psi$.

Also, in [9] we can find a representation theorem for *AGM* revision operators:

Definition 4. *A* faithful assignment *is a function that maps each $\varphi \in \mathcal{L}$ to a pre-order $\leq_\varphi$ over $\mathcal{W}$ such that:*

(C1) If $I, I' \in [\![\varphi]\!]$, then $I \not<_\varphi I'$.
(C2) If $I \in [\![\varphi]\!]$ and $I' \notin [\![\varphi]\!]$, then $I <_\varphi I'$.
(C3) If $\varphi \equiv \psi$, then $\leq_\varphi = \leq_\psi$.

Theorem 5. *A revision operator $\circ$ satisfies postulates (R1)–(R6) if and only if there exists a faithful assignment that maps each φ to a* total *pre-order $\leq_\varphi$ such that:*

$$[\![\varphi \circ \mu]\!] = \min([\![\mu]\!], \leq_\varphi).$$

3.2 Distance Measure

The classical way of model-based revision operators to define a pre-order $\leq_\varphi$ is measuring the distance from worlds to φ.

Definition 6. *A* distance function $d : \mathcal{W} \times \mathcal{W} \to \mathbb{N}$ *satisfies following conditions:*

- $d(I, J) = 0$ *if and only if* $I = J$.
- $d(I, J) = d(J, I)$.

Definition 7. *Given a distance function d, the pre-order $\leq_\varphi^d$ over $\mathcal{W}$ associated with φ is defined as:*

- $I \leq_\varphi^d J$ *if and only if* $d(I, \varphi) \leq d(J, \varphi)$.
- $d(I,\varphi) = \begin{cases} \min\{d(I,J) \mid J \in [\![\varphi]\!]\} & \text{if } \varphi \not\equiv \bot; \\ 1 & \text{if } \varphi \equiv \bot. \end{cases}$

We write $\circ^d$ for the revision operator s.t. $[\![\varphi \circ^d \mu]\!] = \min([\![\mu]\!], \leq_\varphi^d)$.

Two distance functions are worth mentioning here. The widely used *Hamming distance* d_H counts the number of atoms on which two worlds differ [5, 14]: $d_H(I, J) = |\{p \mid I(p) \neq J(p)\}|$. And the *drastic distance* d_D is simply defined as: $d_D(I, J) = 0$ if $I = J$; $d_D(I, J) = 1$ if $I \neq J$.

4 Satisfaction Measure for Revision

4.1 Partial Satisfaction of Terms

To define a total pre-order $\leq_\varphi$ that meets conditions (C1)–(C3), the primary task is to decide how to sort the worlds that do not satisfy φ. This paper aims to define $\leq_\varphi$ by measuring to what extent a world $J \notin [\![\varphi]\!]$ can partially satisfy φ. When φ is a consistent term D, an intuition is that the more the literals of D that are satisfied by J, the more satisfaction D will get in the world J. Specifically, we propose a monotonic condition for $\leq_D$:

(C4) If $D \not\equiv \bot$ and $J \cap D \subseteq I \cap D$, then $I \leq_D J$.

To characterize how (C4) affects the revision behavior of $D \circ \mu$, we introduce two postulates (R7) and (R8), which use IP_μ as an effective tool to compare the overlaps between D and each model of μ. The relation between (C4) and (R7),(R8) is exhibited by including them in an extension of Theorem 5.

(R7) If $D \not\equiv \bot, \mu \not\equiv \bot$, there exists a $D_i \in IP_\mu$ s.t. $D_i \cup (D - \overline{D_i}) \models D \circ \mu$.
(R8) If $D \not\equiv \bot, \mu \not\equiv \bot$, then for any $D_i, D_j \in IP_\mu$ s.t. $D - \overline{D_j} \subseteq D - \overline{D_i}$, $D_j \cup (D - \overline{D_j}) \models D \circ \mu \implies D_i \cup (D - \overline{D_i}) \models D \circ \mu$.

Theorem 8. *A revision operator $\circ$ satisfies postulates (R1)–(R8) if and only if there exists an assignment that maps each φ to a total pre-order $\leq_\varphi$ over $\mathcal{W}$ that satisfies conditions (C1)–(C4) and:*

$$[\![\varphi \circ \mu]\!] = \min([\![\mu]\!], \leq_\varphi).$$

Proof. This theorem is extended from Theorem 5, which was finely proved by Katsuno and Mendelzon in [9]. We follow their approach and omit the repeating part. The main task here is to prove the corresponding relation between (C4) and (R7),(R8).

($\Longrightarrow$) Suppose $\circ$ satisfies (R1)–(R8), we define the total pre-order $\leq_\varphi$ over $\mathcal{W}$ as $I \leq_\varphi J$ if and only if $I \in [\![\varphi]\!]$ or $I \in [\![\varphi \circ \mu]\!]$ where $[\![\mu]\!] = \{I, J\}$. Now we prove that $\leq_D$ satisfies (C4).

Let I, J be two worlds s.t. $J \cap D \subseteq I \cap D$, there are two cases to consider.

(1) $IP_\mu = \{I, J\}$. Since $J \cap D \subseteq I \cap D$, we have $D - \overline{J} \subseteq D - \overline{I}$. Notice that for any world K, $K \cup (D - \overline{K}) = K$, by (R8) we have $J \in [\![D \circ \mu]\!] \implies I \in [\![D \circ \mu]\!]$. By (R1),(R3), we know that $[\![D \circ \mu]\!]$ is a non-empty subset of $[\![\mu]\!]$, so $I \in [\![D \circ \mu]\!]$. By definition, we get $I \leq_D J$.
(2) $IP_\mu = \{D'\}$, and there is a literal $l \notin D'$ s.t. $I = D' \cup l$ and $J = D' \cup \bar{l}$. Since $J \cap D \subseteq I \cap D$, we have $\bar{l} \notin D$. By the fact that $var(D') \cup var(l) = P$, we know that for any literal $k \in D$, if $k \neq l$, then $k \in D'$ or $k \in \overline{D'}$. So $D' \cup (D - \overline{D'}) \subseteq D' \cup l = I$. By (R7), we have $I \in [\![D \circ \mu]\!]$ and $I \leq_D J$ holds.

($\Longleftarrow$) Suppose there is an assignment that maps each φ to a total pre-order $\leq_\varphi$ over $\mathcal{W}$ that satisfies conditions (C1)–(C4), we define a revision operator as $[\![\varphi \circ \mu]\!] = \min([\![\mu]\!], \leq_\varphi)$. Now we prove that $\circ$ satisfies postulates (R7) and (R8).

Let $D_i \in IP_\mu$, for all $I \in [\![\mu]\!]$ s.t. $I \supseteq D_i$, we have:

- $I \cap D = D - \overline{I} \subseteq D - \overline{D_i}$.
- $I \cap D = D - \overline{D_i} \iff D - \overline{D_i} \subseteq I \iff D_i \cup (D - \overline{D_i}) \subseteq I$.

So we can conclude that for all $I, J \supseteq D_i$, if $D_i \cup (D - \overline{D_i}) \subseteq I$, then $J \cap D \subseteq I \cap D$ and by (C4) $I \leq_D J$ holds, which means:

- $I \models D_i \cup (D - \overline{D_i}) \implies I \in \min(\llbracket D_i \rrbracket, \leq_D)$.

Then by the definition of $\circ$, for any $D_i \in IP_\mu$, if $D_i \cup (D - \overline{D_i}) \not\models D \circ \mu$, then for all $I \in \llbracket D_i \rrbracket$, $I \not\models D \circ \mu$.

For (R7), suppose there is no $D_i \in IP_\mu$ s.t. $D_i \cup (D - \overline{D_i}) \models D \circ \mu$, then there is no $I \in \llbracket \mu \rrbracket$ s.t. $I \models D \circ \mu$. This is contradictory with the fact that $\min(\llbracket \mu \rrbracket, \leq_D)$ is a non-empty subset of $\llbracket \mu \rrbracket$, so (R7) holds.

For (R8), suppose $D_i, D_j \in IP_\mu$ and $D - \overline{D_j} \subseteq D - \overline{D_i}$. For any I, J s.t. $I \supseteq D_i \cup (D - \overline{D_i})$ and $J \supseteq D_j \cup (D - \overline{D_j})$, we have $I \cap D = D - \overline{D_i} \supseteq D - \overline{D_j} = J \cap D$. By (C4) we get $I \leq_D J$, then by the definition of $\circ$, $J \in \llbracket D \circ \mu \rrbracket \implies I \in \llbracket D \circ \mu \rrbracket$. So, (R8) holds. □

4.2 Satisfaction Function

In compliance with (C4), we define the pre-order $\leq_\varphi$ by introducing a function that measures the satisfaction of terms. The set of all consistent terms in $\mathcal{L}$ is denoted by $\mathcal{D}$.

Definition 9. *A* satisfaction function $s : \mathcal{W} \times \mathcal{D} \to [0, 1]$ *satisfies following conditions:*

- $s(I, D) = 1$ *if and only if* $I \models D$.
- $s(J, D) \leq s(I, D)$ *if* $J \cap D \subseteq I \cap D$.

Definition 10. *Given a satisfaction function s, the pre-order $\leq^s_\varphi$ over $\mathcal{W}$ associated with φ is defined as:*

- $I \leq^s_\varphi J$ *if and only if* $s(I, \varphi) \geq s(J, \varphi)$.
- $s(I, \varphi) = \begin{cases} \max\{s(I, D) \mid D \in IP_\varphi\} & \text{if } \varphi \not\equiv \bot; \\ 0 & \text{if } \varphi \equiv \bot. \end{cases}$

We write $\circ^s$ for the revision operator s.t. $\llbracket \varphi \circ^s \mu \rrbracket = \min(\llbracket \mu \rrbracket, \leq^s_\varphi)$.

The value of $s(I, D)$ represents to what extent the term D is satisfied by the world I. The first condition of Definition 9, corresponding to (C1) and (C2), states that D is fully satisfied only by its models. The second condition, concerning the case where D is partially satisfied, admits the monotonicity demanded by (C4). In Definition 10, by taking advantage of the uniqueness of IP_φ, the principle of syntax irrelevance, i.e., the condition (C3), is promised, and the function s is extended to measure the satisfaction of all formulae in $\mathcal{L}$. A world I is recognized to be closer to φ in comparison to another world J if and only if φ's satisfaction in I is higher than that in J.

$\leq^s_\varphi$ is properly defined since the following proposition is easily true.

Proposition 11. *Let s be a satisfaction function, then for any $\varphi \in \mathcal{L}$, $\leq^s_\varphi$ satisfies conditions (C1)–(C4).*

5 Comparison

5.1 Satisfaction vs. Distance

The notion of satisfaction measure is clear and specific in the sense that it acknowledges the *true*/*false* state of atoms in a world, the very identity of that world, to be the primary factor for measuring the closeness from that world to formulae, while this is not necessarily true for the distance measure. At the same time, however, the definition of satisfaction function is broad and flexible enough that the most common distance functions can be simulated by satisfaction functions. So, the revision operators based on distance such as $\circ^{d_D}$ and $\circ^{d_H}$ can be redefined in terms of satisfaction functions.

Proposition 12. *Let s_D, s_H be two satisfaction functions such that:*

$$s_D(I, D) = \begin{cases} 1 & if I \models D; \\ 0 & if I \not\models D. \end{cases} \qquad s_H(I, D) = \frac{|P| - |\overline{I} \cap D|}{|P|}.$$

For any φ, μ, we have $\varphi \circ^{d_D} \mu \equiv \varphi \circ^{s_D} \mu$ and $\varphi \circ^{d_H} \mu \equiv \varphi \circ^{s_H} \mu$.

Proof. It is easy to check that s_D and s_H meet the conditions in Definition 9. We prove that for all $\varphi \in \mathcal{L}$ and all $I, J \in \mathcal{W}$, both $I \leq_{\varphi}^{d_D} J \iff I \leq_{\varphi}^{s_D} J$ and $I \leq_{\varphi}^{d_H} J \iff I \leq_{\varphi}^{s_H} J$ hold. The former one is obvious. For the latter, the non-trivial case is when $\varphi \not\equiv \bot$ and we have that:

$$\begin{aligned} & I \leq_{\varphi}^{s_H} J \\ \iff & s_H(I, \varphi) \geq s_H(J, \varphi) \\ \iff & \min\{|\overline{I} \cap D| \mid D \in IP_{\varphi}\} \leq \min\{|\overline{J} \cap D| \mid D \in IP_{\varphi}\} \\ \iff & \min\{d_H(I, K) \mid K \in [\![\varphi]\!]\} \leq \min\{d_H(J, K) \mid K \in [\![\varphi]\!]\} \\ \iff & I \leq_{\varphi}^{d_H} J. \end{aligned}$$

□

5.2 An Example with Five Operators

Firstly, we define a *normal satisfaction function* s_N which serves as a paradigm for satisfaction functions:

$$s_N(I, D) = \frac{|I \cap D|}{|D|}.$$

The principle of s_N is as simple as that each atom within a term is of equal importance for accumulating the term's satisfaction. For instance, two out of three literals within the term $D = p_1 \wedge p_2 \wedge \neg p_3$ are satisfied by the world $I = \{p_1, \neg p_2, \neg p_3, \cdots\}$, hence from the perspective of s_N the satisfaction of D in I should just be $\frac{2}{3}$.

Moreover, we introduce the revision operator $\circ_{\text{ps}}$ in [4] which is based on the function of *normal partial satisfiability*. Generally, $\varphi \circ_{\text{ps}} \mu$ is syntax sensitive w.r.t. φ since its function applies to all DNF and CNF formulae. As a result, $\circ_{ps}$

satisfies conditions (R1)–(R6), especially (R4), if and only if the belief base φ is compiled into a unique syntactical from. In [4], the authors have no preference for this choice of φ's form. Actually, if φ is in the form of *canonical* DNF, i.e., $\bigvee [\![\varphi]\!]$, $\circ_{\mathrm{ps}}$ would coincide with $\circ^{d_H}$; and if φ is in the form of $\bigvee IP_\varphi$, we have $\varphi \circ_{\mathrm{ps}} \mu \equiv \varphi \circ^{s_N} \mu$. For comparison, we study the case where φ is in the form of $\bigwedge PI_\varphi$ and rephrase the operator as $\circ^{PI}_{\mathrm{ps}}$ such that:

- $[\![\varphi \circ^{PI}_{\mathrm{ps}} \mu]\!] = \min([\![\mu]\!], \leq^{\mathrm{ps}}_\varphi)$.
- $I \leq^{\mathrm{ps}}_\varphi J$ if and only if $\mathrm{ps}(I, \varphi) \geq \mathrm{ps}(J, \varphi)$.
- $\mathrm{ps}(I,\varphi) = \begin{cases} \dfrac{1}{|PI_\varphi|} \sum\limits_{C_j \in PI_\varphi} \mathrm{ps}(I, C_j) & \text{if } \varphi \not\equiv \top; \\ 1 & \text{if } \varphi \equiv \top. \end{cases}$
- $\mathrm{ps}(I, C) = \begin{cases} 1 & \text{if } I \models C; \\ 0 & \text{if } I \not\models C. \end{cases}$

In [15], J. Marchi et al. proposed two revision operators: $\circ^{D}_{IP}$ that based on the Hamming distance between terms which are induced by prime implicants of formulae, and $\circ^{\hat{D}}_{IP}$ that based on the notion of *exclusive coordinates* which relates to both prime implicants and prime implicates. Without digging into details of those two operators, we use the example in [15] to compare the revision results of those two and $\circ^{d_H}, \circ^{s_N}, \circ^{PI}_{\mathrm{ps}}$.

Example 13. Let φ be the formula we studied in Example 3 s.t.:

$\bigvee IP_\varphi = D_1 \vee D_2 \vee D_3$ where $D_1 = \neg p_2 \wedge p_4$, $D_2 = \neg p_2 \wedge \neg p_3$,

$D_3 = \neg p_1 \wedge \neg p_3 \wedge p_4$.

and μ be a formula s.t. $\mu \wedge \varphi \models \bot$ and $\bigvee IP_\mu = (p_1 \wedge p_2) \vee (p_3 \wedge \neg p_4)$.

The authors of [15] have shown that:

$\bigwedge PI_\varphi = C_1 \wedge C_2 \wedge C_3 \wedge C_4$ where $C_1 = \neg p_2 \vee \neg p_3$, $C_2 = \neg p_2 \vee p_4$,

$C_3 = \neg p_2 \vee \neg p_1$, $C_4 = \neg p_3 \vee p_4$.

and the revision results of their operators are:

- $\varphi \circ^{D}_{IP} \mu = (p_1 \wedge p_2 \wedge \neg p_3) \vee (\neg p_2 \wedge p_3 \wedge \neg p_4) \vee (p_1 \wedge p_2 \wedge p_4)$
- $\varphi \circ^{\hat{D}}_{IP} \mu = (\neg p_2 \wedge p_3 \wedge \neg p_4) \vee (p_1 \wedge p_2 \wedge \neg p_3 \wedge p_4)$.

Now, we enumerate all worlds in $[\![\mu]\!]$ and exhibit how $\circ^{d_H}$, $\circ^{s_N}$ and $\circ^{PI}_{\mathrm{ps}}$ operate for the revision of $\varphi \circ \mu$:

	p_1	p_2	p_3	p_4	$d_H(J,\varphi)$ $I_1, I_2, I_3, I_4, I_5, I_6, I_7$	$s_H(J,\varphi)$ D_1, D_2, D_3	$\mathrm{ps}(J,\varphi)$ C_1, C_2, C_3, C_4
J_1	T	T	T	T	$\min\{1,2,3,2,2,3,4\}$	$\max\{\frac{1}{2},\frac{0}{2},\frac{1}{3}\}$	$\frac{0+1+0+1}{4}$
J_2	T	T	T	F	$\min\{2,3,2,3,3,4,3\}$	$\max\{\frac{0}{2},\frac{0}{2},\frac{0}{3}\}$	$\frac{0+0+0+0}{4}$
J_3	T	T	F	T	$\min\{2,1,2,1,3,2,3\}$	$\max\{\frac{1}{2},\frac{1}{2},\frac{2}{3}\}$	$\frac{1+1+0+1}{4}$
J_4	T	T	F	F	$\min\{3,2,1,2,4,3,2\}$	$\max\{\frac{0}{2},\frac{1}{2},\frac{1}{3}\}$	$\frac{1+0+0+1}{4}$
J_5	T	F	T	F	$\min\{1,2,1,4,2,3,2\}$	$\max\{\frac{1}{2},\frac{1}{2},\frac{0}{3}\}$	$\frac{1+1+1+0}{4}$
J_6	F	T	T	F	$\min\{3,4,3,2,2,3,2\}$	$\max\{\frac{0}{2},\frac{0}{2},\frac{1}{3}\}$	$\frac{0+0+1+0}{4}$
J_7	F	F	T	F	$\min\{2,3,2,3,1,2,1\}$	$\max\{\frac{1}{2},\frac{1}{2},\frac{1}{3}\}$	$\frac{1+1+1+0}{4}$

Then we can conclude that:

- $[\![\varphi \circ^{d_H} \mu]\!] = \{J_1, J_3, J_4, J_5, J_7\} = [\![\varphi \circ^{D}_{IP} \mu]\!]$
- $[\![\varphi \circ^{PI}_{\mathrm{ps}} \mu]\!] = \{J_3, J_5, J_7\} = [\![\varphi \circ^{\hat{D}}_{IP} \mu]\!]$
- $[\![\varphi \circ^{s_N} \mu]\!] = \{J_3\}$

5.3 Remarks

(1) The fact that $\circ^{d_H}$ coincides with $\circ^{D}_{IP}$ is admitted by the Theorem 4 in [15].[1] Both operators are applying the Hamming distance, one on models and the other on prime implicants. Since each prime implicant of a formula φ represents a subset of $[\![\varphi]\!]$, $\circ^{D}_{IP}$ can be regarded as a cross-check for computing $\circ^{d_H}$ without bothering deep into the level of models, and therefore it is not surprising that $\circ^{D}_{IP}$ has a better complexity performance as J. Marchi et al. have asserted in [15].

(2) For the formulae φ, μ in Example 13, we have $[\![\varphi \circ^{PI}_{\mathrm{ps}} \mu]\!] = [\![\varphi \circ^{\hat{D}}_{IP} \mu]\!]$. We wonder whether this equation is universally true since both operators are taking the prime implicates (of φ) as the basis for the notion of minimal change. However, it turns out that there are counterexamples. By checking the revision results of $\varphi \circ \mu_1$ and $\varphi \circ \mu_2$ where:

- $\mu_1 = \neg p_2 \wedge \neg p_4$
- $\mu_2 = (\neg p_2 \wedge \neg p_4) \vee (\neg p_1 \wedge \neg p_2)$
- $IP_\varphi = (p_3 \wedge p_2 \wedge p_5) \vee (p_3 \wedge p_1 \wedge p_4) \vee (p_3 \wedge p_2 \wedge p_4)$
- $PI_\varphi = p_3 \wedge (p_1 \vee p_2) \wedge (p_2 \vee p_4) \wedge (p_4 \vee p_5)$

we can see that neither $[\![\varphi \circ^{\hat{D}}_{IP} \mu]\!] \subseteq [\![\varphi \circ^{PI}_{\mathrm{ps}} \mu]\!]$ nor $[\![\varphi \circ^{PI}_{\mathrm{ps}} \mu]\!] \subseteq [\![\varphi \circ^{\hat{D}}_{IP} \mu]\!]$ holds.[2] So $\circ^{\hat{D}}_{IP}$ does not always coincide with $\circ^{PI}_{\mathrm{ps}}$. The difference between them is that

[1] The operator $\circ^{d_H}$ is named as $\circ^{D}$ in [15].

[2] See [15] for the details of $\circ^{\hat{D}}_{IP}$.

while $\circ^{PI}_{\rm ps}$ directly picks the models of μ that can satisfy most $C \in PI_\varphi$, the operator $\circ^{\hat{D}}_{IP}$, roughly speaking, picks the ones that can satisfy most $C \in PI_\varphi$ from the terms which are generated from prime implicants of μ by complying with some $D \in IP_\varphi$. From this perspective, we argue that the operator $\circ^{\hat{D}}_{IP}$ is not a successful application of prime implicants since it is not as accurate as $\circ^{PI}_{\rm ps}$ for sorting the models in $[\![\mu]\!]$ according to the satisfaction of $\bigwedge PI_\varphi$, which, as we perceive, is the genuine intention of [15].

$\circ^{s_N}$ vs. $\circ^{d_H}$. The difference between them is that each atom in $var(\varphi)$ is of equal importance w.r.t. $d_H(J, \varphi)$, while this kind of equality is only locally true within each $D \in IP_\varphi$ w.r.t. $s_N(J, D)$.

For example, J_1 is one atom away from a model (I_1) of D_1, but the atoms that really count for $s_N(J_1, D_1)$ are only p_2, p_4 for the fact that there is always a world in $\{I_1, I_2, I_5, I_6\} = [\![D_1]\!]$ to match J_1 regarding p_1, p_3 no matter how J_1 is defined on them. So, we have $s_N(J_1, D_1) = \frac{1}{2}$. Since $|D_1| = |D_2| = 2$ and $|D_3| = 3$, each atom in D_3 is of less importance for $s_N(J, D_3)$ than that of D_1 or D_2 ($\frac{1}{3} < \frac{1}{2}$). By checking each $d_H(J, \varphi)$ in the above table, we can find that among all five worlds in $[\![\varphi \circ^{d_H} \mu]\!]$, i.e., the models of μ which are one atom distance from φ, only J_3 has claimed this minimal distance from $\{I_4, I_6\} = [\![D_3]\!]$ ($d_H(J_3, I_4) = 1$). And that is why J_3 was differentiated from others and became the sole selection in $[\![\varphi \circ^{s_N} \mu]\!]$ ($\frac{2}{3} > \frac{1}{2}$).

$\circ^{s_N}$ vs. $\circ^{PI}_{\rm ps}$. The difference between them is basically the difference between prime implicants and prime implicates. As shown in Definition 9 and 10, we in this paper preferred to take the belief base φ in the form of $\bigvee IP_\varphi$.

Firstly, by Theorem 5, the purpose of defining a pre-order $\leq_\varphi$ is to make sure that the revision result is as close as possible to φ, and of course the best scenario is where the revision result could satisfy φ. So it is a natural approach to measure how close a world is to certain conditions that satisfy φ, as a $D \in IP_\varphi$ or an $I \in [\![\varphi]\!]$. From this perspective, the satisfaction measure, to which $\circ^{s_N}$ pertains, is on the same page with the classical distance measure, while $\circ^{PI}_{\rm ps}$ is not.

Secondly, it is true that the prime implicates altogether could make up a condition that satisfies φ since $\bigwedge PI_\varphi \equiv \varphi$, but each $C \in PI_\varphi$ individually is rather about a condition that falsifies φ since by definition $\neg C \models \neg\varphi$. Actually, for each $C \in PI_\varphi$, there is a $D \in IP_{\neg\varphi}$ s.t. $D \equiv \neg C$. Moreover, although $\circ^{s_N}$ and $\circ^{PI}_{\rm ps}$ are sharing the same idea that each conjunct in a conjunction is of equal importance for accumulating the truth of that conjunction, it sounds less plausible for $\circ^{PI}_{\rm ps}$. Unlike the atoms within a term which are independent individually, the conjuncts of $\bigwedge PI_\varphi$ are more complex and correlate to each other, and therefore can not be accumulated one by one. For example, if a world I satisfies the C_1 in the Example 13, then it must satisfy either $C_2 \wedge C_3$ or C_4.

That being said, both $\circ^{s_N}$ and $\circ^{PI}_{\rm ps}$ are proper AGM operators, and the choice between them is debatable.

6 Conclusion and Future Work

The main motivation of this paper is to study the model-based revision operators with satisfaction measure, which should be a fair alternative to the classical distance measure. The idea of satisfaction measure was declared in a condition (C4), which was later characterized by two postulates (R7),(R8) in an extension of the AGM representation theorem. By introducing the notion of satisfaction function, we developed a method to generate total pre-orders that meet the condition (C4) along with (C1)–(C3) required by the faithful assignments.

There are other revision operators motivated by similar ideas, e.g., the $\circ_B$ in [2] is defined as: $[\![\varphi \circ_B \mu]\!] = \bigcup_{I \in [\![\varphi]\!]} \min([\![\mu]\!], \leq_I)$ where $J_1 \leq_I J_2 \iff J_2 \cap I \subseteq J_1 \cap I$. But those operators are not proper AGM operators because the inclusion relation $\subseteq$ is a partial order and their methods failed to integrate $\leq_I$ of different $I \in [\![\varphi]\!]$ into a total pre-order as required. In contrast, an operator $\circ^s$ based on satisfaction measure can avoid this problem since the function s places $\leq_D$ of all $D \in IP_\varphi$ within a unified measurement, i.e., $[0, 1]$.

Another feature of this paper is that we chose to compile the information from a belief base φ into the form of $\bigvee IP_\varphi$, the disjunction of all prime implicants of φ. There are two reasons for this choice. For one, the satisfaction of a formula φ can be naturally reduced to the satisfactions of its prime implicants since each $D \in IP_\varphi$ is a minimal condition to satisfy φ. The other reason is that we need a unique syntactical form to neutralize the sensitivity of syntax caused by the definition of satisfaction function.

As regards the future work, we intend to characterize a specific type of AGM operators. Notice that the conditions (C1)–(C3) of faithful assignment are defined for each individual $\leq_\varphi$, which means the pre-orders associated with different formulae are independent of each other. However, we are generally interested in the operators, as all of those mentioned in this paper, that stick with one method for assigning pre-orders to all formulae in $\mathcal{L}$. To characterize this kind of *internal coherence* of an operator, we need conditions about how the pre-orders of different formulae relate to each other. For now, we have two conditions under consideration:

(C5) If $I \leq_{\varphi_1} J$ and $I \leq_{\varphi_2} J$, then $I \leq_{\varphi_1 \vee \varphi_2} J$.
(C6) If $D_1 \subseteq D_2$ and for all $L \in D_2 - D_1$, $L \in I \iff L \in J$ holds, then $I \leq_{D_1} J \iff I \leq_{D_2} J$.

Unfortunately, neither of them is an ideal option. The condition (C5) is too weak, on the one hand, for the distance measure since no matter how irregular a distance function d is, the associated pre-order $\leq^d_\varphi$ by Definition 7 could meet it for the fact that $[\![\varphi_1 \vee \varphi_2]\!] = [\![\varphi_1]\!] \cup [\![\varphi_2]\!]$. On the other hand, (C5) is not suitable for the satisfaction measure because it rules out most pre-orders $\leq^s_\varphi$ in Definition 10 including $\leq^{s_N}_\varphi$. As for the condition (C6), it is similar to the elimination property in social choice theory [7], and it describes a coherent relation among the pre-orders $\leq_D$. (C6) is respected by both $\circ^{d_H}$ and $\circ^{s_N}$. The problem is that (C6) has nothing to say about $\leq_\varphi$ when φ is not a term. To get a full understanding of such coherence, we should consider more conditions and check whether they are suitable for the operators of other kinds in the literature.

Also, we are interested to extend satisfaction measure to other related issues. For example, the study of belief merging [11] would benefit from satisfaction measure by a standard of fairness. A merging operator can be regarded as a revision operator for multiple belief bases. For a merging operator based on Hamming distance, the belief bases could have different influence on the result since the largest distance number of each base could be different from each other. A merging operator with satisfaction measure, in contrast, will normalize the influence of each base into $[0, 1]$ then output an impartial merging result. With the method in [11], we can define a merging operator Δ^s from a revision operator $\circ^s$. All the IC postulates of belief merging will be respected by Δ^s except (IC4), and that should not be a problem since the symmetrical property in (IC4) is rather a customized condition for the distance measure.

References

1. Alchourrón, C., Gärdenfors, P., Makinson, D.: On the logic of theory change: partial meet contraction and revision functions. J. Symbolic Logic **50**(2), 510–530 (1985)
2. Borgida, A.: Language features for flexible handling of exceptions in information systems. ACM Trans. Database Syst. (TODS) **10**(4), 565–603 (1985)
3. Borja-Macías, V., Pozos-Parra, P.: Model-based belief merging without distance measures. In: Proceedings of the 6th International Joint Conference on Autonomous Agents and Multiagent Systems, pp. 1–3 (2007). https://doi.org/10.1145/1329125.1329312
4. Chávez-Bosquez, O., Pozos-Parra, P., Ma, J.: Implementing Δ_{ps} (ps-merge) belief merging operator for belief revision. Computación y Sistemas **21**(3), 419–434 (2017)
5. Dalal, M.: Investigations into a theory of knowledge base revision: preliminary report. In: Proceedings of the 7th National Conference on Artificial Intelligence (AAAI-88), pp. 475–479 (1988)
6. Fermé, E., Hansson, S.O.: AGM 25 years: twenty-five years of research in belief change. J. Philos. Log. **40**(2), 295–331 (2011)
7. Fine, B., Fine, K.: Social choice and individual ranking I. Rev. Econ. Stud. **41**(3), 303–322 (1974)
8. Hamilton, A.G.: Logic for Mathematicians. Cambridge University Press, Cambridge (1988)
9. Katsuno, H., Mendelzon, A.O.: Propositional knowledge base revision and minimal change. Artif. Intell. **52**(3), 263–294 (1991)
10. Kean, A., Tsiknis, G.: An incremental method for generating prime implicants/implicates. J. Symb. Comput. **9**(2), 185–206 (1990)
11. Konieczny, S., Pérez, R.P.: Merging information under constraints: a logical framework. J. Log. Comput. **12**(5), 773–808 (2002)
12. Lehmann, D., Magidor, M., Schlechta, K.: Distance semantics for belief revision. J. Symbolic Logic **66**(1), 295–317 (2001)
13. Liberatore, P., Schaerf, M.: Arbitration: a commutative operator for belief revision. In: Proceedings of the 2nd World Conference on the Fundamentals of Artificial Intelligence (WOCFAI 1995), pp. 217–228 (1995)

14. Lin, J., Mendelzon, A.O.: Knowledge base merging by majority. In: Pareschi, R., Fronhöfer, B. (eds.) Dynamic Worlds: From the Frame Problem to Knowledge Management, vol. 12, pp. 195–218. Springer, Dordrecht (1999). https://doi.org/10.1007/978-94-017-1317-7_6
15. Marchi, J., Bittencourt, G., Perrussel, L.: Prime forms and minimal change in propositional belief bases. Ann. Math. Artif. Intell. **59**(1), 1–45 (2010)
16. Ramesh, A., Becker, G., Murray, N.V.: CNF and DNF considered harmful for computing prime implicants/implicates. J. Autom. Reason. **18**(3), 337–356 (1997)
17. Rodrigues, O., Gabbay, D., Russo, A.: Belief revision. In: Gabbay, D., Guenthner, F. (eds.) Handbook of Philosophical Logic, vol. 16, pp. 1–114. Springer, Dordrecht (2011)
18. Schlechta, K.: Non-prioritized belief revision based on distances between models. Theoria **63**(1–2), 34–53 (1997)
19. Tison, P.: Generalization of consensus theory and application to the minimization of boolean functions. IEEE Trans. Electron. Comput. EC-16(4), 446–456 (1967). https://doi.org/10.1109/PGEC.1967.264648

A Formal Analysis of Hollis' Paradox

Thomas Ågotnes[1,2(✉)] and Chiaki Sakama[3]

[1] University of Bergen, Bergen, Norway
thomas.agotnes@uib.no
[2] Shanxi University, Taiyuan, China
[3] Wakayama University, Wakayama, Japan
sakama@wakayama-u.ac.jp

Abstract. In Hollis' paradox, A and B each chose a positive integer and whisper their number to C. C then informs them, jointly, that they have chosen different numbers and, moreover, that neither of them are able to work out who has the greatest number. A then reasons as follows: B cannot have 1, otherwise he would know that my number is greater, and by the same reasoning B knows that I don't have 1. But then B also cannot have 2, otherwise he would know that my number is greater (since he knows I don't have 1). This line of reasoning can be repeated indefinitely, effectively forming an inductive proof, ruling out any number – an apparent paradox. In this paper we formalise Hollis' paradox using public announcement logic, and argue that the root cause of the paradox is the wrongful assumption that A and B assume that C's announcement necessarily is *successful*. This resolves the paradox without assuming that C can be untruthful, or that A and B are not perfect reasoners, like other solutions do. There are similarities to the surprise examination paradox. In addition to a semantic analysis in the tradition of epistemic logic, we provide a syntactic one, deriving conclusions from a set of premises describing the initial situation – more in the spirit of the literature on Hollis' paradox. The latter allows us to pinpoint which assumptions are actually necessary for the conclusions resolving the paradox.

Keywords: Epistemic Logic · Hollis' Paradox · Public Announcement Logic

1 Introduction

In *A paradoxical train of thought* [9], Martin Hollis describes the following situation.

A thinks of a number and whispers it privately to C. B does the same. C tells them, 'You have each thought of a different positive whole number. Neither of you can work out whose is the greater'.... Sitting alone in his homebound train, A muses as follows. 'I picked 157 and have no idea what B picked. So, assuming that he indeed chose a different positive whole number, C is right.... Well, clearly B did not choose 1, as he would then be able to work out that mine is greater; and by the same token he knows that I did not choose 1. So he did not choose 2, since he could then use the previous reasoning to prove that my number is greater. Similarly, he can know that I did not choose 2 either. With 2 out of the way, I

N. Alechina et al. (Eds.): LORI 2023, LNCS 14329, pp. 306–321, 2023.
https://doi.org/10.1007/978-3-031-45558-2_24

infer that he did not choose 3; and he can infer that I did not choose 3. ... I can keep this up for ever. But that is absurd. It means that I cannot have picked 157, which I certainly did'.

Several solutions attempting to resolve the apparent paradox have been proposed [11, 14, 17] (see also Hollis' response to the two first in [10]). What they have in common is that they argue that the announcement by C might not be truthful, and even if it were A and B might not have justified belief in that. Like most well known epistemic puzzles, Hollis' paradox leaves many assumptions implicit or ambiguous, so let us in this paper assume the following: (a) all agents always tell the truth (if they say something it is true and they know that it is true) and (b) this is common knowledge among all agents. Thus, we will be modelling knowledge rather than belief, and at any point in time an agent's knowledge is a result of the information she has received. We also assume that it is common knowledge that everyone is a perfect reasoner[1].

As far as we are aware, no *formal* analysis of Hollis' paradox appears in the literature, unlike most other well known epistemic or doxastic puzzles or paradoxes which have been studied using dynamic epistemic logic – see [18, 19] for an overview and references. Indeed, the precision and clarity of formal logic have been crucial in understanding these puzzles and clarify hidden premises (and these puzzles have again been a driving force as case studies in the development of dynamic epistemic logic).

In this paper we use public announcement logic [15] to model and analyse Hollis' paradox. This allows us to untangle subtleties in the alleged paradox, and in particular to be precise about the distinction between truth *before* an announcement and *after*, a distinction often lost in other analyses of the paradox. We argue that the root cause of the paradoxical situation is a wrongful assumption that the announcements by C always are *successful*, i.e., that they always remain true after they are announced. In Hollis' argument, this assumption is used as a premise in the inductive "proof". This has, as far as we know, not been pointed out in other studies of the paradox, and we believe this is the first solution to the paradox that does not rely on weakening the assumptions outlined above. However, it should come as no surprise. As pointed out already in [14], Hollis' paradox is similar[2] to *the surprise examination paradox*[3] which was first analysed using dynamic epistemic logic by Gerbrandy [7, 8]. Gerbrandy pointed out that the root cause of that paradox is the same phenomenon that lies behind many other epistemic puzzles with counter-intuitive solutions, the *muddy children (or three wise men) problem* [5] being the most well known, namely that announcements can become false as a result of being announced[4] – they are not necessarily successful. Olin [14] also points out that there are still "important differences" between the two paradoxes. We discuss the connection further in the last section of the paper.

In addition to arguing why, under the assumptions outlined above, Hollis' paradox is actually not a paradox, we shed light on other epistemic aspects of the puzzle, such

[1] Hollis [9] already hints at this assumption: "...each of us has to assume that the other is not stupid...".

[2] Olin [14] claims that it is "a version of surprise examination"; Hollis [10] on the other hand argues that his paradox is "wider".

[3] See [12] for an overview of different variants and a discussion of historic origins.

[4] In muddy children, that happens in the last joint announcement by the children.

as whether common knowledge must be assumed (it must not) or how many layers of nested knowledge are relevant (two). We provide two alternative and complementary analyses: a semantic analysis (in the style of Gerbrandy) where we give a single model of the initial situation described in the story and show that it has certain logical properties (Sect. 3), as well as a syntactic analysis (more in the style of Hollis and his respondents in *Analysis*, but formalised) where we describe the situation using a set of logical formulas and show that the same properties can be derived (Sect. 4). First, we give a brief technical introduction to epistemic logic and the logic of public announcements (see [19] for more details).

2 Background

2.1 Epistemic Logic

The most popular *epistemic logic* (i.e., logic for reasoning about knowledge) is modal propositional epistemic logic [5]. It extends propositional logic over a set of primitive propositions P with modalities K_a, where a is one of the agents in a given finite set Ag of agents. Intuitively, $K_a\phi$ means that agent a knows ϕ. Formally, the language is defined by the following grammar:

$$\varphi ::= p \mid \neg\varphi \mid \varphi_1 \wedge \varphi_2 \mid K_a\varphi$$

where $p \in P$ and $a \in Ag$. It is interpreted in *(epistemic) models* $M = (S, \sim, V)$ where S is a non-empty set of *states* (or worlds); $\sim$ gives an equivalence relation $\sim_a$ on S for each $a \in Ag$, a's *accessibility relation*; and $V : P \to \wp(S)$ is a *valuation function*, saying which primitive propositions are true in which states. Intuitively, $s \sim_a t$ models that agent a cannot discern between the states s and t; if the state of the world is s she considers it possible that it is actually t, and vice versa.

We write $M, s \models \phi$ to denote the fact that formula ϕ is true in state s of model M, defined recursively as follows:

$$\begin{array}{ll} M, s \models p \;\Leftrightarrow s \in V(p) & M, s \models K_a\varphi \;\Leftrightarrow (\forall t \in S)(s \sim_a t \Rightarrow M, t \models \varphi) \\ M, s \models \neg\varphi \Leftrightarrow M, s \not\models \varphi & M, s \models \varphi \wedge \psi \Leftrightarrow M, s \models \varphi \;\&\; M, s \models \psi \end{array}$$

Thus, $K_a\varphi$ is true if and only if φ is true in all indiscernible (for a) states. We use the usual derived propositional connectives, in addition to $\hat{K}_a\phi$ for $\neg K_a\neg\phi$, intuitively meaning that agent a considers that ϕ *possible*, i.e., that φ is true in at least one indiscernible state.

2.2 Public Announcement Logic

Public announcement logic (PAL) [15] extends epistemic logic in order to be able to reason about *change* in agents' knowledge and ignorance, resulting from a specific type of events: public announcements (such as the ones made by C in Hollis' paradox). Syntactically PAL extends epistemic logic with modalities of the form $[\phi]$ where ϕ is a formula. A formula $[\phi]\psi$ intuitively means that *after ϕ is truthfully*[5] *and publicly announced, ψ becomes true*. Formally, the language is defined by the following grammar:

$$\varphi ::= p \mid \neg\varphi \mid \varphi_1 \wedge \varphi_2 \mid K_a\varphi \mid [\varphi_1]\varphi_2$$

where $p \in P$ and $a \in Ag$. This language is also interpreted in epistemic models, extending the interpretation of the epistemic language with a clause for the public announcement operators. Informally, $[\phi]\psi$ is true in a state s in a model M ($M, s \models [\phi]\psi$) if ψ is true in state s ($M', s \models \psi$) in the model (call it M') resulting from *removing all states t in M where ϕ is false* ($M, t \models \neg\phi$). This captures the epistemic effects of a public announcement of ϕ: after the announcement, no-one considers it possible that $\neg\phi$ was[6] true, no-one considers it possible that anyone considered $\neg\phi$ possible, and so on (it becomes *common knowledge* that ϕ was true, capturing the word "publicly" above). Formally:

$$M, s \models [\varphi]\psi \quad \Longleftrightarrow \quad (M, s \models \varphi \Rightarrow M|\varphi, s \models \psi)$$

where $M|\varphi = (S', \sim', V')$ is a model such that for any $a \in Ag$ and $p \in P$, $S' = \{t \in S \mid M, t \models \varphi\}$, $\sim'_a = \sim_a \cap (S' \times S')$, and $V'(p) = V(p) \cap S'$.

The precondition $M, s \models \phi$ in the interpretation of $[\phi]\psi$ is needed because without it the definition would not be well-defined: if ϕ is false in s then s itself would be removed in the model update. This captures the "truthful" in the informal reading "*after ϕ is truthfully and publicly announced, ψ becomes true*" - or, alternatively, "*if ϕ is true then ψ will become true after ϕ is publicly announced*". The *dual*, $\langle\phi\rangle\psi = \neg[\phi]\neg\psi$, means that *$\phi$ is true and ψ will become true after ϕ is publicly announced.*

We write $M \models \phi$ to denote the fact that ϕ is true in all states in model M. A formula ϕ is *valid* if $M \models \phi$ for all models M. When Γ is a set of formulas, $\Gamma \models \phi$ means that for all M, s, if $M, s \models \Gamma$ then $M, s \models \phi$ (ϕ is logically entailed by Γ).

We say that a formula ϕ is an *(un)successful update* in M, s iff $M, s \models \langle\phi\rangle\phi$ ($M, s \models \langle\phi\rangle\neg\phi$); ϕ is a *successful formula* iff $[\phi]\phi$ is valid and an *unsuccessful formula* if not.

2.3 Axioms

Axiomatisations of epistemic logic and Public Announcement Logic are shown in Table 1. These axiomatisations are sound and complete [15, 19], in the sense that any formula is valid if and only if it can be derived using these axioms and rules.

We write $\vdash \phi$ to denote that formula ϕ is derivable (is a theorem), i.e., that there is a finite sequence of formulas ending with ϕ where every formula is either an instance of an axiom schema or the result of applying an inference rule to formulas earlier in the sequence. When Γ is a set of formulas, $\Gamma \vdash \phi$ ("ϕ can be derived from Γ") means that there is a finite subset $\{\gamma_1, \ldots, \gamma_k\}$ of Γ such that $\vdash \bigwedge_{1 \leq i \leq k} \gamma_i \rightarrow \phi$.

[5] Here and in the following we mean "truthful" in the strong sense that the announcement is in fact true (rather than only believed to be true).

[6] If ϕ is, e.g., a primitive proposition, then "was true" is the same as "is true". However, this is not the case in general: it could be that ϕ was true in a certain state before the announcement, but became false in the same state *as a result of the announcement*. The canonical example of the latter is the so-called Moore sentence $\phi = p \wedge \neg K_a p$.

3 A Semantic Analysis

Table 1. Axiomatisation of epistemic logic (left) and PAL (left and right).

Propositional tautology instances	Prop		
$K_a(\phi \to \psi) \to (K_a\phi \to K_a\psi)$	KD	$[\phi]p \leftrightarrow (\phi \to p)$	APerm
$K_a\phi \to \phi$	T	$[\phi]\neg\psi \leftrightarrow (\phi \to \neg[\phi]\psi)$	ANeg
$K_a\phi \to K_aK_a\phi$	4	$[\phi](\psi \wedge \chi) \leftrightarrow ([\phi]\psi \wedge [\phi]\chi)$	AConj
$\neg K_a\phi \to K_a\neg K_a\phi$	5	$[\phi]K_a\psi \leftrightarrow (\phi \to K_a[\phi]\psi)$	AKnow
From ϕ and $\phi \to \psi$, infer ψ	MP	$[\phi][\psi]\chi \leftrightarrow [\phi \wedge [\phi]\psi]\chi$	AComp
From ϕ, infer $K_a\phi$	Nec		

Hollis' paradox is well suited to a semantic (model theoretic) analysis, because the story intuitively and implicitly completely describes a single epistemic model. Figure 1 shows the epistemic model of the agents' knowledge after they have chosen their numbers but *before* C makes any announcement. A state corresponds to each agent having selected a number, we will refer to the combination as a *selection*. We only model A and B (as agents a and b respectively); C's knowledge is not relevant for the paradox beyond the assumption that his two announcements are actually true when they are made. Let 7_a be an atomic proposition meaning that agent a has chosen the number 7, and similarly for other numbers and for agent b. Also, let p_a mean that agent a's number is strictly greater than agent b's, and p_b that agent b's number is strictly greater than agent a's. We can now formalise the two announcements "you have each thought of a different number" and "neither of you can work out whose is the greater", respectively as:

$$ann1 = p_a \vee p_b \qquad ann2 = \neg K_a p_a \wedge \neg K_a p_b \wedge \neg K_b p_a \wedge \neg K_b p_b$$

While $ann1$ is straightforward, the formalisation $ann2$ of the second announcement deserves comment. In this formalisation we interpret "work out" as "deduce". "Work out" doesn't seem to imply, e.g., waiting for further information, asking questions, or guessing. Indeed, this is a common interpretation: informal descriptions of Hollis' paradox that have appeared after the original statement [9] explicitly use "deduce" instead of "work out"; e.g., [16] ("Neither of you can deduce which number is greatest"). It is worth noting that this formalisation is similar to Gerbrandy's formalisation of the the announcement in the surprise exam paradox [8], and that it has been argued [2] that the latter does not capture the intended meaning and that a stronger *self-referential* proposition is needed. In Sect. 5 we discuss why the same argument does not apply to our case. Also note that this formalisation is made in the context of the assumptions made in the introduction (common knowledge of truthfulness, perfect reasoners). A formula $K_a\phi$ holds iff ϕ follows from the information agent a currently has, and can thus be deduced by a perfect reasoner. $\neg K_a\phi$ holds if a cannot deduce ϕ (work out that ϕ holds).

In this initial model, agents a and b each only know their own number and consider any possibility for the other agent's number. They have no additional information (yet). For example, we have that $M_1, (2, 3) \models \neg K_a p_b$: if A has selected 2 and B has selected

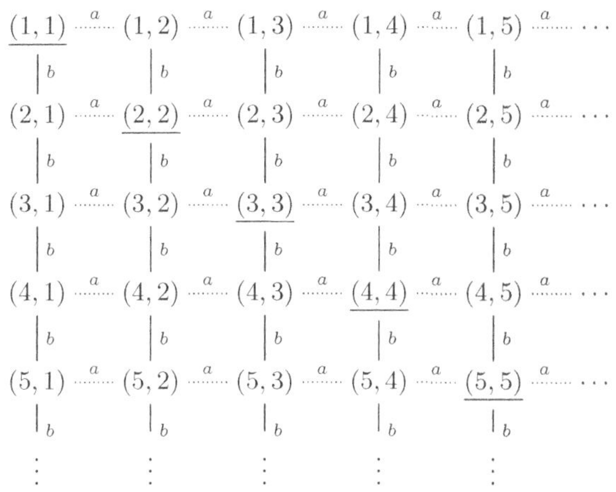

Fig. 1. Initial model M_1. In state $(2, 3)$ agent a has chosen the number 2 and agent b has chosen the number 3, and so on for the other states. The accessibility relation for agent a is depicted using dotted lines. Reflexive loops and transitive "jumps" are not shown; the actual accessibility relation is the reflexive, transitive closure of the relation in the picture. More intuitively: agent a cannot discern between states on the same row. Similarly for agent b, solid lines, and the same column. Atom $p_a$is true in all states to the left of the underlined diagonal; p_b in all states to the right of the diagonal. States where $ann1$ is *false* are underlined. $ann2$ is true in all states.

3, then A does not know that B's number is highest. In fact, in *all* states, i.e., no matter what the selection is, it holds that none of the agents know which number is greatest: $M_1 \models ann2$. However, note that if the selection, e.g., is $(1, 1)$, A knows that her number cannot be strictly greater than B's: $M_1, (1, 1) \models K_a \neg p_a$.

Let us now consider the situation immediately after C makes the announcement $ann1$. This announcement is *informative* for A and B; they learn something from it. Thus we have to update the model M_1 with the new information $ann1$ which is jointly received by a and b. We do that by removing the states in model M_1 where $ann1$ is false. The resulting model, M_2, is shown in Fig. 2.

As mentioned, the agents' knowledge has now changed, and in particular we have that $M_2 \models 1_a \rightarrow K_a p_b$. Similarly, $M_2 \models 1_b \rightarrow K_b p_a$. In words: if A has chosen the number 1, she *now* knows that she has a strictly lower number than B. Written another way: $M_1 \models 1_a \rightarrow [ann1] K_a p_b$.

As a consequence, we now (after the first announcement) have that the statement $ann2$ is not true in, e.g., state $(1, 3)$: $M_2, (1, 3) \models \neg ann2$.

Consider now the announcement of $ann2$ by C. The consequence of this announcement is that no one no longer considers states where $ann2$ was false (at the moment the announcement was made) possible (i.e., the bold states in the figure), and we update model M_2 by removing those states. The resulting model, M_3, is also illustrated in Fig. 2. Observe that we now have that, e.g., $M_3, (2, y) \models K_a p_b$ for all $y > 2$, and $M_3, (x, 2) \models K_b p_a$ for all $x > 2$. In other words, $M_2 \models 2_a \rightarrow [ann2] K_a p_b$, or: $M_1 \models 2_a \rightarrow [ann1][ann2] K_a p_b$ – no matter what the selection is, if A's number is 2 then she will know that B's number is highest after both announcements.

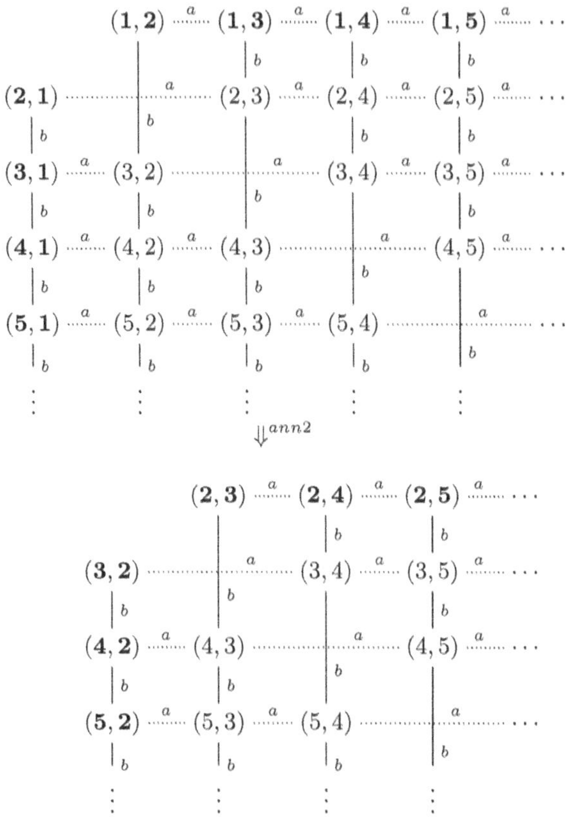

Fig. 2. M_2 (top), the result of announcing $ann1$ in M_1. M_3 (bottom), the result of announcing $ann2$ in M_2. States where $ann2$ is *false* are in **bold**. $ann1$ is true in all states.

Let us consider the claims in the statement of the paradox. "clearly B did not choose 1": this is true; $M_1 \models [ann1][ann2]K_a\neg 1_b$. "...and by the same token he knows that I did not choose 1": also true; $M_1 \models [ann1][ann2]K_b\neg 1_a$. "So he did not choose 2, since he could then use the previous reasoning to prove that my number is greater": no, this is in fact not true. In fact, no matter what the selection is, each of the two agents considers it possible that the other agent has 2 unless she has it herself:

$$M_1 \models \neg 2_a \rightarrow [ann1][ann2]\neg K_a\neg 2_b$$

and similarly with a and b swapped. This shows that the inductive argument in the "proof" of the paradox halts. The reason that the argument in the "proof" of the paradox doesn't work is that while the announcement $ann2$ might have been successful, A and B cannot know that: $M_1 \models (\neg 2_a \wedge \neg 2_b) \rightarrow [ann1][ann2](ann_2 \wedge \neg K_a ann_2 \wedge \neg K_b ann_2)$.

So why do we still have states $(2, y)$ and $(x, 2)$ in the model after the second announcement? Observe that the second announcement removed all states $(1, y)$ and $(x, 1)$. What enabled this was that $ann1$ was announced first – without that $ann2$ would not have removed those states. $ann2$ plays a similar role for the states $(2, y)$ and $(x, 2)$: after the announcement of $ann2$, $ann2$ becomes *false* in those states. However, it was true in the same states before the announcement, which is why they are not removed.

It could perhaps be argued that C implicitly meant something stronger than just that $ann2$ was true at the moment it was announced, for example that it would also stay true after the announcement (and that this was clear to A and B)[7]. That would be modelled explicitly by the announcement $ann2' = ann2 \wedge [ann2]ann2 = \langle ann2 \rangle ann2$. The effect of that announcement would in fact be identical to the effect of announcing $ann2$ twice in a row. As argued above, the third announcement (announcing $ann2$ a second time) would remove the $(2, y)$ and $(x, 2)$ states. Now, after this annoucement, $ann2$ becomes false in all $(3, y)$ and $(x, 3)$ states. Formally: $M_1 \models 3_a \rightarrow [ann1][ann2][ann2]K_a p_b$ (while $M_1 \models 3_a \rightarrow [ann1][ann2]\neg K_a p_b$). We can continue this argument: repeating the announcement "Neither of you can deduce which number is greatest" removes more and more states. It is only in this sense that "you can extend this line of reasoning to include any number you like" is true: extending this line of reasoning implies that the announcement has to be made *again* to exclude the number 2, and again for the number 3, and so on. If repeated enough times, we will reach a point where either A or B has learned who has the greatest number, and the announcement is unsuccessful and cannot be repeated any more[8]. In the statement of the paradox, the announcement is only made once, which explains why the reasoning cannot be extended beyond the number 1. This resolves the paradox.

4 A Syntactic Analysis

We now turn to analyse the paradox *syntactically*, by describing the situation as a set of formulas Γ, and deriving conclusions from them. In particular, we will show, similarly to in the model theoretic analysis, that

$$\Gamma \vdash [ann1][ann2]K_a \neg 1_b$$

– after the two announcements A knows that B does not have 1, but

$$\Gamma \vdash 157_a \rightarrow [ann1][ann2]\neg K_a \neg 2_b$$

– she does *not* know that B does not have 2 (in the case that A has 157 as in the description of the paradox), stopping the inductive train of thought in its tracks.

4.1 Describing the Initial Situation

We start by defining Γ, describing A's and B's initial knowledge and ignorance. For the purpose of the two derivations mentioned above we basically only need two premises (more discussion on this perhaps surprising fact below).

---

[7] Gerbrandy [8, pp. 27–29] discusses the same point in the context of surprise examination.

[8] This can be expressed elegantly by the iterated announcement operator in [13]: $M_1 \models \langle ann2^* \rangle \neg ann2$, which is true iff $M_1 \models \underbrace{\langle ann2 \rangle \cdots \langle ann2 \rangle}_{n} \neg ann2$ for some $n \geq 1$. See also [20] for a further discussion of this and related operators.

The first is that *everyone knows their own number*. For any $i \in \{a, b\}$:

$$x_i \to K_i x_i \tag{A0}$$

and furthermore that *this is known by both A and B*. For any $i, j \in \{a, b\}$:

$$K_j(x_i \to K_i x_i) \tag{A1}$$

Since (A1) implies (A0) (see epistemic logic axiom T), we actually only need (A1). We will use axiom T in the same way implicitly in the following.

The second is that initially (before any announcements) *each agent considers it possible that the other has chosen any number* (and this is known by both). For any $i, j \in \{a, b\}$ and any number y, we write $\bar{i}$ for "the other agent", i.e., $\bar{a} = b$ and $\bar{b} = a$:

$$K_j \hat{K}_i y_{\bar{i}} \tag{A2}$$

In addition to these two[9] premises we need some bookkeeping: the logic of the linear order of the natural numbers and the agents' knowledge of that. This is captured by the following three premises.

First, the relationship between p_b and p_a. *If i's number is greatest, then the other agent's number is not* (and this is known). For any $i, j \in \{a, b\}$:

$$K_j(p_i \to \neg p_{\bar{i}}) \tag{A3}$$

Second, we need two premises describing the relationship between atoms of the form 156_a and p_a. The first says that *one is the lowest number* (and anyone knows this, and anyone knows that anyone knows this[10]). The second is that *if agent i has the greatest number then p_i holds* (and this is known). For any $i, j, k \in \{a, b\}$ and numbers $x > y$:

$$K_j K_k(1_i \to \neg p_i) \tag{A4}$$

$$K_j K_k((x_i \wedge y_{\bar{i}}) \to p_i) \tag{A5}$$

Thus, we let Γ be all instances of (A1)–(A5):

$$\Gamma = \left\{ \begin{array}{ll} K_j(x_i \to K_i x_i), & \\ K_j \hat{K}_i y_{\bar{i}}, & \\ K_j(p_i \to \neg p_{\bar{i}}), & : i, j, k \in \{a, b\}, x, y \in \mathbb{N}, x > y \\ K_j K_k(1_i \to \neg p_i), & \\ K_j K_k((x_i \wedge y_{\bar{i}}) \to p_i) & \end{array} \right\}$$

where $\mathbb{N}$ is the set of natural numbers. Note that, while Γ is an infinite set of premises, any derivation $\Gamma \vdash \phi$ of ϕ from Γ can only use a finite number of those premises.

[9] There are two *schemas* but actually infinitely many formulas.

[10] We could assume that these premises are *common knowledge*, writing e.g., $C_{\{a,b\}}(1_i \to \neg p_i)$. However, it turns out that assuming common knowledge is *not needed*, and it is of interest to illucidate exactly how many levels of nested knowledge are sufficient: e.g., two levels for (A4).

4.2 Simplifying Announcements

It is a straightforward exercise in PAL to show that, for any ϕ,

$$\vdash [ann1][ann2]\phi \leftrightarrow [\beta]\phi \tag{1}$$

where

$$\beta = (p_b \vee p_a) \wedge \\ \neg K_b(p_a \rightarrow p_b) \wedge \neg K_b(p_b \rightarrow p_a) \wedge \neg K_a(p_a \rightarrow p_b) \wedge \neg K_a(p_b \rightarrow p_a)$$

From $K_b(p_b \rightarrow \neg p_a) \in \Gamma$ (A3) and similarly for the other combinations, we also have[11]:

$$\Gamma \vdash \alpha \leftrightarrow \beta \tag{2}$$

where

$$\alpha = (p_b \vee p_a) \wedge \hat{K}_b p_a \wedge \hat{K}_b p_b \wedge \hat{K}_a p_a \wedge \hat{K}_a p_b$$

4.3 I Know that She Does Not Have 1

We now show that $\Gamma \vdash [ann1][ann2]K_a\neg 1_b$. Here and in the following we often combine several proof steps. In particular, we liberally use known epistemic logic and PAL theorems – referred to as "S5" and "PAL" respectively.

$$\begin{array}{lll}
1 & \Gamma \vdash K_a(1_b \rightarrow K_b 1_b) & (A1) \\
2 & \Gamma \vdash K_a K_b(1_b \rightarrow \neg p_b) & (A4) \\
3 & \Gamma \vdash K_a(K_b 1_b \rightarrow K_b \neg p_b) & 2, S5 \\
4 & \Gamma \vdash K_a(1_b \rightarrow K_b \neg p_b) & 1, 3, S5 \\
5 & \Gamma \vdash K_a(1_b \rightarrow \neg\alpha) & 4, Prop \\
6 & \Gamma \vdash K_a(\alpha \rightarrow \neg(\alpha \rightarrow 1_b)) & 5, Prop \\
7 & \Gamma \vdash K_a(\alpha \rightarrow \neg[\alpha]1_b) & 6, APerm \\
8 & \Gamma \vdash K_a[\alpha]\neg 1_b & 7, ANeg \\
9 & \Gamma \vdash \alpha \rightarrow K_a[\alpha]\neg 1_b & 8, Prop \\
10 & \Gamma \vdash [\alpha]K_a\neg 1_b & 9, AKnow \\
11 & \Gamma \vdash [ann1][ann2]K_a\neg 1_b & 10, Eq.(1), Eq.(2), Prop
\end{array}$$

4.4 But I Don't Know that She Does Not Have 2

We show that $\Gamma \vdash 157_a \rightarrow [ann1][ann2]\neg K_a\neg 2_b$. "$x/y$" means "replace x with y".

[11] Observe that α expresses that (1) the two numbers are different, and (2) both agents consider each of the numbers to be the greatest (α implies $ann1 \wedge ann2$ but not the other way around).

1	$\Gamma \vdash \hat{K}_a 2_b$	(A2), S5
2	$\Gamma \vdash 157_a \rightarrow K_a 157_a$	(A1), S5
3	$\Gamma \vdash K_a((157_a \wedge 2_b) \rightarrow p_a)$	(A5), S5
4	$\Gamma \vdash K_a 157_a \rightarrow K_a(2_b \rightarrow p_a)$	3, S5, Prop
5	$\Gamma \vdash 157_a \rightarrow \hat{K}_a(2_b \wedge (2_b \rightarrow p_a))$	1, 2, 4, S5
6	$\Gamma \vdash 157_a \rightarrow \hat{K}_a(2_b \wedge (p_b \vee p_a))$	5, Prop
7	$\Gamma \vdash K_a \hat{K}_a 2_b$	(A2)
8	$\Gamma \vdash K_a 157_a \rightarrow K_a \hat{K}_a(2_b \wedge 157_a)$	7, S5
9	$\Gamma \vdash K_a((2_b \wedge 157_a) \rightarrow p_a)$	(A5), S5
10	$\Gamma \vdash K_a 157_a \rightarrow K_a \hat{K}_a p_a$	8, 9, S5
11	$\Gamma \vdash 157_a \rightarrow K_a \hat{K}_a p_a$	2, 10, Prop
12	$\Gamma \vdash 157_a \rightarrow K_a \hat{K}_a p_b$	Like $7{-}11:\ 2_b/158_b$ and p_a/p_b
13	$\Gamma \vdash K_a \hat{K}_b 3_a$	(A2)
14	$\Gamma \vdash K_a(2_b \rightarrow K_b 2_b)$	(A1)
15	$\Gamma \vdash K_a(2_b \rightarrow \hat{K}_b(3_a \wedge 2_b))$	13, 14, S5
16	$\Gamma \vdash K_a K_b(3_a \wedge 2_b \rightarrow p_a)$	(A5)
17	$\Gamma \vdash K_a(2_b \rightarrow \hat{K}_b p_a)$	15, 16, S5
18	$\Gamma \vdash K_a(2_b \rightarrow \hat{K}_b p_b)$	Like $13{-}17:\ 3_a/1_a$ and p_a/p_b
19	$\Gamma \vdash 157_a \rightarrow \hat{K}_a(2_b \wedge \alpha)$	6, 11, 12, 17, 18, S5
20	$\Gamma \vdash 157_a \rightarrow (\alpha \rightarrow \neg K_a(\alpha \rightarrow \neg 2_b))$	19, Prop
21	$\Gamma \vdash 157_a \rightarrow (\alpha \rightarrow \neg K_a(\alpha \rightarrow \neg(\alpha \rightarrow 2_b)))$	20, Prop
22	$\Gamma \vdash 157_a \rightarrow (\alpha \rightarrow \neg K_a(\alpha \rightarrow \neg[\alpha]2_b))$	21, APerm
23	$\Gamma \vdash 157_a \rightarrow (\alpha \rightarrow \neg K_a[\alpha]\neg 2_b)$	22, ANeg
24	$\Gamma \vdash 157_a \rightarrow (\alpha \rightarrow (\alpha \wedge \neg K_a[\alpha]\neg 2_b))$	23, Prop
25	$\Gamma \vdash 157_a \rightarrow (\alpha \rightarrow \neg(\alpha \rightarrow K_a[\alpha]\neg 2_b))$	24, Prop
26	$\Gamma \vdash 157_a \rightarrow (\alpha \rightarrow \neg[\alpha]K_a \neg 2_b)$	25, AKnow
27	$\Gamma \vdash 157_a \rightarrow [\alpha]\neg K_a \neg 2_b$	26, ANeg
28	$\Gamma \vdash 157_a \rightarrow [ann1][ann2]\neg K_a \neg 2_b$	29, Eq. (1), Eq.(2)

4.5 Dealing with Infinite Disjunction

In the previous section we showed how to derive $\Gamma \vdash 157_a \rightarrow [ann1][ann2]\neg K_a \neg 2_b$, and which assumptions were sufficient for that derivation. The number 157, taken from the original formulation of the paradox, is of course arbitrary – it could be replaced with 15 or 1570 or indeed any number different from 2 itself. So we get $\Gamma \vdash 15_a \rightarrow [ann1][ann2]\neg K_a \neg 2_b$ and so on in the same way. By this reasoning, it seems that we should be able to get the more general $\Gamma \vdash \neg 2_a \rightarrow [ann1][ann2]\neg K_a \neg 2_b$. However, this does in fact not hold – the assumptions in Γ turn out to not be strong enough to make $\neg 2_a \rightarrow [ann1][ann2]\neg K_a \neg 2_b$ derivable. To see this, consider the model M_4 and its transformations as a result of the two announcements in Fig. 3. It is easy to see that $M_4, (-, 3) \models \Gamma$, but since $M_6, (-, 3) \models K_a \neg 2_b$ we have that $M_4, (-, 3) \not\models [ann1][ann2]\neg K_a \neg 2_b$. In other words, $\Gamma \not\models \neg 2_a \rightarrow [ann1][ann2]\neg K_a \neg 2_b$, and thus $\Gamma \nvdash \neg 2_a \rightarrow [ann1][ann2]\neg K_a \neg 2_b$.

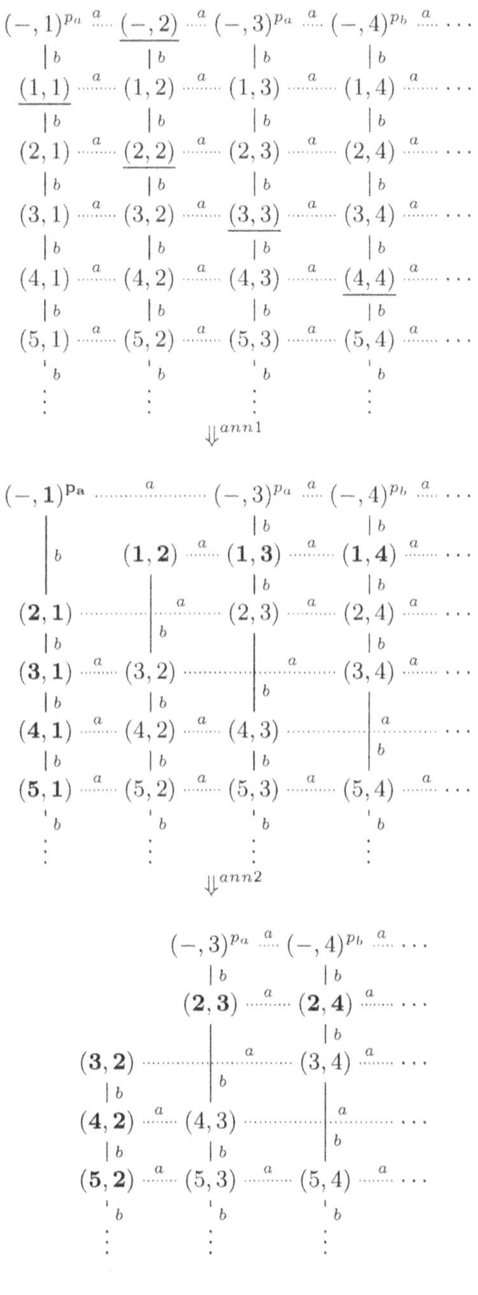

Fig. 3. Models M_4 (top), as well as M_5 (middle) and M_6 (bottom) – the results of announcing $ann1$ in M_4 and $ann2$ in M_5, respectively. States where $ann1$/$ann2$ is false are underlined/in bold. The valuation is the same as in M_1 for corresponding states. For the "new" states (first row), the valuation is as follows: x_b is given by the state, e.g., 3_b is true in state $(-, 3)$; x_a is *false* in all these states; the truth values of p_a and p_b are indicated in the figure.

So, Γ must be strengthened if we want to derive $\neg 2_a \rightarrow [ann1][ann2]\neg K_a \neg 2_b$, so that models like M_4 are ruled out. That model contains states where one agent (a) has not chosen any number, clearly conflicting with the description of the puzzle[12]. However, the assumption that A has chosen some number corresponds to an infinite disjunction of the form $\bigvee_{x \geq 1} x_a$, which cannot be written as a formula.

It turns out, however, that a weaker assumption is sufficient. Notice that if we have that $\neg 1_a$ and $\neg 2_a$ and 2_b, it follows that p_a – if A doesn't have 1 or 2 she must have a number greater than B's number 2. $\neg 1_a \wedge \neg 2_a \wedge 2_b \rightarrow p_a$ does not follow from Γ (to see this observe that it is false in state $(-, 2)$ in M_4). We now strengthen Γ with a generalisation of that assumption, namely, for any $i, j \in \{a, b\}$, $k \geq 1$ and $m \leq k$:

$$K_j(\neg 1_i \wedge \neg 2_i \wedge \cdots \wedge \neg k_i \wedge m_{\bar{i}} \rightarrow p_i) \tag{A6}$$

We will also need a negative variant of A1 (everyone knows their own number), saying that if I have *not* chosen x then I know that. For any $i, j \in \{a, b\}$:

$$K_j(\neg x_i \rightarrow K_i \neg x_i) \tag{A1'}$$

Finally, we will need to assume the following as a first principle (any $i \in \{a, b\}$):

$$K_i(\hat{K}_a p_b \wedge \hat{K}_b p_a) \tag{A7}$$

– in the initial situation (before any announcements), A considers it possible that B has chosen a greater number, and conversely for B (note that we cannot assume, e.g., $\hat{K}_a p_a$ – because if A has 1 she does not consider it possible that her number is greater than B's).

Let Γ' be Γ extended with premises (A6), (A1') and (A7), i.e., $\Gamma' = \Gamma \cup \{K_j(\neg 1_i \wedge \neg 2_i \wedge \cdots \wedge \neg k_i \wedge m_{\bar{i}} \rightarrow p_i), K_j(\neg x_i \rightarrow K_i \neg x_i), K_i(\hat{K}_a p_b \wedge \hat{K}_b p_a) : i, j \in \{a, b\}, k \geq 1, m \leq k\}$. We now show that $\Gamma' \vdash \neg 2_a \rightarrow [ann1][ann2]\neg K_a \neg 2_b$.

In the following, by "L. 4.4:x-y" we mean "like in lines x to y in the proof in Sect. 4.4".

[12] Nevertheless, that was no problem for $157_a \rightarrow [ann1][ann2]\neg K_a \neg 2_b$, which happens to hold in those models too.

1	$\Gamma' \vdash (\neg 1_a \wedge \neg 2_a) \rightarrow \hat{K}_a(2_b \wedge (p_b \vee p_a))$	L. 4.4:1-6; $157_a/(\neg 1_a \wedge \neg 2_a)$, (A1)/(A1'), (A5)/(A6)
2	$\Gamma' \vdash (\neg 1_a \wedge \neg 2_a) \rightarrow K_a \hat{K}_a p_a$	L. 4.4:7-11; $157_a/(\neg 1_a \wedge \neg 2_a)$ (A5)/(A6)
3	$\Gamma' \vdash \hat{K}_a p_b$	(A7)
4	$\Gamma' \vdash (\neg 1_a \wedge \neg 2_a) \rightarrow K_a \hat{K}_a p_b$	3, Prop
5	$\Gamma' \vdash K_a \hat{K}_b p_a$	(A7)
6	$\Gamma' \vdash K_a(2_b \rightarrow \hat{K}_b p_a)$	5, S5
7	$\Gamma' \vdash K_a(2_b \rightarrow \hat{K}_b p_b)$	L. 4.4:13-18; $3_a/1_a$, p_a/p_b
8	$\Gamma' \vdash (\neg 1_a \wedge \neg 2_a) \rightarrow \hat{K}_a(2_b \wedge \alpha)$	1, 2, 4, 6, 7, S5
9	$\Gamma' \vdash (\neg 1_a \wedge \neg 2_a) \rightarrow [\alpha]\neg K_a \neg 2_b$	L. 4.4:19-27
10	$\Gamma' \vdash (\neg 1_a \wedge \neg 2_a) \rightarrow (\alpha \rightarrow \langle\alpha\rangle\neg K_a \neg 2_b)$	9, PAL
11	$\Gamma' \vdash (\neg 1_a \wedge \neg 2_a \wedge \alpha) \rightarrow \langle\alpha\rangle\neg K_a \neg 2_b$	10, Prop
12	$\Gamma' \vdash \alpha \rightarrow \neg 1_a$	as in Sec. 4.3
13	$\Gamma' \vdash (\neg 2_a \wedge \alpha) \rightarrow \langle\alpha\rangle\neg K_a \neg 2_b$	11, 12, Prop
14	$\Gamma' \vdash \neg 2_a \rightarrow (\alpha \rightarrow \langle\alpha\rangle\neg K_a \neg 2_b$	13, Prop
15	$\Gamma' \vdash \neg 2_a \rightarrow [\alpha]\neg K_a \neg 2_b$	14, PAL ([19, Prop. 4.13])
16	$\Gamma' \vdash \neg 2_a \rightarrow [ann1][ann2]\neg K_a \neg 2_b$	15, Eq. (1), Eq. (2)

5 Discussion

We have argued that under assumptions about common knowledge of truthfulness and perfect reasoners, Hollis' paradox can be resolved by observing that the second announcement is not neccessarily successful. Note that it will *actually* be a *successful update* – except in the cases that either A or B has chosen 1 or 2. Thus, a more precise explanation is that the agents *don't know whether the announcement was successful.* As is well known in dynamic epistemic logic, an announcement can be unsuccessful yet informative, a likely source of the confusion behind the so-called paradox.

As mentioned in the introduction, there are similarities between Hollis' paradox and the surprise examination paradox. In particular, they are built on the same fallacy: that announcements always are successful. This was first pointed out for the surprise examination paradox by Gerbrandy [8], using a variant of public announcement logic, in a similar way to the semantic analysis in this paper. Several other logical analyses have since appeared [2–4, 12]. While it can be argued that the root cause behind the two "paradoxes" is the same (unsuccessful formulas), the logical modelling is quite different. Gerbrandy's formalisation has in common with our formalisation of Hollis' paradox that there is an initial announcement that eliminates some states in the model and that the (false) assumption that the initial announcement would stay true after that initial elimination would eliminate yet more states and that this can be repeated in several steps eventually leading to a paradoxical situation where all states have been eliminated. In both cases the "paradox" can be seen as an inductive "proof" that actually fails after the first step due to the false premise that the initial announcement is successful. A significant difference is that in the surprise examination paradox the state space is finite, while in Hollis' paradox the state space is infinite and an inductive argument is crucial.

Other significant differences is that the state space in Gerbrandy's solution is very simple, consisting of only three states, and our model of Hollis' paradox is more complex, while on the other hand the announcement of "the exam date will be a surprise" in the former[13], is more complex than $ann2$. The reason for the latter is the iterative opening of the doors which has no correspondent in Hollis' paradox. In fact, from a modelling perspective Hollis' paradox has more in common with *Sum and Product* [6], with a state space that is a (in that case finite) subset of the cartesian product of the natural numbers and where states are eliminated in a sequence of announcements. In that case the announcements are given explicitly and there is no paradox.

It has however, been forcefully argued [2] that Gerbrandy's *non-self-referential* formalisation of the announcement is not a very natural interpretation of the sentence "the exam date will be a surprise" nor is it indeed the interpretation most commenters on the paradox agree with. This argument hinges on the word "will" which refers to the future and in particular, it is argued, to the actual future immediately after the announcement is made, and thus that a *self-referential* interpretation of the statement to mean "you will not know in advance the exam day (i.e., after hearing *this* very announcement")[14]. This is indeed convincing, but we argue that the same argument does not apply to Hollis' paradox where the announcement is "neither of you can work it out" (or "neither of you can deduce it" [16]). Granted, "can work out" (or "deduce") seems to refer to the future as well, but a perfect reasoner has at any point already "worked out" (deduced) all possible consequences of her knowledge. The operative word here is "*can*", referring to the present, the announcement is not "neither of you *will* be able to work out".

Our formalisation hinges on the two assumptions of common knowledge of truthfulness and perfect reasoners, both of which it would be interesting to relax in future work on formalisations. Modeling non-perfect reasoners (see, e.g., [1]) might seem particularly relevant since it gives more meaning to the phrase "can work out", but there are no clues in the description of the paradox how the agents abilities to "work out" things are limited (indeed, on the contrary, as mentioned in the introduction Hollis hinted at joint knowledge of good reasoning abilities).

While the semantic modelling of the initial situation in Hollis' paradox allowed us to pinpoint exactly where the inductive argument breaks down, existing discourse on Hollis' paradox [9–11, 14, 17] typically employ (informal) derivations of conclusions from premises in some implicit epistemic/doxastic logic. In keeping with this tradition we also provided a "syntactic" analysis where we modelled the initial situation as a set of premises and derived our conclusions from them – albeit in a more detailed, formal way. This furthermore allowed us to pinpoint which of the facts in the initial situation were sufficient for the conclusions. It turned out that we did not need to completely describe the grid model from the semantic analysis. Furthermore, while it can clearly be argued that it is implicitly assumed that it is common knowledge that A and B each know their own number, the derivation of the fact that none of the agents can rule out

[13] $(we \wedge \neg Kwe) \vee (th \wedge [\neg we]\neg Kth) \vee (fr \wedge [\neg we][\neg th]\neg Kfr) \vee K\bot$. Note that Gerbrandy assumes that the knowledge modalities are K45 rather than S5.

[14] Note that this kind of self-reference is not the same as saying that "you don't know it now and you still don't know it after it is announced that you don't know it now" as briefly discussed at the end of Sect. 3.

that the other one has 2 only relies on *general* knowledge (everybody-knows) of that fact. That conclusion only relies on up to 2 levels of nested knowledge of any of the premises (everybody knows that everybody knows).

The fact that we don't need to assume common knowledge of the premises has an interesting corollary. Intuitively, a "static" epistemic or doxastic logic seems to be insufficient to deal with the paradox, because we need to be able to reason about knowledge/beliefs at different time points – in particular "before" and "after" announcements. Indeed, failure to make that distinction is exactly what lies behind the original paradox as well as other attempts to resolve it. However, the fact that we don't need common knowledge means that the premises, conclusions and the whole derivation can be translated into pure (static!) epistemic logic [15]! So, Hollis' paradox can be resolved by pure "static" epistemic reasoning about the initial situation after all.

Acknowledgments. The first author was supported by L. Meltzers Høyskolefond.

References

1. Ågotnes, T.: A logic of Finite Syntactic Epistemic States. Ph.D. thesis, Department of Informatics, University of Bergen, Norway (2004)
2. Baltag, A., Bezhanishvili, N., Fernández-Duque, D.: The topology of surprise. In: Proceedings of the International Conference on Principles of Knowledge Representation and Reasoning, vol. 19, pp. 33–42 (2022)
3. Baltag, A., Smets, S.: Surprise?! an answer to the hangman, or how to avoid unexpected exams!. In: Logic and Interactive Rationality Seminar (LIRA), Slides (2009)
4. Van Ditmarsch, H., Kooi, B.: The secret of my success. Synthese **151**(2), 201–232 (2006)
5. Fagin, R., Halpern, J.Y., Moses, Y., Vardi, M.: Reasoning about Knowledge. MIT, Cambridge (1995)
6. Freudenthal, H.: Solution to problem no. 223. Nieuw Archief voor Wiskunde **3**(18), 102–106 (1970)
7. Gerbrandy, J.: Bisimulations on Planet Kripke. University of Amsterdam, Amsterdam (1999)
8. Gerbrandy, J.: The surprise examination in dynamic epistemic logic. Synthese **155**, 21–33 (2007)
9. Hollis, M.: A paradoxical train of thought. Analysis **44**(4), 205–206 (1984)
10. Hollis, M.: More paradoxical epistemics. Analysis **46**(4), 217–218 (1986)
11. Kinghan, M.: A paradox derailed: reply to Hollis. Analysis **46**(1), 20–24 (1986)
12. Marcoci, A.: The surprise examination paradox in dynamic epistemic logic. Master's thesis, Universiteit van Amsterdam (2010)
13. Miller, J.S., Moss, L.S.: The undecidability of iterated modal relativization. Stud. Logica. **79**, 373–407 (2005)
14. Olin, D.: On a paradoxical train of though. Analysis **46**(1), 18–20 (1986)
15. Plaza, J.A.: Logics of public communications. In: Proceedings of the ISMIS 1989, pp. 201–216 (1989)
16. Poundstone, W.: Labyrinths of Reason: Paradox, Puzzles, and the Frailty of Knowledge. Anchor (2011)
17. Rea, G.: A variation of Hollis' paradox. Analysis **47**(4), 218–220 (1987)
18. van Ditmarsch, H., Kooi, B.: In: One Hundred Prisoners and a Light Bulb, pp. 83–94. Springer, Cham (2015). https://doi.org/10.1007/978-3-319-16694-0_9
19. Van Ditmarsch, H., van Der Hoek, W., Kooi, B.: Dynamic Epistemic Logic, volume 337 of Synthese Library. Springer, Cham (2007)
20. van Ditmarsch, H.: To be announced. Inf. Comput. **292**, 105026 (2023)

Author Index

N. Alechina et al. (Eds.): LORI 2023, LNCS 14329, p. 323, 2023.
https://doi.org/10.1007/978-3-031-45558-2

The manufacturer's authorised representative in the EU is Springer Nature Customer Service Centre GmbH, Europaplatz 3, 69115 Heidelberg, Germany. If you have any concerns regarding our products, please contact ProductSafety@springernature.com

Printed and bound by CPI Group (UK) Ltd, Croydon, CR0 4YY
06/05/2026
02104301-0002